Agroecology and Pollination Management

Agroecology and Pollination Management

Edited by Molly Ismay

SYRAWOOD
PUBLISHING HOUSE

New York

Published by Syrawood Publishing House,
750 Third Avenue, 9th Floor,
New York, NY 10017, USA
www.syrawoodpublishinghouse.com

Agroecology and Pollination Management
Edited by Molly Ismay

© 2017 Syrawood Publishing House

International Standard Book Number: 978-1-68286-390-9 (Hardback)

Cataloging-in-publication Data

Agroecology and pollination management / edited by Molly Ismay.
    p. cm.
Includes bibliographical references and index.
ISBN 978-1-68286-390-9
1. Agricultural ecology. 2. Pollen management. 3. Pollination. I. Ismay, Molly.
S589.7 .A37 2017
577.55--dc23

Printed in the United States of America.

TABLE OF CONTENTS

PREFACE

This book unfolds the innovative aspects of agro-ecology and pollination management, which will be crucial for the progress of this field in the future. It attempts to understand the multiple branches that fall under this discipline and how such concepts have practical applications. Agro-ecology refers to the study of the various ecological processes related to agriculture. It specifically uses these processes to innovate new agriculture management techniques. Pollination management is a part of agro-ecology. It refers to the practice of enhancing crop pollination, in order to improve yield and crop quality. This text includes some of the vital pieces of work being conducted across the world, on various topics related to agro-ecology and pollination management. It aims to serve as a resource guide for students and experts alike and contribute to the growth of the discipline.

This book was inspired by the evolution of our times; to answer the curiosity of inquisitive minds. Many developments have occurred across the globe in the recent past which has transformed the progress in the field.

This book was developed from a mere concept to drafts to chapters and finally compiled together as a complete text to benefit the readers across all nations. To ensure the quality of the content we instilled two significant steps in our procedure. The first was to appoint an editorial team that would verify the data and statistics provided in the book and also select the most appropriate and valuable contributions from the plentiful contributions we received from authors worldwide. The next step was to appoint an expert of the topic as the Editor-in-Chief, who would head the project and finally make the necessary amendments and modifications to make the text reader-friendly. I was then commissioned to examine all the material to present the topics in the most comprehensible and productive format.

I would like to take this opportunity to thank all the contributing authors who were supportive enough to contribute their time and knowledge to this project. I also wish to convey my regards to my family who have been extremely supportive during the entire project.

Editor

Plant Volatile Analogues Strengthen Attractiveness to Insect

Yufeng Sun[1], Hao Yu[2], Jing-Jiang Zhou[3], John A. Pickett[3], Kongming Wu[1]*

1 State Key Laboratory for Biology of Plant Diseases and Insect Pests, Institute of Plant Protection, Chinese Academy of Agricultural Sciences, Beijing, People's Republic of China, **2** Department of Entomology, Henan Institute of Science and Technology, Xinxiang, People's Republic of China, **3** Department of Biological Chemistry and Crop Protection, Rothamsted Research, Harpenden, Hertfordshire, United Kingdom

Abstract

Green leaf bug *Apolygus lucorum* (Meyer-Dür) is one of the major pests in agriculture. Management of *A. lucorum* was largely achieved by using pesticides. However, the increasing population of *A. lucorum* since growing Bt cotton widely and the increased awareness of ecoenvironment and agricultural product safety makes their population-control very challenging. Therefore this study was conducted to explore a novel ecological approach, synthetic plant volatile analogues, to manage the pest. Here, plant volatile analogues were first designed and synthesized by combining the bioactive components of β-ionone and benzaldehyde. The stabilities of β-ionone, benzaldehyde and analogue **3 g** were tested. The electroantennogram (EAG) responses of *A. lucorum* adult antennae to the analogues were recorded. And the behavior assay and filed experiment were also conducted. In this study, thirteen analogues were acquired. The analogue **3 g** was demonstrated to be more stable than β-ionone and benzaldehyde in the environment. Many of the analogues elicited EAG responses, and the EAG response values to **3 g** remained unchanged during seven-day period. **3 g** was also demonstrated to be attractive to *A. lucorum* adults in the laboratory behavior experiment and in the field. Its attractiveness persisted longer than β-ionone and benzaldehyde. This indicated that **3 g** can strengthen attractiveness to insect and has potential as an attractant. Our results suggest that synthetic plant volatile analogues can strengthen attractiveness to insect. This is the first published study about synthetic plant volatile analogues that have the potential to be used in pest control. Our results will support a new ecological approach to pest control and it will be helpful to ecoenvironment and agricultural product safety.

Editor: Youjun Zhang, Institute of Vegetables and Flowers, Chinese Academy of Agricultural Science, China

Funding: This study was supported by the National Natural Science Funds (No. 31321004) from National Natural Science Foundation of China and the Special Fund for Agro-scientific Research in the Public Interest (201103012) from the Chinese Ministry of Agriculture. The funders had no role in study design, data collection and analysis, decision to publish, or preparation of the manuscript.

Competing Interests: The authors have declared that no competing interests exist.

* E-mail: kmwu@ippcaas.cn

Introduction

The green leaf bug, *Apolygus lucorum* (Meyer-Dür) (Hemiptera: Miridae), is a major pest of many cultivated plants including cotton, cereals, vegetables and fruit trees. With the reduced application of insecticides on Bt cotton for the control of lepdopitera pests, the abundance of *A. lucorum* has substantially increased in China [1]. This species can easily attain outbreak densities, switch hosts, and wilder spread because of its environmental adaptability [1–4], high population growth rate [2–3], and strong dispersal capacity [5–6]. These characteristics make *A. lucorum* difficult to control.

Currently, various agronomic (soil tillage, removing weeds before sowing seeds) and chemical (spraying with organophosphates, pyrethroids and nicotinoid) measures are applied to control *A. lucorum*. To reduce large yield losses caused by the pest and insecticide use which are usually harmful to human beings and the environment, it is necessary to develop new approach in integrated pest management (IPM) schemes for this pest.

Insect ecology involves the relationship between insect and its surroundings that seek both to proceed with IPM and to protect the ecological environment [7]. Research on insect ecology is helpful to ecoenvironment and agricultural product safety. Plant volatiles, a factor of insect ecology, emitted by plant, in response to mechanical or herbivore damage, may achieve this aim and can be applied at the farm or landscape level. They involve mediating the behavior of insects [8–9], natural enemies [10–12] or neighboring plant [13–15]. Their potential value in pest population control has been recognized [16–18].

The current study concerns the use of plant volatiles, β-ionone and benzaldehyde, as attractants for *A. lucorum*. According to previous reports, β-ionone, a volatile released from cotton, tomato, and other plants [19–23], attracts brown planthoppers [24] and repels phytophagous mites [25]. Benzaldehyde, a common component of plant volatiles [26], attracts many pest species [27–28]. A recent study has indicated that β-ionone and benzaldehyde can be recognized by adult *A. lucorum* and can affect *A. lucorum* behavior under laboratory conditions [29]. However, the two chemicals cannot be efficiently used as *A. lucorum* attractants because of their low stability and mediocre ability to attract *A. lucorum* in field conditions.

Indeed, so far there has been no publish about plant volatile analogues synthesis and their use. In the current study, we try to explore novel ecological approach to control pest based on synthetic plant volatile analogues. Here we first hypothesized that the analogues synthesized with substructure combination strategy by combining the bioactive components of β-ionone and benzaldehyde would contribute to achieving our objectives: i) to increase stability by changing chemical functional groups, for example, aldehyde group, which is associated with low oxidative stability and (ii) to enhance attractiveness to *A. lucorum* by combining the active groups of the two compounds. Based on such working hypothesis, we designed and synthesized 13 analogues of β-ionone and benzaldehyde. We tested the stabilities of β-ionone, benzaldehyde and analogue **3 g**. And we then conducted EAG test, behavior assay and filed experiment to evaluate the attractivities of the analogues to *A. lucorum*.

Materials and Methods

Ethics statement

We captured the insects in Xinxiang Experiment Station (Henan) of Institute of Plant Protection, Chinese Academy of Agricultural Sciences. The wild-captured *A. lucorum* used in this study was serious pest in China. Therefore, no specific permits were required for the described insect collection and experimentation.

Insects

A culture of *A. lucorum* was maintained on fresh ears of corn (*Zea mays* L.) in a climate chamber at $29\pm1°C$, $60\pm5\%$ RH, and 14:10 L:D at the Institute of Plant Protection, Chinese Academy of Agricultural Sciences, Beijing. Adults were used in the laboratory experiment.

Synthesis of β-ionone and benzaldehyde analogues 3a−m

Both β-ionone (0.88 g, 4.6 mmol) and benzene formaldehyde (**2a−m**, 6.0 mmol) were dissolved in ethanol (10 mL) in a round-bottom flask (50 mL) before 15 mL of a sodium hydrate solution (1 mmol/mL) was added dropwise. The reaction mixture was stirred at room temperature for 5 h, brought to pH 7.0 with 10% hydrochloric acid, and subsequently extracted with ether (50 mL ×2). The organic layer was dried with anhydrous Na_2SO_4, and the solvent was removed under reduced pressure. The residue was purified by column chromatography on silica gel using ethyl acetate-petroleum (60–90°C) at a ratio of 1:13–1:25 as the eluent to afford 13 analogues, which were designated **3a−m** (Figure 1) and R of the structures are illustrated in Table 1. They were characterized by melting point, IR, ^{13}C NMR, ^1H NMR and high-resolution mass spectrometry (HRMS). Melting points of chemicals were determined with an X-4 binocular microscope (Yuhua Instrument Co., Goyi, China) with thermometer. IR spectra were recorded on neat samples with a Nicolet 6700 FT-IR spectrometer (Thermo Fisher Scientific Inc., USA). ^{13}C NMR and ^1H NMR spectra were recorded with a Bruker Avance DPX300 spectrometer (Bruker-Spectrospin AG, Swiss). Chemical shifts were described in δ (ppm) relative to the signal of an internal standard (tetramethylsilane) and using $CHCl_3$-$d1$ or DMSO-$d6$ as the solvent. Coupling constants were given in Hz. HRMS spectra were displayed under electron impact (150 eV) condition using a Bruker APEX IV spectrometer (Bruker Instruments Co. Ltd., USA).

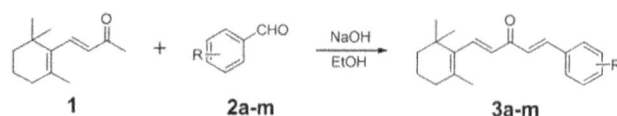

Figure 1. General route for the synthesis of 13 β-ionone and benzaldehyde analogues (3a−m). 1 = β-ionone, 2a−m = 13 forms of benzaldehyde, 3a−m = 13 analogues of β-ionone and benzaldehyde. R is listed in Table 1.

Stability of β-ionone, benzaldehyde and analogue (3 g)

GC-MS analyses were used for testing the stability of β-ionone, benzaldehyde and one representative analogue, **3 g** (Figure 2), whose structure is quite similar to other synthesized analogues. GC-MS analyses were performed with a Thermo Trace GC Ultra gas chromatograph coupled to a Thermo ISQ mass spectrometer. The pure compounds (20 mg) were added into colorless, transparent vials (2-cm) individually. Then the vials were covered with 100 mesh gauzes and exposed to air and sunlight. 1 mg of each compound was sampled and stored in sealed brown, transparent vial every day. The sampling was last for seven days. Each compound was represented by three replicates. The GC was operated in splitless injection mode and fitted with a TG-5SILMS column (30 m×0.25 mm×0.25 μm). For β-ionone and benzaldehyde, firstly the oven was programmed from 40–130°C at 3°C/min after an initial delay of 1 min and held at 130°C for 1 min, and then the oven was programmed from 130–250°C at 10°C/min and held at 250°C for 5 min. For **3 g**, firstly the oven was programmed from 60–200°C at 10°C/min after an initial delay of 1 min and held at 200°C for 1 min, then the oven was programmed from 200–250°C at 10°C/min and held at 250°C for 10 min. Injector temperature was 250°C; MS quadrupole temperature was 150°C; MS source temperature was 250°C; and transfer line temperature was 250°C. The sampling and analyses were performed twice (in May 2013 and in September 2013) to confirm that whether the stability were consistent under different time horizons.

Laboratory EAG experiment

The responses of *A. lucorum* females and males to the 13 analogues of β-ionone and benzaldehyde (**3a−m**) were measured by EAG experiment. The EAG responses of *A. lucorum* were recorded with Syntech GC/EAD interface temp controller TC-02 and stimulus controller CS-55 (Hilversum, The Netherlands). Laboratory EAG experiment was conducted to evaluate whether the test compounds stimulate *A. lucorum*. Five concentrations (0.01, 0.1, 1, 10, 100 μg/μL) in dichloromethane were checked in preliminary experiment. The concentration, 10 μg/μL, was suitable to each test compound. The compounds elicited relatively high EAG responses by *A. lucorum* at this concentration. Therefore we measured the EAG responses of the herbivore to chemicals only using the concentration, 10 μg/μL. And it was enough to evaluate whether the test compounds stimulate *A. lucorum*.

The test compounds were dissolved in distilled dichloromethane, and the solutions (20 μL, 10 μg/μL) were then added to a piece of folded filter paper (0.5 cm×4 cm). After evaporation for 30 s, the filter paper was inserted into a glass Pasteur pipette. The antennae of *A. lucorum* adults were excised and mounted between electrodes [30]. The stimuli were delivered to the antennae in a constant airstream of 150 mL/min at 30–40 s intervals, and the EAG signals were recorded. β-ionone and benzaldehyde were tested as positive controls. Dichloromethane and ethyl benzoate were also included as a background and standard stimulus,

Table 1. Physical properties, IR, HRMS and ^{13}C NMR data for compounds 3a–m.

Compd	R	Mp(°C)	State	Yield (%)	IR (cm^{-1})	HRMS [M+H], (calcd)	^{13}C NMR (75 MHz, δ, ppm)
3a	H	80–81	yellow crystal	61.3	2930, 2858, 1651, 1593, 1567, 1447, 1345	281.1905 (281.1900)	189.7, 144.3, 142.1, 137.9, 137.4, 134.8, 133.0, 130.5, 127.1, 125.4, 40.8, 35.1, 34.7, 29.8, 22.8, 19.8
3b	4-F	69–70	yellow crystal	64.2	2933, 2865, 1667, 1598, 1572, 1455, 1339	299.1809 (299.1806)	189.9, 163.2, 144.2, 142.3, 137.8, 137.4, 132.1, 131.1, 130.4, 126.3, 117.1, 116.8, 40.8, 35.1, 34.6, 29.8, 22.8, 22.4, 19.8
3c	4-Cl	87–89	yellow crystal	63.2	2930, 2859, 1654, 1597, 1567, 1410, 1352	315.1508 (315.1510)	189.7, 144.2, 142.0, 137.8, 137.4, 137.0, 134.4, 130.4, 130.3, 130.1, 127.0, 40.8, 35.1, 34.6, 29.8, 22.8, 19.8
3d	4-Br	95–97	yellow crystal	68.0	2928, 2858, 1654, 1596, 1564, 1489, 1350	359.1011 (359.1005)	190.0, 144.0, 143.6, 137.5, 135.9, 131.2, 130.5, 129.8, 129.2, 126.6, 40.8, 35.1, 34.6, 29.8, 22.8, 19.8
3e	4-Me	92–94	yellow crystal	76.2	2930, 2862, 1651, 1587, 1569, 1455, 1350	295.2055 (295.2056)	190.1, 143.8, 143.6, 141.6, 137.5, 137.2, 133.2, 130.6, 129.2, 125.8, 40.8, 35.1, 34.6, 29.8, 22.8, 22.4, 19.8
3f	4-Et	62–64	yellow crystal	32.8	2931, 2863, 1650, 1614, 1588, 1452, 1347	309.2215 (309.2213)	189.3, 147.1, 142.9, 136.5, 132.4, 130.0, 129.6, 128.5, 124.8, 39.8, 34.2, 33.7, 28.9, 21.9, 18.9, 15.4
3g	4-OMe	62–63	yellow crystal	78.5	2930, 2866, 1667, 1602, 1581, 1465, 1339	311.2009 (311.2006)	190.0, 162.4, 143.5, 143.4, 137.5, 137.0, 130.9, 130.7, 128.6, 124.5, 115.3, 56.3, 40.8, 35.1, 34.6, 29.8, 22.8, 19.9
3h	4-OEt	91–92	yellow crystal	42.3	2927, 2866, 1664, 1597, 1573, 1478, 1337	325.2163 (325.2162)	190.0, 161.8, 143.5, 137.5, 137.0, 132.9, 130.9, 130.7, 128.4, 124.4, 116.8, 115.6, 64.5, 40.8, 35.1, 34.6, 29.8, 22.7, 19.9, 15.6
3i	4-NO$_2$	112–114	yellow crystal	68.7	2926, 2863, 1669, 1616, 1598, 1455, 1340	326.1752 (326.1751)	189.2, 149.4, 145.1, 142.1, 140.3, 139.0, 137.4, 130.1, 130.0, 129.7, 125.1, 40.8, 35.1, 34.8, 29.8, 22.8, 19.8
3j	3- Me	99–100	yellow crystal	53.2	2930, 2863, 1651, 1591, 1453, 1338	295.2062 (295.2056)	190.1, 143.9, 143.8, 139.5, 137.5, 137.4, 135.9, 132.0, 130.5, 129.8, 129.7, 126.5, 126.4, 40.8, 35.1, 34.6, 29.8, 22.8, 22.2, 19.8
3k	2- OMe	/	yellow oil	70.1	2930, 1647, 1622, 1592, 1562, 1487, 1350	311.2006 (311.2006)	190.8, 159.5, 137.5, 136.8, 132.7, 132.4, 131.1, 130.7, 129.7, 124.9, 121.6, 112.1, 56.4, 40.7, 34.5, 29.7, 22.7, 19.9
3l	3, 4-Me$_2$	85–87	yellow crystal	75.1	2930, 2863, 1651, 1612, 1589, 1452, 1344	309.2217 (309.2213)	190.2, 143.9, 143.7, 140.4, 138.0, 137.5, 137.1, 133.6, 131.1, 130.6, 130.4, 126.8, 125.7, 40.8, 35.1, 34.6, 29.8, 22.8, 20.7, 20.6, 19.9
3m	2, 4-Cl$_2$	109–110	yellow crystal	61.9	2938, 2865, 1667, 1610, 1565, 1468, 1384	349.1126 (349.1121)	189.7, 144.6, 138.1, 138.0, 137.4, 137.1, 136.7, 132.8, 130.9, 129.6, 129.3, 128.4, 40.8, 35.1, 34.7, 29.8, 22.8, 19.8

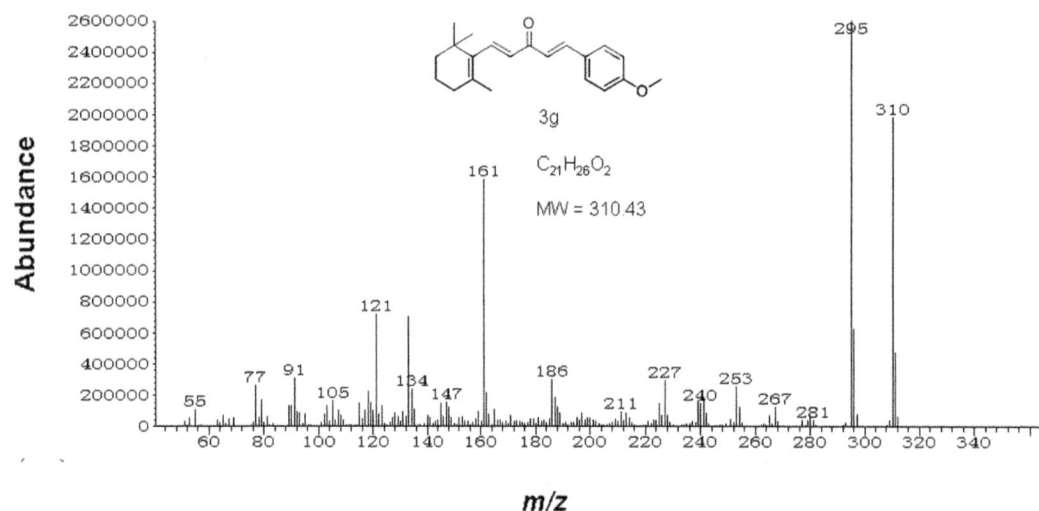

Figure 2. EI mass spectrum of analogue 3 g, also showing its structure, deduced molecular formula and molecular weight.

respectively, and these were applied before and after stimulation with each test compound. EAG responses were obtained from three replicate females and three replicate males per test compound.

The EAG responses of *A. lucorum* females and males to the samples of β-ionone, benzaldehyde, β-ionone + benzaldehyde and **3 g** exposed to air and sunlight and prepared in "**Stability of β-ionone, benzaldehyde and the analogue (3 g)**" were also measured.

The above laboratory EAG experiments were also performed twice (in May 2012 and in September 2013 for the first EAG experiment, and in July 2013 and in September 2013 for the second EAG experiment) to confirm that whether the responses of adult *A. lucorum* to the samples were consistent under different time horizons.

Laboratory behavior experiment

Behavioural responses of adult *A. lucorum* to the pure and treated samples (exposed to air and sunlight for periods up to one and/or seven days) of β-ionone, benzaldehyde, β-ionone + benzaldehyde and **3 g** were investigated with a glass Y-tube olfactometer (2.8 cm uniform diameter, 21.8 cm main body length, and 18.8 cm branch length). An airflow (0.2 mL/min) was introduced into each arm of the olfactometer through glass stimulus chamber (an odour source adapter), attached to each of the two ending arms. In this way, two well-separated laminar air flows were generated in the olfactometer. As mentioned above, the concentration, 10 µg/µL in dichloromethane, was suitable to each test compound in the EAG experiment. The compounds elicited relatively high EAG responses by *A. lucorum* at this concentration. So here we used the same concentration, 10 µg/µL, for behavior assay. In each test 20 µL of dichloromethane solution of each chemical (10 µg/µL) was placed in the glass stimulus chamber of the "treatment" arm. As a control, 20 µL of dichloromethane was placed in the glass stimulus chamber of the "CK' arm of the olfactometer. Experiments were performed at room temperature. The olfactometer was washed with water and ethanol before each experiment. Adult *A. lucorum* was introduced at the bottom of the olfactometer individually and let free to walk. After 5 minutes, the *A. lucorum* in the treatment and control arms of the olfactometer was recorded. The insect that did not move and remained at the base of the Y tube was recorded as not reaction.

Behavioral responses were obtained from 60 replicate females and 60 replicate males per test compound.

Field experiment

An experiment was conducted in a 3-month-old alfalfa (*Medicago sativa* L.) field at the Xinxiang Experiment Station (35°18′ N, 113°54′ E) in Henan Province during 16–27 July 2012. The field had been tilled so that it was devoid of vegetation before alfalfa was planted; the alfalfa had not been sprayed with pesticide. For preparation of lures, the analogues of β-ionone and benzaldehyde (200 µL, 10 µg/µL in dichloromethane) were added into red rubber septa (Enoy Technology, Zhangzhou, China), and the dichloromethane was allowed to evaporate. The lures were positioned in the centers of white sticky cards (28 cm×22 cm), which were horizontally hung below ship-type traps (Enoy Technology, Zhangzhou, China) attached by wires to a bamboo stake; the traps were flush with the top of alfalfa plants (Figure S1). The trapping devices were placed ~10 m apart, and were randomly assigned to contain one of the 13 analogues or the controls. Lures containing β-ionone, lures containing benzaldehyde and lures containing β-ionone + benzaldehyde prepared as above were used as positive controls. Red rubber septa without any compound were used as blank controls. Each compound or control was represented by three replicate traps. Sticky cards were replaced every 3 days, and the trapped *A. lucorum* were counted and sex was determined.

As with above method, another experiment was also conducted in this field during 15–22 July 2013. Lures containing one of β-ionone, benzaldehyde, β-ionone + benzaldehyde, **3 g** or blank controls were used. Sticky cards were replaced every day, and the trapped *A. lucorum* were counted and sex was determined.

Statistical analysis

The relative EAG response value of *A. lucorum* antennae to each test compound was calculated using the following equation: Relative EAG response value = (EAG response value to the test compound − mean EAG response value to the background stimulus)/(mean EAG response value to the standard stimulus − mean EAG response value to the background stimulus). The mean relative EAG response values of female or male *A. lucorum* antennae to each test compound and to each positive control were compared using Student's t-tests. Student's t-tests were also used to compare the mean relative EAG response of female and male *A. lucorum* antennae to each test compound.

For the behavior research, the percent responses of *A. lucorum* were used for analysis and differences between "treated" and "CK" were compared with nonparametric tests followed by chi-square statistical.

For the field study, the cumulative numbers of *A. lucorum* collected in each sticky trap were used for analysis. One-way ANOVA was carried out for comparisons between the treatments.

All the statistic tests were conducted using SPSS (version 12.0).

Results

Thirteen analogues (**3a−m**) were produced by the aldol reaction between β-ionone and benzene formaldehyde (**2a−m**) with minimal amounts of by-products (Fig. 1). This reaction was simple to perform, did not require dangerous conditions, and did not produce toxic substances. The 13 analogues were easily purified. Their identities were confirmed by IR, high-resolution mass spectrometry (HRMS), ^{13}C NMR and ^{1}H NMR. The physical properties, IR, HRMS and ^{13}C NMR data for compounds **3a–m** are listed in **Table 1**. ^{1}H NMR data for compounds **3a−m** are listed in **Table 2**.

The stability of β-ionone, benzaldehyde and one representative analogue, **3 g** (Fig. 2), were checked after exposing them to air and

sunlight for a period up to seven days. In such conditions, the stability tests performed in May 2013 indicated that samples of pure β-ionone degraded completely and samples of pure benzaldehyde declined to 5.5% a.i. on the first day, whereas samples of pure **3 g** declined to 94.3% a.i. on the seventh day (Fig. 3A and Fig. S2). And the stability tests performed in September 2013 showed a similar result (Fig. 3B). The experiments confirmed that the stabilities of the samples were consistent under different time horizons and **3 g** was much more stable in the environment than β-ionone and benzaldehyde.

A previous study showed that β-ionone and benzaldehyde elicited EAG responses from adult *A. lucorum* antennae, suggesting that the compounds were recognized as signals by *A. lucorum* adults [29]. To test the attractiveness of the analogues to *A. lucorum*, we first measured the EAG responses elicited by the analogues from adult *A. lucorum*. In the case of the experiment performed in May 2012, the mean EAG response value to the background stimulus, dichloromethane, was -35.74 ± 21.10 µV for *A. lucorum* females and -39.08 ± 19.80 µV for males. The mean response values to the standard stimulus, ethyl benzoate (20 µL, 10 µg/µL), was -80.03 ± 62.55 µV for females and -94.8 ± 50.15 µV for males, which were significantly higher than to the background (both P< 0.001). The difference between females and males in their responses to the standard stimulus was not significant (P = 0.083).

β-ionone, benzaldehyde and β-ionone + benzaldehyde were included as positive controls in our EAG response experiment.

Table 2. ^{1}H NMR data for compounds 3a−m.

Compd	^{1}H NMR (300 MHz, δ, ppm)
3a	1.11 (s, 6H, 5-Me, 5-Me), 1.48–1.52 (m, 2H, 4-H, 4-H), 1.61–1.67 (m, 2H, 3-H, 3-H), 1.83 (d, 3H, J = 0.7 Hz, 1-Me), 2.10 (t, 2H, J = 6.2 Hz, 2-H, 2-H), 6.48 (d, 1H, J = 16.1 Hz, 8-H), 7.00 (d, 1H, J = 15.9 Hz, 10-H), 7.39–7.42 (m, 3H, 14-H, 15-H, 16-H), 7.51 (d, 1H, J = 16.1 Hz, 7-H), 7.58–7.61 (m, 2H, 13-H, 17-H), 7.67 (d, 1H, J = 16.0 Hz, 11-H)
3b	1.11 (s, 6H, 5-Me, 5-Me), 1.48–1.52 (m, 2H, 4-H, 4-H), 1.60–1.65 (m, 2H, 3-H, 3-H), 1.83 (d, 3H, J = 0.7 Hz, 1-Me), 2.09 (d, 2H, J = 6.2 Hz, 2-H, 2-H), 6.46 (d, 1H, J = 16.1 Hz, 8-H), 6.93 (d, 1H, J = 15.9 Hz, 10-H), 7.09 (t, 2H, J = 8.7 Hz, 14-H, 16-H), 7.48–7.66 (m, 4H, 13-H, 17-H, 7-H, 11-H)
3c	1.11 (s, 6H, 5-Me, 5-Me), 1.48–1.51 (m, 2H, 4-H, 4-H), 1.61–1.67 (m, 2H, 3-H, 3-H), 1.83 (d, 3H, J = 0.7 Hz, 1-Me), 2.11 (t, 2H, J = 6.1 Hz, 2-H, 2-H), 6.46 (d, 1H, J = 16.1 Hz, 8-H), 6.97 (d, 1H, J = 15.9 Hz, 10-H), 7.36–7.54 (m, 5H, 7-H, 13-H, 14-H, 16-H, 17-H), 7.61 (d, 1H, J = 16.0 Hz, 11-H)
3d	1.11 (s, 6H, 5-Me, 5-Me), 1.48–1.52 (m, 2H, 4-H, 4-H), 1.61–1.68 (m, 2H, 3-H, 3-H), 1.83 (s, 3H, 1-Me), 2.10 (t, 2H, J = 6.2 Hz, 2-H, 2-H), 6.46 (d, 1H, J = 16.1 Hz, 8-H), 6.98 (d, 1H, J = 15.9 Hz, 10-H), 7.44–7.62 (m, 6H, 7-H, 11-H, 13-H, 14-H, 16-H, 17-H)
3e	1.11 (s, 6H, 5-Me, 5-Me), 1.48–1.52 (m, 2H, 4-H, 4-H), 1.60–1.68 (m, 2H, 3-H, 3-H), 1.82 (d, 3H, J = 0.3 Hz, 1-Me), 2.09 (t, 2H, J = 6.2 Hz, 2-H, 2-H), 2.38 (s, 3H, 15-Me), 6.47 (d, 1H, J = 16.1 Hz, 8-H), 6.96 (d, 1H, J = 15.9 Hz, 10-H), 7.12 (d, 1H, J = 8.0 Hz, 14-H), 7.50 (t, 2H, J = 6.8 Hz, 16-H, 7-H), 7.64 (d, 1H, J = 15.9 Hz, 11-H)
3f	1.11 (s, 6H, 5-Me, 5-Me), 1.25 (t, 3H, J = 7.6 Hz, 15-C-Me)1.48–1.52 (m, 2H, 4-H, 4-H), 1.63–1.66 (m, 2H, 3-H, 3-H), 1.83 (d, 3H, J = 0.6 Hz, 1-Me), 2.10 (t, 2H, J = 6.1 Hz, 2-H, 2-H), 2.68 (q, 2H, J = 7.6 Hz, 15-C-H, 15-C-H), 6.47 (d, 1H, J = 16.1 Hz, 8-H), 6.96 (d, 1H, J = 15.9 Hz, 10-H), 7.25 (d, 2H, J = 4.5 Hz, 14-H, 16-H), 7.50–7.53 (m, 3H, 7-H, 13-H, 17-H), 7.65 (d, 1H, J = 15.9 Hz, 11-H)
3g	1.11 (s, 6H, 5-Me, 5-Me), 1.48–1.52 (m, 2H, 4-H, 4-H), 1.60–1.66 (m, 2H, 3-H, 3-H), 1.82 (d, 3H, J = 0.7 Hz, 1-Me), 2.09 (t, 2H, J = 6.1 Hz, 2-H, 2-H), 3.85 (s, 3H, 15-O-Me), 6.47 (d, 1H, J = 16.0 Hz, 8-H), 6.86–6.93 (m, 3H, 14-H, 16-H, 10-H), 7.45–7.56 (m, 3H, 7-H, 13-H, 17-H), 7.64 (d, 1H, J = 15.9 Hz, 11-H)
3h	1.11 (s, 6H, 5-Me, 5-Me), 1.43 (t, 3H, J = 7.0 Hz, 15-O-C-Me)1.48–1.52 (m, 2H, 4-H, 4-H), 1.62–1.66 (m, 2H, 3-H, 3-H), 1.82 (s, 3H, 1-Me), 2.09 (t, 2H, J = 6.1 Hz, 2-H, 2-H), 4.07 (q, 2H, J = 7.0 Hz, 15-O-C-H, 15-O-C-H), 6.46 (d, 1H, J = 16.0 Hz, 8-H), 6.85–6.92 (m, 3H, 10-H, 14-H, 16-H), 7.45–7.55 (m, 3H, 7-H, 13-H, 17-H), 7.64 (d, 1H, J = 15.9 Hz, 11-H)
3i	1.13 (s, 6H, 5-Me, 5-Me), 1.49–1.53 (m, 2H, 4-H, 4-H), 1.62–1.67 (m, 2H, 3-H, 3-H), 1.85 (d, 3H, J = 0.6 Hz, 1-Me), 2.12 (t, 2H, J = 6.2 Hz, 2-H, 2-H), 6.48 (d, 1H, J = 16.3 Hz, 8-H), 7.10 (d, 1H, J = 15.9 Hz, 10-H), 7.55–7.75 (m, 4H, 7-H, 11-H, 13-H, 17-H), 8.24–8.28 (m, 2H, 14-H, 16-H)
3j	1.12 (s, 6H, 5-Me, 5-Me), 1.48–1.52 (m, 2H, 4-H, 4-H), 1.63–1.67 (m, 2H, 3-H, 3-H), 1.83 (d, 3H, J = 0.7 Hz, 1-Me), 2.09 (t, 2H, J = 6.2 Hz, 2-H, 2-H), 2.39 (s, 3H, 14-Me), 6.48 (d, 1H, J = 16.1 Hz, 8-H), 6.98 (d, 1H, J = 16.0 Hz, 10-H), 7.19–7.32 (m, 2H, 13-H, 15-H), 7.40 (d, 2H, J = 5.8 Hz, 16-H, 17-H), 7.48 (d, 1H, J = 0.8 Hz, 16-H, 17-H), 7.53 (dd, 1H, J = 0.8 and 0.8 Hz, 7-H), 7.64 (d, 1H, J = 15.4 Hz, 11-H)
3k	1.11 (s, 6H, 5-Me, 5-Me), 1.48–1.52 (m, 2H, 4-H, 4-H), 1.60–1.68 (m, 2H, 3-H, 3-H), 1.82 (d, 3H, J = 0.7 Hz, 1-Me), 2.09 (t, 2H, J = 5.9 Hz, 2-H, 2-H), 3.89 (s, 3H, 13-O-Me), 6.50 (d, 1H, J = 16.1 Hz, 8-H), 6.91–7.09 (m, 3H, 10-H, 14-H, 16-H), 7.33–7.36 (m, 1H, 15-H), 7.45 (dd, 1H, J = 0.7 and 0.8 Hz, 7-H), 7.59 (q, 1H, J = 3.1 Hz, 17-H), 7.99 (d, 1H, J = 16.2 Hz, 11-H)
3l	1.11 (s, 6H, 5-Me, 5-Me), 1.48–1.52 (m, 2H, 4-H, 4-H), 1.60–1.66 (m, 2H, 3-H, 3-H), 1.83 (s, 3H, 1-Me), 2.10 (t, 2H, J = 6.1 Hz, 2-H, 2-H), 2.29 (s, 6H, 14-Me, 15-Me), 6.49 (d, 1H, J = 16.0 Hz, 8-H), 6.95 (d, 1H, J = 15.9 Hz, 10-H), 7.16 (d, 1H, J = 7.6 Hz, 13-H), 7.34 (q, 2H, J = 4.3 Hz, 16-H, 17-H), 7.49 (d, 1H, J = 15.7 Hz, 7-H), 7.63 (d, 1H, J = 15.9 Hz, 11-H)
3m	1.12 (s, 6H, 5-Me, 5-Me), 1.48–1.52 (m, 2H, 4-H, 4-H), 1.63–1.67 (m, 2H, 3-H, 3-H), 1.83 (d, 3H, J = 0.7 Hz, 1-Me), 2.11 (t, 2H, J = 6.2 Hz, 2-H, 2-H), 6.50 (d, 1H, J = 16.1 Hz, 8-H), 6.93 (d, 1H, J = 16.0 Hz, 10-H), 7.27–7.30 (m, 1H, 16-H), 7.45–7.55 (m, 2H, 7-H, 14-H), 7.62 (d, 1H, J = 8.5 Hz, 17-H), 7.97 (d, 1H, J = 16.0 Hz, 11-H)

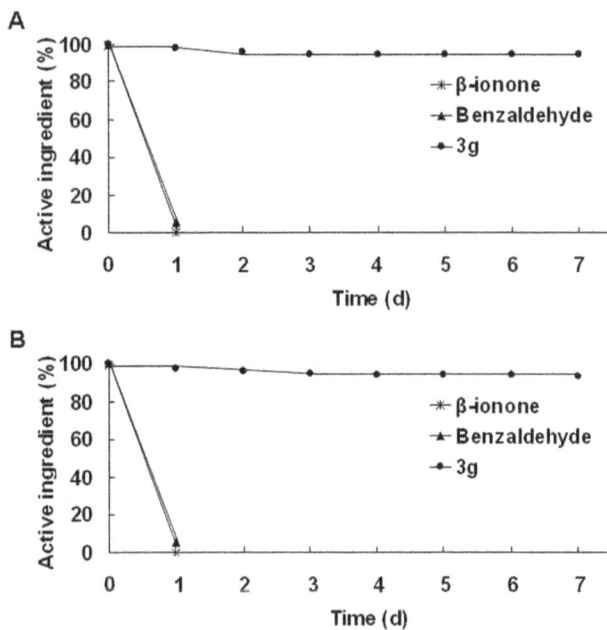

Figure 3. Active ingredient content of β-ionone, benzaldehyde and 3 g on different day after exposing them to air and sunlight for a period up to seven days. (A) Stability tests performed in May 2013, (B) Stability tests performed in September 2013.

The mean relative EAG response values to β-ionone (**β**) was 1.05 ± 0.02 for female *A. lucorum* and 1.16 ± 0.34 for males, and the difference between females and males was not significant ($P=0.632$). In contrast, the response to benzaldehyde (**B**) was stronger for females (4.74 ± 0.59) than for males (1.30 ± 0.17) (P<

0.01), and the response to β-ionone + benzaldehyde (**β+B**) was also stronger for females (3.69 ± 0.79) than for males (1.23 ± 0.39) (P< 0.01) (Fig. 4A).

Among the 13 synthesized compounds, only three (**3b**, **3g**, and **3j**) elicited stronger EAG responses from female *A. lucorum* antennae than from male antennae; for all other compounds, male antennae exhibited stronger responses than female antennae (student's *t*-tests P<0.05) (Fig. 4A). Compounds **3 h** and **3 m** elicited stronger responses from both female and male antennae than other analogues (Fig. 4A).

The EAG responses of female antennae were stronger to compounds **3h**, **3j**, and **3m** than to the positive control β-ionone (P<0.05), while responses of female antennae were weaker to all 13 analogues than to the positive control benzaldehyde and β-ionone + benzaldehyde (P<0.05). The responses of male antennae were stronger to **3d**, **3h**, **3i**, **3k**, **3l**, and **3m** than to β-ionone, benzaldehyde and β-ionone + benzaldehyde (P<0.05).

The EAG experiment performed in September 2013 showed a similar pattern as above (Fig. 4B). Our results demonstrate that most analogues synthesized in this study elicit responses from *A. lucorum* and the responses of adult *A. lucorum* to the samples were consistent under different time horizons.

Samples of β-ionone, benzaldehyde, β-ionone + benzaldehyde and **3g** that were exposed to air and sunlight for different periods were tested for their EAG responses with antennae of adult female and male *A. lucorum*. In the case of the experiment performed in July 2013, all samples of β-ionone, benzaldehyde and **3g** elicited EAG responses from female and male *A. lucorum* antennae (Fig. 5A). The relative EAG response values to β-ionone, benzaldehyde and β-ionone + benzaldehyde decreased day by day. On the contrary, the relative EAG response values to **3g** changed very little over seven days. The EAG experiments were performed again in September 2013 and showed a similar result (Fig. 5B).

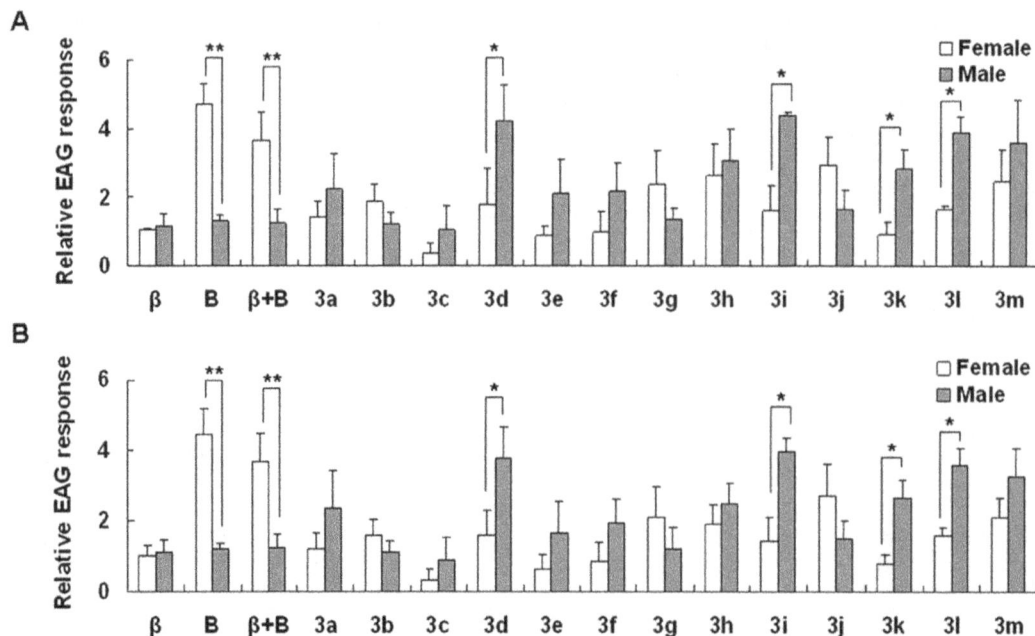

Figure 4. Relative EAG responses (mean ± SD) of female and male *A. lucorum* to synthetic analogues of β-ionone and benzaldehyde. (A) EAG experiment performed in May 2012, (B) EAG experiment performed in September 2013. β = β-ionone and B = benzaldehyde. Asterisks indicate significant differences in EAG responses between female and male antennae: * P<0.05, ** P<0.01.

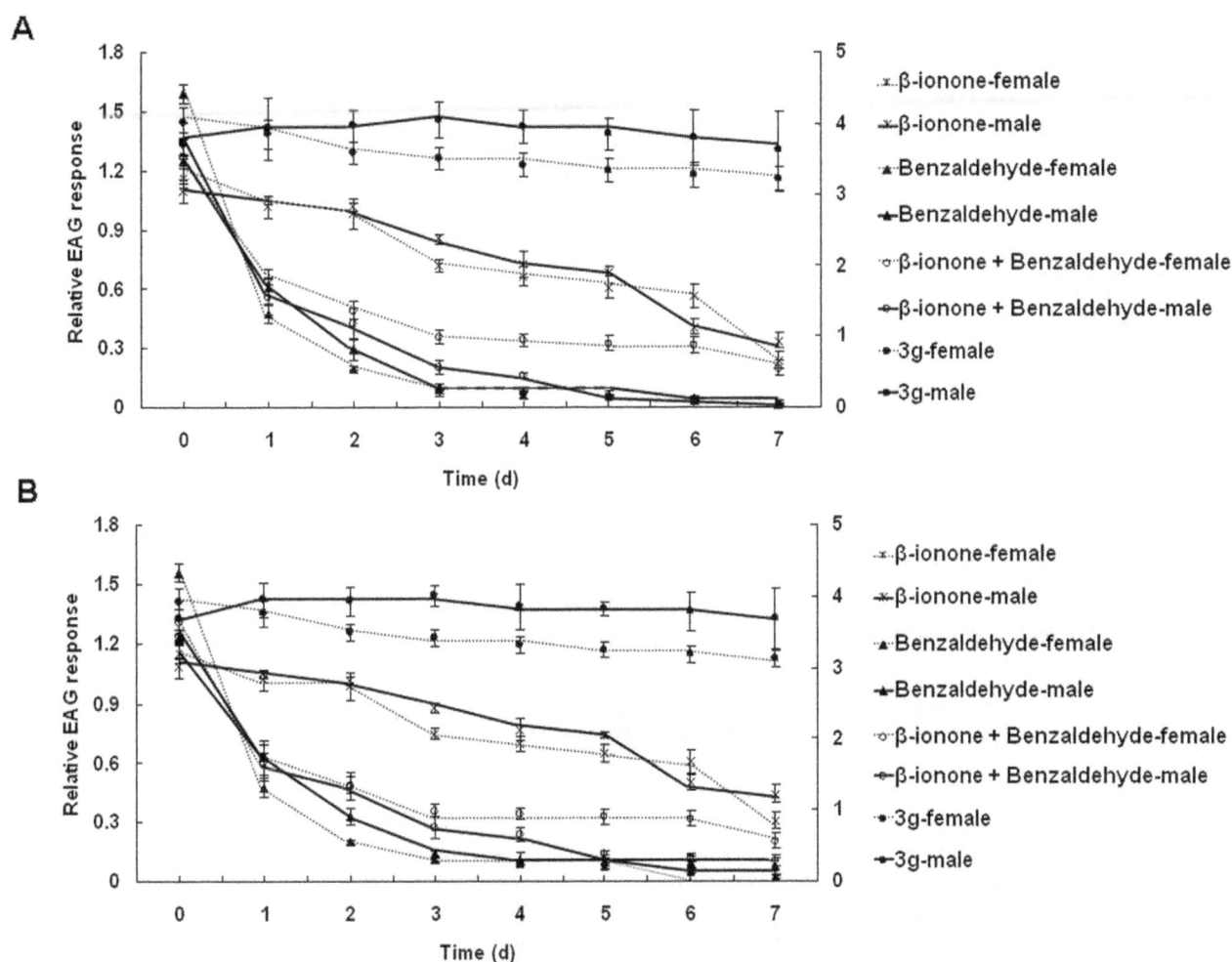

Figure 5. Relative EAG responses (mean ± SE) of female and male *A. lucorum* to *β*-ionone, benzaldehyde, *β*-ionone + benzaldehyde and 3 g. (A) EAG experiment performed in July 2013, (B) EAG experiment performed in September 2013. Secondary axis in the chart showed relative EAG responses of female *A. lucorum* to benzaldehyde and to β-ionone + benzaldehyde.

In behavior experiment, *A. lucorum* showed preferences for the pure β-ionone, benzaldehyde, β-ionone + benzaldehyde and **3 g** when tested against solvent dichloromethane (CK). For female *A. lucorum*, the differences were only statistically significant for pure benzaldehyde, β-ionone + benzaldehyde and **3 g** (Fig. 6A), but for male *A. lucorum*, the differences were statistically significant for all compounds (Fig. 6B). The differential attractiveness between odours of β-ionone[1d] (pure β-ionone left exposing to air and sunlight for one day, the same below) and CK was not pronounced in the experiments where they were offered together as choices. The results obtained in benzaldehyde[1d]–CK and (β-ionone[1d] + benzaldehyde[1d])–CK were similar to β-ionone[1d]–CK foregoing. Whereas, female and male *A. lucorum* showed a significant preference for **3g**[1d] and **3g**[7d] when they were offered next to CK respectively (Fig. 6C and 6D). Our results demonstrated that **3 g** was attractive to adult *A. lucorum* and its attractiveness persisted longer than β-ionone, benzaldehyde and β-ionone + benzaldehyde.

We then determined whether the analogues of β-ionone and benzaldehyde could attract *A. lucorum* in an alfalfa field. The results showed that the volatiles β-ionone, benzaldehyde and β-ionone + benzaldehyde (positive controls) attracted significantly more *A. lucorum* (females and males) than dichloromethane (blank control, CK).

CK). The number of *A. lucorum* in traps containing analogue lures varied with the analogue. Compounds **3a**, **3c**, **3d**, **3j**, **3k**, and **3m** trapped a moderate number of *A. lucorum*, while **3b**, **3e**, **3f**, **3g**, **3h**, **3i**, and **3l** trapped a high number *A. lucorum*. In particular, **3g** trapped significantly more *A. lucorum* than positive control (Fig. 7).

The compounds that were most attractive in the field were not the same compounds that elicited the strongest EAG responses in the laboratory. For example, **3d**, **3k**, and **3m** elicited strong EAG responses in the laboratory but did not result in the trapping of a high number of *A. lucorum* in the field. Laboratory EAG experiment was conducted in an unnatural environment. The odours in the experimental area, interference from instrument and human activities may influence the result of EAG experiment. In addition, field experiment was carried out in a natural environment. Insect physiological condition, environmental condition and other factors can affect the efficiency of the tested compounds used in the fields [11]. Therefore EAG response values may only represent whether the test compounds stimulate insects, but not reflect positive correlation with the insects behavior [31–33].

We also determined whether β-ionone, benzaldehyde, β-ionone + benzaldehyde or **3 g** attracted *A. lucorum* constantly. The result showed that β-ionone, benzaldehyde and β-ionone + benzaldehyde attracted significantly more *A. lucorum* than dichloromethane

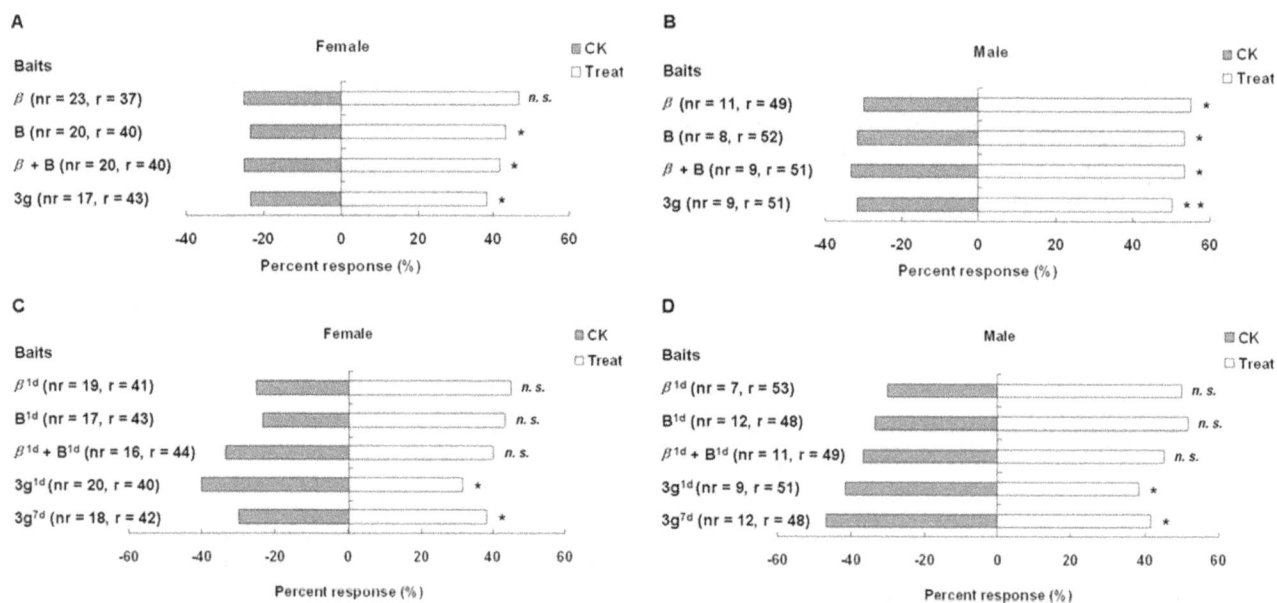

Figure 6. Choices of adult *A. lucorum* in the Y-tube olfactometer. (A) female towards pure β-ionone, benzaldehyde, β-ionone + benzaldehyde and 3 g, (B) male towards pure β-ionone, benzaldehyde, β-ionone + benzaldehyde and 3 g, (C) female towards β-ionone[1d], benzaldehyde[1d], β-ionone[1d] + benzaldehyde[1d], 3g[1d] and 3g[7d], (D) male towards β-ionone[1d], benzaldehyde[1d], β-ionone[1d] + benzaldehyde[1d], 3g[1d] and 3g[7d]. The bars represent the percentage of tested insects that made a particular choice. The asterisks with the choice bars indicate significant preferences. *P<0.05, **P<0.01, n.s. = not significant, nr = not reacting, r= reacting. Superscript characters of compounds represent the periods that the chemicals left exposing to air and sunlight.

(blank control, **Ctrl**) in the first two days and then attracted very little *A. lucorum* in the last five days. However, the analogue **3 g** trapped sustained and balanced numbers of *A. lucorum* during seven days (Fig. 8). This indicated that **3 g** possessed persistent attractiveness to *A. lucorum*.

Discussion

Plants synthesize and release volatile organic compounds. These signals can be detected by phytophagous insects via olfactory sensilla on the antennae and used to locate hosts and avoid non-host plants [34–36]. Plant volatiles have been artificially synthesized and successfully used as pest attractants in IPM. For example, methyl anthranilate has been used as an attractant for the thrips *Thrips hawaiiensis* Morgan and *Thrips coloratus* Schmutz [17]. A blend of *cis*-3-hexene acetate, linalol, and methyl

jasmonate was shown to be attractive to the Colorado potato beetle, *Leptinotarsa decemlineata* (Say) [18].

β-ionone and benzaldehyde are common plant volatiles released by many kinds of crops [20–23]. Although a recent study indicated that β-ionone and benzaldehyde can alter the behavior of *A. lucorum* adults under laboratory conditions [29], they could not be efficiently used for pest control in the field because of their low stability and insufficient attractiveness.

So far there has been no publish about plant volatile analogues synthesis and their use. In the current study, we try to explore novel ecological approach to control pest based on synthetic plant volatile analogues. We first hypothesized and synthesized analogues of β-ionone and benzaldehyde that combined moieties of the chemicals in order to produce a compound with increased stability and enhanced attractiveness to *A. lucorum* adults.

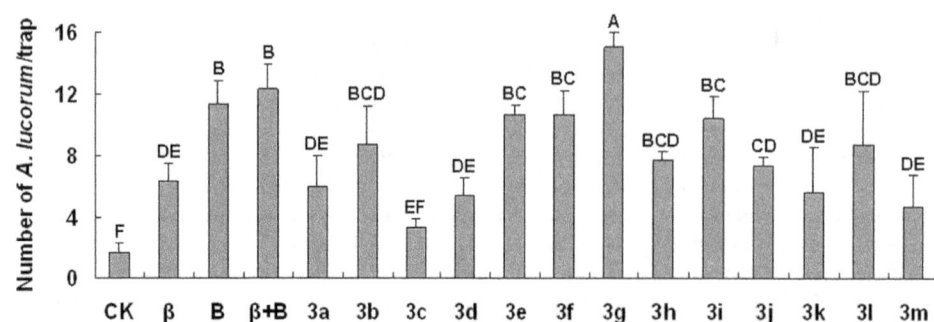

Figure 7. Number of *A. lucorum* captured in sticky traps (mean ± SD per trap) baited with synthetic analogues of *β*-ionone and benzaldehyde during 16–27 July 2012. CK = blank control, β= β-ionone, and B = benzaldehyde, β+B= β-ionone + benzaldehyde. Means with the same letter are not significantly different.

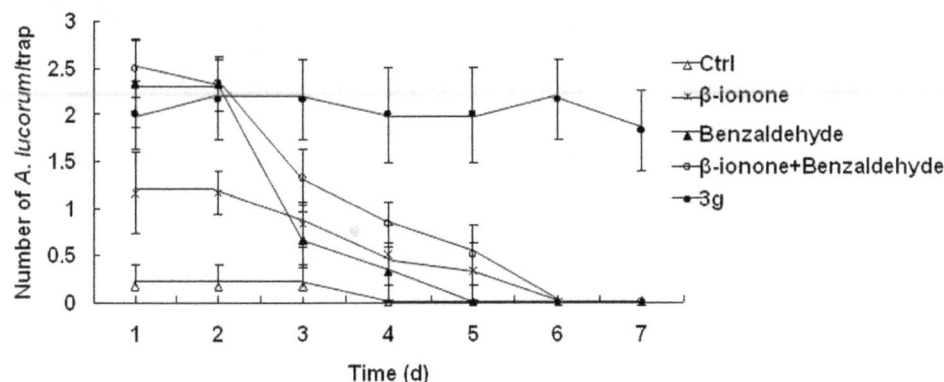

Figure 8. Number of *A. lucorum* captured in sticky traps (mean ± SE per trap) baited with *β*-ionone, benzaldehyde, *β*-ionone + benzaldehyde and 3 g from different day during 15–22 July 2013.

In this study, the analogues synthesized were demonstrated to have high stability compared to the original plant volatiles *β*-ionone or benzaldehyde. The laboratory EAG experiment showed that most of the analogues elicited responses from *A. lucorum* adult antennae. The laboratory behavior experiment displayed that analogue **3 g** was attractive to *A. lucorum* and its attractiveness persisted longer than the original plant volatiles. The field experiment indicated that most of the analogues were attractive to *A. lucorum* adults. The high stability and persistent attractiveness of analogue **3 g** in particular make it suitable for field use and potential as an attractant. The results support our hypothesis that designing and synthesizing analogues by combining the bioactive components of *β*-ionone and benzaldehyde would contribute to increasing stability and enhancing attractiveness.

Now it's necessary to increase the number of studies like this to design, synthesize plant volatile analogues and evaluate their bioactivities to insect. Once this is accomplished, there is the possibility of supporting it as a widely-accepted ecological approach that may extend its use in pest control in the field.

Acknowledgments

We thank Professor Yunhe Li from the Institute of Plant Protection, Chinese Academy of Agricultural Sciences for valuable comments on the manuscript. We thank the graduate trainees at Department of Entomology, Henan Institute of Science and Technology in 2012 and 2013 for assistance with the field surveys.

Author Contributions

Conceived and designed the experiments: KW YS. Performed the experiments: YS HY. Analyzed the data: KW YS HY. Contributed reagents/materials/analysis tools: KW YS HY. Contributed to the writing of the manuscript: KW JP JZ YS.

References

1. Lu YH, Wu KM, Jiang YY, Xia B, Li P, et al. (2010) Mirid bug outbreaks in multiple crops correlated with wide-scale adoption of Bt cotton in China. Science 328: 1151–1154.
2. Lu YH, Wu KM (2008) Biology and Control of Cotton Mirids. Golden Shield Press. 151 p.
3. Ting YQ (1963) Studies on the ecological characteristics of cotton mirids. I. effect of temperature and humidity on the development and distribution of the pests. Acta Phytophy Sin 2: 285–296.
4. Ting YQ (1964) Studies on the population fluctuations of cotton mirids in the cotton cultivation region of Kwanchung, Shensi, China. Acta Entomol Sin 13: 298–310.
5. Lu YH, Wu KM, Guo YY (2007) Flight Potential of *Lygus lucorum* (Meyer-Dür) (Heteroptera: Miridae). Environ Entomol 36: 1007–1013.
6. Lu YH, Wu KM, Wyckhuys KAG, Guo YY (2009) Comparative flight performance of three important pest Adelphocoris species of Bt cotton in China. Bull Entomol Res 99: 543–550.
7. Price PW, Denno RF, Eubanks MD, Finke DL, Kaplan I (2011) *Insect ecology: behavior, populations and communities.* Cambridge University Press, Cambridge.
8. Takabayashi J, Dicke M (1996) Plant-carnivore mutualism through herbivore-induced carnivore attractants. Trends Plant Sci 1: 109–113.
9. Kessler A, Baldwin IT (2001) Defensive function of herbivore-induced plant volatile emissions in nature. Science 291: 2141–2144.
10. Pare PW, Tumlinson JH (1999) Plant volatiles as a defense against insect herbivores. Plant Physiol 121: 325–332.
11. James DG (2005) Further field evaluation of synthetic herbivore-induced plant volatiles as attractants for beneficial insects. J Chem Ecol 31: 481–495.
12. Simpson M, Gurr GM, Simmons AT, Wratten SD, James DG, et al. (2011) Attract and reward: combining chemical ecology and habitat manipulation to enhance biological control in field crops. J Appl Ecol 48: 580–590.
13. Farmer EE, Ryan CA (1990) Interplant communication: airborne methyl jasmonate induces synthesis of proteinase inhibitors in plant leaves. Proc Natl Acad Sci 87: 7713–7716.
14. Shulaev V, Silverman P, Raskin I (1997) Airborne signalling by methyl salicylate in plant pathogen resistance. Nature 385: 718–721.
15. Baldwin IT, Halitschke R, Paschold A, Von Dahl CC, Preston CA (2006) Volatile signaling in plant-plant interactions: "talking trees" in the genomics era. Science 311: 812–815.
16. Cruz-López L, Malo EA, Toledo J, Virgen A, Mazo AD, et al. (2006) A new potential attractant for *Anastrepha obliqua* from *Spondias mombin* fruits. J Chem Ecol 32: 351–365.
17. Murai T, Imai T, Maekawa M (2000) Methyl anthranilate as an attractant for two thrips species and the thrips parasitoid ceranisus menes. J Chem Ecol 26: 2557–2565.
18. Martel JW, Alford AR, Dickens JC (2005) Laboratory and greenhouse evaluation of a synthetic host volatile attractant for Colorado potato beetle, *Leptinotarsa decemlineata* (Say). Agr Forest Entomol 7: 71–78.
19. Hedin PA, Thompson AC, Gueldner RC, Minyard JP (1971) Constituents of the cotton bud. Phytochemistry 10: 3316–3318.
20. Loughrin JH, Hamilton-Kemp TR, Andersen RA, Hildebrand DF (1990) Volatiles from flowers of Nicotiana sylvestris, N. otophora and Malus × domestica: headspace components and day/night changes in their relative concentrations. Phytochemistry 29: 2473–2477.

21. Baldwin EA, Scott JW, Shewmaker CK, Schuch W (2000) Flavor trivia and tomato aroma: biochemistry and possible mechanisms for control of important aroma components. HortScience 35: 1013–1021.

22. Kolosova N, Gorenstein N, Kish CM, Dudareva N (2001) Regulation of circadian methyl benzoate emission in diurnally and nocturnally emitting plants. Plant Cell 13: 2333–2347.

23. Underwood BA, Tieman DM, Shibuya K, Loucas HM, Dexter RJ, et al. (2005) Ethylene-regulated floral volatile synthesis in petunia corollas. Plant Physiol 138: 255–266.

24. Obata T, Koh HS, Kim M, Fukami H (1981) Planthopper atractants in rice plant. Appl Entomol Zool 25: 47–51.

25. Wang SF, Ridsdill-Smith TJ, Ghisalberti EL (2005) Chemical defenses of Trifolium glanduliferum against redlegged earth mite Halotydeus destructor. J Agric Food Chem 53: 6240–6245.

26. Knudsen JT, Eriksson R, Gershenzon J, Stahl B (2006) Diversity and distribution of floral scent. Bot Rev 72: 1–120.

27. Blum MS, Padovani F, Curley A, Hawk RE (1969) Benzaldehyde: defensive secretion of a harvester ant. Comp Biochem Phys 29: 461–465.

28. Lee BH, Choi WS, Lee SE, Park BS (2001) Fumigant toxicity of essential oils and their constituent compounds towards the rice weevil, *Sitophilus oryzae* (L.). Crop Prot 20: 317–320.

29. Tian WH (2012) Exploration and functioonal analysis of odorant binding protein related genes in green plant bug, *Apolygus lucorum*. Master dissertation, Shanxi Normal University, Shanxi.

30. Roelofs WL (1984) Electroantennogram assays: rapid and convenient screening procedures for pheromones. In: Hummel HE, editors. Techniques in pheromone research. Springer, New York, pp. 131–159.

31. Xiao C, Du JW, Zhang ZN (2000) Electroantennogram responses of cotton bollworm (*Helicoverpa armigera*) to several plant volatiles. Acta Agriculturae Jiangxi, 12: 27–31.

32. Liu XH (2011) The development of phyto-attractant of *Chlorophorus* sp. Master dissertation, Beijing Forestry University, Beijing.

33. Chen W (2012) Electroantennogram response of *Nilaparvata lugens* (Stal) to rice volatiles and cloning odorant binding protein genes. Master dissertation, Huazhong Agricultural University, Wuhan.

34. Pickett JA, Wadhams LJ, Woodcock CM (1998) Insect supersense: mate and host location by insects as model systems for exploiting olfactory interactions. Biochemist 20: 8–13.

35. Bruce TJA, Wadhams LJ, Woodcock CM (2005) Insect host location: a volatile situation. Trends Plant Sci 10: 269–174.

36. Halitschke R, Stenberg JA, Kessler D, Kessler A, Baldwin IT (2008) Shared signals 'alarm calls' from plants increase apparency to herbivores and their enemies in nature. Eco Lett 11: 24–34.

Impaired Olfactory Associative Behavior of Honeybee Workers Due to Contamination of Imidacloprid in the Larval Stage

En-Cheng Yang[1,2]*, Hui-Chun Chang[1], Wen-Yen Wu[1], Yu-Wen Chen[3]

1 Department of Entomology, National Taiwan University, Taipei, Taiwan, **2** Graduate Institute of Brain and Mind Sciences, National Taiwan University, Taipei, Taiwan, **3** Department of Animal Science, National Ilan University, Ilan, Taiwan

Abstract

The residue of imidacloprid in the nectar and pollens of the plants is toxic not only to adult honeybees but also the larvae. Our understanding of the risk of imidacloprid to larvae of the honeybees is still in a very early stage. In this study, the capped-brood, pupation and eclosion rates of the honeybee larvae were recorded after treating them directly in the hive with different dosages of imidacloprid. The brood-capped rates of the larvae decreased significantly when the dosages increased from 24 to 8000 ng/larva. However, there were no significant effects of DMSO or 0.4 ng of imidacloprid per larva on the brood-capped, pupation and eclosion rates. Although the sublethal dosage of imidacloprid had no effect on the eclosion rate, we found that the olfactory associative behavior of the adult bees was impaired if they had been treated with 0.04 ng/larva imidacloprid in the larval stage. These results demonstrate that a sublethal dosage of imidacloprid given to the larvae affects the subsequent associative ability of the adult honeybee workers. Thus, a low dose of imidacloprid may affect the survival condition of the entire colony, even though the larvae survive to adulthood.

Editor: Nicolas Chaline, Université Paris 13, France

Funding: This study was part of the interdisciplinary research on colony collapse disorder (CCD) of honeybee (Apis mellifera) (Project No. 97R0533-3 and 98R0529-3), and financially supported by the Excellence Research Program of National Taiwan University to ECY and HCC. The funders had no role in study design, data collection and analysis, decision to publish, or preparation of the manuscript.

Competing Interests: The authors have declared that no competing interests exist.

* E-mail: ecyang@ntu.edu.tw

Introduction

Honeybees play critical roles in agriculture and the global ecosystem by pollinating plants while at the same time producing bee products with a high economic value [1,2,3]. However, the rather recent phenomenon of colony collapse disorder (CCD) involving the sudden and massive disappearance of bee colonies around the world is worrisome [4]. This phenomenon manifests itself with the en masse disappearance of adult bees with only a few adult bee bodies being found around the beehives. The reason for this tends to be that these bees died while away from the hive, collecting pollen and nectar in the field, and were unable to navigate back home, thus leading to CCD. The severe loss of bee products and agricultural products caused by CCD [5,6,7] is of great concern to academics as well as to farmers. Recent studies show that RNA-induced acute paralysis virus, chronic paralysis virus, Kashmir bee virus (KBV), deformed wing virus (DWV), Israeli acute paralysis disease (IAPV), and disease caused by *Nosema ceranae* all contribute to CCD [8,9,10,11]. However, the large scale use of systemic insecticides, such as imidacloprid and thiamethoxam, has also been considered as being a major contributing factor to CCD. Thiamethoxam has recently been demonstrated to induce homing failure that could potentially cause colony collapse as a result of a sublethal dosage as low as 1.34 ng/bee (i.e. an acute oral exposure to 20 µL of liquid containing 67 µg/L of the insecticide) [12]. An increasing number of studies imply that

imidacloprid is also associated with colony disorder. Therefore, this study investigated the effect of imidacloprid on honeybees.

Imidacloprid is a neonicotinoid neurotoxic insecticide. Its major effect is acting on the nicotinic acetylcholine receptors (nAChR) of insects [13,14], causing death of the nervous system due to hyper-excitation followed by paralysis [15,16]. It has been considered an ideal insecticide because its toxicity is higher to insects than to mammals [17]. Imidacloprid is a systematic insecticide with contact toxicity and gastric toxicity. It can be applied either by foliar spray application, granular spray application, irrigation or seed treatment to protect plants from pest invasions [18]. In 1994, an event of obvious weakening of bee colonies, and a reduction in bee products was reported after the suspected consumption of sunflower nectar by the bees resulted in abnormal honeybee behavior to a point of the bees failing to return home [5,6]. This event is likely attributed to the fact that sunflower Gaucho® seeds were used in that area of France. Gaucho® seed is a seed treated with imidacloprid. Imidacloprid spreads to every part of the plant through the vascular system (i.e. xylem) and kills insects that feed from the plant. The sunflower Gaucho® seeds event increased the awareness of the important influence the use of imidacloprid has on honeybees, and resulted in an increased interest in studying the median lethal dosage (LD_{50}) and median lethal concentration (LC_{50}) of imidacloprid on honeybees [5,6,18,19,20,21,22,23]. Although LD_{50} and LC_{50} are the prime toxicity indicators for insecticides, the sublethal dosage of imidacloprid on honeybees needs to be considered as well. Previous studies have shown that after feeding the honeybees imidacloprid, their activity level decreased at an imidacloprid concentration of 100 μg/kg [24], and in addition their collection activities changed at a concentration above 6–100 μg/kg [25,26,27,28,29,30] while their olfactory learning and memory abilities were impaired at a concentration above 12–24 μg/kg [22,27]. The bees' medium-term olfactory memory was impaired with an imidacloprid dose of 12 ng/bee [31]. Recently, Eiri and Nieh reported that the ingestion of imidacloprid of 0.21 ng/bee could increase the threshold of sucrose concentration that a forager would accept and reduce waggle dancing at different time scale which consequently diminished the fitness of the bee colony [32].

Honeybee foragers collect food outside the hives and thus have frequent contact with plants that are contaminated with imidacloprid [33], and thus they bring honey and pollen containing imidacloprid back to their hives for storage [34,35,36]. Nurse bees then feed the imidacloprid-containing honey and pollen to the bee larvae, which could possibly affect their development as a result of accumulating the influences of imidacloprid during their growth period. However, previous studies failed to demonstrate a reduced honeybee mortality rate, collection activities, beeswax production, larvae capped-brood rate, colony activities, homing rate, and pollen carrying rate after feeding imidacloprid to honeybees [6,37]. It was therefore concluded that a low concentration of imidacloprid does not affect the bee colony and the larvae capped-brood rate. In the natural environment, imidacloprid concentration is less than 10 μg/kg in the soil, nectar and pollen [5,6,31,34,38,39,40,41]. Nevertheless, this does not exclude the possibility of accumulative intoxication through the repetitive consumption of honey and pollen containing only a low concentration of imidacloprid. In addition to the flowering products of a plant, the resin sources for propolis are assumed to transfer this systemic insecticide into the bee colony as well [42]. The relatively high residue of imidacloprid was observed in the honeycomb and propolis of depopulated beehives [42], indicating that the insecticide could be accumulated in these materials, resulting in the larvae of the colonies being continuously exposed to the contamination before the hives were depopulated. However, previous studies focused on the investigation of adult bees and bee colonies rather than on larvae. Therefore, our study focused on providing honeybee larvae with various doses of imidacloprid and to investigate and observe if the larvae, capped-brood, pupation and eclosion rates as well as the olfactory associative behavior change after having been exposed to imidacloprid.

Materials and Methods

1. Source of the test insects

Apis mellifera L. served as the test insect for this study. Honeybee colonies were raised in the apiaries of National Ilan University. Each test colony had a population working on about 9 frames of honeycomb in a Langstroth hive, and the test population included a queen bee spawning normally. The hives were checked every week to ensure that the normal function of the honeybee colony was being maintained.

2. Preparation of chemicals

Imidacloprid (95%, TG, Bayer Cropscience AG, Monheim am Rhein, Germany) is in powder form, thus it has to be dissolved in an organic solvent such as dimethylsulfoxide (DMSO) or acetone. We did not examine if there were any negative influences on the honeybees from the organic solvents because the amount of

organic solvent was trivial in the end test solution (0.1 or 1%, v/v) [19,20,22,27,31,37]. Previous studies showed that the feeding behavior of honeybees is not influenced by 0.1% DMSO treatment; however, the feeding activity of honeybees reduces significantly with acetone treatment [30]. We therefore chose DMSO (MP Biomedicals LLC., Solon, OH, USA) as the solvent for imidacloprid.

The imidacloprid stock solutions were prepared with concentrations of 8 and 200 g/L (imidacloprid in DMSO), respectively. Test solutions of imidacloprid were freshly prepared just prior to each application. The intermediate-concentration solutions of 100 and 6000 mg/L were prepared in advance by diluting the 8 g/L stock solution with DMSO. Then 2 μL of each of the above prepared intermediate-concentration solutions was added into 1998 μL of distilled deionized water (DDW), resulting in test solutions with imidacloprid concentrations of 0.1 and 6 mg/L. These test solutions contained 0.1% (v/v) DMSO. Another set of imidacloprid test solutions with higher concentration were prepared from the 200 g/L stock solution by adding 0.5, 5, 10, 15 and 20 μL of this stock solution to 1999.5, 1995, 1990, 1985 and 1980 μL of DDW respectively, resulting in an imidacloprid concentration of 50, 500, 1000, 1500 and 2000 mg/L with a DMSO concentration ≤1% in the imidacloprid test solution (the DMSO concentrations were 0.025, 0.25, 0.5, 0.75 and 1% respectively). In addition, a set of imidacloprid test solutions with sublethal concentration (according to the result of this study) were prepared for testing the effect of imidacloprid on olfactory association. The intermediate-concentration solutions of 0.01, 0.1, 1 and 10 mg/L were prepared by diluting the 8 mg/mL stock solution with DMSO. Fresh test solutions of imidacloprid were prepared by adding 20 μL of each of the above prepared intermediate-concentration solutions into 1980 μL of DDW respectively, resulting in imidacloprid concentrations of 0.1, 1, 10 and 100 μg/L with an uniform DMSO concentration of 1% (v/v) in the imidacloprid test solution.

3. Testing the effect of imidacloprid on larval survival

Four honeybee colonies were selected. The queens were restricted to depositing their eggs in empty frames of honeycomb using vertical and horizontal queen excluders for 24 hours in the test hives. The frames with eggs were left inside the test hives and checked daily. On the 3rd day, the frames were removed and transparent slides were placed on the honeycombs to mark the relative locations of the cells occupied by 1-day-old (emerged within 24 hours) larvae with brood food. More than three hundred cells were marked in each colony. The larvae in the cells were divided into 10 groups, each consisting of 30 to 40 larvae, and each marked with a different color on the transparent slide. The larvae from 7 of these 10 groups were treated with different doses of imidacloprid by adding 1 μL of imidacloprid test solution each at the respective concentrations of 0.1, 6, 50, 500, 1000, 1500 and 2000 mg/L into their cells. The larvae from the other 3 groups were set as control groups in order to evaluate the effect of the solvents. For the control groups, the larvae were treated with 1 μL each of DDW and 0.1 and 1% (v/v) DMSO solutions respectively using the same procedures of the imidacloprid treatment groups. After the drug application, the marked transparent slides were removed and kept as a reference for the applications of the following days. The frames of honeycomb were then put back into the respective colony, and the larvae were allowed to be reared inside the colonies. The drug application was conducted once a day for 4 consecutive days (from 1 to 4 days old), and the total doses of imidacloprid added into the nest cells were 0.4, 24, 200, 2000, 4000, 6000 and 8000 ng/larva (nominal doses) respectively.

Chemicals can be absorbed by the larvae orally or by contact. The capped-brood rate was calculated starting from day 7. On day 15, the pupae were removed from their colonies and placed on 24-hole plates to calculate the pupation rate. The pupae were later placed in a dark incubator at 34°C and 70% relative humidity without other bees for eclosion rate observation.

4. Treatment for testing the effect of imidacloprid after eclosion

An additional three honeybee colonies were selected for testing the effect of imidacloprid at a sublethal dosage on the contaminated larvae in adulthood. Seven groups of 100 one-day-old larvae each were obtained by the same procedures as described above. The other corresponding procedures of adding 1 μL of DDW, solution of 1% DMSO and solutions of 0.1, 1, 10 and 100 μg/L imidacloprid respectively once a day to each cell of the larvae among the 6 groups for 4 consecutive days (from 1 to 4 days old) were performed as well. The total amounts of imidacloprid added over 4 days were 0, 0, 0.0004, 0.004, 0.04 and 0.4 ng/larva (nominal doses) respectively. A group of 100 larvae was left intact as a control. On day 15, the pupae were removed from the colonies and placed on 24-hole plates and then transferred into the dark incubator under the conditions described above. The eclosion rate was about 90% for each group. Color labels were affixed to the dorsum of the thorax of the honeybees after eclosion so as to differentiate the honeybees of each group. After having been color-labeled, the honeybees were released into their original colony. The color-labeled honeybees were randomly selected to test their olfactory associative behavior 15 days after eclosion. Although worker bees turn into foragers 20 days after eclosion [43], our pretest showed that sampling difficulties occurred if the olfactory association was to be tested on day 20. This difficulty was possibly due to the honeybees' homing failure similar to the phenomenon that was demonstrated in the adulthood of a forager when subjected to acute contamination by neonicotinoids [12]. It has been shown that worker bees can associate on day 15 after eclosion [44]. Therefore, we choose the color-labeled honeybees on day 15 after eclosion to test the olfactory associative behavior of the worker bees.

5. Proboscis extension reflex (PER) test

Conventional conditioning of the proboscis extension reflex (PER), as shown in many previous studies, was used to test the honeybees' associative ability [22,45,46,47,48,49]. We applied the principles of classical conditioning [50] with odor as the conditioned stimulus (CS) and sugar water as the unconditioned stimulus (US) to test the olfactory associative behavior of the honeybees by observing their PER. Honeybees were placed in a dark incubator at 25°C and 40% relative humidity where they were starved for 4 hours before the test. After the 3rd hour, the honeybees were anesthetized by low temperature by being placed in a 4°C ice bucket for 5–10 minutes. After anesthesia, honeybees were fixed at 1000 μL pipette tips by beeswax/resin mixture, left there for 1 hour until their physiological conditions recovered. Cotton swabs dipped with 50% (w/v) sucrose solution were applied over the honeybees' antenna (precautions were taken not to let the honeybees take in any sugar water). Honeybees with a normal PER were tested for their olfactory associative behavior. Those that could not produce a PER were eliminated. Groups of approximately thirty honeybees per group were tested for their olfactory associative behavior after various treatments.

When training the bee to associate, the bees were offered 50% sucrose solution for 3 seconds. Then the odorous stimulation was applied for 6 seconds by placing the bees 1 cm away from the

blow hole of a pneumatic PicoPump (PV380, World Precision Instruments, Inc., Sarasota, FL, USA) connected to a bottle filled with citral (≥96%, FG, W230308, Sigma-Aldrich, Inc., St. Louis, MO, USA). At the 3rd second the bees were simultaneously offered sugar water while continuing to experience the odor.

The test for the PER by providing odorous stimulation only was conducted to observe if the honeybees associated this stimulation with sugar water. Each honeybee went through 4 associative tests (Conditioning 1–4) and each test was held 20 minutes apart. The successful rate of the PER response rate was calculated according to Ray and Ferneyhough [44]:

PER response rate =

(counts of honeybees with response/counts of honeybees tested) × 100%

6. Statistical analysis and calculation of medial lethal dose (LD$_{50}$)

SPSS software (IBM Corp., Armonk, NY, USA) was used for the statistical analysis in this study. The effects of organic solvents and imidacloprid on the larvae capped-brood rate, pupation rate, eclosion rate as well as honeybee's olfactory associative behavior after eclosion were analyzed by the two-tailed Kruskal-Wallis H tests, because the data were not distributed normally, and a two-tailed Mann-Whitney U test with the Dunn-Šidák correction at the 95% confidence level was performed as a post-hoc test. In this study each data was shown as the mean ± standard deviation. The CalcuSyn software (Biosoft, Ferguson, MO, USA) was used to calculate the LD$_{50}$ of imidacloprid for the honeybee larvae capped-brood rate.

Results

1. Influence of DMSO on larvae capped-brood rate, pupation rate and eclosion rate

Larvae capped-brood rates were 100±0, 98.75±2.50 and 91.04±9.36% for larvae treated with DDW, 0.1 and 1% DMSO, respectively. The analysis showed there was a variation among the larvae capped-brood rates (two-tailed Kruskal-Wallis H test, $\chi^2 = 18.646$, $df = 2$, $P < 0.001$). The capped-brood rate of the group of 1% DMSO was slightly lower than the rates of the groups of DDW (two-tailed Mann-Whitney U test with Dunn-Šidák correction, $U = 10320$, adjusted $P < 0.001$) and 0.1% DMSO (two-tailed Mann-Whitney U test with Dunn-Šidák correction, $U = 10137.5$, adjusted $P = 0.018 < 0.05$). Nevertheless, the pupation rates were 96.25±3.23, 97.50±3.54 and 91.04±9.36%, showing no significant difference among the three treatment groups (two-tailed Kruskal-Wallis H test, $\chi^2 = 5.061$, $df = 2$, $P = 0.08$). Eclosion rates were 89.38±6.57, 89.82±10.10 and 89.17±7.26%, showing no significant difference (two-tailed Kruskal-Wallis H test, $\chi^2 = 0.031$, $df = 2$, $P = 0.984$) as well. We summarized these results, and concluded that the DMSO applied in this study had an ignorable influence on the development of the larvae.

2. Influence of imidacloprid on the larvae capped-brood rate

Larvae capped-brood rates were 97.50±2.04, 90.83±7.39 and 87.50±6.16% for 0.4, 24 and 200 ng/larva imidacloprid treatment, while the rates were 98.75±2.50 and 91.04±9.36% for 0.1 and 1% DMSO treatment (control group). With the dose raised to 2000, 4000, 6000 and 8000 ng/larva, most larvae were removed

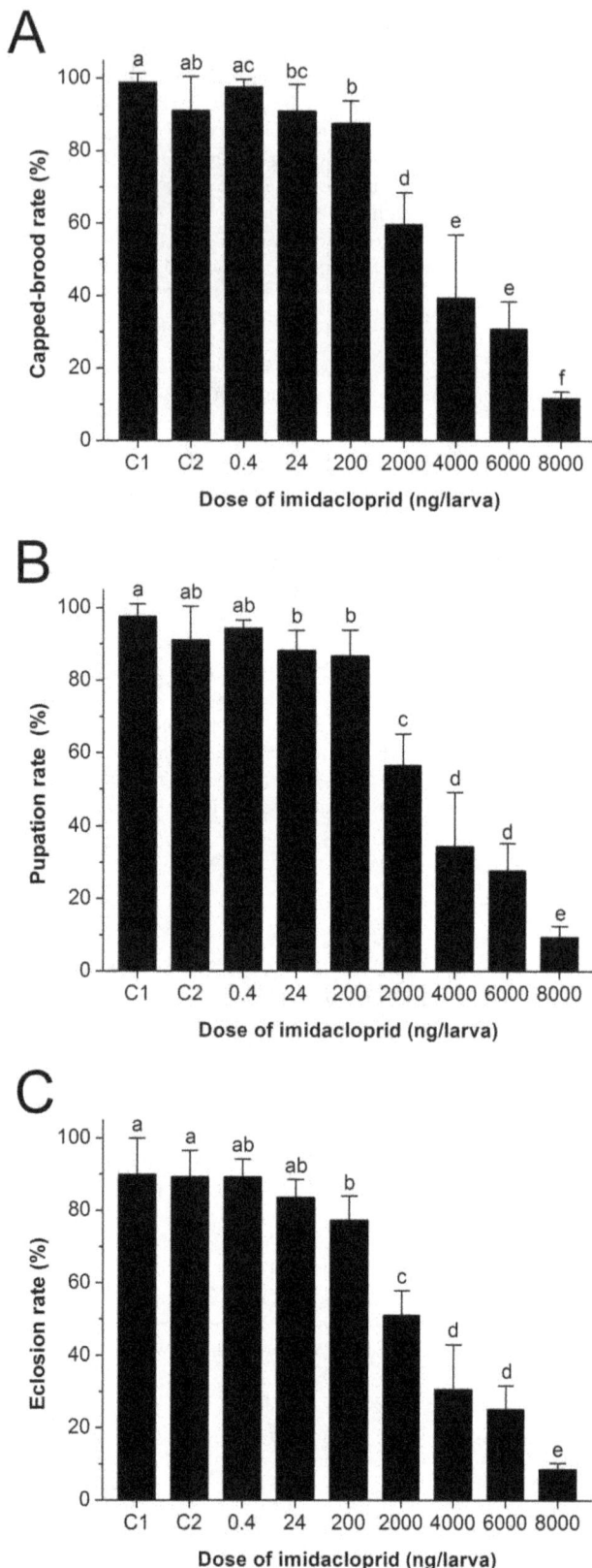

Figure 1. Lethal effect of imidacloprid on the honeybee larvae.
The subfigures show the effects of imidacloprid on the capped-brood rate (A), pupation rate (B) and the eclosion rate (C) of honeybee larvae under field conditions (Tested larvae were obtained from 4 colonies, $N_{0.1\% \text{ DMSO}} = 40+40+35+40$, $N_{1\% \text{ DMSO}} = 30+40+40+30$, $N_{0.4 \text{ ng}} = 40+40+35+40$, $N_{24 \text{ ng}} = 30+30+40+30$, $N_{200 \text{ ng}} = 30+40+40+30$,

$N_{2000 \text{ ng}} = 40+40+30+30$, $N_{4000 \text{ ng}} = 40+40+30+30$, $N_{6000 \text{ ng}} = 40+40+30+30$, $N_{8000 \text{ ng}} = 30+40+30+30$ larvae). The doses of imidacloprid treatments are 0.4, 24, 200, 2000, 4000, 6000 and 8000 ng/larva. C1 and C2 are the control groups (0.1 and 1% DMSO). In each subfigure, any two effects of the experimental treatments without any same letter above the columns are significantly different (two-tailed Kruskal-Wallis H test, $P<0.001$, compared by two-tailed Mann-Whitney U tests with Dunn-Šidák correction at the 95% confidence level).

by the nurse bees by day 2 or day 3 with larvae capped-brood rates of 59.58 ± 8.83, 39.38 ± 17.37, 30.83 ± 7.55 and $11.67\pm1.92\%$ respectively. Larvae capped-brood rates were significantly different among the experimental treatments (two-tailed Kruskal-Wallis H test, $\chi^2 = 560.317$, $df=8$, $P<0.001$). There was no significant difference between the control group of 0.1% DMSO and the group that received low dose imidacloprid treatment at a dose of 0.4 ng/larva (two-tailed Mann-Whitney U test with Dunn-Šidák correction, $U= 11857.5$, adjusted $P= 0.991$), but the groups that received a higher dose above or equal to 24 ng/larva differed significantly from the control group (two-tailed Mann-Whitney U tests with Dunn-Šidák corrections, adjusted $P<0.05$) (Figure 1A). The LD_{50} of imidacloprid is about 1400 ng/larva as analyzed by the CalcuSyn software.

3. Influence of imidacloprid on the pupation rate

Pupation rates were 94.29 ± 2.24, 88.13 ± 5.54 and $86.67\pm7.07\%$ with 0.4, 24 and 200 ng/larva imidacloprid while the rates were 97.50 ± 3.54 and $91.04\pm9.36\%$ for the control group with 0.1 and 1% DMSO. With the imidacloprid dose raised to 2000, 4000, 6000 and 8000 ng/larva, the pupation rates were 56.46 ± 8.72, 34.38 ± 14.77, 27.71 ± 7.56 and $9.38\pm2.99\%$ respectively. Pupataion rates were significantly different among the experimental treatments (two-tailed Kruskal-Wallis H test, $\chi^2 = 565.454$, $df=8$, $P<0.001$). The pupation rates were not different between the control group of 0.1% DMSO and the group that received low dose imidacloprid treatment at a dose of 0.4 ng/larva (two-tailed Mann-Whitney U test with Dunn-Šidák correction, $U= 11625$, adjusted $P= 0.785$), but the groups that received a higher dose above or equal to 24 ng/larva were significantly different from the control group (two-tailed Mann-Whitney U tests with Dunn-Šidák corrections, adjusted $P<0.05$) (Figure 1B).

4. Influence of imidacloprid on the eclosion rate

Eclosion rates were 89.20 ± 4.90, 83.54 ± 4.92 and $77.29\pm6.64\%$ with 0.4, 24 and 200 ng/larva imidacloprid while the rates were 89.82 ± 10.10 and $89.17\pm7.26\%$ for the control group. With imidacloprid dose raised to 2000, 4000, 6000 and 8000 ng/larva, the eclosion rates were 51.04 ± 6.78, 30.63 ± 12.31, 25.21 ± 6.47 and $8.54\pm1.72\%$, respectively. Eclosion rates were significantly different among the experimental treatments (two-tailed Kruskal-Wallis H test, $\chi^2 = 486.874$, $df=8$, $P<0.001$). The eclosion rates were not different between the control group of 0.1% DMSO and the groups that received low dose imidacloprid treatments at a dose of 0.4 ng/larva (two-tailed Mann-Whitney U test with Dunn-Šidák correction, $U= 11935$, adjusted $P= 1$) and 24 ng/larva (two-tailed Mann-Whitney U test with Dunn-Šidák correction, $U= 9487.5$, adjusted $P= 0.756$), respectively, but the groups that received a higher dose above or equal to 2000 ng/larva differed significantly from the control group (two-tailed Mann-Whitney U tests with Dunn-Šidák corrections, adjusted $P<0.05$)(Figure 1C).

5. Influence of DMSO on the olfactory associative behavior of honeybees

The olfactory associative behavior of the larvae treated with DDW and 1% DMSO 15 days after eclosion is shown in Figure 2A. For the DDW treated group, the PER rate was 0% for conditioning trial 1 (T1), $33.02 \pm 5.52\%$ for conditioning trial 2 (T2), $47.78 \pm 1.92\%$ for conditioning trial 3 (T3) and $59.05 \pm 3.72\%$ for conditioning trial 4 (T4). For the control group, the PER rate was 0% for T1, $36.67 \pm 5.77\%$ for T2, $52.63 \pm 2.35\%$ for T3 and $61.21 \pm 3.94\%$ for T4. For the 1% DMSO treated group, PER rate was 0% for T1, $32.92 \pm 9.75\%$ for T2, $50.86 \pm 8.24\%$ for T3 and $63.17 \pm 3.34\%$ for T4. The PER responses of these groups showed no significant difference (two-tailed Kruskal-Wallis H tests, T2: $\chi^2 = 0.317$, $df = 2$, $P = 0.853$; T3:

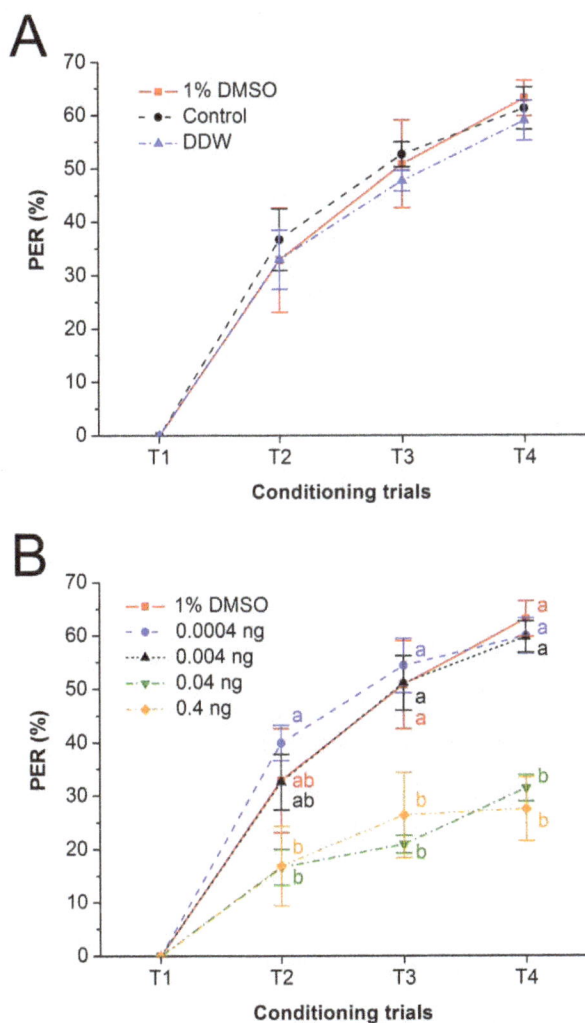

$\chi^2 = 0.46$, $df = 2$, $P = 0.794$; T4: $\chi^2 = 0.336$, $df = 2$, $P = 0.845$). These results indicate that the influence from 1% DMSO on the olfactory associative behavior after eclosion can be ignored.

6. Influence of imidacloprid on the olfactory associative behavior of honeybees

The curve of larvae treated with 0.0004–0.4 ng/larva imidacloprid 15 days after eclosion is shown in Figure 2B. For larvae treated with a low dose of 0.0004 ng imidacloprid, the response rates were 0, 40.00 ± 3.33, 54.44 ± 5.09 and $60.00 \pm 3.33\%$ for T1–T4. With the dose raised to 0.004 ng, the response rates were 0, 32.57 ± 5.24, 51.11 ± 5.09 and $59.72 \pm 2.93\%$ for T1–T4. With the relatively high dose of 0.04 ng, the response rates were 0, 16.67 ± 3.33, 20.95 ± 1.65 and $31.43 \pm 2.47\%$ for T1–T4. With 0.4 ng imidacloprid treatment, the response rates were 0, 16.90 ± 7.45, 26.42 ± 7.95 and $27.57 \pm 5.96\%$ for T1–T4. Results of the two-tailed Kruskal-Wallis H tests revealed that there were significant differences of PER rates among 1% DMSO and imidacloprid treated groups in T2 ($\chi^2 = 18.348$, $df = 4$, $P = 0.001 < 0.05$), T3 ($\chi^2 = 35.976$, $df = 4$, $P < 0.001$) and T4 ($\chi^2 = 43.413$, $df = 4$, $P < 0.001$), respectively. The bees belonging to the 1% DMSO and 0.0004 and 0.004 ng imidacloprid treated groups showed better olfactory associative ability than the bees treated with 0.04 and 0.4 ng imidacloprid in T3 and T4 (two-tailed Mann-Whitney U tests with Dunn-Šidák corrections, adjusted $P < 0.05$), and the bees belonging to the 0.0004 ng imidacloprid treated group showed better olfactory associative ability than the bees treated with 0.04 and 0.4 ng imidacloprid in T2 (two-tailed Mann-Whitney U tests with Dunn-Šidák corrections, adjusted $P < 0.05$)(Figure 2B).

Discussion

In our field test with larvae, the larvae were most of the time raised by the colony except when they were treated with imidacloprid by the experimenters. This method is as close as possible to the natural condition of the honeybee colony, which the larval survival was affected both by its development and the hygienic behavior, which ejected unhealthy larva, of the nurse bee. Thus, we applied it to evaluate the effects of exposing the larvae to imidacloprid contamination in the field. Our results showed that the capped-brood, pupation and eclosion rates as well as the subsequent olfactory associative ability in adulthood were affected by imidacloprid. The relative high survival rates of control groups and low dosage (below and equal 200 ng/larva) treatment groups against the dosage-related decline of the survival rates of high dosage (above and equal 2000 ng/larva) treatment groups indicates the cross contamination of treated imidacloprid was not occurred in this study. In addition, the high dosage of imidacloprid would knock down and kill a nurse bee (NOEC of the knockdown effect = 0.94 ng/bee [51,52], $LD_{50} = 3$–81 ng/bee [5,6,18,19,20,21,22,23]) and would result in larval deaths due to starvation. However, this was not observed in the high dose (above and equal 2000 ng/larva) treatment groups of this study, indicates the contaminated brood food should not be removed by nurse bees when the larva alive. Nevertheless, a damage of nursing task may not be excluded when the nursing behavior was not examined in this study.

Imidacloprid is a systematic insecticide and their residue could be found in the plant tissue and in the soil after spraying or seed-coating treatment. Previous studies have shown that the concentration of imidacloprid residue is below 10 µg/kg in the soil, nectar and pollen in an argo-environment [5,6,34,38,39,40,41,53]. Because a honeybee larva can consume 160 µL of brood food before its pupation [54], it is quite possible that honeybee larvae

Figure 2. Impaired olfactory associative behavior in adulthood caused by the larval contamination of imidacloprid. The subfigures show the effect of DMSO (A) and imidacloprid (B) on the olfactory associative behavior of the honeybee (Tested honeybees were obtained from 3 colonies, $N_{1\% \text{ DMSO}} = 35 + 25 + 30$, $N_{\text{Control}} = 33 + 30 + 30$, $N_{\text{DDW}} = 30 + 30 + 28$, $N_{0.0004 \text{ ng}} = 30 + 30 + 30$, $N_{0.004 \text{ ng}} = 32 + 30 + 30$, $N_{0.04 \text{ ng}} = 30 + 30 + 35$, $N_{0.4 \text{ ng}} = 26 + 29 + 32$ bees). T1, T2, T3 and T4 are conditioning trials 1–4, respectively. In figure 2B, any two effects of the experimental treatments among T2, T3 or T4 without any same letter beside the data points are significantly different (two-tailed Kruskal-Wallis H test, $P \leq 0.001$, compared by two-tailed Mann-Whitney U tests with Dunn-Šidák correction at the 95% confidence level).

are affected by the environmental residue of imidacloprid. In this study, we tested the dose at 0.0004, 0.004, 0.04 and 0.4 ng/larva, which corresponds to expose the larvae to an imidacloprid concentration of approximately 0.0025, 0.025, 0.25 and 2.5 µg/L respectively, which represents the level that is very likely present in an argo-environment. The result revealed that the adult olfactory associative behavior was impaired if the brood-food was contaminated by the imidacloprid treatment at a concentration of 0.25 µg/L or more. This is strong evidence and indicates that a honeybee larva could remain exposed to the residual imidacloprid in an agro-environment. Because honeybee larvae do not consume raw nectar or pollen, we presumed that they were protected from the contamination of a bee colony, or at least that they were protected by the repellent effect of imidacloprid on the forager [25] and the detoxification abilities of a nectar-collecting forager and a larva food-preparing nurse bee [55]. Nevertheless, because the detoxification gene is deficient in a honeybee [56], this protection may break down under the synergy of other stresses, such as malnutrition, disease and the intoxication by insecticides of adult workers, and result in colony disorder.

According to our results, the larva capped-brood rate was not influenced by imidacloprid below 0.4 ng/larva. However, the larvae capped-brood rate was significantly reduced when the imidacloprid dose was increased to 24 ng/larva and more (Figure 1A). The calculated imidacloprid LD_{50} is about 1400 ng/larva. Previous studies show that imidacloprid LD_{50} is 3–81 ng/bee for adult bees [5,6,18,19,20,21,22,23]. Comparing our results with previous studies, it is obvious that larvae have a higher tolerance to imidacloprid than adult bees. Imidacloprid acts mainly on a specific nAChR rather than on all nAChRs presented in a victim [31]. A honeybee possesses as many as 11 members of insect nAChR subunits [57], and these subunits are expressed differently in the different stages of honeybee development [58,59]. Furthermore, many structures are absent in the early stage of a larva such as the Kenyon cells in the mushroom bodies (MBs) [60,61] which are the primary target of nAChR for imidacloprid [62,63,64]. The high tolerance for imidacloprid therefore may be related to the fact that the larvae lack nAChR, which has a high agonistic affinity for imidacloprid in adults. In addition, it has been demonstrated that feeding laboratory reared larva with artificial food contaminated with imidacloprid at a rate as high as 400 mg/kg will induce apoptotic cell death in the tissue of a larva's midgut [65]. We therefore assume that the lethal effect on larvae treated with a high dose of imidacloprid may be attributed to an induced apoptosis rather than neural toxicity.

The understanding of the sublethal effect of imidacloprid on honeybee larvae is relatively poorer than what has been reported for the adult bee. Decourtye et al. reported a delayed development of honeybee larvae that were fed food contaminated with imidacloprid at 5 µg/kg [23,66]. This delayed-development effect was also observed in the honeybee larvae reared in honeycomb that contained residue of insecticides, including imidacloprid at 45 µg/kg [67]. A similar effect on larvae exposed to imidacloprid at 30–300 µg/kg was confirmed as well for another pollinator, *Osmia lignaria* [68]. In the present study, we report an effect that can be induced by a tiny dosage, as low as 0.04 ng/larva (or about 0.25 µg/L of the exposed concentration for 4 days), to impair the olfactory associative behavior in adulthood. Imidacloprid has a metabolic half-life ranging between 4.5–5 hour [69] in the adult honeybee, but the metabolites of imidacloprid, 5-hydroxy-imidacloprid, olefin and 4,5-dihydroxy-imidacloprid remain toxic to the honeybee [20,21,22,69,70,71]. Since the metabolism of the larvae remains unclear, we presumed that the imidacloprid and its metabolites could accumulate in the larvae and have a negative

effect on the larvae's development. The imidacloprid's primary target, nAChR, could be linked to the development of the brain and the neural plasticity [72,73]. In an adult honeybee's brain, the nAChRs are mostly distributed among the MBs, antennal lobes (ALs), antennal nerves, visual ganglions, central body and the suboesophageal ganglion [74,75,76]. Functional cytochrome oxidase histochemical studies have shown that the application of imidacloprid could significantly increase the neural metabolic activity among the lip and basal ring of the MBs rather than among the α-lobe of the MBs and ALs [31,77]. When honeybees are treated with nicotinic antagonist—mecamylamine, muscarinic and imidacloprid, then these chemicals bind with the nAChR of the MBs in the brain causing an impaired memory and learning ability [78,79]. The learning ability of honeybees is significantly reduced when they are fed a sucrose solution containing imidacloprid at 12, 24, 48, or 96 µg/kg [22]. In addition, the mid-term memory of honeybees is impaired when they are fed a dose of imidacloprid of 12 ng/bee [31]. Furthermore, the ingestion of imidacloprid at 0.21 ng/bee could increase the threshold of acceptable sucrose concentration of a forager after 1 hour and reduce waggle dancing 24 hours later [32]. Due to the fact that the symptoms of these above-mentioned acute intoxications of adults and the impairment of the olfactory associative behavior observed in this study seem to be related, it also is possible that the brain functions affected in these events may be connected with each other. We expect that the accumulated imidacloprid and its metabolites may interfere with the development of the functional regions of the brain such as in the MBs, ALs as well as the antennal nerves.

Besides the effect on individual larva, a sublethal dosage of imidacloprid (contaminant concentration at 0.7–10 µg/kg) also contributes to reducing the fertility of the queen and the colony growth of bumble bee colonies [80,81]. However, due to the experimental period was as short as the lifespan of a worker, it is very likely that the sublethal effect observed in this study is an original factor that affects the colony yet does not show the response from the colony.

In summary, this study demonstrated that the honeybee larvae are more tolerant to imidacloprid than the adult bees, but that their development, at least that of the MBs, ALs and antennal nerves may be very easily interfered with by imidacloprid contamination. Honeybees depend on the MBs and ALs in the brain to learn and memorize food location as well as their homing routes when they are out collecting [74,78,79,82,83,84,85]. Our results infer that although imidacloprid does not kill the larvae, when these honeybees with both learning and memory impairments go out collecting, it is highly likely that they cannot learn and memorize food locations and homing routes and that therefore they fail to return to their hives, causing a reduction of bee products and getting even worse to induce CCD. Because honeybee larvae could be affected by a contamination of imidacloprid contamination as low as 0.04 ng/larva, neonicotinoid insecticides should be applied very carefully.

Acknowledgments

We thank the anonymous reviewers for insightful corrections and improvements, and also thank SEC, NTU for statistical consulting services.

Author Contributions

Conceived and designed the experiments: ECY. Performed the experiments: ECY HCC. Analyzed the data: ECY HCC WYW YWC. Contributed reagents/materials/analysis tools: ECY HCC WYW YWC. Wrote the paper: ECY HCC WYW.

References

1. Free JB (1993) Insect pollination of crops. San Diego: Academic Press. 544 p.
2. Kevan PG (1999) Pollinators as bioindicators of the state of the environment: species, activity and diversity. Agricult Ecosys Environ 74: 373–393.
3. Porrini C, Sabatini AG, Girotti S, Fini F, Monaco L, et al. (2003) The death of honey bees and environmental pollution by pesticides: the honey bees as biological indicators. Bull Insectol 56: 147–152.
4. Stokstad E (2007) The case of the empty hives. Science 316: 970–972.
5. Schmuck R (1999) No causal relationship between Gaucho® seed dressing in sunflowers and the French bee syndrome. Pflanzenschutz-Nachr Bayer 52: 257–299.
6. Schmuck R, Schöning R, Stork A, Schramel O (2001) Risk posed to honeybees (Apis mellifera L, Hymenoptera) by an imidacloprid seed dressing of sunflowers. Pest Manag Sci 57: 225–238.
7. Swinton SM, Lupi F, Robertson GP, Hamilton SK (2007) Ecosystem services and agriculture: cultivating agricultural ecosystems for diverse benefits. Ecol Econ 64: 245–252.
8. Blanchard P, Schurr F, Celle O, Cougoule N, Drajnudel P, et al. (2008) First detection of Israeli acute paralysis virus (IAPV) in France, a dicistrovirus affecting honeybees (Apis mellifera). J Invertebr Pathol 99: 348–350.
9. Wang C-H, Lo C-F, Nai Y-S, Wang C-Y, Chen Y-R, et al. (2009) Honey bee colony cllapse disorder. Formosan Entomol 29: 119–138.
10. Higes M, Martín-Hernández R, Garrido-Bailón E, González-Porto AV, García-Palencia P, et al. (2009) Honeybee colony collapse due to Nosema ceranae in professional apiaries. Env Microbiol Rep 1: 110–113.
11. Maori E, Paldi N, Shafir S, Kalev H, Tsur E, et al. (2009) IAPV, a bee-affecting virus associated with colony collapse disorder can be silenced by dsRNA ingestion. Insect Mol Biol 18: 55–60.
12. Henry M, Béguin M, Requier F, Rollin O, Odoux J-F, et al. (2012) A common pesticide decreases foraging success and survival in honey bees. Science 336: 348–350.
13. Buckingham S, Lapied B, Corronc H, Sattelle F (1997) Imidacloprid actions on insect neuronal acetylcholine receptors. J Exp Biol 200: 2685–2692.
14. Matsuda K, Buckingham SD, Kleier D, Rauh JJ, Grauso M, et al. (2001) Neonicotinoids: insecticides acting on insect nicotinic acetylcholine receptors. Trends Pharmacol Sci 22: 573–580.
15. Tomizawa M, Yamamoto I (1992) Binding of nicotinoids and the related compounds to the insect nicotinic acetyicholine receptor. J Pestic Sci 17: 231–236.
16. Matsuda K, Shimomura M, Ihara M, Akamatsu M, Sattelle DB (2005) Neonicotinoids show selective and diverse actions on their nicotinic receptor targets: electrophysiology, molecular biology, and receptor modeling studies. Biosci Biotechnol Biochem 69: 1442–1452.
17. Liu MY, Casida JE (1993) High affinity binding of [³H]imidacloprid in the insect acetylcholine receptor. Pestic Biochem Physiol 46: 40–46.
18. Elbert A, Becker B, Hartwig J, Erdelen C (1991) Imidacloprid—a new systemic insecticide. Pflanzenschutz-Nachr Bayer 44: 113–136.
19. Suchail S, Guez D, Belzunces LP (2000) Characteristics of imidacloprid toxicity in two Apis mellifera subspecies. Environ Toxicol Chem 19: 1901–1905.
20. Suchail S, Guez D, Belzunces LP (2001) Discrepancy between acute and chronic toxicity induced by imidacloprid and its metabolites in Apis mellifera. Environ Toxicol Chem 20: 2482–2486.
21. Nauen R, Ebbinghaus-Kintscher U, Schmuck R (2001) Toxicity and nicotinic acetylcholine receptor interaction of imidacloprid and its metabolites in Apis mellifera (Hymenoptera: Apidae). Pest Manag Sci 57: 577–586.
22. Decourtye A, Lacassie E, Pham-Delègue M-H (2003) Learning performances of honeybees (Apis mellifera L) are differentially affected by imidacloprid according to the season. Pest Manag Sci 59: 269–278.
23. Decourtye A, Devillers J (2010) Ecotoxicity of neonicotinoid insecticides to bees. In: Thany SH, editor. Insect Nicotinic Acetylcholine Receptors. 1st ed. New York: Springer. pp. 85–95.
24. Medrzycki P, Montanari R, Bortolotti L, Sabatini AG, Maini S, et al. (2003) Effects of imidacloprid administered in sub-lethal doses on honey bee behaviour. Laboratory tests. Bull Insectol 56: 59–62.
25. Bortolotti L, Montanari R, Marcelino J, Medrzycki P, Maini S, et al. (2003) Effects of sub-lethal imidacloprid doses on the homing rate and foraging activity of honey bees. Bull Insectol 56: 63–67.
26. Colin ME, Bonmatin JM, Moineau I, Gaimon C, Brun S, et al. (2004) A method to quantify and analyze the foraging activity of honey bees: relevance to the sublethal effects induced by systemic insecticides. Arch Environ Contam Toxicol 47: 387–395.
27. Decourtye A, Devillers J, Cluzeau S, Charreton M, Pham-Delègue M-H (2004) Effects of imidacloprid and deltamethrin on associative learning in honeybees under semi-field and laboratory conditions. Ecotoxicol Environ Saf 57: 410–419.
28. Kirchner WH (1999) Mad-bee-disease? Sublethal effects of imidacloprid (Gaucho®) on the behaviour of honeybees. Apidologie 30: 422.
29. Ramirez-Romero R, Chaufaux J, Pham-Delègue M-H (2005) Effects of Cry1Ab protoxin, deltamethrin and imidacloprid on the foraging activity and the learning performances of the honeybee Apis mellifera, a comparative approach. Apidologie 36: 601–611.
30. Yang EC, Chuang YC, Chen YL, Chang LH (2008) Abnormal foraging behavior induced by sublethal dosage of imidacloprid in the honey bee (Hymenoptera: Apidae). J Econ Entomol 101: 1743–1748.
31. Decourtye A, Armengaud C, Renou M, Devillers J, Cluzeau S, et al. (2004) Imidacloprid impairs memory and brain metabolism in the honeybee (Apis mellifera L.). Pestic Biochem Physiol 78: 83–92.
32. Eiri DM, Nieh JC (2012) A nicotinic acetylcholine receptor agonist affects honey bee sucrose responsiveness and decreases waggle dancing. J Exp Biol 215: 2022–2029.
33. Koch H, Weißer P (1997) Exposure of honey bees during pesticide application under field conditions. Apidologie 28: 439–447.
34. Wallner K. Test regarding effects of imidacloprid on honey bees. In: Belzunces LP, Pélissier C, Lewis GB, editors. Les Colloques de l'INRA 98: Hazards of Pesticides to Bees; 1999 September 7–9; Avignon, France. INRA Editions. pp. 91–94.
35. Chauzat M-P, Faucon J-P, Martel A-C, Lachaize J, Cougoule N, et al. (2006) A survey of pesticide residues in pollen loads collected by honey bees in France. J Econ Entomol 99: 253–262.
36. Chauzat M-P, Carpentier P, Martel A-C, Bougeard S, Cougoule N, et al. (2009) Influence of pesticide residues on honey bee (Hymenoptera: Apidae) colony health in France. Environ Entomol 38: 514–523.
37. Faucon J-P, Aurières C, Drajnudel P, Mathieu L, Ribière M, et al. (2005) Experimental study on the toxicity of imidacloprid given in syrup to honey bee (Apis mellifera) colonies. Pest Manag Sci 61: 111–125.
38. Wallner K, Schur A, Stürz B (1999) Test regarding the danger of the seed disinfectant, Gaucho® 70WS, for honeybees. Apidologie 30: 423.
39. Curé G, Schmidt HW, Schmuck R (1999) Results of a comprehensive field research programme with the systemic insecticide imidacloprid (Gaucho®). In: Belzunces LP, Pélissier C, Lewis GB, editors. Les Colloques de l'INRA 98: Hazards of Pesticides to Bees; 1999 September 7–9; Avignon, France. INRA Editions. pp. 49–59.
40. Bonmatin JM, Moineau I, Charvet R, Fleche C, Colin ME, et al. (2003) A LC/APCI-MS/MS method for analysis of imidacloprid in soils, in plants, and in pollens. Anal Chem 75: 2027–2033.
41. Bonmatin JM, Marchand PA, Charvet R, Moineau I, Bengsch ER, et al. (2005) Quantification of imidacloprid uptake in maize crops. J Agric Food Chem 53: 5336–5341.
42. Pareja L, Colazzo M, Pérez-Parada A, Niell S, Carrasco-Letelier L, et al. (2011) Detection of pesticides in active and depopulated beehives in Uruguay. Int J Environ Res Public Health 8: 3844–3858.
43. Seeley TD (1982) Adaptive significance of the age polyethism schedule in honeybee colonies. Behav Ecol Sociobiol 11: 287–293.
44. Ray S, Ferneyhough B (1999) Behavioral development and olfactory learning in the honeybee (Apis mellifera). Dev Psychobiol 34: 21–27.
45. Taylor KS, Waller GD, Crowder LA (1987) Impairment of a classical conditioned response of the honey bee (Apis mellifera L.) by sublethal doses of synthetic pyrethroid Insecticides. Apidologie 18: 243–252.
46. Mamood AN, Waller GD (1990) Recovery of learning responses by honeybees following a sublethal exposure to permethrin. Physiol Entomol 15: 55–60.
47. Stone JC, Abramson CI, Price JM (1997) Task-dependent effects of dicofol (Kelthane) on learning in the honey bee (Apis mellifera). Bull Environ Contam Toxicol 58: 177–183.
48. Abramson CI, Aquino IS, Ramalho FS, Price JM (1999) The effect of insecticides on learning in the africanized honey bee (Apis mellifera L.). Arch Environ Contam Toxicol 37: 529–535.
49. Decourtye A, Devillers J, Genecque E, Le Menach K, Budzinski H, et al. (2005) Comparative sublethal toxicity of nine pesticides on olfactory learning performances of the honeybee Apis mellifera. Arch Environ Contam Toxicol 48: 242–250.
50. Takeda K (1961) Classical conditioned response in the honey bee. J Insect Physiol 6: 168–179.
51. Halm M-P, Rortais A, Arnold G, Taséi JN, Rault S (2006) New risk assessment approach for systemic insecticides: the case of honey bees and imidacloprid (Gaucho). Environ Sci Technol 40: 2448–2454.
52. Wilhelmy H (2000) Substance A - Acute Effects on the Honeybee Apis mellifera (Hymenoptera, Apidae), Non-GLP. Sarstedt, Germany: Dr. U. Noack-Laboratorium für angewandte Biologie.
53. Laurent FM, Rathahao E (2003) Distribution of [¹⁴C]Imidacloprid in sunflowers (Helianthus annuus L.) following seed treatment. J Agric Food Chem 51: 8005–8010.
54. Aupinel P, Fortini D, Dufour H, Tasei J-N, Michaud B, et al. (2005) Improvement of artificial feeding in a standard in vitro method for rearing Apis mellifera larvae. Bull Insectol 58: 107–111.
55. Smirle MJ, Robinson GE (1989) Behavioral status and detoxifying enzyme activity are related in worker honey bees. J Insect Behav 2: 285–289.
56. Claudianos C, Ranson H, Johnson RM, Biswas S, Schuler MA, et al. (2006) A deficit of detoxification enzymes: pesticide sensitivity and environmental response in the honeybee. Insect Mol Biol 15: 615–636.
57. Jones AK, Raymond-Delpech V, Thany SH, Gauthier M, Sattelle DB (2006) The nicotinic acetylcholine receptor gene family of the honey bee, Apis mellifera. Genome Res 16: 1422–1430.

58. Thany SH, Lenaers G, Crozatier M, Armengaud C, Gauthier M (2003) Identification and localization of the nicotinic acetylcholine receptor alpha3 mRNA in the brain of the honeybee, *Apis mellifera*. Insect Mol Biol 12: 255–262.

59. Thany SH, Crozatier M, Raymond-Delpech V, Gauthier M, Lenaers G (2005) Apisα2, Apisα7-1 and Apisα7-2: three new neuronal nicotinic acetylcholine receptor α-subunits in the honeybee brain. Gene 344: 125–132.

60. Malun D (1998) Early development of mushroom bodies in the brain of the honeybee *Apis mellifera* as revealed by BrdU incorporation and ablation experiments. Learn Memory 5: 90–101.

61. Farris SM, Robinson GE, Davis RL, Fahrbach SE (1999) Larval and pupal development of the mushroom bodies in the honey bee, *Apis mellifera*. J Comp Neurol 414: 97–113.

62. Bicker G, Kreissl S (1994) Calcium imaging reveals nicotinic acetylcholine receptors on cultured mushroom body neurons. J Neurophysiol 71: 808–810.

63. Goldberg F, Grünewald B, Rosenboom H, Menzel R (1999) Nicotinic acetylcholine currents of cultured Kenyon cells from the mushroom bodies of the honey bee *Apis mellifera*. J Physiol 514: 759–768.

64. Déglise P, Grünewald B, Gauthier M (2002) The insecticide imidacloprid is a partial agonist of the nicotinic receptor of honeybee Kenyon cells. Neurosci Lett 321: 13–16.

65. Gregorc A, Ellis JD (2011) Cell death localization in situ in laboratory reared honey bee (*Apis mellifera* L.) larvae treated with pesticides. Pestic Biochem Physiol 99: 200–207.

66. Decourtye A, Tisseur M, Taséi J-N, Pham-Delègue M-H (2005) Toxicité et risques liés à l'emploi de pesticides chez les pollinisateurs : cas de l'abeille domestique. In: Regnault-Roger C, editor. Enjeux phytosanitaires pour l'agriculture et l'environnement. Paris: Tec et Doc Lavoisier. pp. 283–299.

67. Wu JY, Anelli CM, Sheppard WS (2011) Sub-lethal effects of pesticide residues in brood comb on worker honey bee (*Apis mellifera*) development and longevity. PLoS ONE 6: e14720.

68. Abbott VA, Nadeau JL, Higo HA, Winston ML (2008) Lethal and sublethal effects of imidacloprid on *Osmia lignaria* and clothianidin on *Megachile rotundata* (Hymenoptera: Megachilidae). J Econ Entomol 101: 784–796.

69. Suchail S, Debrauwer L, Belzunces LP (2004) Metabolism of imidacloprid in *Apis mellifera*. Pest Manag Sci 60: 291–296.

70. Guez D, Belzunces LP, Maleszka R (2003) Effects of imidacloprid metabolites on habituation in honeybees suggest the existence of two subtypes of nicotinic receptors differentially expressed during adult development. Pharmacol Biochem Behav 75: 217–222.

71. Suchail S, De Sousa G, Rahmani R, Belzunces LP (2004) *In vivo* distribution and metabolisation of ^{14}C-imidacloprid in different compartments of *Apis mellifera* L. Pest Manag Sci 60: 1056–1062.

72. Dwyer JB, McQuown SC, Leslie FM (2009) The dynamic effects of nicotine on the developing brain. Pharmacol Ther 122: 125–139.

73. Dupuis J, Louis T, Gauthier M, Raymond V (2012) Insights from honeybee (*Apis mellifera*) and fly (*Drosophila melanogaster*) nicotinic acetylcholine receptors: From genes to behavioral functions. Neurosci Biobehav Rev 36: 1553–1564.

74. Bicker G (1999) Histochemistry of classical neurotransmitters in antennal lobes and mushroom bodies of the honeybee. Microsc Res Tech 45: 174–183.

75. Kreissl S, Bicker G (1989) Histochemistry of acetylcholinesterase and immunocytochemistry of an acetylcholine receptor-like antigen in the brain of the honeybee. J Comp Neurol 286: 71–84.

76. Scheidler A, Kaulen P, Brüning G, Erber J (1990) Quantitative autoradiographic localization of [^{125}I]α-bungarotoxin binding sites in the honeybee brain. Brain Res 534: 332–335.

77. Armengaud C, Causse N, Aït-Oubah J, Ginolhac A, Gauthier M (2000) Functional cytochrome oxidase histochemistry in the honeybee brain. Brain Res 859: 390–393.

78. Lozano VC, Bonnard E, Gauthier M, Richard D (1996) Mecamylamine-induced impairment of acquisition and retrieval of olfactory conditioning in the honeybee. Behav Brain Res 81: 215–222.

79. Lozano V, Armengaud C, Gauthier M (2001) Memory impairment induced by cholinergic antagonists injected into the mushroom bodies of the honeybee. J Comp Physiol A 187: 249–254.

80. Tasei JN, Ripault G, Rivault E (2001) Hazards of imidacloprid seed coating to *Bombus terrestris* (Hymenoptera: Apidae) when applied to sunflower. J Econ Entomol 94: 623–627.

81. Whitehorn PR, O'Connor S, Wackers FL, Goulson D (2012) Neonicotinoid pesticide reduces bumble bee colony growth and queen production. Science 336: 351–352.

82. Menzel R, Erber J, Masuhr T (1974) Learning and memory in the honeybee. In: Barton-Browne L, editor. Experimental Analysis of Insect Behaviour. Berlin: Springer. pp. 195–217.

83. Erber J, Masuhr TH, Menzel R (1980) Localization of short-term memory in the brain of the bee, *Apis mellifera*. Physiol Entomol 5: 343–358.

84. Dacher M, Lagarrigue A, Gauthier M (2005) Antennal tactile learning in the honeybee: effect of nicotinic antagonists on memory dynamics. Neuroscience 130: 37–50.

85. Thany SH, Gauthier M (2005) Nicotine injected into the antennal lobes induces a rapid modulation of sucrose threshold and improves short-term memory in the honeybee *Apis mellifera*. Brain Res 1039: 216–219.

Lack of Evidence for an Association between *Iridovirus* and Colony Collapse Disorder

Rafal Tokarz[1,9], Cadhla Firth[1,9], Craig Street[1], Diana L. Cox-Foster[2], W. Ian Lipkin[1]*

1 Center for Infection and Immunity, Mailman School of Public Health, Columbia University, New York, New York, United States of America, **2** Department of Entomology, The Pennsylvania State University, University Park, Pennsylvania, United States of America

Abstract

Colony collapse disorder (CCD) is characterized by the unexplained losses of large numbers of adult worker bees (*Apis mellifera*) from apparently healthy colonies. Although infections, toxins, and other stressors have been associated with the onset of CCD, the pathogenesis of this disorder remains obscure. Recently, a proteomics study implicated a double-stranded DNA virus, invertebrate iridescent virus (Family *Iridoviridae*) along with a microsporidium (*Nosema* sp.) as the cause of CCD. We tested the validity of this relationship using two independent methods: (i) we surveyed healthy and CCD colonies from the United States and Israel for the presence of members of the *Iridovirus* genus and (ii) we reanalyzed metagenomics data previously generated from RNA pools of CCD colonies for the presence of *Iridovirus*-like sequences. Neither analysis revealed any evidence to suggest the presence of an *Iridovirus* in healthy or CCD colonies.

Editor: Robin F. A. Moritz, Martin-Luther-Universität Halle, Germany

Funding: Funding was provided by the U. S. Department of Defense, Google.org, and the Pennsylvania Department of Agriculture (ME 447514, "Colony Collapse disorder in Honey Bees: Defining associated pathology and enabling reuse of equipment"). The funders had no role in study design, data collection and analysis, decision to publish, or preparation of the manuscript.

Competing Interests: Please note that Google.org is not the same as Google.com. Whereas Google.com is a commercial source, Google.org is a non-profit foundation. Nonetheless, the authors confirm that there is no conflict of interest regarding employment, consultancy, patents, products in development or marketed products.

* E-mail: wil2001@columbia.edu

⑨ These authors contributed equally to this work.

Introduction

Honey bee decline has become an increasingly important problem worldwide, and has been attributed to multiple underlying causes [1,2]. A significant contributor to colony losses is Colony Collapse Disorder (CCD), which is characterized by the sudden absence of adult honey bees (*Apis mellifera*) in hives, while both resources and brood remain. Since the initial description of CCD in 2006, annual colony losses in the United States have exceeded 30%, a number significantly greater than previous average losses [1]. The percentage of collapsed CCD colonies has declined in the last five years; however, approximately one quarter of beekeepers continue to report CCD symptoms in more than 60% of their annual colony losses [3].

Although CCD continues to be associated with an annual colony loss of approximately 10% in the United States [1], attempts to identify the cause or combination of causes that result in CCD have not been successful. Although initial efforts focused on identifying a single pathogen, a consensus has recently emerged that multiple pathogens in concert with increasing environmental stressors are likely to be the driving factors behind CCD [4,5,6,7]. Several pathogens in particular have repeatedly been identified in concert with the appearance of CCD, including the microsporidium *Nosema ceranae* [8] and members of the *Picornavirales* such as Israeli acute paralysis virus [9].

Recently, a double-stranded DNA virus (dsDNA) belonging to the *Iridovirus* genus (Family *Iridoviridae*) has been proposed to act synergistically with *N. ceranae* to cause the development of CCD

[8]. In an attempt to use an unbiased proteomics approach for the detection of CCD markers, Bromenshenk et al. [8] identified 139 peptide fragments with homology to invertebrate iridescent virus 6 (IIV-6), the only member of the genus for which full genome sequences and proteomic analysis are available [10]. IIV-6 naturally infects a range of lepidopteran species, but has not been shown to infect any hymenopteran in the wild [11]. However, another invertebrate iridescent virus (IIV-24) has previously been associated with disease in the Eastern honey bee (*A. cerana*) [12]. Unfortunately, only 355 bp of IIV-24 sequence is publicly available, making it difficult to determine if the peptides identified as IIV-6 by Bromenshenk et al. [8] could be more closely related to IIV-24. Notably, in a re-interpretation of this proteomic analysis, Foster [13] suggested instead that the peptides identified as IIV-6 have higher similarity to *A. mellifera* protein sequences. To reconcile this discrepancy and test the intriguing hypothesis proposed by Bromenshenk et al., we surveyed CCD and healthy *A. mellifera* colonies from the United States and Israel for the presence of several invertebrate iridescent viruses, and re-analyzed previously generated metagenomic data from CCD colonies for *Iridovirus* sequences.

Methods

One hundred and sixty-three bees were surveyed for the presence of members of the *Iridovirus* genus. These were collected in 2007–2009 from 35 colonies distributed across nine commercial apiaries, seven of which have been previously described [5,9]. All samples were obtained under the auspices of the USDA and the Pennsylvania Department of Agriculture. Four colonies appeared healthy at the time of collection and the remaining 31 were in various stages of collapse, as described in Cox-Foster et al. [9]. Many of these apiaries were also represented by the colonies sampled by Bromenshenk et al. [8]. Samples were stored at −70°C from the time of collection.

Invertebrate iridescent viruses have been shown to replicate primarily in the gut walls, Malpighian tubules and fat-bodies of bees [12]; therefore, total nucleic acid was extracted from individual bee abdomens using the EasyMag Extraction platform (Biomerieux). Four assays were developed to detect the presence of invertebrate iridescent viruses. First, a quantitative SYBR Green PCR assay targeting conserved regions of the DNA-dependent RNA polymerase of IIV-6 (based on GenBank accession number NC_003038) was performed in duplicate (Table 1). Two nested PCR assays targeting conserved regions of the DNA Polymerase gene were also designed. The first assay was specific for IIV-6 (Table 1). In an attempt to capture more of the diversity present in the *Iridovirus* genus, a second nested assay was developed based on an alignment of IIV-6 and IIV-31 (GenBank accession number AJ279821), the only members of the *Iridovirus* genus with available DNA polymerase sequence data. Finally, an assay targeting the only sequence available for IIV-24 (GenBank accession number AF042340) was developed in a highly conserved region of the capsid. PCR assays were performed in 20 μl reactions containing 0.25 μM of each primer, 0.125 mM dNTPs, and 1 mM MgCl$_2$. The primer annealing temperatures for each assay are given in Table 1. Plasmids containing the relevant portion of IIV-6 resuspended in 25 ng/μl of bee genomic DNA and DNA extracted from IIV-31-positive isopods (*Armadillidium vulgare*) were used as positive controls. The IIV-24-specific assay was also able to detect the IIV-6 positive control. The sensitivity of each assay was evaluated by serial dilution of the IIV-6 plasmid standard. To verify the quality of our extracted nucleic acid, we also determined the copy number of the bee nuclear genome present using a *β-actin* quantitative assay [14].

Following the report of an IIV-6-like virus present in samples from the same colonies surveyed by Cox-Foster et al. in 2007 [9], we reanalyzed the metagenomic data generated in the 2007 study for the presence of invertebrate iridescent viruses. The purported identification of IIV-6 in these samples was based upon proteomic analysis, suggesting the presence of mRNA transcripts that should be detectable by a metagenomics approach. A searchable database was first created using all 1204 protein-coding sequences for the *Iridovirus* genus (GenBank Tax ID: 10487) and a homology search with blastx [15] was performed using individual reads previously generated from four pools of RNA [9].

Results and Discussion

PCR assays were used to test for the presence of members of *Iridovirus* in healthy and CCD colonies. All four assays were optimized to detect 1–10 DNA copies. We found no evidence of *Iridovirus* in any of the 163 bees, despite detecting an average of 10^5–10^6 DNA copies of *β-actin* per microliter in these samples. During the initial characterization of IIV-24 in diseased bees (*A. cerana*), it was estimated that each bee contained an average of 10^{10}–10^{11} viral particles [12], a quantity likely to be well within the range of detection of our PCR assays. The design of our assays was limited by both the paucity of *Iridovirus* sequences publicly available and the absence of any nucleic acid or amino acid sequences linked to the *Iridovirus* identified by mass spectrometry [8]. To address these deficiencies, we targeted highly conserved regions of the genome and used multiple PCR assays designed to detect IIV-6 or an IIV-6 related virus, as well as IIV-24. We focused on IIV-6-related viruses because of the reported presence of 139 IIV-6 peptides in 100% of CCD and 75% of healthy colonies [8]. We also targeted IIV-24, as it is the only *Iridovirus* previously associated with disease in bees [12,16]. Nevertheless, none of the assays revealed IIV-6 or IIV-24-like viruses in bees from healthy or CCD colonies.

The results of our reanalysis of the metagenomic data from earlier studies of CCD colonies [9] also does not suggest the presence of an *Iridovirus* in CCD bees. Using an e-value cut-off set at 0.0001, two of the four RNA pools (representing pools of CCD positive samples) did not reveal any sequences with homology to any *Iridovirus*. The third and fourth pools, comprised of samples from healthy colonies and royal jelly, contained seven reads with low similarity to IIV-6 (e-value $> 1e^{-18}$). Blasting these individual reads against the entire non-redundant protein-coding database revealed a higher similarity to bacterial or bee genomic sequence

Table 1. Sequences of the primers used for the detection of members of the genus *Iridovirus*.

Primer name	Primer sequence (5′ to 3′)	Product length (nt)	Annealing Temperature (°C)
IIV6.DdRp.F*	CCCAGCATACTATAACATGTCTGCAA	94	60
IIV6.DdRp.R*	GAATATCCTCTGACGACATCATTCC		
IIV6.DNAPol1.F	GGTGGAATGGGATGATCATG	270	57
IIV6.DNAPol1.R	CCCCCTATATAACCAGGAGTTTG		
IIV6.DNAPol2.F	GATGTGTTCGTATCTTGTGATGG	120	53
IIV6.DNAPol2.R	GGGCCAGGAGTTTGTTTATTAAC		
IIV24.CP.F	GGATATGACAATATGATTGGAAAT	177	50
IIV24.CP.R	CGTTAATTTGCATTTCATTGTA		
IIV.DNAPol1.F	GGGGARTTYTWCCSACAATC	283	50
IIV.DNAPol1.R	GGCTWCCCATRTAWGTKGTRCACA		
IIV.DNAPol2.F	GGGATCTTYTRGAYGCCAGAA	227	54
IIV.DNAPol2.R	GGCATAAAYGGTAAMGCTCCAAC		

*Primers for SYBR Green Assay.

in all cases (e-values from 2–88 orders of magnitude lower than when the search was restricted to *Iridovirus*). Although these results do not exclude the possibility that *Iridovirus* sequences may have been present at extremely low concentrations in the 2007 sample set, they do not lend support to a linkage between invertebrate iridescent viruses and CCD.

Typical signs of infection with invertebrate iridescent viruses include lethargy, early and rapid mortality and in bees, the loss of flight [17,18]. The absence of adult worker bees in otherwise healthy colonies that characterizes CCD does not appear to be consistent with signs of invertebrate iridescent virus infection. Furthermore, a primary characteristic of pathogenic invertebrate iridescent virus infections is the presence of a purple-blue iridescence that can be observed both in centrifuged viral particles and directly in infected tissues by microscopy (fat body, Malpighian tubules and the gut wall), even in mild infections [12,17,18,19]. Presumably, any iridovirus present at a level high enough to initiate colony collapse (even if the presence of *N. ceranae* is also required) would also be present at levels high enough for the detection of iridescence. However, during the examination of bee tissues from a large number of CCD colonies, no iridescent qualities were observed (data not shown).

Admittedly, the possibility exists that the invertebrate iridescent virus detected by mass spectrometry is unrelated to those targeted here, although we consider this to be unlikely. Indeed, it has been suggested that the peptides identified as IIV-6-like by Bromenshenk et al. [8] actually show higher similarity to *A. mellifera* proteins when the host proteome is included in the analysis [13]. When our data is considered along with that of Foster [13] it therefore seems unlikely that an invertebrate iridescent virus related to IIV-6 or IIV-24 is a significant contributor to CCD.

Acknowledgments

We thank İkbal Agah İnce and Monique M. van Oers for the gift of purified and characterized IIV-6 virions, Brian A. Federici for the gift of IIV-31 in isopods, and James Ng for technical assistance.

Author Contributions

Conceived and designed the experiments: RT CF CS DLC-F WIL. Performed the experiments: RT CF CS. Analyzed the data: RT CF CS WIL. Contributed reagents/materials/analysis tools: DLC-F WIL. Wrote the paper: RT CF CS DLC-F WIL.

References

1. United Nations Environment Programme (2010) Global honey bee colony disorders and other threats to insect pollinators, UNEP Emerging Issues, (UNEP, Nairobi).
2. Williams GR, Tarpy DR, vanEngelsdorp D, Chauzat MP, Cox-Foster DL, et al. (2010) Colony Collapse Disorder in context. Bioessays 32: 845–846.
3. vanEngelsdorp D, Hayes J, Underwood RM, Caron D, Pettis JS (2011) A Survey of managed honey bee colony losses in the U.S., fall 2009 to winter 2010. J Apicult Res 50: 1–10.
4. Cox-Foster D, vanEngelsdorp D (2009) Saving the honeybee. Sci Am 300: 40–47.
5. vanEngelsdorp D, Evans JD, Saegerman C, Mullin C, Haubruge E, et al. (2009) Colony collapse disorder: a descriptive study. PLoS One 4: e6481.
6. vanEngelsdorp D, Speybroeck N, Evans JD, Nguyen BK, Mullins C, et al. (2010) Weighing risk factors associated with bee colony collapse disorder by classification and regression tree analysis. J Econ Entomol 103: 1517–1523.
7. Ratnieks FLW, Carreck NL (2010) Clarity on honey bee collapse? Science 327: 152–153.
8. Bromenshenk JJ, Henderson CB, Wick CH, Stanford MF, Zulich AW, et al. (2010) Iridovirus and microsporidian linked to honey bee colony decline. PLoS One 5: e13181.
9. Cox-Foster DL, Conlan S, Holmes EC, Palacios G, Evans JD, et al. (2007) A metagenomic survey of microbes in honey bee colony collapse disorder. Science 318: 283–287.
10. Ince IA, Boeren SA, van Oers MM, Vervoort JJ, Vlak JM (2010) Proteomic analysis of Chilo iridescent virus. Virology 405: 253–258.
11. Williams T (2008) Natural invertebrate hosts to iridoviruses (Iridoviridae). Neotrop Entomol 37: 615–632.
12. Bailey L, Ball BV, Woods RD (1976) An iridovirus from bees. J Gen Virol 31: 459–461.
13. Foster LJ (2011) Interpretation of data underlying the link between CCD and an invertebrate iridescent virus. Mol Cell Proteomics doi: 10.1074/mcp.O110.006387.
14. Lourenço AP, Mackert A, Cristino AS, Simões ZLP (2008) Validation of reference genes for gene expression studies in the honey bee, *Apis mellifera*, by quantitative real-time RT-PCR. Apidologie 39: 372–385.
15. Altschul SF, Madden TL, Schäffer AA, Zhang J, Zhang Z, et al. (1997) Gapped BLAST and PSI-BLAST: A new generation of protein database search programs. Nucleic Acids Res 25: 3389–3402.
16. Webby R, Kalmakoff J (1998) Sequence comparison of the major capsid protein gene from 18 diverse iridoviruses. Arch Virol 143: 1949–1966.
17. Bailey L, Ball BV (1978) *Apis* iridescent virus and "clustering disease" of *Apis cerana*. J Invert Pathol 31: 368–371.
18. Kleespies RG, Tidona CA, Darai G (1999) Characterization of a new iridovirus isolated from crickets and investigations on the host range. J Invert Pathol 73: 84–90.
19. Williams T (1996) The iridoviruses. Adv Viral Res 46: 345–412.

Predictive Markers of Honey Bee Colony Collapse

Benjamin Dainat[1]*, Jay D. Evans[2], Yan Ping Chen[2], Laurent Gauthier[1], Peter Neumann[1,3]

1 Swiss Bee Research Centre, Agroscope Liebefeld-Posieux Research Station ALP, Bern, Switzerland, **2** Bee Research Laboratory, United States Department of Agriculture-Agricultural Research Service, Beltsville, Maryland, United States of America, **3** Department of Zoology and Entomology, Rhodes University, Grahamstown, South Africa

Abstract

Across the Northern hemisphere, managed honey bee colonies, *Apis mellifera*, are currently affected by abrupt depopulation during winter and many factors are suspected to be involved, either alone or in combination. Parasites and pathogens are considered as principal actors, in particular the ectoparasitic mite *Varroa destructor*, associated viruses and the microsporidian *Nosema ceranae*. Here we used long term monitoring of colonies and screening for eleven disease agents and genes involved in bee immunity and physiology to identify predictive markers of honeybee colony losses during winter. The data show that DWV, *Nosema ceranae*, *Varroa destructor* and *Vitellogenin* can be predictive markers for winter colony losses, but their predictive power strongly depends on the season. In particular, the data support that *V. destructor* is a key player for losses, arguably in line with its specific impact on the health of individual bees and colonies.

Editor: Patricia V. Aguilar, University of Texas Medical Branch, United States of America

Funding: This work was supported by a grant of the Swiss Federal Veterinary Office. The funders had no role in study design, data collection and analysis, decision to publish, or preparation of the manuscript.

Competing Interests: The authors have declared that no competing interests exist.

* E-mail: benjamin.dainat@alp.admin.ch

Introduction

Agricultural pollination should integrate wild species, which provide pollination as an ecosystem service, and managed pollinator introduction as crop management practices [1]. Amongst the managed pollinators, the Western honey bee, *Apis mellifera*, is clearly a cornerstone, because pollination of many crops in most parts of the world relies on this species [1]. However, managed honey bee colonies are currently affected by a syndrome corresponding to an abrupt depopulation during winter [2]. Many biotic and abiotic factors are suspected to be involved in this condition, either alone or in combination [2–3]. Among them, parasites contribute to weakening colony health, leaving room to secondary infections. In particular, the ectoparasitic mite *Varroa destructor* [4] is now considered to be the main candidate involved in winter colony losses in Europe [5–8]. This parasite originates from South-East Asia and has now become widespread across most of the continents [4]. It has been shown that *V. destructor* or its associated microbes can affect the immune system of parasitized bees [9–12]. In addition, viral infections linked with *V. destructor* are generally considered as a major cause of bee losses. Indeed, *V. destructor* plays a central role as a mechanical and biological vector of several viruses [12–17]. In addition *V. destructor* appears to accelerate the replication of latent viral infections [12,17–20]. Although more than 19 different viruses have been detected in *A. mellifera*, only three have been associated with winter losses on a large scale [5,21–23], namely Deformed wing virus [24], Acute bee paralysis virus [25,26] and Israeli acute bee paralysis virus [27]. These viruses are positive stranded RNA viruses belonging to the *Iflaviridae* and *Dicistroviridae* families and are suspected to enhance the deleterious action of *V. destructor* on bee colonies by their strong association with this mite [20,28,29]. In the United States, IAPV was first identified as a predictive factor for producing CCD symptoms [21]. However, subsequent surveys indicate that this virus was not the main factor responsible for losses but only one of multiple possible factors involved [30,31]. In Europe, DWV and ABPV are generally suspected to be involved in winter colony losses [5,22,23]. Both are transmitted by the mite after feeding on bee pupae or adults [20,29]. ABPV is highly virulent for bees when injected directly into the hemolymph [25,29,32]. On the contrary, DWV is much less virulent and generates typical symptoms of deformed wings only in bees from colonies highly infested with *Varroa* mites [33]. Despite its low virulence for bees, DWV infects a large range of bee tissues and can produce high titers in infected bees, suggesting a potential impact on bee physiology [34,35].

Another potential candidate involved in colony losses is the microsporidian *Nosema ceranae* [36,37] although the impact of this parasite on colony health in Europe still remains controversial [5,38,39].

Despite the fact that several studies have pointed out the potential involvement of pathogens on colony losses, no common pattern has yet emerged. This is probably due in part to the different parameters present in these studies such as climate, bee races, beekeeping practices or sampling methods. Indeed sampling often consists of bees collected once from either healthy, weak or dead colonies. In this context long term monitoring appears crucial, especially because pathogens causing colony death may have disappeared leaving room for opportunistic infections.

In this study, we aimed to identify predictive markers of winter honey bee colony losses. Although the up- or down-regulation of a marker in a sample may not be related to the principal cause(s) of the disease, such markers would help beekeepers or bee inspectors to set up a reliable diagnostic tool and to standardize bee colony monitoring all around the world. For this purpose we performed a survey of bee colonies in Switzerland over six months and checked for the presence and loads of eleven honey bee pathogens in the samples, as well as the levels of expression of three *A. mellifera* genes involved in bee immunity.

Materials and Methods

1. Experimental design

In summer 2007, 29 queenright colonies were selected from our local bee stocks (predominantly *A. m. carnica*) of similar strength (~14'000 workers each) at the Swiss Bee Research Centre in Bern, Switzerland. In order to get sufficiently different parasitism levels with the ectoparasitic mite *V. destructor*, 18 colonies were adequately and timely treated in winter 2006–2007 and in summer and fall 2007 against this mite using organic acids following the Liebefelder alternative treatment (twice with formic acid by evaporation using the FAM diffuser in summer and late summer and oxalic acid by droplets in Fall) [40], while the others were left untreated. Both groups were otherwise managed in exactly the same way prior to and during the experiment. In the experiment, the two groups were physically separated by 250 m and one four-storied building to minimize mite movement between colonies (e.g. via drifting and/or robbing; [4,41]), but nevertheless had similar foraging conditions. Mite infestation levels were monitored weekly in each colony using the natural mite fall method [4,42], distinguishing colonies with high and low mite infestations. Pooled worker samples (N = 100) were collected alive in the brood nest in summer (18-08-2007) at the beginning of the experiment and in fall (22/11/2007). These two samplings were called "Summer" and "Fall". Another sampling, "Winter", was ultimately performed during winter but at different times according to the destiny of the colony: in colonies which survived winter, bees were collected on 28/01/2008 while for the colonies that did not survive, this sampling was done just before collapsing from 30/11/2007 to 28/01/2008. Colony-level traits (areas covered with bees, number of cells with open, sealed brood, honey and pollen) were estimated in dm^2 every 3 weeks from September until the end of October and at the beginning of March until May using the Liebefelder standard method [43].

2. Molecular approaches

Pools of 100 workers were collected alive from the brood nests of each colony and immediately frozen at $-20°C$. For total RNA extraction, bees were first homogenized in 20 ml of Tris-NaCl buffer (Tris 10 mM; NaCl 400 mM; pH 7.5). An aliquot of 50 µl of the homogenate was used for RNA extraction with the NucleoSpin RNA II Kit® (Macherey-Nagel) following the recommendations of the supplier. Then, cDNA was immediately processed using M-MLV reverse transcriptase (Invitrogen) and random hexamers [44]. These samples were checked for the presence of eight honey bee viruses (DWV, ABPV, IAPV, Chronic bee paralysis virus, Kashmir bee virus, Black queen cell virus, Sacbrood virus and Slow bee paralysis virus), using a qualitative PCR assay [31,44,45] with viral cDNA for positive control and water for the negative control. DWV, BQCV and ABPV positive samples were further processed for the quantification of viral titers using quantitative PCR (qPCR). The microsporidians *N. ceranae* and *N. apis* were quantified as well in each sample using qPCR [31]. In parallel, the expression levels of *A. mellifera* transcripts including *vitellogenin*, *eater* and *hymenoptaecin* were monitored using qPCR [46]. In order to normalize the data according to the amount of RNA in the sample, analysis of the *β-actin* gene was performed in parallel for each sample [47]. For all the targets except DWV, normalization was done using the comparative quantification method (delta CT method) [48]. DWV loads in samples were quantified by absolute quantification method using

standard curves made of serial dilutions of known amounts of the amplicons [49] and presented as equivalent copies of DWV genome. All qPCR reactions were conducted using a thermal profile of: 50°C (2 min) then 95°C (10 min) followed by 40 cycles of 95°C (15 s), 60°C (1 min). Because qPCR assays were performed using SYBR-Green (Eurogentec), a melting curve was performed at the end of each run to ascertain the amplification of the target.

3. Data analyses

The bee colony samples were divided into two groups according to their winter survival. The first group (DC: dead colonies) consisted of 13 colonies which died during winter; most of these (11 out of 13) belonged to the set of colonies that received no treatment against *V. destructor*. The second group (SC: surviving colonies) included the 16 colonies that survived winter and coincided with the colonies that received a treatment against the mite. The variation estimates of transcript abundance of the studied variable between different groups were evaluated by using two tailed t-tests and non-parametric Kruskal–Wallis test as appropriate. Since DWV values covered a wide range, the data were transformed as the log10 DWV. For survival analysis, a Kaplan–Meier survival analysis was performed and used to compare the groups using Mantel-Hansel tests. Linear Models were also performed using Standard Least squares fitting in the Fit Model platform of JMP software to test the effects of colony status when nested within season. To visualize the results, regression diagnostics were performed with the model leverage plot. This allowed us to test if the variables are predictive on overwintering abilities and also to check the influence of each point on each hypothesis test. Multivariate Spearman correlations were performed between the variables (Pathogens and physiological markers). *P*-values below 0.05 were considered significant. The analyses were performed using Systat 12® and JMP® software.

Results

1. Bee population measurements and timing of colony collapse

While the colony size did not differ significantly in September between SC and DC groups (Figure 1; Mean: 12743.75 bees and 13376.92 respectively, Mann-Whitney test P = 0.07), a significant decrease in population was observed in October (29/10/2007) in the group of colonies that died during winter (Mann-Whitney test, P<0.01, Figure 1 A). In the untreated group most of the colonies collapsed during December 2007 (N = 9) and two out of 11 died before February 2008. Two colonies properly treated against *V. destructor* mites died during winter but later in the season (end of February) and obviously because of queen failure (these colonies became drone layers). The rest of the colonies did develop with an increased population as expected in this season (Figure 1 A). The *V. destructor* infestation levels and timing of treatments are indicated in Figure 1 B. After each mite treatment, we had considerable treatment Varroa fall confirming the treatment efficacy. From August to September, the natural daily mite fall in the surviving colonies decreased from [average ± SE] 8.13±1.95 to 5.21±0.5 per day with 0.377±0.089 after the last oxalic acid treatment.

2. Parasites and Pathogens

The number of *V. destructor* mites collected during the course of the experiment in the DC group exceeded those collected in the SC group (Figure 1 B and 2 C). The mite level in the DC increased constantly until October before decreasing slightly thereafter. Mite

A

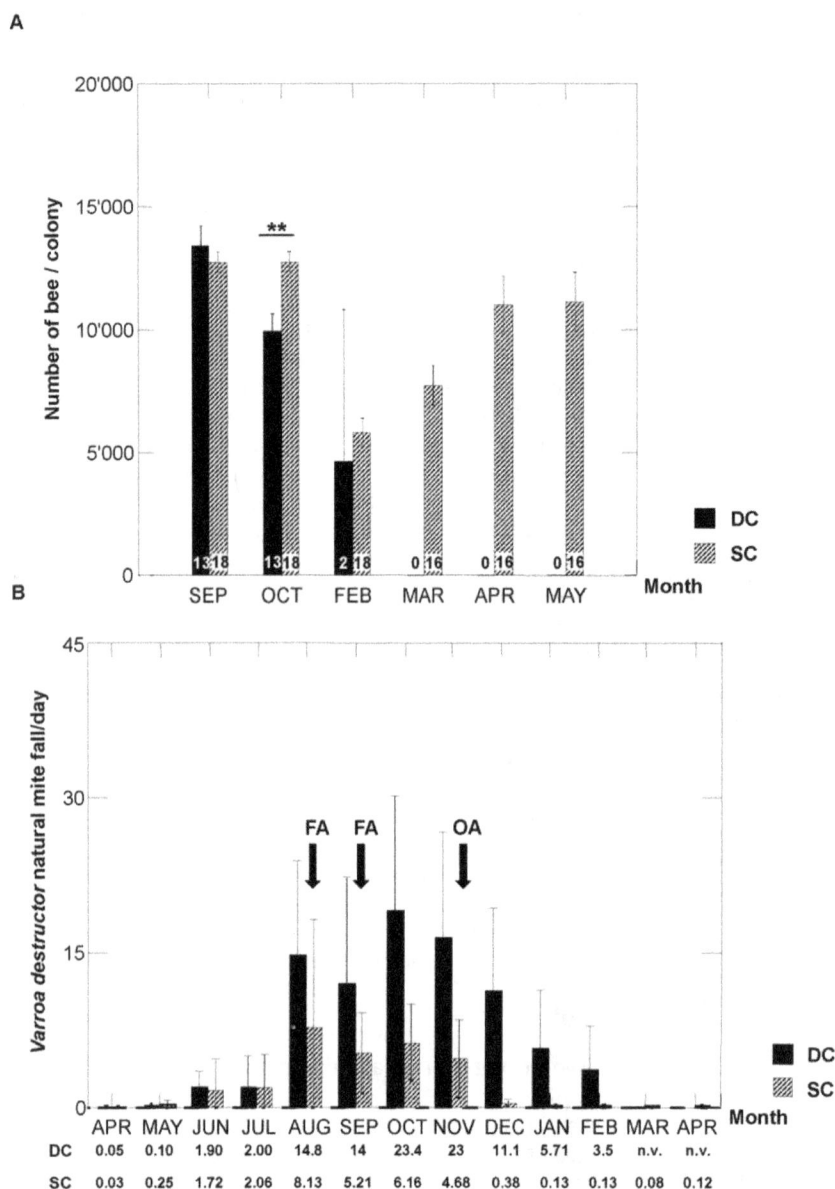

B

Figure 1. Colony strength and *V. destructor* natural mite fall. A) Colony strength (Y-axis) by sampling month in 2007/2008 (X-axis) for the colonies that died during winter (DC, black boxes) and the ones that survived (SC, grey boxes). P value is indicated (Mann-Whitney test): **P<0.01. Number of live colonies is given (N). B) Natural average *V. destructor* mite fall in 2007/2008 on the hive bottom boards per day per colony (Y-axis) for the two groups Dying Colonies (DC) and Surviving Colonies (SC) over the experimental period. (X-axis) Infestation levels [average ± SD] are shown for each month and timing of treatments (OA = Oxalic Acid, FA = Formic acid) is indicated by arrows (n.v. = no value).

loads in the surviving colonies dropped substantially from October to the winter sample.

Only five honey bee viruses (DWV, BQCV, ABPV, SBV and SPV) were detected in the experimental colonies (Table 1). The viruses CBPV, KBV and IAPV were not detected. DWV had 56.25% prevalence for SC and 61.5% for DC in summer. Almost all colonies were positive for this virus in fall (93.75% for SC and 100% for DC). Conversely, BQCV and SBV displayed a lower prevalence during the cold season, with no detectable SBV sample in the fall sample. We failed to detect SPV in summer and this virus was identified in less than 20% of the samples collected in fall in both DC and SC groups. No SPV could be detected in the bees that survived winter although the virus was detected in three

colonies of the DC just before colony collapse. Although all colonies were positive for *N. ceranae*, *N. apis* could not be detected. The prevalence of *N. ceranae* was equivalent between the groups in all seasons and ranged between 18.75% and 50% (Table 1). No distinction could be made in terms of numbers of the detected investigated pathogens between the SC and DC groups in any season as shown in table 1 (Mann-Whitney U-tests, Summer: P = 0.729; Fall: P = 0.854, Winter: P = 0.359).

3. Expression levels of bee pathogens and genes related to the bee immunity

As none of the qualitative PCR analysis showed significant differences between surviving colonies and colonies that died

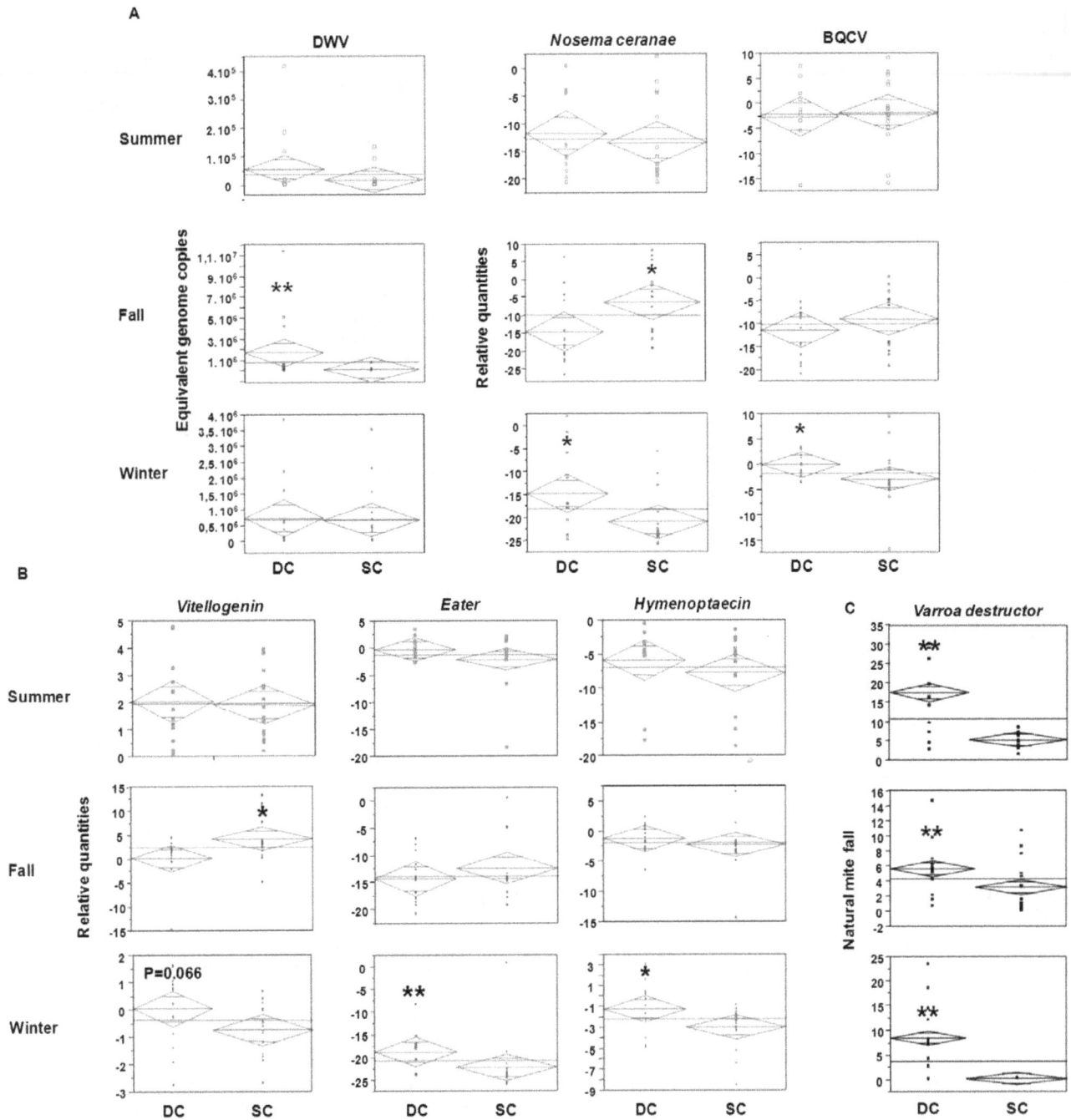

Figure 2. Seasonal variability in Summer, Fall and Winter. A) Pathogens B) Gene expression profiles C) *V. destructor*. The Y-axis shows the relative quantities except for DWV, where it shows equivalent genome copies and the X-axis displays the groups Dying Colonies DC (N = 13) and Surviving Colonies SC (N = 16). Significant differences (Two tailed t-test) are indicated with * = P<0.05 and ** = P<0.01.

during winter (Table 1), we conducted quantitative analyses from samples collected in summer, fall and winter in order to point out potential differences. Because winter samples from the DC group were collected just before collapsing and therefore at different time points during winter, data analyses are presented separately for DC and SC groups (Fig. 2). We measured the expression levels of three pathogens (DWV, BQCV and *N. ceranae*) based on their high prevalence in our samples and their potential pathogenic effects at the colony level. ABPV was excluded from these assays because only three samples were found infected by this virus. In addition,

we completed these analyses by measuring in parallel the expression levels of three *A. mellifera* genes involved in bee immune defenses (*hymenoptaecin* and *eater*) or bee life expectancy (*vitellogenin*).

Seasonal variation. Without considering the outcome of the colonies (death or survival during winter), seasonal variation was observed between summer and fall (Figure 3). DWV expression was higher in fall than in summer (P<0.001), but conversely BQCV titers were lower in fall than in summer (P<0.001). *N. ceranae* quantification showed equivalent titers between summer and fall (P = 0.479). *Vitellogenin* mRNA titers remained stable

Table 1. Proportion of colonies with detectable levels of pathogens as measured by PCR and qPCR in surviving (SC) and dying (DC) colonies.

Season	Status	Colonies	DWV (%)	BQCV (%)	ABPV (%)	SBV (%)	SPV (%)	N. ceranae (%)	Nr Path
Summer	DC	13	61.5	84.6	7.7	15.4	0	38.5	2.08±0.64
	SC	16	56.3	87.5	0	43.8	0	31.3	2.19±1.17
Fall	DC	13	100	53.8	7.7	0	15.4	30.8	2.08±1.04
	SC	16	93.8	50	0	0	18.8	50	1.94±0.85
Winter	DC	13	100	100	0	7.7	23.1	38.5	2.69±0.95
	SC	16	100	93.8	0	25	0	18.8	2.38±0.62

Total numbers of pathogens (Nr Path) are summarized in the last column (±SD). No significant differences in pathogen numbers were found between Live and Dead in any season (Mann-Whitney test, Summer P = 0.729; Fall P = 0.854, Winter P = 0.359). CBPV, KBV, IAPV and N. apis were not found in any colony.

between summer and fall (P = 0.544) *while hymenoptaecin* mRNA levels were higher in fall than in summer (P<0.001). Conversely, *eater* expression was significantly reduced in fall compared to summer (P<0.001).

Such comparisons between fall and winter were more difficult to address, especially for colonies that died during winter, because the sampling was done just before the population decline and thus for this group, winter sampling encompassed several weeks, from December, when most of these colonies died, to January. In contrast, all samples were collected by the end of January in colonies that survived winter. The results are presented in table 2. In the group that died during winter, significant differences could be observed between samples collected in fall and those collected upon collapsing but only for BQCV which displayed increasing levels (P<0.001) and the *eater* gene for which expression conversely decreased (P<0.05). The levels of *N. ceranae* remained stable (P = 0.94). In the surviving group, significant variations were observed between fall and winter: *N. ceranae* levels decreased (P<0.001) as well as *eater* (P<0.001) and *vitellogenin* (P<0.001) mRNA levels. In contrast, DWV increased (P<0.05) as well as BQCV (P<0.05). Levels of *hymenoptaecin* were stable (P = 0.981).

Comparisons between DC and SC groups. Significant variations were observed in the fall between DC and SC groups, but only for DWV, *N. ceranae* and *vitellogenin* mRNA (Figure 2). No differences in any of the targets analyzed here were observed in summer between the two groups. DWV loads were higher in fall in the collapsing colonies than in the surviving ones (P<0.01). In fall, colonies that collapsed during winter displayed lower levels of *N. ceranae* than surviving ones (P<0.05).while in the summer *N. ceranae* levels were similar in both DC and SC groups (P = 0.483).Variations in BQCV titers between SC and DC groups were only observed in winter (P<0.05). While no *vitellogenin* differences could be observed in summer between both SC and DC groups, this gene displayed significantly higher expression levels in fall in the surviving colonies than in the DC group (P<0.05).

4 Correlations between markers

Correlations between the six markers were observed (Figure 4). In summer, DWV *vs. hymenoptaecin* ($r_s = 0.50$, P<0.01), *hymenoptaecin vs. eater* ($r_s = 0.46$; P<0.05), *vitellogenin vs. eater* ($r_s = 0.88$, P<0.001) and *vitellogenin vs. hymenoptaecin* ($r_s = 0.37$, P<0.05) showed significantly positive correlations. In fall, these correlations were not observed anymore but BQCV showed a positive correlation with *N. ceranae* ($r_s = 0.44$, P<0.05). In fall, the number of *V. destructor* mites collected on bottom boards was significantly correlated with both DWV (positively) and *vitellogenin* (negatively) expression levels

($r_s = 0.57$, P<0.05 and $r_s = -0.37$, P<0.05, respectively). In contrast, *V. destructor* correlated positively in winter with *N. ceranae* ($r_s = 0.41$, P<0.05), Eater ($r_s = 0.58$, P<0.01) and *vitellogenin* ($r_s = 0.45$, P<0.05). Full data are presented in Figure 4.

5. Identification of predictive markers for colony collapse

Using a linear model (see M&M) the results show that over all seasons, DWV (P<0.05), *V. destructor* (P<0.001), *N. ceranae* (P<0.001) and *vitellogenin* (P<0.001) could be considered as predictive markers for winter losses. This was not the case for BQCV (P = 0.467), *eater* (P = 0.173) and *hymenoptaecin* (P = 0.376) which displayed few variations between DC and SC groups either in summer or in fall. However, the model showed that there is a significant seasonal impact on the expression of these markers (DWV: P<0.05; *V. destructor*: P<0.01, *N. ceranae*: P<0.05 and *vitellogenin*: P<0.01). Then in summer only *V. destructor* (P<0.01) could be considered as a significant predictive marker while in fall, DWV (increased; P<0.01), *V. destructor* (increased; P<0.01), *N. ceranae* (decreased; P<0.05) and Vitellogenin (decreased; P<0.05) could be considered as significant predictors of colony collapse.

Discussion

Here we showed that both *V. destructor* and DWV are strong predictive markers for honey bee colony death during winter.

Among the high number of microorganisms which are coexisting with honey bee colonies, most are opportunistic and induce troubles under as-yet undefined environmental conditions. It is then crucial for establishing a proper diagnosis of bee diseases to be able to distinguish between a normal situation and a pathogenic one. This can be partly achieved by measuring the expression levels of pathogens in honey bees using quantitative techniques, although data can only be recorded when clinical signs are detected or after colony collapse. Here we present a novel approach, which consists of identifying markers which could predict the destiny of the colony during winter. These markers may help to establish reliable diagnostic tools in relation with field observations, and ultimately to identify the causes of colony mortality.

1. *Varroa destructor* is strongly associated with colony collapse during winter

In our assay, all of the colonies which were left untreated died during winter, while only two colonies collapsed despite proper treatment. These two colonies collapsed late in the season (in February) in comparison with the majority of the other colonies which died before the end of December. These two colonies had

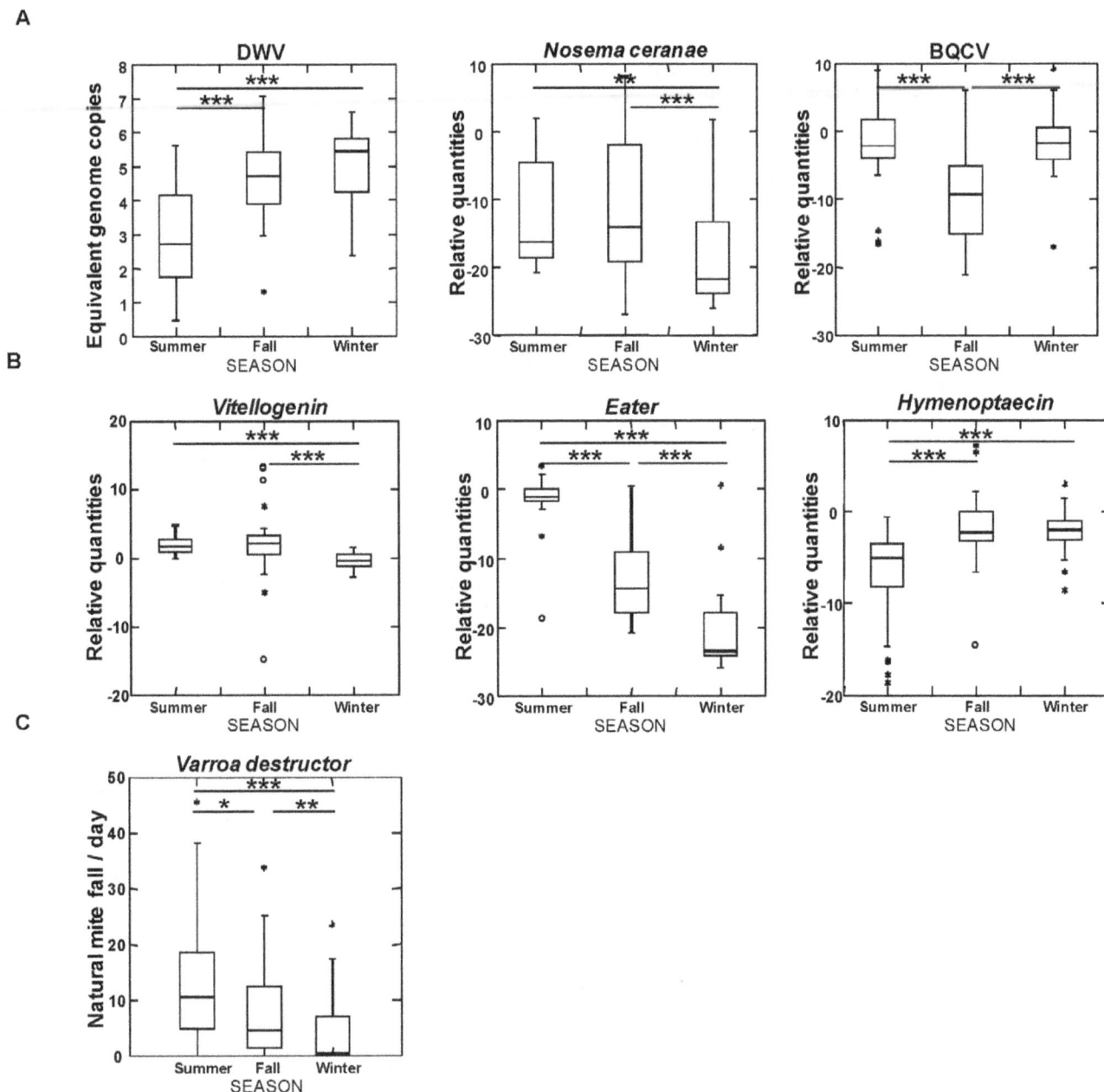

Figure 3. Overall seasonal variability of colonies (N = 29) from Summer to Winter. A) Pathogens B) Genes C) *V. destructor*. The Y-axis show the relative quantities except for DWV, where it shows equivalent genome copies [log10]. The seasons summer, fall and winter are shown on the X-axis. (Mann-Whitney test with Bonferoni correction,* P<0.05; **P<0.01; ***P<0.001).

an over-abundance of male bees, arguably a sign of failing queen fecundity. Colonies that died during winter had significantly more mites than the surviving group either in summer, fall or winter. *V. destructor* loads were thus very predictive of colony death and, in fact, provided the only predictive marker in summer. The mite level in the colonies which died during winter increased to reach a peak in October and decreased thereafter following the drop in colony size.

2. DWV is a predictive marker of colony collapse

Several bee viruses were identified. Apart from SPV which was only detected in highly mite-infested colonies here and rarely in previous surveys of European bees [45,50], DWV, SBV and

BQCV are commonly detected in western honey bee colonies all around the world, most of the time in the absence of clinical signs, although these viruses can be pathogenic under favorable environmental circumstances [20]. We found a significant increase of DWV titers between summer and fall. This result is consistent with previous reports showing an increase in DWV titers in fall [49], which probably reflects the close link between DWV and *V. destructor* since mite numbers climb rapidly from summer to fall. It is also illustrated with the significant positive correlation in fall between *V. destructor* and DWV. This mite is indeed an efficient vector of DWV, transmitting it upon feeding on bee pupae or during the phoretic phase of its biological cycle on adult bees [14,51,52]. The mite seems also competent for replication of the

Table 2. Seasonal variability for all colonies together from Summer to Winter for pathogen loads, *Varroa destructor* (Vd) and expression levels for *vitellogenin* and immune genes.

TARGET	SUMMER TO FALL		FALL TO WINTER	
	DC	SC	DC	SC
Vd	→	→	→	↓↓↓
DWV	↑↑↑	↑	→	↑
BQCV	↓↓	↓↓	↑↑↑	↑
Nosema ceranae	→	↑	→	↓↓↓
Vitellogenin	→	→	→	↓↓↓
Eater	↓↓↓	↓↓↓	↑	↓↓↓
Hymenoptaecin	↑↑↑	↑↑	→	→

Data are grouped by colony status after Winter (SC = Surviving colonies, DC = Dying colonies). (Mann-Whitney test : ↑ = P<0.05; ↑↑ or ↓↓ = P<0.01; ↑↑↑ or ↓↓↓ = P<0.001; → = No significant difference).

virus thereby increasing DWV prevalence and titers in mite infested colonies [33,53]. Proper treatments of bee colonies against *V. destructor* drastically reduced DWV titers in colonies supporting arguments that this virus is not efficiently horizontally transmitted in the absence of mites [54] and that vertical transmission routes are unlikely to generate heavy loads in bee progeny. We found higher DWV titers in collapsing colonies than in the surviving group suggesting that this virus might be involved in the process of collapsing. DWV was shown to replicate in various bee tissues including fat body, a key tissue involved in many physiological processes including immune defenses [34,55]. In particular, the fat body is the site for production of the egg yolk protein vitellogenin [34,56]. Vitellogenin is involved in immunity and ageing through hormonal regulatory pathways and is therefore a common molecular marker for the overall health and lifespan of individual bees [57]. One can hypothesize that DWV replication impairs the expression of vitellogenin in fat body cells, which could explain why we found significantly less *vitellogenin* mRNA titers in collapsing colonies than in surviving ones. In addition, *V. destructor* infestation levels and *vitellogenin* were significant negatively correlated in fall. The data agree well with a previous study showing a reduction of Vitellogenin titers in mite infested workers [58], although DWV quantification was not addressed in the prior study. Put together, this reinforces the hypothesis of an impact of DWV on fat body function.

3. *Nosema ceranae* titers are higher in healthy colonies

N. ceranae is a microsporidian intracellular parasite suspected to have replaced its common relative *N. apis* during the last decade [59,60], although there are some reports showing that *N. ceranae* has been present in western honey bees for greater than twenty years [59,61]. This parasite replicates in the bee gut and is therefore a potent pathogen of bee colonies even if its pathogenicity has only been observed in experimental conditions. These infections are very common in the absence of clinical symptoms and recent reports have shown that this parasite is not involved in bee colony collapses in Europe [5,39] except in Spain [36,62]. Our results support that *N. ceranae* is not involved in colony collapse, because no significant differences were observed between colonies that died during winter and those which survived. Likewise, we did not observe an increase of *N. ceranae* titers from summer to fall, even in December when colonies were

Summer:

Variable	by Variable	Spearman ρ	P value	Plot
Hym	DWV	0.4969	**0.0061**	
Hym	Eater	0.4665	**0.0107**	
Vg	Eater	0.8823	**<.0001**	
Vg	Hym	0.3680	**0.0495**	

Fall:

Variable	by Variable	Spearman ρ	P value	Plot
BQCV	NCer	0.4444	**0.0157**	
Vd	DWV	0.5739	**0.0011**	
Vd	Vg	-0.3783	**0.0430**	

Winter:

Variable	by Variable	Spearman ρ	P value	Plot
Eater	NCer	0.5310	**0.0030**	
Eater	BQCV	0.3789	**0.0427**	
Hym	BQCV	0.5013	**0.0056**	
Hym	DWV	0.5862	**0.0008**	
Vg	NCer	0.6345	**0.0002**	
Vg	Eater	0.6163	**0.0004**	
Vd	NCer	0.4084	**0.0309**	
Vd	Eater	0.5788	**0.0013**	
Vd	Vg	0.4457	**0.0175**	

Figure 4. Spearman rank correlations for the different variables who showed significant P values (P<0.05, indicated in bold) for the three seasons summer, fall and winter (N = 29 colonies). The variables shown are: Hym = Hymenoptaecin, Vg = Vitellogenin, Eater, NCer = *Nosema ceranane*, DWV, BQCV and Vd = *Varroa destructor*.

about to die. No *N. apis* were detected in any sample in spite of the fact that this species has a lower temperature preference than *N. ceranae* and might therefore be favored in temperate climates such as in Switzerland [60,63]. In a case study in Switzerland, *N. apis* was detected only in mixed infections with *N. ceranae* [64], consistent with these results.

4. Immune genes are upregulated in dying colonies

We monitored the expression of three *A. mellifera* genes involved in humoral and cellular immune defenses to identify physiological responses that may occur before colony collapse. In workers, in addition to being the precursor of royal jelly proteins produced by nurses to feed the larvae, the egg yolk glycoprotein vitellogenin displays multiple functions affecting important physiological pathways [65]. Among these, it has been shown that vitellogenin plays a role in bee immunity as a zinc transporter [66]. Hymenoptaecin is an antimicrobial peptide, which is highly expressed in the bee hemolymph after challenge with bacterial infections [48,67]. Eater is a major phagocytic receptor for a broad range of bacterial pathogens in *Drosophila* [68] and its homolog was identified in the honey bee genome [48,69].

The analysis of such markers from pooled individuals is complex, especially for samples collected in summer, which contained a mix of workers of different ages, because the expression of these genes might also vary according to the age-related task of worker bees. In foragers it has been shown that the vitellogenin titers as well as the number of hemocytes were strongly reduced compared to the nurse bees [70,71]. Furthermore, no data are available concerning the gene expression patterns of winter bees. Despite that, we observed significant seasonal variations among these genes. While *vitellogenin* expression remained stable from summer to fall, *hymenoptaecin* and *eater* displayed opposite expression patterns in both groups of colonies (surviving and non surviving colonies). The rapid decrease of *eater* from summer to fall suggests that hemocyte numbers are reduced in winter bees while humoral immune responses are activated. Data analyses from fall to winter in samples collected from colonies that did not survive winter (mostly collected in December before collapsing) showed no variation in either *vitellogenin* or *hymenoptaecin* mRNA levels while *eater* displayed a slight increase. In contrast, samples collected from colonies that survived winter showed a very high increase of both *vitellogenin* and *eater* transcripts while *hymenoptaecin* kept stable. Since these samples were collected at the end of January, these results may point out a different physiological state of bees collected in winter clusters two months apart. One can hypothesize that *V. destructor* – DWV – *N. ceranae* may induce different immune signaling transduction pathways than the pathway, leading to the immune transcript hymenoptae-cin [48]. As such, none of the immune genes could be identified as significant predictive markers, while vitellogenin does seem to be a viable marker.

5. *Varroa destructor* and DWV are strong predictive markers for colony collapse during winter

From the putative investigated markers, four of them were shown to be good predictive ones but seasonal. Among them, *V. destructor* and DWV were already identified during summer as strong predictive markers for collapsing during winter. As winter bees are reared as soon as mid-summer until mid-September in Switzerland [72], these data suggest that *V. destructor* infestations levels in colonies should be estimated by beekeepers as soon as summertime in order to anticipate winter colony losses. Therefore, recording of DWV loads might not be as suitable since DWV was shown to be a strong marker of winter colony collapse in fall but not in summer.

In general, the results presented here are in line with those of several studies showing that *V. destructor* and DWV are associated with colony losses in winter [5,6,8,22], and in contrast to analyses of the enigmatic Colony Collapse Disorder in the U.S., for which mite numbers were a poor correlate with CCD risk [31].

6. Conclusion

This study provides evidence that *Varroa destructor* is a key player for winter colony losses and highlights the urgent need for efficient treatments against this parasite. The data suggest an indirect effect of mite infestation on honeybee overwintering abilities through the promotion of opportunistic viral infections, which eventually lead to the impairment of critical physiological functions. The knowledge gathered in this work will help to improve our understanding of bee losses, standardize methods for biomarkers of disease and finally to mitigate causes of bee declines.

Acknowledgments

We would like to address our thanks to Hélène Berthoud, Monika Haueter, Alexandra Roetschi and Rolf Kuhn for their valuable assistance. Our thanks are also addressed to Dawn Lopez for technical assistance in USA and Virginia Williams for editorial comments. Use of commercial names in this paper is for informational purposes only.

Author Contributions

Conceived and designed the experiments: BD PN. Performed the experiments: BD. Analyzed the data: BD JDE YPC. Contributed reagents/materials/analysis tools: BD LG PN. Wrote the paper: BD JDE YPC LG PN.

References

1. Aebi A, Vaissière BE, vanEngelsdorp D, Delaplane K, Roubik DW, et al., editors. Back to the future: *Apis* vs. non-*Apis* pollination. Trends Ecol Evol 27: In press.
2. Neumann P, Carreck NL (2010) Honey bee colony losses. J Apic Res 49: 1–6. doi:10.3896/IBRA.1.49.1.01.
3. Potts SG, Biesmeijer JC, Kremen C, Neumann P, Schweiger O, et al., editors. Global pollinator declines: drivers and impacts. Trends Ecol Evol 25: 345–353.
4. Rosenkranz P, Aumeier P, Ziegelmann B (2010) Biology and control of *Varroa destructor*. J Invertebr Pathol 103: 96–119. doi:10.1016/j.jip.2009.07.016.
5. Genersch E, von der Ohe W, Kaatz H, Schroeder A, Otten C, et al., editors. The German bee monitoring project: a long term study to understand periodically high winter losses of honey bee colonies. Apidologie 41: 332–352. doi:10.1051/apido/2010014.
6. Schäfer MO, Ritter W, Pettis JS, Neumann P (2010) Winter losses of honeybee colonies (Hymenoptera: Apidae): The role of infestations with *Aethina tumida* (Coleoptera: Nitidulidae) and *Varroa destructor* (Parasitiformes: Varroidae). J Econ Entomol 103: 10–16.
7. Dahle B (2010) The role of *Varroa destructor* for honey bee colony losses in Norway. J Apic Res 49: 124–125. doi:10.3896/IBRA.1.49.1.26.
8. Guzman-Novoa E, Eccles L, Calvete Y, Mcgowan J, Kelly PG, et al., editors. *Varroa destructor* is the main culprit for the death and reduced populations of overwintered honey bee (*Apis mellifera*) colonies in Ontario, Canada. Apidologie 41: 443–450. doi:10.1051/apido/2009076.
9. Ball BV (1988) The impact of secondary infections in honey-bee colonies infested with the parasitic mite *Varroa jacobsoni*. In: Needham GR, Page RE, Delfinado-Baker M, Bowman CE, eds. Africanized Honeybees and Bee Mites. Chichester, London, UK: Ellis Horwood. pp 457–461.
10. Yang XL, Cox-Foster DL (2005) Impact of an ectoparasite on the immunity and pathology of an invertebrate: Evidence for host immunosuppression and viral amplification. Proc Natl Acad Sci U S A 102: 7470–7475.
11. Gregory PG, Evans JD, Rinderer T, de Guzman L (2005) Conditional immune-gene suppression of honeybees parasitized by Varroa mites. J Insect Sci 5: 7.
12. Yang X, Cox-Foster D (2007) Effects of parasitization by *Varroa destructor* on survivorship and physiological traits of *Apis mellifera* in correlation with viral incidence and microbial challenge. Parasitology 134: 405–412.
13. Ball BV, Allen MF (1988) The prevalence of pathogens in honey bee *Apis mellifera* colonies infested with the parasitic mite *Varroa jacobsoni*. Ann Appl Biol 113: 237–244.
14. Bowen-Walker PL, Martin SJ, Gunn A (1999) The transmission of deformed wing virus between honeybees (*Apis mellifera* L.) by the ectoparasitic mite *Varroa jacobsoni* Oud. J Invertebr Pathol 73: 101–106.
15. Chen YP, Pettis JS, Evans JD, Kramer M, Feldlaufer MF (2004) Transmission of Kashmir bee virus by the ectoparasitic mite *Varroa destructor*. Apidologie 35: 441–448.
16. Shen MQ, Cui LW, Ostiguy N, Cox-Foster D (2005) Intricate transmission routes and interactions between picorna-like viruses (Kashmir bee virus and sacbrood virus) with the honeybee host and the parasitic varroa mite. J Gen Virol 86: 2281–2289.
17. Shen MQ, Yang XL, Cox-Foster D, Cui LW (2005) The role of varroa mites in infections of Kashmir bee virus (KBV) and deformed wing virus (DWV) in honey bees. Virology 342: 141–149.
18. Dall DJ (1985) Inapparent infection of honey bee pupae by Kashmir and sacbrood bee viruses in Australia. Ann Appl Biol 106: 461–468.
19. Anderson DL, Gibbs AJ (1988) Inapparent virus infections and their interactions in pupae of the honey bee (*Apis mellifera* Linnaeus) in Australia. J Gen Virol 69: 1617–1625.
20. Chen YP, Siede R (2007) Honey bee viruses. Advances in Virus Research Vol 70: 33–80.
21. Cox-Foster DL, Conlan S, Holmes EC, Palacios G, Evans JD, et al., editors. A metagenomic survey of microbes in honey bee colony collapse disorder. Science 318: 283–287.
22. Highfield AC, El Nagar A, Mackinder LCM, Noel LMLJ, Hall MJ, et al., editors. Deformed wing virus implicated in overwintering honeybee colony losses. Appl Environ Microbiol 75: 7212–7220.
23. Berthoud H, Imdorf A, Haueter M, Radloff S, Neumann P (2010) Virus infections and winter losses of honey bee colonies (*Apis mellifera*). J Apic Res 49: 60–65. doi:10.3896/IBRA.1.49.1.08.
24. Lanzi G, de Miranda JR, Boniotti MB, Cameron CE, Lavazza A, et al., editors. Molecular and biological characterization of deformed wing virus of honeybees (*Apis mellifera* L.). J Virol 80: 4998–5009.
25. Bailey L, Gibbs AJ, Woods RD (1963) Two viruses from adult honey bees (*Apis mellifera* Linnaeus). Virology 21: 390–395.
26. Govan VA, Leat N, Allsopp M, Davison S (2000) Analysis of the complete genome sequence of acute bee paralysis virus shows that it belongs to the novel group of insect-infecting RNA viruses. Virology 277: 457–463.
27. Maori E, Lavi S, Mozes-Koch R, Gantman Y, Peretz Y, et al., editors. Isolation and characterization of Israeli acute paralysis virus, a dicistrovirus affecting honeybees in Israel: evidence for diversity due to intra- and inter-species recombination. J Gen Virol 88: 3428–3438.
28. Di Prisco G, Pennacchio F, Caprio E, Boncristiani HF, Evans JD, et al., editors. *Varroa destructor* is an effective vector of Israeli acute paralysis virus in the honeybee, *Apis mellifera*. J Gen Virol 92: 151–155. doi:10.1099/vir.0.023853-0.

29. Ribière M, Ball BV, Aubert M (2008) Natural history and geographic distribution of honey bee viruses. In: Aubert M, Ball BV, Fries I, Moritz R, Milani N, et al., editors. Virology and the honey bee. Luxembourg: European Communities. pp 15–84.

30. vanEngelsdorp D, Hayes J, Underwood RM, Pettis J (2008) A survey of honey bee colony losses in the US, fall 2007 to spring 2008. Plos One 3: e4071.

31. vanEngelsdorp D, Evans JD, Saegerman C, Mullin C, Haubruge E, et al., editors. Colony collapse disorder: A descriptive study. Plos One 4: e6481.

32. de Miranda JR, Cordoni G, Budge G (2010) The acute bee paralysis virus - Kashmir bee virus - Israeli acute paralysis virus complex. J Invertebr Pathol 103: 30–47. doi:10.1016/j.jip.2009.06.014.

33. de Miranda JR, Genersch E (2010) Deformed wing virus. J Invertebr Pathol 103: S48–S61. doi:10.1016/j.jip.2009.06.012.

34. Fievet J, Tentcheva D, Gauthier L, de Miranda J, Cousserans F, et al., editors. Localization of deformed wing virus infection in queen and drone *Apis mellifera* L. Virology Journal 3: 1–5.

35. Boncristiani HF, Di Prisco G, Pettis JS, Hamilton M, Chen YP (2009) Molecular approaches to the analysis of deformed wing virus replication and pathogenesis in the honey bee, *Apis mellifera*. Virology Journal 6: 1–9. doi:10.1186/1743-422X-6-221.

36. Higes M, Martin-Hernandez R, Botias C, Bailon EG, Gonzalez-Porto AV, et al., editors. How natural infection by *Nosema ceranae* causes honeybee colony collapse. Environ Microbiol 10: 2659–2669.

37. Higes M, Martin-Hernandez R, Martinez-Salvador A, Garrido-Bailon E, Gonzalez-Porto AV, et al., editors. A preliminary study of the epidemiological factors related to honey bee colony loss in Spain. Environ Microbiol Rep 2: 243–250. doi:10.1111/j.1758-2229.2009.00099.x.

38. Chauzat MP, Martel AC, Zeggane S, Drajnudel P, Schurr F, et al., editors. A case control study and a survey on mortalities of honey bee colonies (*Apis mellifera*) in France during the winter of 2005-6. J Apic Res 49: 40–51. doi:10.3896/IBRA.1.49.1.06.

39. Gisder S, Hedtke K, Mockel N, Frielitz MC, Linde A, et al., editors. Five-Year Cohort Study of *Nosema* spp. in Germany: Does Climate Shape Virulence and Assertiveness of *Nosema ceranae*? Appl Environ Microbiol 76: 3032–3038. doi:10.1128/AEM.03097-09.

40. Imdorf A, Charrière JD, Kilchenmann V, Bogdanov S, Fluri P (2003) Alternative strategy in central Europe for the control of *Varroa destructor* in honey bee colonies. Apiacta 38: 258–278.

41. Goodwin RM, Taylor MA, McBrydie HM, Cox HM (2006) Drift of *Varroa destructor*-infested worker honey bees to neighbouring colonies. J Apic Res 45: 155–156.

42. Imdorf A, Charrière JD (1998) What is the Varroa population in my colonies ? Bee Biz 7: 37.

43. Imdorf A, Bühlmann G, Gerig L, Kilchenmann V, Wille H (1987) Überprüfung der Schätzmethode zur Ermittlung der Brutfläche und der Anzahl Arbeiterinnen in freifliegenden Bienenvölkern. Apidologie 18: 137–146.

44. Tentcheva D, Gauthier L, Zappulla N, Dainat B, Cousserans F, et al., editors. Prevalence and seasonal variations of six bee viruses in *Apis mellifera* L. and *Varroa destructor* mite populations in France. Appl Environ Microbiol 70: 7185–7191.

45. de Miranda JR, Dainat B, Locke B, Cordoni G, Berthoud H, et al., editors. Genetic characterization of slow paralysis virus of the honeybee (*Apis mellifera* L.). J Gen Virol 91: 2524–2530. doi:10.1099/vir.0.022434-0.

46. Evans JD (2006) Beepath: An ordered quantitative-PCR array for exploring honey bee immunity and disease. J Invertebr Pathol 93: 135–139.

47. Lourenco AP, Mackert A, Cristino AD, Simoes ZLP (2008) Validation of reference genes for gene expression studies in the honey bee, *Apis mellifera*, by quantitative real-time RT-PCR. Apidologie 39: 372-U33.

48. Evans JD, Aronstein K, Chen YP, Hetru C, Imler JL, et al., editors. Immune pathways and defence mechanisms in honey bees *Apis mellifera*. Insect Mol Biol 15: 645–656.

49. Gauthier L, Tentcheva D, Tournaire M, Dainat B, Cousserans F, et al., editors. Viral load estimation in asymptomatic honey bee colonies using the quantitative RT-PCR technique. Apidologie 38: 426–436.

50. Ball BV, Bailey L (1997) Viruses. In: Morse RA, Flottum K, eds. Honey bee pests, predators & diseases. MedinaOH: A.I. Root Company. pp 11–31.

51. Santillan-Galicia MT, Ball BV, Clark SJ, Alderson PG (2010) Transmission of deformed wing virus and slow paralysis virus to adult bees (*Apis mellifera* L.) by *Varroa destructor*. J Apic Res 49: 141–148. doi:10.3896/IBRA.1.49.2.01.

52. Mockel N, Gisder S, Genersch E (2011) Horizontal transmission of deformed wing virus: pathological consequences in adult bees (*Apis mellifera*) depend on the transmission route. J Gen Virol 92: 370–377. doi:10.1099/vir.0.025940-0.

53. Gisder S, Aumeier P, Genersch E (2009) Deformed wing virus: replication and viral load in mites (*Varroa destructor*). J Gen Virol 90: 463–467.

54. Sumpter DJT, Martin SJ (2004) The dynamics of virus epidemics in Varroa-infested honey bee colonies. J Anim Ecol 73: 51–63.

55. Trenczek T, Faye I (1988) Synthesis of immune proteins in primary cultures of fat body from *Hyalophora cecropia*. Insect Biochem 18: 299–312.

56. Tufail M, Takeda M (2008) Molecular characteristics of insect vitellogenins. J Insect Physiol 54: 1447–1458.

57. Munch D, Amdam GV, Wolschin F (2008) Ageing in a eusocial insect: molecular and physiological characteristics of life span plasticity in the honey bee. Funct Ecol 22: 407–421.

58. Amdam GV, Hartfelder K, Norberg K, Hagen A, Omholt SW (2004) Altered physiology in worker honey bees (Hymenoptera : Apidae) infested with the mite *Varroa destructor* (Acari : Varroidae): A factor in colony loss during overwintering? J Econ Entomol 97: 741–747.

59. Paxton RJ, Klee J, Korpela S, Fries I (2007) *Nosema ceranae* has infected *Apis mellifera* in Europe since at least 1998 and may be more virulent than *Nosema apis*. Apidologie 38: 558–565.

60. Fries I (2010) *Nosema ceranae* in European honey bees (*Apis mellifera*). J Invertebr Pathol 103: S73–S79. doi:10.1016/j.jip.2009.06.017.

61. Chen Y, Evans JD, Smith IB, Pettis JS (2008) *Nosema ceranae* is a long-present and wide-spread microsporidian infection of the European honey bee (*Apis mellifera*) in the United States. J Invertebr Pathol 97: 186–188.

62. Higes M, Martin-Hernandez R, Garrido-Bailon E, Gonzalez-Porto AV, Garcia-Palencia P, et al., editors. Honeybee colony collapse due to *Nosema ceranae* in professional apiaries. Environ Microbiol Rep 1: 110–113. doi:10.1111/j.1758-2229.2009.00014.x.

63. Higes M, Garcia-Palencia P, Botias C, Meana A, Martin-Hernandez R (2010) The differential development of microsporidia infecting worker honey bee (*Apis mellifera*) at increasing incubation temperature. Environ Microbiol Rep 2: 745–748.

64. Dainat B, vanEngelsdorp D, Neumann P (2012) CCD in Europe?. Environ Microbiol Rep;doi:10.1111/j.1758-2229.2011.00312.x.

65. Amdam GV, Norberg K, Hagen A, Omholt SW (2003) Social exploitation of vitellogenin. Proc Natl Acad Sci U S A 100: 1799–1802.

66. Amdam GV, Simoes ZLP, Hagen A, Norberg K, Schroder K, et al., editors. Hormonal control of the yolk precursor vitellogenin regulates immune function and longevity in honeybees. Exp Gerontol 39: 767–773.

67. Chan QWT, Melathopoulos AP, Pernal SF, Foster LJ (2009) The innate immune and systemic response in honey bees to a bacterial pathogen, *Paenibacillus larvae*. Bmc Genomics 10: 387.

68. Kocks C, Cho JH, Nehme N, Ulvila J, Pearson AM, et al., editors. Eater, a transmembrane protein mediating phagocytosis of bacterial pathogens in drosophila. Cell 123: 335–346.

69. Simone M, Evans JD, Spivak M (2009) Resin Collection and Social Immunity in Honey Bees. Evolution 63: 3016–3022.

70. Wilson-Rich N, Dres ST, Starks PT (2008) The ontogeny of immunity: Development of innate immune strength in the honey bee (*Apis mellifera*). J Insect Physiol 54: 1392–1399.

71. Schmid MR, Brockmann A, Pirk CWW, Stanley DW, Tautz J (2008) Adult honeybees (*Apis mellifera* L.) abandon hemocytic, but not phenoloxidase-based immunity. J Insect Physiol 54: 439–444.

72. Merz R, Gerig L, Wille H, Leuthold R (1979) Das Problem der Kurz- und Langlebigkeit bei der Ein- und Auswinterung im Bienenvolk (*Apis mellifica* L.): eine Verhaltensstudie. Rev Suisse Zool 86: 663–671.

5

Pathogen Webs in Collapsing Honey Bee Colonies

R. Scott Cornman[1], David R. Tarpy[2], Yanping Chen[1], Lacey Jeffreys[2], Dawn Lopez[1], Jeffery S. Pettis[1], Dennis vanEngelsdorp[3], Jay D. Evans[1]*

1 Bee Research Laboratory, Agricultural Research Service, United States Department of Agriculture, Beltsville, Maryland, United States of America, 2 Department of Entomology, North Carolina State University, Raleigh, North Carolina, United States of America, 3 Department of Entomology, University of Maryland, College Park, Maryland, United States of America

Abstract

Recent losses in honey bee colonies are unusual in their severity, geographical distribution, and, in some cases, failure to present recognized characteristics of known disease. Domesticated honey bees face numerous pests and pathogens, tempting hypotheses that colony collapses arise from exposure to new or resurgent pathogens. Here we explore the incidence and abundance of currently known honey bee pathogens in colonies suffering from Colony Collapse Disorder (CCD), otherwise weak colonies, and strong colonies from across the United States. Although pathogen identities differed between the eastern and western United States, there was a greater incidence and abundance of pathogens in CCD colonies. Pathogen loads were highly covariant in CCD but not control hives, suggesting that CCD colonies rapidly become susceptible to a diverse set of pathogens, or that co-infections can act synergistically to produce the rapid depletion of workers that characterizes the disorder. We also tested workers from a CCD-free apiary to confirm that significant positive correlations among pathogen loads can develop at the level of individual bees and not merely as a secondary effect of CCD. This observation and other recent data highlight pathogen interactions as important components of bee disease. Finally, we used deep RNA sequencing to further characterize microbial diversity in CCD and non-CCD hives. We identified novel strains of the recently described Lake Sinai viruses (LSV) and found evidence of a shift in gut bacterial composition that may be a biomarker of CCD. The results are discussed with respect to host-parasite interactions and other environmental stressors of honey bees.

Editor: Sarah K. Highlander, Baylor College of Medicine, United States of America

Funding: This work was supported by the National Honey Board, the Pennsylvania Department of Agriculture, the North American Pollinator Protection Campaign, the Apiary Inspectors of America and the USDA-CSREES Coordinated Agricultural Project for Pollinator Health. The funders had no role in study design, data collection and analysis, decision to publish, or preparation of the manuscript.

Competing Interests: The authors have declared that no competing interests exist.

* E-mail: Jay.evans@ars.usda.gov

Introduction

In addition to producing hive products such as honey, bee pollen, and propolis, managed colonies of the European honey bee, *Apis mellifera*, are in increasing demand for commercial crop pollination [1,2] Yet in the midst of this demand, beekeepers on multiple continents have suffered severe losses in recent years [3,4]. Since 2006, a substantial fraction of honey bee losses in the United States have been ascribed to Colony Collapse Disorder (CCD), an enigmatic sudden disappearance of adult worker bees [5]. 'Disappearing diseases' similar to CCD have long been described in honey bees, and are apparently a recurring feature of domesticated honey bee populations. Historically, these declines have not shown recognized pathologies [6] and have generally gone unresolved for years following their occurrence [7,8].

Current research on this phenomenon has focused on three general, non-exclusive factors: (1) environmental contaminants, especially agricultural pesticides; (2) poor nutrition and subsequent developmental disorders; and (3) novel or resurgent pathogens. While numerous additional hypotheses have been raised to explain CCD, including genetic homogeneity, breakdowns in social cues, a failure in colony thermoregulation, and the impacts of genetically modified or toxic pollen [9,10], these hypotheses have not found broad support in studies to date.

Current evidence for a chemotoxic basis of CCD is equivocal. Honey bees have been exposed for many years to diverse anthropogenic chemicals, primarily agricultural applications aimed at reducing pest plants or arthropods. Chemical residues, including known insecticides, have been detected in bees and in hive materials (mostly wax and pollen) [11]. Recent evidence suggests the effects of low-level exposure to such chemicals range from impaired behavior (Henry et al., 2012) to lowered disease resistance (Alaux et al., 2012, Pettis et al., 2012), and further study of agrochemical toxicity is warranted. Nevertheless, neither individual chemicals nor overall chemical loads have been tied to increased risk of CCD; in fact, levels of the pesticides coumaphos and Esfenvalerate have been found at higher levels in control colonies as compared to CCD colonies [5,10]. The interpretation of this finding is complicated by the fact that coumaphos is itself directly applied to honey bee hives to reduce levels of the parasitic mite, *Varroa destructor*, and thus the apparent positive correlation between coumaphos level and colony health is confounded with *Varroa* management practices. Even so, genes presumed to be involved in pesticide detoxification have not been detected as differentially expressed in bees from CCD versus non-CCD colonies [12].

While nutritional resources certainly affect honey bee longevity, including survival over the stressful winter (when CCD has been most prevalent), there is no direct evidence linking food resources to colony collapses. Bees from collapsed colonies showed typical body weights, protein complements, and lipid stores when compared to temporal controls [5].

There are many microorganisms that affect honey bees, ranging from viruses to bacteria, fungi, trypanosomes, and amoebae [13,14]. The roles of many of these microbes on individual and colony health remain unclear, and even less understood are the interactions and relationships among pathogens. In an earlier microbial survey in the U.S., declining honey bees colonies showed an especially high prevalence of two dicistroviruses, Israeli acute paralysis virus (IAPV) and Kashmir bee virus (KBV), and two microsporidian species in the genus *Nosema* when compared to healthy controls [15]. Of these, IAPV was most strongly linked to colony collapse, a trend that has not been supported with deeper sampling effort [5]. Nosemosis has since been associated with collapsing hives in additional studies [16], but other work has not found *Nosema* to be a predictor of CCD [5] or general colony loss [17] and the broad distribution of *Nosema* in apparently strong colonies [18,19] contradicts a unifying role for these pathogens in CCD. Moreover, no consistent differences were observed in the expression of honey bee immune genes between CCD and non-CCD samples [12], indicating the absence of a characteristic immune response associated with this syndrome. However, CCD colonies did have more pathogen species present than did non-CCD colonies in a recent survey and there was evidence that the condition is contagious [5]. Furthermore, some viruses have been found to be predictors of overwinter colony loss [17], and an increase in ribosomal RNA fragments among transcripts from CCD samples was interpreted as implicating one or more of a group of honey bee RNA viruses [12]. Taken collectively, current data suggest that multiple factors underlie CCD, some of which may be interchangeable or dispensable but which may interact synergistically to cause disease. Thus, while a prominent role for pathogens seems likely, the causes of CCD remain elusive.

Here we further explore the connections between pathogens and CCD via a country-wide survey of pathogens in collapsed and healthy colonies. This survey includes but expands upon samples previously collected and analyzed [5,15]. A retrospective strategy is an efficient approach to identifying potential pathogen interactions associated with CCD and a necessary prelude to laborious and costly prospective studies specifically targeting this syndrome, given the erratic nature of its occurrence. For example, a recent, large-scale prospective study of honey bee colonies over a ten-month period provided an invaluable catalog of pathogens in managed colonies but did not encounter any unexplained colony losses [19].

Our survey was based on two approaches. We first used quantitative real-time PCR (qPCR) assays for the major honey bee pathogens to provide a fine-scale analysis of pathogen levels in individual bees, colonies, apiaries, and the U.S. as a whole. Our objectives were to 1) further quantify pathogen loads in CCD colonies (relative to non-CCD colonies) across a broad geographic area; 2) identify significant covariation among pathogens and determine if such correlations are greater in CCD colonies. We then used deep sequencing to identify novel microbial strains and species, and to compare levels of the predominant gut microbiota in bees from CCD and healthy colonies. As adult honey bees have a relatively simple and consistent gut microbiome [15,20], deviations in this flora could be a biomarker for, or directly related to, disease.

We found that CCD colonies exhibited a higher prevalence, abundance, and positive covariance of pathogens. In marked contrast, otherwise weak colonies lacking CCD traits did not have increased pathogen loads relative to strong colonies, and non-CCD colonies in general exhibited little pathogen covariance. It remains unknown whether these statistical associations reflect an actual synergy among pathogens in CCD hives or are instead ancillary to some other variable, but we show that positive correlations among pathogens develop at the level of individual bees and are not contingent on the pre-existence of CCD. We also found important heterogeneity in pathogen distributions in our samples, which were collected from across the United States. While our data supports the view that pathogen interactions contribute to CCD, they also indicate that there may be multiple routes to the same phenotype or that particular combinations are not deterministic in their effects. Finally, we take a metagenomic step toward elucidating other biotic components of CCD by identifying novel virus strains and finding evidence of a CCD-associated shift in gut bacterial composition.

Materials and Methods

Colony Censuses and Collection of Material

CCD and some non-CCD colonies were sampled from the same apiaries in late 2006 and early 2007 as described in [5,15]. CCD cases were drawn either from temporary migratory commercial beekeeping apiaries on the East Coast (n = 24) or temporary commercial apiaries established in California for almond pollination (n = 37). The latter had previously been stationed in Minnesota, Pennsylvania, Washington, Nebraska, and Montana. We sought additional non-CCD colonies that were far from regions with CCD, out of concern that some non-CCD colonies could potentially have developed CCD at a later date (this was not monitored). The non-CCD samples can therefore be subdivided into 'sympatric' colonies that were in or near CCD outbreaks (n = 37) and additional 'non-sympatric' colonies (n = 26) that were taken from California apiaries with no record of CCD in January, 2008, and similarly healthy Maryland apiaries in July, 2004, and July, 2008. As a result, there is both temporal and geographic breadth in our sampling but we were not able to explicitly pair CCD and non-CCD colonies for each sampling time and location. All sampled colonies were 'overwintered', i.e., hives that were, in the view of their managers, healthy the previous summer and provisioned adequately for winter. A detailed analysis of how CCD and non-CCD hives differ at diagnosis in terms of population size and age structure is given in [5]. For each colony, we collected over 200 live adult bees from the nest interior and shipped these in 50 ml centrifuge tubes on dry ice. After shipping, samples were stored at −80°C until analysis.

Estimates of Pathogen Abundance: Among-colony Analysis

Abdomens were cut from eight bees from each colony and ground together in 4 ml RNAqueous buffer (Ambion). Whole bees were not used because of the potential for PCR inhibition [21]. A 700 µl aliquot of this homogenate was extracted according to the manufacturer's protocol. cDNA was generated with Superscript II (Invitrogen) and primed by an oligo-dT cocktail 12–18 nt in length.

qPCR reactions were carried out in 96-well plates using a Bio-Rad iCycler (Bio-Rad Corp). Reactions consisted of 1.5 µg template, 1 U Taq with proscribed buffer (Roche Applied Sciences), 1 mM dNTP mix, 2 mM $MgCl_2$, 0.2 µM of each primer, 1X concentration SYBR-Green I dye (Applied Biosystems), and 10 nM fluorescein in a 25 µl reaction volume. All reactions were carried out with a thermal protocol consisting of 5 min at 95°C, then 40 cycles of a four-step protocol consisting of

94°C for 20 s, 60°C for 30 s, 72°C for 1 min, and 78°C for 20 s. Fluorescence measurements were taken repeatedly during the 78°C step to reduce the contribution of primer artifacts to the inferred concentration of the target. Dissociation kinetics were monitored to verify the product melt temperature, and a subset of products for each targeted pathogen was sequenced to confirm primer specificity [22]. Positive and negative control reactions were run on each plate.

We used published primers (Table 1 of [5]) to survey for nine known honey-bee pathogens: KBV, IAPV, deformed wing virus (DWV), acute bee paralysis virus (ABPV), black queen cell virus (BQCV), sacbrood virus (SBV), *Nosema ceranae*, *N. apis*, and the trypanosome *Crithidia mellificae*. Given the past importance of IAPV as an indicator of bee disease, three primer pairs [5] were used to confirm the sensitivity of this assay. (One primer listed in [5] was discovered to be incorrect; the actual sequence for IAPV-PW-R17 is GCAGGACATTAATGTACTATATCCAG). The mean amplification efficiency of each qPCR primer pair (**File S1**) was estimated by dilution-series analysis. The geometric mean of three honey-bee genes – actin, ribosomal protein S5 (RPS5), and microsomal glutathione-S-transferase (MGST) – was used to calculate the normalized abundance of each target in each sample (ΔC_T), following the recommendation of [23]. Since the mean of multiple efficiency estimates for each control gene was close to one, we assumed equal and perfect efficiencies of the reference genes. However, we used the actual estimated efficiency for each target primer pair to calculate fold change in abundance, using the $\Delta\Delta C_T$ method [23]. That is, differential abundance equals $(1+\text{efficiency})^{\Delta\Delta C_T}$, where $\Delta\Delta C_T$ is the difference in mean ΔC_T between two sample populations.

Whether the mean difference ($\Delta\Delta C_T$) in pathogen abundances was significant was determined by Analysis of Variance of ΔC_T without conversion to linear relative abundance for each sample, because converted values can have strongly non-normal distributions, particularly for viruses. In addition to comparing CCD colonies to non-CCD colonies, we performed an additional analysis that decoupled the effects of colony size (an indicator of colony strength [5] from CCD *per se*. Colonies from apiaries that had no report of CCD were split into 'weak' colonies with six or fewer frames of bees (n = 15) and 'strong' colonies with seven or more frames of bees (n = 29). These thresholds correspond to a natural break in colony size distribution, while retaining

sufficient samples in each bin for statistical analysis. Partial correlations of abundance of each pathogen were estimated using Spearman's rho statistic with a Bonferroni correction for multiple tests.

Pathogen Covariance in Individual Bees

We investigated whether covariation in pathogen abundance extended to individual bees with natural infections of common pathogens. We quantified pathogen loads of individual workers taken from otherwise strong colonies known to contain *N. ceranae*. This analysis was deliberately removed from the context of CCD, and focused on *Nosema* and several RNA viruses because prior work had already suggested synergistic interactions among these common pathogens [15]. We sampled 17–24 adult workers from each of four colonies located in Raleigh, North Carolina (n = 77 bees). qPCR was performed for Chronic bee paralysis virus (CBPV), BQCV, DWV, *N. apis*, and *N. ceranae*. (Note that CBPV was not included in the survey of CCD and non-CCD colonies because initial work and subsequent RNA sequencing indicated a very low incidence/abundance of this pathogen in those samples.) Total RNA was isolated from the thorax using the RNeasy Mini kit and cDNA synthesis performed using 3.0 µl buffer, 3.0 µl 2.5 mM dNTP, 0.75 µl RNaseOUT, 0.3 µl (0.18 µg) of random primer cocktail, 0.75 µl Superscript III, 2.2 µl H2O, and 5 µl (approximately 2.5 µg) RNA. qPCR measurements were performed on an ABI 7900 Fast Real-Time PCR system with Sequence Detection Systems software version 2.3. qPCR reactions included SYBR Green Master Mix, 10 picomoles of each primer, and 2 µl of cDNA in a 10 µl volume. Product specificity was evaluated by dissociation curve. Total loads and partial correlations were calculated as above, except with RPS5 as the single normalization gene.

Deep RNA Sequencing of Healthy and Collapsed Colonies for Microbe Discovery

RNA was pooled by combining equal aliquots from each CCD or non-CCD colony described above. Five µg of RNA from the "CCD−" pool was used to generate cDNA using a cocktail of random heptamer primers. cDNA was size-selected from agarose and end-polished with End Repair Enzyme (Illumina) following manufacturer protocols. A 3′ polyadenine tract was then added

Table 1. Honey-bee pathogen incidence by colony status.

Pathogen	Present, non-CCD colony	Absent, non-CCD colony	Proportion present, non-CCD colony	Present, CCD colony	Absent, CCD colony	Proportion present, CCD colony	P-value
ABPV	30	33	0.48	31	30	0.51	0.722
BQCV	53	10	0.84	55	6	0.90	0.314
DWV	26	37	0.41	38	23	0.62	0.019
IAPV	10	53	0.16	15	46	0.25	0.225
KBV	8	55	0.13	23	38	0.38	0.001
SBV	9	54	0.14	16	45	0.26	0.096
NC	36	27	0.57	41	20	0.67	0.247
NA	6	57	0.10	20	41	0.33	0.001
Crithidia	49	14	0.78	53	8	0.87	0.182

Values are the number of colony samples (n = 61 for CCD and n = 63 for non-CCD) in which the pathogen was detected at any level. The likelihood ratio chi-square test of contingency was used to compute the probability of equal pathogen incidence in CCD and non-CCD colonies. ABPV = acute bee paralysis virus; DWV = deformed wing virus; SBV = sacbrood virus; BQCV = black queen cell virus; IAPV = Israeli acute paralysis virus; KBV = Kashmir bee virus; NC = *Nosema ceranae*; NA = *Nosema apis*.

with Klenow fragment (Invitrogen) and the products purified with a Qiaquick DNA purification column (Qiagen). Illumina adapters were ligated to cDNA with T4 DNA ligase and the products were amplified under the following thermocycler conditions: an initial denaturing step at 98°C for 30 seconds, followed by 14 cycles at 98°C for 30 seconds, 65°C for 30 seconds, and 72°C for 30 seconds. Final products of 100–300 bp were size-selected from agarose and sequenced on an Illumina Genome Analyzer by the Institute for Genome Sciences, University of Maryland, Baltimore.

Equivalently prepared cDNA from the "CCD+" pool was sequenced using a paired-end strategy with a 350-bp fragment size. A paired-end approach facilitates the assembly of longer contigs, and therefore may provide more diagnostic sequences for annotation, but at a cost of reduced read length (67 bp). Both sequencing runs were quality-trimmed by retaining only the longest contiguous sequence of each read with a minimum (Phred-equivalent) quality score of 15, excepting at most one ambiguous base. Reads less than 50 bp after this trimming step were discarded. A small number of reads were removed because they matched Illumina primer sequence in the Univec database (www. ncbi.nlm.nih.gov/VecScreen/UniVec.html).

Reads were assembled into contigs using the Velvet assembly package [24]. CCD− reads were assembled into contigs using multiple iterations of Velvet with successive hash lengths of 21, 31, 41, 51, or 61. Contigs of less than 100 bp or with less than 3X coverage were discarded. This assembly strategy was chosen to accommodate the broad spectrum of RNA sources in the sample (viruses, a diverse bacterial community, and eukaryotic pathogens as well as the host genome) that are likely to have different optimal hash lengths for assembly. CCD+ reads were assembled in a similar fashion without read-pair information; in addition, a single paired-end assembly was performed with Velvet using a hash length of 21 and an expected fragment length of 350. Contigs from all intermediate assemblies were then merged using the BlastClust component of BLAST at 98% identity and 90% nonreciprocal overlap. Because there was substantial redundancy of contigs remaining after this step, we input the contigs to CAP3 [25] for more aggressive assembly, requiring a 60-bp overlap with 92% identity. Raw reads are available as accessions SRX028143 and SRX028145 of the NCBI Short Read Archive, however, the resulting contigs were not submitted because of an NCBI policy against hosting assemblies from mixed sources. The contigs are included here as **File S2**.

Contigs were searched against the GenBank nr database with BLASTN and BLASTX, requiring an expectation of ≤1.0E-10. All ribosomal matches to trypanosomes (e.g., *Leptomonas* and *Leishmania*) were considered to be *Crithidia mellificae* for this analysis, although the taxonomy of bee trypanosomes is not well established [26] and there is evidence of substantial genetic divergence among isolates from honey bee (R. Schwarz and J. DeRisi, unpublished data). Contigs with BLASTN hits to bacterial ribosomal sequence were submitted to the Classifier tool [27] for taxonomic evaluation. In addition to these bacterial contigs, we also used GenBank accessions of honey bee gut bacteria for read mapping (see below) because they are cloned 16S amplicons rather than short-read assemblies and are also longer on average than our contigs.

We used Bowtie [28] to map reads to reference sequences for quantitative comparisons between CCD− and CCD+. To avoid an ascertainment bias between the two sequence samples, which have different maximum lengths, we trimmed all reads to a maximum of 65 bp for mapping (chosen because it resulted in more similar mean lengths than did the 67-bp maximum length of CCD reads). Reads were mapped sequentially to a series of reference databases, allowing a defined number of mismatches (see Results) and normalized to the total number of reads in each library. We did not report counts with an additional normalization for reference length because the assembled contigs are generally fragments of larger molecules and, particularly for RNA viruses, are unlikely to include weakly expressed or noncoding regions. Since, for feasibility, the two sequenced samples were pooled from various sources as described, we cannot calculate technical or biological components of variance and thus make no statistical test of differential abundance between CCD+ and CCD−.

We performed an initial cull of all reads that mapped to known honey bee sequences, including the reference genome version Amel_4.0 [29] (GenBank accession PRJNA13343), GenBank accessions of mitochondrial and ribosomal sequence of that species (NC_001566, AY703484, AY703551), and contigs derived from this study that were BLASTN matches to the previous. Reads were then mapped to a database containing 1) representative GenBank accessions of known honey bee viruses: IAPV (NC_009025.1), KBV (NC_004807.1), acute bee paralysis virus (ABPV, NC_002548.1), chronic bee paralysis virus (CBPV, NC_010711.1 and NC_010712.1), DWV (NC_004830.2), BQCV (NC_003784.1), sacbrood virus (SBV, NC_002066.1), and slow bee paralysis virus (NC_014137.1); 2) whole genome sequences of eukaryotic pathogens and commensals, specifically the fungi *Ascosphaera apis* [30] (PRJNA17285) and *Saccharomyces cerevisiae* [31] (PRJNA128), the microsporidia *N. ceranae* [32] (PRJNA48321) and *N. apis* (Y.-P. Chen, unpublished data), the trypanosome *Crithidia mellificae* (R.S. Schwarz, unpublished data), the mite *Varroa destructor* [33] (PRJNA33465); and 3) GenBank accessions of bacterial ribosomal sequence that were classified as the dominant gut phylotypes by [20] (accessions listed below). After culling reads that matched this database, the remaining reads were mapped to the assembled contigs themselves. Residual reads were then mapped to ribosomal sequence of the SILVA database [34]. This last mapping was done only to identify the number of residual reads that were recognizably ribosomal, not to identify their taxonomic source (for which there is little power with short reads). After this first pass, unmatched reads were re-assembled with Velvet in a manner analogous to the original assembly, but with hash lengths of 23, 37, and 51, respectively. The resulting contigs were annotated with BLAST in the same manner as above and the mapping procedure re-iterated to produce the final read counts. After all Bowtie mapping steps, we used Mega BLAST to the genome and gene set of honey bee to identify residual bee sequence. Reads mapping to the Kakugo variant of DWV [35] were not treated separately due to the high nucleotide identity between the two. In contrast, Varroa destructor virus 1 [36] is more distinct from DWV, but only a single read mapped to this reference, such that we chose for simplicity to combine this read with the DWV counts.

The goals of the present study with respect to the honey bee microbiome were to further characterize what species were present and to identify any changes in species distributions between CCD− and CCD+ that are suggestive of physiological state or pathogenicity. However, this study was not designed specifically for metagenomic analysis of microbial community structure or gene content. It is inherently difficult to assemble short ribosomal reads from a diverse pool into contigs of sufficient length for phylogenetic assessment, and protein-coding transcripts from a source other than honey bee are expected to be poorly represented in total RNA. Uneven representation of phylogenetic groups in public databases and non-uniform criteria for their annotation are other sources of bias limiting our ability to accurately classify ribosomal sequence. There may also be

Table 2. Contrasts in honey-bee pathogen abundance by colony status.

Target	All CCD colonies vs. all non-CCD colonies				Weak vs. strong colonies in non-CCD apiaries			
	$\Delta\Delta C_T$	SE	P-value	Fold change	$\Delta\Delta C_T$	SE	P-value	Fold change
ABPV	+2.23	0.96	0.02	4.57	+0.84	0.87	0.34	1.77
BQCV	+2.81	1.08	0.01	6.67	−1.70	1.59	0.29	0.32
DWV	+3.90	1.15	<0.01	14.26	−0.07	0.61	0.96	0.95
IAPV	+0.22	0.60	0.72	1.15	+1.64	0.98	0.11	2.83
KBV	+2.58	0.78	<0.01	5.49	+0.67	0.67	0.32	1.56
SBV	+0.28	0.64	0.66	1.27	+0.36	1.52	0.81	1.36
N. ceranae	+1.19	1.22	0.33	1.85	+1.70	1.59	0.29	2.41
N. apis	+3.94	1.07	<0.01	20.97	−0.03	0.14	0.81	0.98
Crithidia	+2.62	1.12	0.02	6.15	−0.79	1.95	0.69	0.58

The difference in mean ΔC_T values (normalized threshold cycle in qPCR reactions) was compared by ANOVA for two non-independent contrasts: all CCD colonies (n = 61) versus all non-CCD colonies (n = 63), and weak (n = 15) versus strong (n = 29) colonies in non-CCD apiaries. Weak colonies had six or fewer frames of bees and strong colonies had seven or more frames (see Materials and Methods). Non-CCD colonies include both sympatric and allopatric colonies, which were combined for increased statistical power. $\Delta\Delta$ is the mean $\Delta$of the non-CCD population minus the mean Δvalue of the CCD population. Thus, positive numbers represent a decrease in mean threshold cycle and an increase in pathogen abundance. Fold change between categories is calculated as $(1+ \text{primer efficiency})^{\Delta\Delta C_T}$. SE = standard error of population mean ΔC_T; P-value = probability of equal mean ΔC_T by ANOVA (i.e., that the true $\Delta\Delta C_T = 0$); ABPV = acute bee paralysis virus; DWV = deformed wing virus; SBV = sacbrood virus; BQCV = black queen cell virus; IAPV = Israeli acute paralysis virus; KBV = Kashmir bee virus.

unknown biological or methodological biases, such as variation among organisms in intrinsic expression level or the efficiency of RNA extraction. Relative abundances *within* a sample should therefore be treated with caution. However, there is no reason these obstacles to microbiotic classification should generate artifactual differences *between* equivalently prepared samples, especially in light of the relatively simple and stable community of honey bee gut bacteria [20]. We therefore mapped reads to GenBank accessions representative of this microbiota that were drawn from the phylogenetic analysis of [20] (AJ971849, AJ971850, AJ971857, AY370183–AY370186, AY370188, AY370191, AY370192, DQ837604, DQ837605, DQ837611, DQ837616, DQ837617, DQ837622–DQ837626, DQ837632–DQ837634, DQ837636, EF187232, EF187235–EF187237, EF187240, EF187242, EF187244, EF187250, EU055544, HM107876, HM108310, HM108312, HM108315, HM108316, HM108318, HM108324, HM108330, HM108332, HM108334, HM108335, HM108337, HM108346, HM108542, HM108563, HM111870, HM111875, HM111880, HM111883, HM111887, HM111901, HM111923, HM111924, HM111973, HM111977, HM112025, HM112033, HM112038, HM112042, HM112050, HM112068, HM112094, HM112104, HM112118, HM112130, HM112858, HM112866, HM113259, HM113300), as well as to contigs of our assembly that were considered bacterial based on BLAST match or the Classifier tool of the RDP database [27].

Characterization of Novel Virus Candidates

Two groups of novel viruses were identified in this study (see Results) for which additional documentation was performed. One group had protein-level similarity to CBPV, now known to be variants of the recently described Lake Sinai Viruses [19], whereas the other group had protein-level similarity to members of the Partitiviridae. We sequenced PCR amplicons of several hundred base pairs to confirm the assembled sequences, for one LSV-related contig and for two partitivirus contigs. We used the following primer pairs: LSV, CATCGCAAATAGGCTGAGCA (forward) and CTCCTGGGTTGGCCTCACTA (reverse); Partitivirus1, TGAAGTCATGGATTGTAGTCTCGCT (forward) and CATCTGGTATGCCATGGTCTC (reverse), Partitivirus2,

AGTCAAGCATCCGTGTTCATTC (forward) and TCGTGATCTGTTACCATCAGACTG (reverse). These amplicons all matched their predicted products and were deposited in GenBank as accessions JF732913–732915.

Results

Incidence, Abundance, and Covariance of Honey Bee Pathogens in CCD

CCD colonies showed moderately higher incidences of pathogens (**Table 1**) than non-CCD colonies. For all nine targets, the proportion of positive colonies was higher among CCD colonies than non-CCD colonies, although only DWV, KBV, and *N. apis* were significant at $\alpha = 0.05$. Proportions of each target species were highly correlated between the two classes of hive (r = 0.97), indicating that common pathogens were common in both CCD and non-CCD hives and rarer pathogens were also rare in both. Concordant with the increased number of positive colonies, the mean number of pathogens present per colony was 4.8 (SE = 0.23) in CCD colonies and 3.6 (SE = 0.23) in non-CCD colonies, a significant difference of means (p = 0.004).

In addition to increased incidence of pathogens, CCD colonies also had higher loads of those pathogens, as measured by qPCR (i.e., a significantly lower mean ΔC_T; **Table 2**). CCD colonies showed higher levels of the viruses BQCV, DWV, KBV, and ABPV as well as the gut parasites *N. apis* and *C. mellificae*. *N. apis* loads were over 20-fold higher in bees from CCD colonies than non-CCD colonies. In contrast to previous work [15], neither IAPV nor *N. ceranae* levels were significantly higher in CCD colonies.

To determine whether increased pathogen loads were dependent on CCD diagnosis or were merely characteristic of weak colonies in general, we also contrasted pathogen loads in 'strong' non-CCD colonies, with seven or more frames of bees, with those in 'weak' non-CCD colonies, with six or fewer frames. No pathogen had higher loads in 'weak' colonies relative to 'strong' (**Table 2**), indicating that higher pathogen loads is a hallmark of CCD rather than of a small colony *per se*.

Covariation in abundance across different pathogen species was widespread in CCD colonies but rare in non-CCD colonies (**Table 3**). For CCD colonies, 11 of 36 pathogen pairs had significantly positively correlated ΔCT at a Bonferroni-corrected p<0.01. The RNA viruses ABPV, BQCV, DWV, and KBV were predominant in the list of significant pairwise interactions. **Figure 1** illustrates these "webs" of pathogen correlations in CCD compared with non-CCD colonies by linking each pair with a line the thickness of which is scaled to the correlation coefficient. SBV is not included in **Figure 1** because it was not significantly correlated with any other pathogen, perhaps because it replicates in larvae and adults are only carriers (although see below).

Samples from geographically distant sources had different pathogen complements. Colonies in the western U.S. tended to show higher incidences of pathogens (**Fig. S1,** panel A) than did colonies sampled at the same time in the eastern U.S. Both *N. apis* and ABPV were far more common in western colonies (**Fig. S1,** panel B), while IAPV trended higher in eastern colonies. KBV was the only pathogen to show higher abundance in CCD colonies in both eastern and western samples (**Fig. S1,** panel B).

Covariance in pathogen abundance was also observed at the level of individual bees drawn from colonies located in a CCD-free apiary (**Table 4**). These colonies were known to be infected with *N. ceranae* but were otherwise strong colonies. Six viruses and both *Nosema* species were measured by qPCR, although in these colonies CBPV was present and KBV was not. *Crithidia* levels were not measured in this cohort. Of the 28 pairwise correlations, 13 were significantly positive after Bonferroni correction. Even species likely to be in direct competition, such as *N. ceranae* and *N. apis*, which both reproduce in the gut epithelium, were positively correlated. Although the complement of pathogens in this single apiary differed somewhat from the colony-level survey as a whole, the data show that statistical interactions among pathogens scale to the level at which biological synergism is expected to be manifested, and that they are not secondary to some other CCD-associated variable.

Metagenomic Analysis of RNA Sequences

Deep sequencing of RNA was performed primarily as a metagenomic strategy to identify novel pathogens that may be associated with CCD. The data were evaluated in two ways, first by annotating assembled contigs and then by classifying each sequence read according to the reference sequence they best matched, if any (see Materials and Methods). After quality-trimming, there were 19.28 million Illumina sequence reads for the non-CCD sample and 41.95 million reads for the CCD sample (counting paired reads separately). Our combined assembly of the two sets of reads produced 2,413 contigs with an N50 contig length of 436 bp. Contig sequences are given in **File S2**. **File S3** contains a spreadsheet of BLAST matches (expectation <1E−10). The number of reads mapping to sequential reference sequences (see Materials and Methods) at each step is summarized in **Table 5**.

The distribution of contigs by best BLAST match is shown in **Table 6**. As expected, the majority of contigs (1,683 or 70%) matched the genus *Apis*. Another 35 contigs had best matches to other insects, principally ribosomal sequence from the genus *Bombus* and other bees. These contigs are presumed to be *A. mellifera* alleles that diverge from the reference genome, rather than genuinely derived from another species. Smaller numbers of contigs were homologous to various pathogens included in the qPCR survey, such as *Crithidia*, *Nosema*, and most of the RNA viruses investigated. Surprisingly, twelve contigs homologous to *Varroa* ribosomal loci were identified; since these are relatively

large ectoparasites that are readily removed, contamination of honey bee RNA by *Varroa* was not expected. However, the possibility that cells or RNA moieties are transferred to bees by feeding mites is suggested by the fact that other investigators have also found *Varroa* ribosomal sequence in *A. mellifera* deep-sequencing reads (e.g., GenBank accession HP469569 from a 454 transcriptome assembly).

Fifty-eight contigs were apparently of plant origin and presumably derive from consumed pollen, as has been observed in other studies (e.g., [20]). Fungi were the next most abundant group of eukaryotes, but none of the top BLAST matches to these contigs were known entomopathogens. Nine of 19 contigs were yeasts related to *Saccharomyces*/*Zygosaccharomyces* and six more were strong matches to other members of the *Saccharomycetaceae*, which is consistent with the known abundance of yeasts in the honey bee gut [37]. Three contigs had greatest similarity to the plant-pathogenic genera *Cronartium*, *Endocronartium*, and *Melamspora* and were likely associated with pollen. The remaining contig had greatest similarity to a common environmental fungus, *Myceliophthora thermophila*. No contigs had best BLAST matches to fungi related to *Penicillium* or *Aspergillus*, which have been reported to be present in honey-bee guts [37].

We identified 303 contigs that had bacterial best BLAST matches. Using the Classifier tool for 16S ribosomal loci, we could assign 67 of these contigs to bacterial orders with 80% bootstrap support (**Table 7**). The identified taxa were consistent with previous studies of the honey bee gut microbiome ([15,20] and references therein), including a diversity of Lactobacillales and Enterobacteriales. The remaining contigs were either phylogenetically ambiguous at this confidence level or not 16S sequences. Among these unclassified contigs, 12 had strong BLASTN matches to the *Melissococcus plutonius* genome (GenBank accession AP012200), the bacterial pathogen underlying European foulbrood disease of honey bees (reviewed by [38]). This pathogen was modestly more abundant in CCD+ by read count (a log2 difference of +0.39, or an increase of 31%). The bacterial pathogen causing American foulbrood, *Paenibacillus larvae* [39], was also detected by read mapping, but was also only moderately more abundant in CCD+ (a log2 differential of +0.10, or a 7% increase). Four contigs had best BLAST matches to the bacterial genus *Arsenophonus*, which is known to occur as an intracellular symbiont in some insect species and has been reported in honey bee [40].

The gut bacteria of honey bees can be clustered by 16S ribosomal sequence into a relatively small number of distinct phylotypes that are numerically predominant [20]. To examine how these major bacterial groups vary between CCD− and CCD+, we compared the relative abundance of reads mapping to 72 GenBank accessions that are representative of these phylotypes (see Materials and Methods). **Figure 2A** shows a strong and consistent pattern in which accessions representative of the Alpha1, Alpha2.1, Alpha2.2, and Bifidobacterium phylotypes of [20] are reduced in CCD+, with log2 differences in the range of −0.5 to −1.5. In contrast, the Betaproteobacteria, Firmicutes, and Gammaproteobacteria phylotypes are consistently increased in CCD+, by more moderate amounts. Since these phylotypes are numerically dominant among honey-bee gut bacteria, changes in their numbers are likely to be autocorrelated, such that the opposing direction of change in these two groups of taxa may well reflect a common underlying cause.

Although short sequence reads lack sufficient resolution for taxonomic quantification when many taxa are plausible matches, in this case the majority of bacterial reads are expected to map to only one phylotype. Furthermore, the results are consistent when

Table 3. Correlations of pathogen abundance within different colony types.

Pathogen pair		CCD (n = 61)		Non-CCD (n = 63)		Non-CCD, sympatric (n = 37)		Non-CCD, non-sympatric (n = 26)	
		Correlation	P-value	Correlation	P-value	Correlation	P-value	Correlation	P-value
BQCV	ABPV	0.606	<0.001*	0.135	0.291	0.214	0.204	0.018	0.928
DWV	ABPV	0.508	<0.001*	0.005	0.968	−0.018	0.914	0.035	0.863
KBV	ABPV	0.506	<0.001*	−0.003	0.981	0.029	0.865	−0.115	0.568
KBV	DWV	0.520	<0.001*	0.049	0.705	0.094	0.580	−0.021	0.918
KBV	IAPV	0.527	<0.001*	0.446	<0.001*	0.667	<0.001*	−0.083	0.681
DWV	BQCV	0.492	<0.001*	0.020	0.876	0.054	0.752	−0.037	0.854
KBV	BQCV	0.460	<0.001*	0.034	0.793	0.176	0.298	−0.295	0.135
Crithidia	NA	0.465	<0.001*	−0.110	0.391	−0.194	0.251	−0.115	0.568
NC	KBV	0.390	0.002*	0.004	0.974	0.084	0.621	−0.082	0.684
NC	DWV	0.361	0.004*	−0.038	0.769	0.142	0.401	−0.269	0.174
NA	BQCV	0.353	0.005*	0.289	0.022	0.361	0.028	0.140	0.485
NC	SBV	0.305	0.016	0.276	0.028	0.121	0.484	0.334	0.089
NA	NC	0.301	0.018	0.095	0.459	0.198	0.240	−0.163	0.415
NC	BQCV	0.278	0.030	0.130	0.310	0.119	0.482	0.126	0.533
IAPV	BQCV	0.273	0.033	0.195	0.126	0.310	0.062	−0.235	0.239
IAPV	DWV	0.233	0.071	0.120	0.350	0.099	0.562	0.181	0.367
SBV	KBV	0.228	0.074	−0.086	0.501	0.007	0.968	−0.145	0.470
Crithidia	BQCV	0.223	0.084	0.030	0.814	0.024	0.890	0.225	0.260
SBV	BQCV	0.212	0.098	0.088	0.495	−0.013	0.941	0.157	0.434
NC	ABPV	0.192	0.137	0.118	0.358	0.095	0.576	−0.044	0.829
NA	SBV	0.139	0.283	−0.080	0.534	−0.066	0.701	−0.089	0.661
SBV	DWV	0.151	0.240	0.099	0.440	0.222	0.192	0.070	0.727
NA	ABPV	0.146	0.260	0.284	0.024	0.259	0.122	0.353	0.071
SBV	ABPV	0.141	0.273	−0.012	0.929	0.314	0.062	−0.275	0.165
IAPV	ABPV	0.115	0.377	0.052	0.684	0.026	0.880	−0.023	0.908
Crithidia	IAPV	−0.096	0.461	−0.080	0.532	−0.174	0.304	0.050	0.806
NC	IAPV	0.093	0.478	−0.066	0.609	−0.185	0.272	0.240	0.228
Crithidia	ABPV	0.091	0.484	0.031	0.810	−0.048	0.778	−0.037	0.854
Crithidia	NC	0.088	0.499	0.215	0.091	0.402	0.014	0.084	0.678
NA	DWV	0.081	0.537	−0.167	0.191	−0.214	0.202	−0.120	0.552
SBV	IAPV	0.073	0.572	0.090	0.483	−0.106	0.538	0.511	0.007*
Crithidia	KBV	−0.056	0.666	−0.036	0.779	0.014	0.935	−0.154	0.442
Crithidia	SBV	0.053	0.683	0.160	0.209	0.032	0.852	0.398	0.040
NA	KBV	−0.035	0.789	−0.094	0.465	−0.125	0.460	−0.063	0.755
NA	IAPV	−0.013	0.920	−0.084	0.515	−0.121	0.477	−0.051	0.802
Crithidia	DWV	−0.000	0.999	0.273	0.031	0.390	0.017	−0.044	0.829

The number of colonies for each category is shown in parentheses. Sympatric non-CCD colonies are those that occurred in the same apiaries as CCD colonies, whereas non-sympatric colonies were sampled from different locations or in different years, or both, and thus were far removed from any diagnosed cases of CCD. The distinction is made because it was not possible to follow non-CCD colonies after sampling to determine if any subsequently experienced CCD. ABPV = acute bee paralysis virus; DWV = deformed wing virus; SBV = sacbrood virus; BQCV = black queen cell virus; IAPV = Israeli acute paralysis virus; KBV = Kashmir bee virus. Asterisk indicates a significant comparison after Bonferroni correction for multiple tests.

mapped either to the 16S references or the assembled contigs. **Fig. 2A** shows that the differential read counts are consistent among the different accessions that constitute each phylotype and are not driven by individual outliers. The 67 contigs that were assigned by Classifier to a bacterial order exhibited a comparable deficit in CCD+ (**Fig. 2B**) for some alpha-proteobacteria as well as actinobacteria that are presumed to be *Bifidobacterium* based on Classifier output and BLAST match. Read mapping to all 303

bacterial contigs again suggests a bimodal distribution of change in relative abundance (**Fig. 2C**). Interestingly, two *Arsenophonus*-related contigs had the highest proportional increase in CCD+ among all contigs with moderate to high read counts (>1,000 mapped reads in either sample). These contigs are highlighted in **Fig. 2C**.

Several novel RNA virus sequences were identified in the assembled contigs. BLASTX matches to LSV1 or LSV2 [19] were

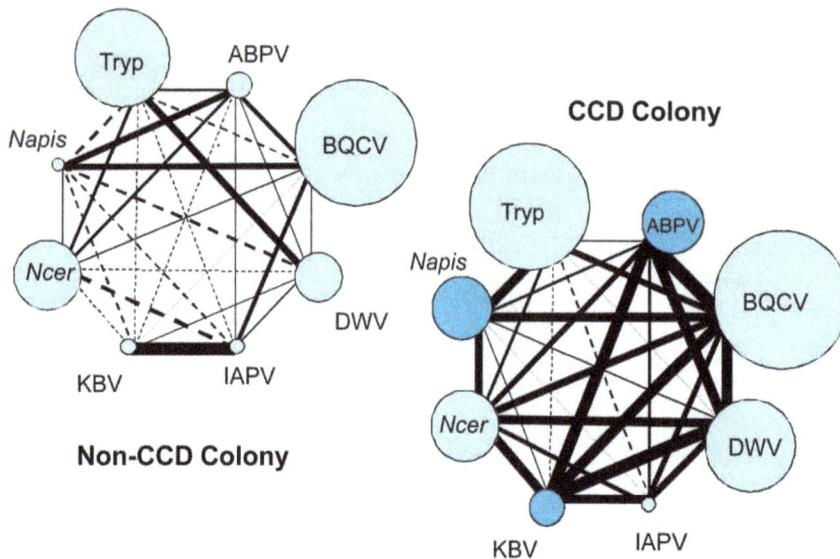

Figure 1. Graphical representation of pairwise correlations between pathogen abundance in CCD and non-CCD colonies. The thickness of lines is scaled to the Spearman's rho correlation coefficient for each pair, the values of which are given in Table 3.

found for 109 contigs, indicating that these recently identified viruses are widespread in the U.S. Interestingly, contigs or reads were not found that matched the reference genome of CBPV itself (GenBank accessions NC_010711 and NC_010712) [41]. Twelve of the contigs with homology to LSV were greater than 1 kb but none covered the full length of the coding sequence of LSV1 and LSV2. We therefore investigated the diversity of LSV sequences by constructing separate nucleotide phylogenies for the five longest contigs aligning with the 5′ end of these viruses (**Fig. 3A**) and the five longest contigs aligning with the 3′ end (**Fig. 3B**). The trees for these two groups of contigs are similar in branch lengths and topology, suggesting a one-to-one correspondence between each 5′-aligning contig and a 3′-aligning contig. This inference is further supported by the distribution of reads mapped to each of these contigs and to LSV1/LSV2 (**Fig. 4**). That is, normalized read counts for each contig in **Fig. 3A** mirror those for a contig in **Fig. 3B** with a similar phylogenetic position (for example, Contig600 and Contig511). Additional sequencing of longer LSV clones beyond the scope of this study is needed to clarify the diversity of this viral taxon, but the weight of evidence suggests that multiple strains intermediate between LSV1 and LSV2 were present in the sampled colonies. Interestingly, while LSV2 and the contigs most closely related to it (Contig876 and Contig762) were comparatively even in abundance in CCD− and CCD+, LSV1 and the other LSV-related contigs showed pronounced deviations between the two samples (**Fig. 4**). This observation suggests a potential association between LSV strain and CCD status that merits further investigation.

An additional three contigs were identified with BLASTX matches (expectation <1.0E-10) to the RNA-dependent RNA polymerase (RDRP) of viruses in the Partitiviridae, which are double-stranded RNA viruses with segmented genomes [42]. The level of amino-acid sequence identity of these matches was 30–36% (**Fig. 5**) and matches to other regions, such as capsid-encoding genes, were not detected in our assembly. As replicating partitiviruses have been identified to date only from fungi and plants [42], it seems probable that these viral sequences derive from fungi or pollen present in the gut rather than honey bee tissue. They are nonetheless noteworthy in that they were found

almost exclusively in CCD+ (1733 reads versus 9 in CCD−, or a normalized, log2 differential of +6.5).

Finally, two short, low-coverage contigs were near exact matches to Tobacco Ringspot Virus, a single-strand RNA virus of the Picornavirales. We are not aware of any published reports of bees infected by this virus and assume that it is associated with pollen.

We did not detect any known DNA virus transcripts in the assembled contigs using BLASTX (best match with an expectation <1.0E-10). However, as it has been recently argued that an insect iridescent virus (IIV) contributes to CCD [43], we performed an additional search for this virus. We identified all ORFs of 20 amino-acids or more from all unmapped reads and used BLASTP to compare these to all GenBank protein accessions of IIV3 and IIV6, the most representative taxa proposed for the putative honeybee IIV [43]. Only a single read in each sample had a match to these reference sequences, and in both cases those reads were better matches to unrelated viral proteins. We conclude that if an IIV was present in these samples, it was not transcriptionally active at detectable levels.

Comparison of Read Mapping to qPCR Estimates of Pathogen Loads

A second objective of RNA sequencing was to corroborate the qPCR estimates of pathogen loads. To do this, we converted the fold change in pathogen abundance inferred from qPCR (**Table 2**) to "expected" log2 differentials in reads and compared these with actual log2 differentials in mapped reads. There was general agreement between the two methods for seven of the nine targets (**Table 8**). For ABPV and SBV, reads were mapped from CCD+ but not CCD−, consistent with the qPCR results for these viruses, but preventing a quantitative estimate of relative change. On the other hand, there was a strong discrepancy in *Crithidia* abundance assessed by the two methods. The decline in trypanosome-mapped reads in CCD+ was similar regardless of whether a GenBank ribosomal accession (GU321196), assembled contigs (**Table 6**), or a draft whole-genome assembly of *C. mellificae* strain ATCC-30254 (R. Schwarz unpublished data) was used as the reference (results not shown). However, since another, highly divergent lineage of

Table 4. Pearson correlation coefficients of pathogen abundance within randomly sampled individual worker bees from colonies with an observable *Nosema* infection but no characteristics of CCD.

Pathogen pair		Correlation	Number of co-infected bees	P-value
CBPV	ABPV	0.547	66	<0.0001*
DWV	NA	0.604	45	<0.0001*
DWV	NC	0.639	35	<0.0001*
DWV	CBPV	0.500	57	<0.0001*
SBV	ABPV	0.528	71	<0.0001*
SBV	CBPV	0.538	67	<0.0001*
DWV	BQCV	0.411	62	0.001*
NC	NA	0.482	34	0.004*
SBV	NC	0.427	42	0.005*
BQCV	NC	0.410	41	0.008*
CBPV	BQCV	0.322	63	0.010*
CBPV	NC	0.381	40	0.015*
IAPV	CBPV	0.283	69	0.018*
IAPV	ABPV	0.238	74	0.041
ABPV	NC	0.304	43	0.048
DWV	ABPV	0.248	62	0.052
BQCV	ABPV	0.219	68	0.073
SBV	DWV	0.229	62	0.074
SBV	BQCV	0.200	68	0.103
SBV	IAPV	0.168	74	0.153
BQCV	NA	0.181	49	0.214
IAPV	NA	−0.159	55	0.247
SBV	NA	0.159	53	0.257
IAPV	BQCV	0.042	71	0.731
CBPV	NA	0.039	49	0.792
IAPV	NC	−0.022	44	0.887
ABPV	NA	−0.009	55	0.950
IAPV	DWV	−0.001	65	0.991

These samples tested whether pathogen covariation occurred at the level of individual *Nosema*-exposed bees, outside of a CCD context. N equals the number of co-infected bees upon which the correlation is calculated for each pathogen pair, out of a total of 77 bees tested. CBPV = Chronic bee paralysis virus; ABPV = acute bee paralysis virus; DWV = deformed wing virus; SBV = sacbrood virus; BQCV = black queen cell virus; IAPV = Israeli acute paralysis virus; NC = *Nosema ceranae*; NA = *Nosema apis*. Asterisk indicates a significant comparison after Bonferroni correction for multiple tests.

Crithida has been isolated and sequenced from infected honey bees (J. De Risi, unpublished data), it remains possible that the strong discrepancy between the two methods reflects an underlying genetic heterogeneity. A second major discrepancy was seen for IAPV abundance. Genotypic variation is an unlikely explanation in this case, however, as three primer pairs were used to detect this virus. Given that cDNA sequencing libraries were generated from RNA that had been pooled from many samples, and that viral abundances are typically skewed (i.e., a few samples have values much higher than the mean), even small stochastic errors at this stage could introduce nontrivial technical variation. IAPV was infrequent generally (**Table 1**), exacerbating this potential random error. Also, our more stringent requirement that three

Table 5. Assignment of Illumina sequencing reads to bins.

Mapping step	Reference	Program	Match criterion	CCD−			CCD+		
				Starting reads (000s)	Mapped reads (000s)	Percent of total reads mapping	Starting reads (000s)	Mapped reads (000s)	Percent of total reads mapping
1	Honey bee reference sequence	Bowtie	V = 2	19,276	16,921	87.8%	41,950	37,692	89.9%
2	Reference sequence for known eukaryotic, prokaryotic, and viral associates of honey bee	Bowtie	V = 2, best	2,355	195	1.0%	4,258	496	1.2%
3	Assembly contigs	Bowtie	V = 2, best	2,160	1,586	8.2%	3,762	2,992	7.1%
4	Silva LSU and SSU ribosomal databases (4/21/2010)	Bowtie	V = 3	574	116	0.6%	770	293	0.7%
5	NCBI Univec database	Mega BLAST	E <0.01	458	41	0.2%	477	11	0.03%
6	Honey bee reference sequence	Mega BLAST	E <0.01	417	63	0.32%	466	69	0.16%

Match criteria are the conditions under which an aligned read is considered a valid mapping. For Bowtie, this column shows the number of mismatches (the parameter 'V') and whether the match was required to be the best match in the reference database. For Mega BLAST, the minimum expectation of the match is shown.

Table 6. Best BLAST match of contigs assembled from deep sequencing of the CCD+ and CCD− cDNA libraries derived from pooled colony samples of total RNA.

Taxon	Number of contigs
Apis mellifera	1683
Other insect*	35
*Crithidia***	10
Nosema apis	10
Nosema ceranae	8
Varroa destructor	12
Fungi	19
Plants	58
Uncultured eukaryote	7
Nematoda	1
Eubacteria	303
Black queen cell virus	5
Deformed wing virus	6
Israeli acute paralysis virus	9
Kashmri bee virus	2
Sacbrood virus	1
Lake Sinai Virus 1	25
Lake Sinai Virus 2	85
Partitiviridae	3
Tobacco Ring Virus	2
No match	129

*Best BLAST match was to ribosomal sequence of another insect species but contig is presumed to derive from *A. mellifera*.
**Includes the trypanosome genera *Leishmania* and *Leptomonas*.

primer pairs amplify for an IAPV sample to be considered positive may have prevented low-level infections from being detected in the qPCR analysis, but would not have biased the sequencing pool. Despite the uncertainty regarding IAPV abundances, the data are at least consistent with there being no significant increase in IAPV incidence or abundance in our CCD samples, contrary to earlier work [15]. More work is needed to clarify the disagreement

between methods for these two pathogens, but we do not believe it materially affects the conclusions of the study.

Discussion

Colonies of the domesticated honey bee have been in decline in the United States for sixty years. This decline has been driven in part by economic forces, including the increased costs of disease management [10]. Nevertheless, honey bee colony losses in the U.S. have reached new highs in the past several years, exceeding 30% country-wide during the vulnerable winter period (an absolute rate of 400,000+ colonies each winter in the United States alone) [10]. Parasites and infectious agents have been posited to play a role in CCD, a syndrome tied to many of these overwinter colony losses. RNA viruses and microsporidia have been implicated in past studies [15,16], but no single pathogen has been identified that is consistently associated with collapse. An emerging hypothesis to explain these findings is that interactions among multiple subclinical infections can lead to the rapid depletion of adult workers that characterizes CCD. Alternatively, CCD as operationally defined could conflate unrelated diseases that produce similar phenotypes, thereby confounding studies of the underlying causes. More extensive studies of biotic correlates with CCD have been needed to clarify these issues.

Here we have presented a retrospective study of pathogen incidence, abundance, and covariance in a large, geographically diverse sample. Our results revealed an increase in pathogen loads and extensive pathogen covariance in CCD colonies that were not observed in weak colonies generally. No single pathogen was uniformly associated with CCD, however, consistent with the body of data on the subject. For example, levels of the microsporidian pathogen *N. apis* were more than an order of magnitude higher in CCD samples overall, but it was completely undetected in eastern cases of CCD. Its congener *N. ceranae* was widespread but not significantly increased in CCD colonies. However, positive correlations between *N. ceranae* and other pathogens were observed at both the colony and individual levels. In CCD colonies, *N. ceranae* loads were significantly correlated with levels of DWV and KBV. Individual bees from *Nosema*-infected colonies that were otherwise strong showed positive correlations between the loads of *N. ceranae* they carried and the level of co-infecting DWV, SBV, CBPV, BQCV, and *N. apis*, demonstrating that these interactions can occur independently of CCD status. These results support other studies that have linked *Nosema* infection with increased

Table 7. Taxonomic distribution of cDNA contigs with homology to 16S ribosomal sequence, by bacterial order.

Class	Order	Number of 16S contigs	Log2 difference in read counts (CCD+/CCD−)
Actinobacteria	Bifidobacteriales	3	−0.82
Alphaproteobacteria	Rhizobiales	7	−1.47
Alphaproteobacteria	Rhodospirillales	6	−0.90
Bacilli (Firmicutes)	Clostridiales	5	+0.19
Bacilli (Firmicutes)	Lactobacillales	24	+0.29
Betaproteobacteria	Burkholderiales	1	N/A (3 reads in CCD−, 2 reads in CCD+)
Betaproteobacteria	Neisseriales	3	+0.32
Gammaproteobacteria	Enterobacteriales	13	+0.42
Gammaproteobacteria	Pseudomonadales	5	N/A (64 reads in CCD−, 1 read in CCD+)

Taxonomy was estimated by the RDP Classifier tool for all 303 contigs with bacterial best BLAST matches. Only the 67 contigs with a minimum 80% bootstrap support at the order level are included here. N/A = not applicable, due to a low number of mapped reads.

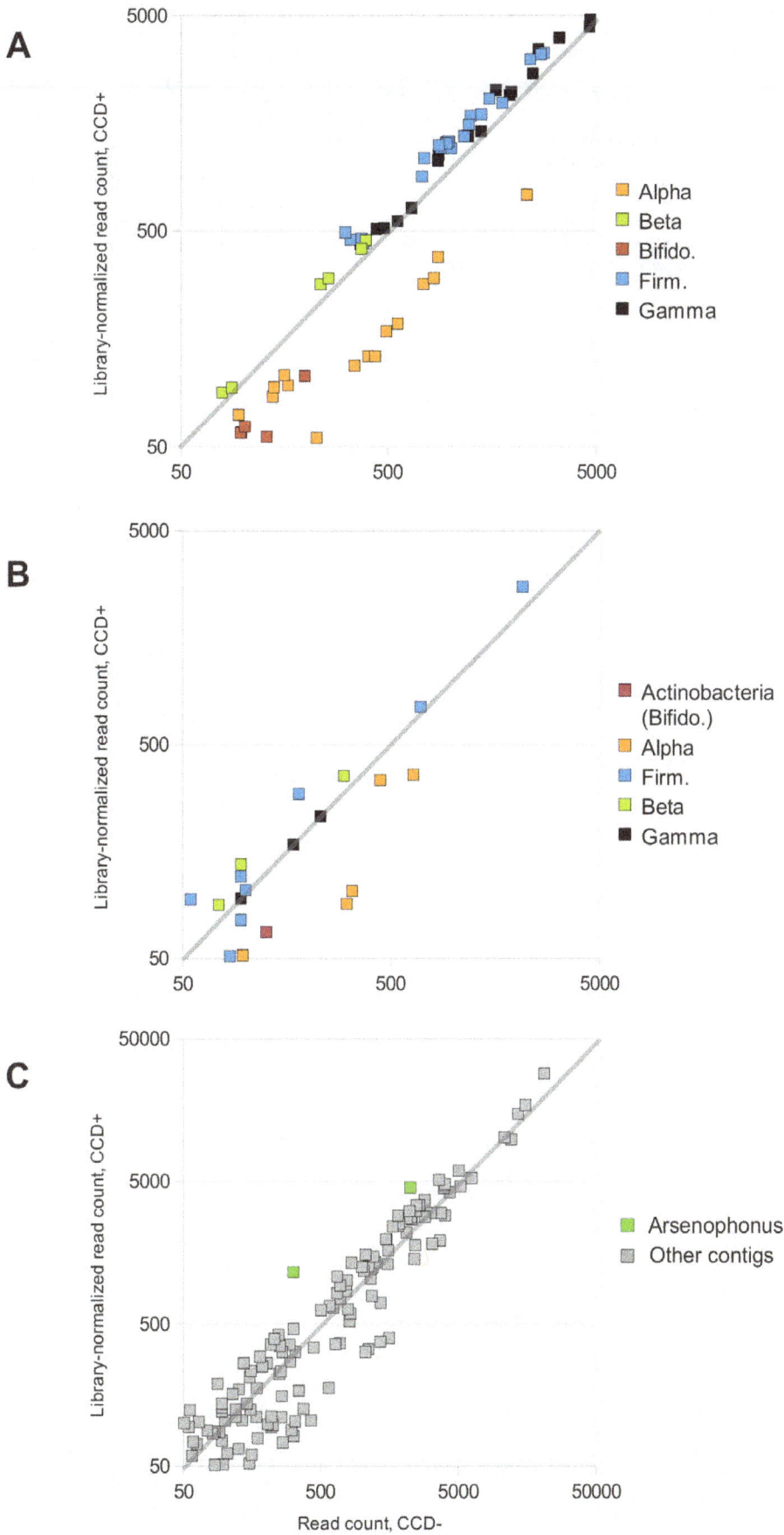

A

Library-normalized read count, CCD+

5000
500
50

Read count, CCD- 50 — 500 — 5000

Legend:
- ■ Alpha
- ■ Beta
- ■ Bifido.
- ■ Firm.
- ■ Gamma

B

Library-normalized read count, CCD+

5000
500
50

Read count, CCD- 50 — 500 — 5000

Legend:
- ■ Actinobacteria (Bifido.)
- ■ Alpha
- ■ Firm.
- ■ Beta
- ■ Gamma

C

Library-normalized read count, CCD+

50000
5000
500
50

Read count, CCD- 50 — 500 — 5000 — 50000

Legend:
- ■ Arsenophonus
- ■ Other contigs

Figure 2. Change in abundance of bacterial taxa inferred from mapping of Illumina reads. In all three panels, the horizontal axis is the number of reads mapping to each reference in the CCD− sample and the vertical axis is reads mapped in CCD+, adjusted for library size. The gray diagonal line in each panel demarcates equal representation in the two samples, and the axes are log10 scale. Only references with normalized read counts greater than 50 in each sample are displayed. **A.** Read counts for 72 GenBank accessions that are representative of the major gut microbial phylotypes of the honey bee. The accessions are drawn from Fig. S1 of [20] and are listed in Materials and Methods. Each accession is color-coded by taxonomy, following the phylotypes of [20]: Alpha = the alpha-proteobacteria clusters Alpha1, Alpha2.1, and Alpha2.2; Beta = beta-proteobacteria cluster, Gamma = the gamma-proteobacteria clusters Gamma1 and Gamma2, Bifido. = *Bifidobacteria*, and Firm. = the firmicutes clusters Firm4 and Firm5. **B.** Read counts of contigs in **File S3** that were assigned to bacterial phyla using the Classifier tool [27]. All three actinobacteria contigs belonged to the genus *Bifidobacteria* based on high Classifier bootstrap support at the genus level (**File S4**) and best BLAST match (**File S3**). Other contigs are color-coded by phylum: Alpha = alpha-proteobacteria, Beta = beta-proteobacteria, Firm = firmicutes, and Gamma = gamma-proteobacteria. **C.** Read counts of all contigs with bacterial BLAST matches. A more diffuse but still bimodal distribution of relative change in read counts is apparent. The two contigs that show the greatest increase in CCD+ relative to other contigs (highlighted in green) both have best BLAST matches to the genus *Arsenophonus* with an expectation at least four orders of magnitude lower than the next closest taxon, but the maximum identity of these matches is only 90%.

susceptibility to other pathogens [15,16]. The strong association of *N. apis* with colony collapse in this study is somewhat unexpected given the apparent decline in its geographic distribution [18] and data indicating a more detrimental effect of *N. ceranae* infection [16], although work in a colder climate found little impact by either species [44].

Most of the known RNA viruses quantified in this study were significantly more abundant in CCD colonies. This general pattern of increased viral loads is consistent with other published data [45,46]. However, given the strong association between IAPV and CCD in one prior survey [15], it is puzzling that we found no positive association between the presence or infection load of this virus and CCD. Our detection strategy and sampling approach were similar to but somewhat broader than the earlier survey. Specifically, we had a stronger focus on the western U.S., where IAPV was generally scarce in both normal and collapsed colonies. While our qPCR and read count data conflicted regarding IAPV

abundance in CCD colonies, we are not likely to be underestimating its frequency with qPCR but rather may be over-estimating it. IAPV remains a bee pathogen of concern, however, given its worldwide distribution [47,48,49] and experimentally demonstrated association with honey bee mortality [47]. We did find strong correlations with disease for the closely related viruses ABPV and KBV, and as such the family Dicistroviridae remains linked to poor bee health.

We found no transcript evidence for an iridescent virus of honey bees. These DNA viruses were recently proposed to have an association with CCD based on proteomic work [43], a result that has since been strongly criticized on methodological grounds [50]. No nucleic acid sequence attributable to a honey bee IIV has been isolated, so a more definitive assessment is not possible with our data. However, our analyses imply that IIV, if present, is unlikely to be a major contributor to CCD in the geographic regions covered by this survey.

The gut microbiota play important roles in host health and nutrition [37], and our survey found evidence of a phylogenetically clustered shift in the honey bee bacterial community involving declines in *Bifidobacterium* and alpha-proteobacteria. Although short-read sequencing provides limited resolution of taxonomic groups, the coherence and magnitude of change in these taxa support their biological relevance. Since CCD colonies have a marked deficit of older workers, age structure *per se* could well contribute to the bacterial pattern observed. The apparent increased abundance in CCD of bacteria related to *Arsenophonus*, an endosymbiont genus identified in numerous insects, is an intriguing observation, but it is not yet clear how often colonies harbor these bacteria (they are not among the predominant phylotypes that have been identified by [20] and others). The phylogenetic relationship of these contigs with other described *Arsenophonus* remains to be clarified, but our results suggest a potential association with bee health that merits further investigation. Yeasts are also important components of the honey bee gut microflora [37] and we found ribosomal sequence related to *Saccharomyces* as expected (**Table 6**). However, we did not detect other fungi that have been reported in the honey bee gut, such as *Aspergillus* and *Penicillium* species, although their distributions are considered more erratic [37]. Metagenomic studies of the interactions among bacteria, fungi, and their host constitute an important future direction of apicultural research.

Honey bees play critical pollination roles in natural and managed ecosystems, and an understanding of the biological causes behind honey bee losses will enable improved management and breeding strategies aimed at improving bee health. Here we describe the most extensive survey to date of microbes associated with CCD colonies. We have decoupled otherwise weak colonies from those diagnosed with CCD and have shown that the latter colonies have substantially heavier pathogen loads (although

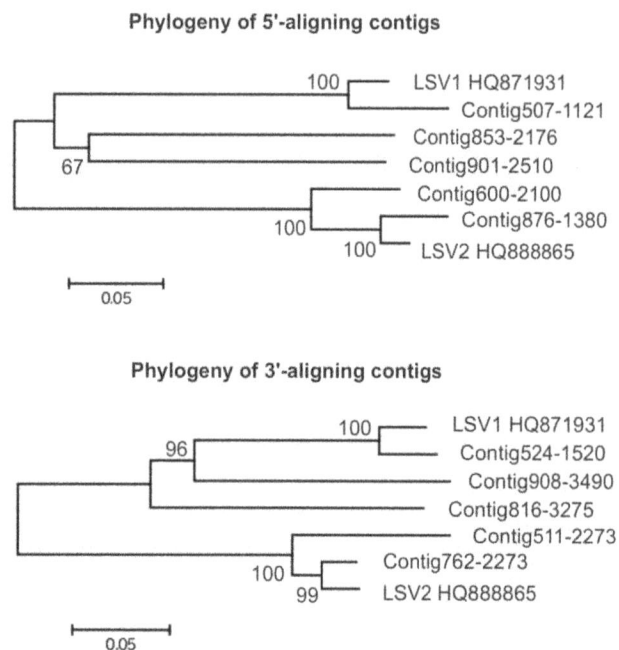

Figure 3. Phylogeny of contigs related to the Lake Sinai Viruses (LSV1 and LSV2). A. Phylogeny of the five longest 5′-aligning contigs with LSV1 and LSV2 (GenBank accessions HQ871931.1 and HQ888865.1) **B**. Phylogeny of the five longest 3′-aligning contigs with LSV1 and LSV2. The two trees have similar branch lengths and topologies, suggesting that a physical linkage between each 5′-aligning contig and a 3′-aligning contig.

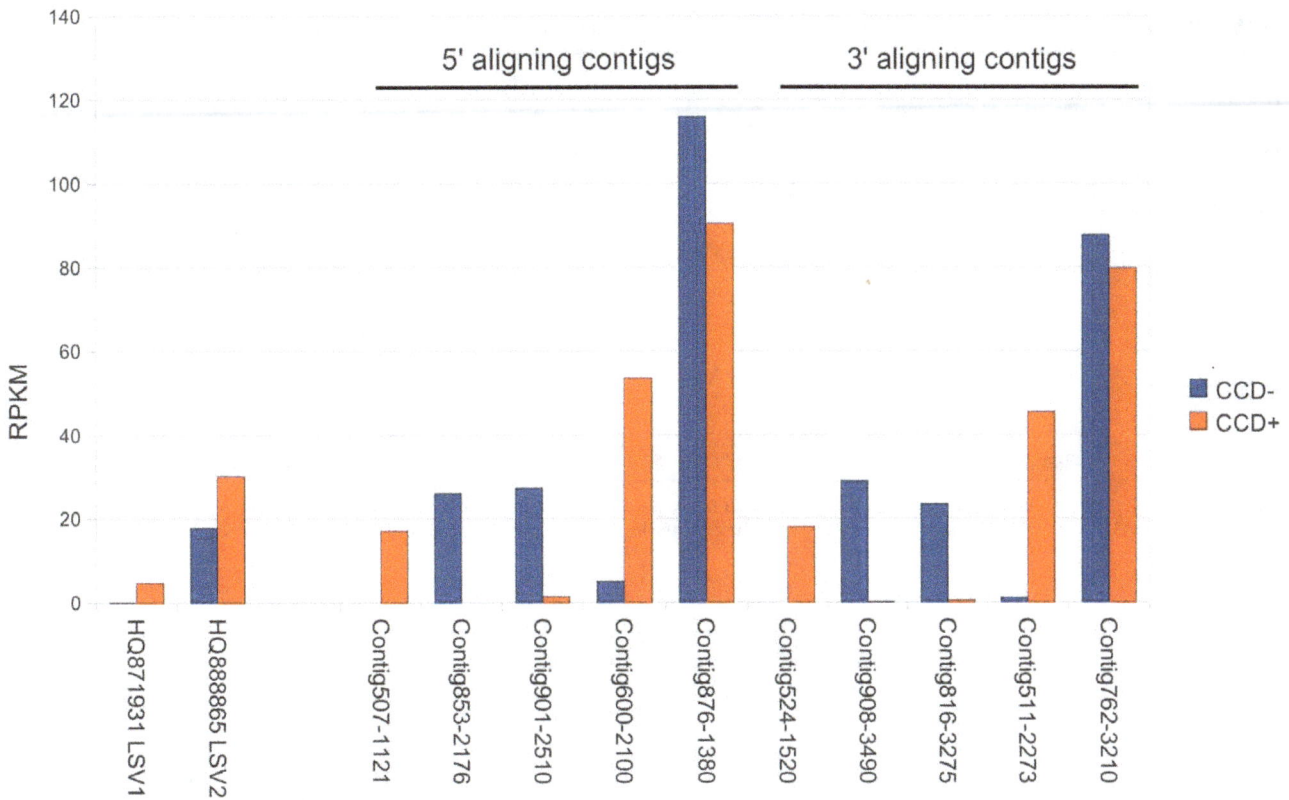

Figure 4. Relative abundance of LSV strains in CCD− and CCD+ samples. Contigs and accessions are the same as in **Figure 3**, with contigs aligning to the 5′ and 3′ regions, respectively, of LSV denoted as such. The frequency of mapped reads for each 5′ aligning contig is mirrored by that of a corresponding 3′ contig, suggesting physical linkage. Here read counts are normalized by contig length (reads per kilobase per million mapped reads, or RPKM) because the frequency of viral fragments of different lengths are being compared.

whether this increase is a cause or an effect of CCD remains unknown). Via *de novo* transcript assembly, we have identified novel RNA viruses of potential importance to bee health that can now be characterized with controlled infections and molecular analyses. The diverse LSV sequences are of particular interest because

outbreaks of the distantly related CBPV have been known to cause workers to die en masse away from the hive, albeit rarely [51].

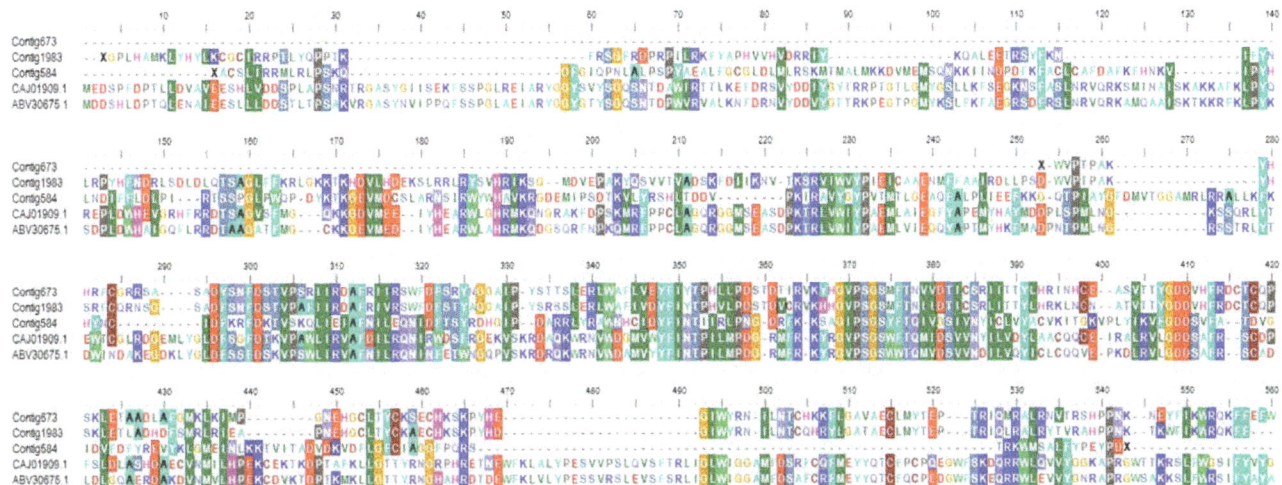

Figure 5. Sequence alignment of three contigs with BLASTX matches to the RDRP of *Penicillium stoloniferum virus S,* **GenBank accessions CAJ01909.1 and AY156521.2.** Shading at each position indicates amino-acid similarity among at least 50% of the residues, based on the BLOSUM62 matrix. Alignment performed with ClustalW using default settings.

Table 8. Comparison of RNA target abundances in Illumina sequence reads compared with qPCR results.

Target	Expected log2 differential abundance in pool based on qPCR	Normalized log2 differential abundance of mapped reads	Comments
ABPV	+2.27	N/A	0 reads in CCD−, 8 reads in CCD+
BQCV	+2.89	+1.72	
DWV	+3.97	+2.14	
IAPV	+0.24	−2.85	The two measures are strongly discordant
KBV	+2.71	+5.12	2 reads in CCD−, 126 reads in CCD+
SBV	+0.22	N/A	0 reads in CCD−, 25 reads in CCD+
Nosema apis	+3.54	+4.10	
Nosema ceranae	+1.60	+0.87	
Crithidia	+2.62	−1.57	The two measures are strongly discordant

The expected log2 differential is taken as the log2 of fold change in CCD colonies from Table 2. For read counts, differential abundance = log2 (CCD+ reads) − log2 (CCD− reads) − log2 (total CCD+ reads/total CCD− reads). N/A indicates that the log2 difference in abundance could not be calculated because no reads matched in CCD−.

Future Directions

An inherent limitation of our approach is the unknown degree to which bees remaining in a CCD hive at the time of sampling can serve as indicators of the events leading to its decline and the physiological status of the missing bees. However, the greater abundance and covariance of pathogens in CCD hives are informative and must be explained by any proposed model of how CCD occurs. A definitive analysis of the causes of CCD will ultimately require its controlled replication through the experimental manipulation of the relevant variables. Given the complexity of natural systems and the number of potential variables, retrospective and prospective observational studies are necessary for narrowing hypotheses to a manageable number. Our results indicate several promising pairings for such tests, in particular, *Nosema* with the RNA viruses DWV, KBV, BQCV, or ABPV. If pathogen webs are indeed precipitators of colony collapse, future work must demonstrate how this occurs at the level of individual bees and the overall hive to produce a rapid loss of foragers without overt disease in the remaining bees. The apparent variation in pathogen distributions, including strain variation, also needs to be better described in order to 1) identify and investigate discrepancies between epidemiological and experimental data, and 2) better inform management and policy decisions, including the possibility of quarantine. Agrochemical exposure also needs to be more fully explored as a contributor to CCD, and we stress that our results do not speak for or against its role in colony loss.

Supporting Information

Figure S1   Differential microbial abundances for colonies sampled in the western and eastern United States. A. Proportional abundance of discitroviruses (red), iflaviruses (blue), bacteria (green), microsporidia (pink) and trypanosome (orange) pathogens in non-CCD (n = 38) and CCD (n = 61) bee samples, as indicated by letter size. A = ABPV, I = IAPV, K = KBV, D = DWV, Q = BQCV, S = SBV, B = bacterial load, C = *Nosema ceranae*, P = *Nosema apis*, T = *Crithidia*. B. Mean relative abundances (ΔC_T) of four viruses and two *Nosema* species in CCD and non-CCD colonies in the two geographic regions. For comparison, the values are scaled by adding a constant such that the minimum value of all samples is zero.

File S1   Estimated efficiencies of qPCR reactions using primers, templates, and reaction conditions described in the text.

File S2   cDNA contigs resulting from our assembly of Illumina sequence reads.

File S3   Best BLAST matches for the assembled contigs.

File S4   Classifier [27] taxonomic assignments for contigs with at least 80% bootstrap support at the level of order.

Acknowledgments

We gratefully acknowledge the substantial improvements to the manuscript suggested by six anonymous reviewers and the reviewing editor.

Author Contributions

Conceived and designed the experiments: JDE JSP DV DT YC. Performed the experiments: DL LJ DT DV JSP. Analyzed the data: RSC JDE. Contributed reagents/materials/analysis tools: YC JDE DV JSP DT. Wrote the paper: RSC JDE.

References

1. Gallai N, Salles JM, Settele J, Vaissière BE (2009) Economic valuation of the vulnerability of world agriculture confronted with pollinator decline. Ecological Economics 68: 810–821.
2. Aizen MA, Harder LD (2009) The global stock of domesticated honey bees is growing slower than agricultural demand for pollination. Current Biology 19: 915–918.
3. Potts SG, Roberts SPM, Dean R, Marris G, Brown MA, et al. (2010) Declines of managed honey bees and beekeepers in Europe. Journal of Apicultural Research 49: 15–22.
4. VanEngelsdorp D, Hayes Jr J, Underwood RM, Pettis JS (2011) A survey of managed honey bee colony losses in the United States, fall 2009 to winter 2010. Journal of Apicultural Research 50: 1–10.

5. vanEngelsdorp D, Evans JD, Saegerman C, Mullin C, Haubruge E, et al. (2009) Colony collapse disorder: a descriptive study. PLoS ONE 4: e6481.

6. Shimanuki H (1997) Synonymy in bee diseases. In: Morse RA, Flottum K, editors. Honey Bee Pests, Predators, and Diseases. Medina, Ohio: A. I. Root. 534–537.

7. Underwood R, VanEngelsdorp D (2007) Colony Collapse Disorder: Have we seen this before? Bee Culture 35: 13–18.

8. Wilson WT, Menapace DM (1979) Disappearing disease of honey bees: a survey of the United States. American Bee Journal 119: 184–186.

9. Oldroyd BP (2007) What's killing American honey bees? PLoS Biol 5: e168.

10. vanEngelsdorp D, Meixner MD (2010) A historical review of managed honey bee populations in Europe and the United States and the factors that may affect them. Journal of Invertebrate Pathology 103: S80–95.

11. Mullin CA, Frazier M, Frazier JL, Ashcraft S, Simonds R, et al. (2010) High levels of miticides and agrochemicals in North American apiaries: implications for honey bee health. PLoS ONE 5: e9754.

12. Johnson RM, Evans JD, Robinson GE, Berenbaum MR (2009) Changes in transcript abundance relating to colony collapse disorder in honey bees (Apis mellifera). Proceedings of the National Academy of Sciences of the United States of America 106: 14790–14795.

13. Morse RA, Flottum K, editors (1997) Honey Bee Pests Predators and Diseases. 3 ed. Medina, Ohio: A.I. Root Co. 718 p.

14. Evans JD, Schwarz RS (2011) Bees brought to their knees: microbes affecting honey bee health. Trends Microbiol 19: 614–620.

15. Cox-Foster DL, Conlan S, Holmes EC, Palacios G, Evans JD, et al. (2007) A metagenomic survey of microbes in honey bee colony collapse disorder. Science 318: 283–287.

16. Higes M, Martin-Hernandez R, Botias C, Bailon EG, Gonzalez-Porto AV, et al. (2008) How natural infection by Nosema ceranae causes honeybee colony collapse. Environmental Microbiology 10: 2659–2669.

17. Genersch E, Von der Ohe W, Kaatz H, Schroeder A, Otten C, et al. (2010) The German bee monitoring project: a long term study to understand periodically high winter losses of honey bee colonies. Apidologie 41: 332–352.

18. Chen Y, Evans JD, Smith IB, Pettis JS (2008) Nosema ceranae is a long-present and wide-spread microsporidian infection of the European honey bee (Apis mellifera) in the United States. Journal of Invertebrate Pathology 97: 186–188.

19. Runckel C, Flenniken ML, Engel JC, Ruby JG, Ganem D, et al. (2011) Temporal analysis of the honey bee microbiome reveals four novel viruses and seasonal prevalence of known viruses, Nosema, and Crithidia. PLoS One 6: e20656.

20. Martinson VG, Danforth BN, Minckley RL, Rueppell O, Tingek S, et al. (2011) A simple and distinctive microbiota associated with honey bees and bumble bees. Mol Ecol 20: 619–628.

21. Boncristiani H, Li J, Evans JD, Pettis J, Chen Y-P (2011) Scientific note on PCR inhibitors in the compound eyes of honey bees, Apis mellifera. Apidologie 42: 457–460.

22. Evans JD (2006) Beepath: An ordered quantitative-PCR array for exploring honey bee immunity and disease. Journal of Invertebrate Pathology 93: 135–139.

23. Vandesompele J, De Preter K, Pattyn F, Poppe B, Van Roy N, et al. (2002) Accurate normalization of real-time quantitative RT-PCR data by geometric averaging of multiple internal control genes. Genome Biol 3: RESEARCH0034.

24. Zerbino DR, Birney E (2008) Velvet: algorithms for de novo short read assembly using de Bruijn graphs. Genome Res 18: 821–829.

25. Huang X, Madan A (1999) CAP3: A DNA sequence assembly program. Genome Res 9: 868–877.

26. Schmid-Hempel R, Tognazzo M (2010) Molecular divergence defines two distinct lineages of Crithidia bombi (Trypanosomatidae), parasites of bumblebees. J Eukaryot Microbiol 57: 337–345.

27. Wang Q, Garrity GM, Tiedje JM, Cole JR (2007) Naive Bayesian classifier for rapid assignment of rRNA sequences into the new bacterial taxonomy. Appl Environ Microbiol 73: 5261–5267.

28. Langmead B, Trapnell C, Pop M, Salzberg SL (2009) Ultrafast and memory-efficient alignment of short DNA sequences to the human genome. Genome Biol 10: R25.

29. Consortium HBGS (2006) Insights into social insects from the genome of the honeybee Apis mellifera. Nature 443: 931–949.

30. Qin X, Evans JD, Aronstein KA, Murray KD, Weinstock GM (2006) Genome sequences of the honey bee pathogens Paenibacillus larvae and Ascosphaera apis. Insect Molecular Biology 15: 715–718.

31. Cherry JM, Ball C, Weng S, Juvik G, Schmidt R, et al. (1997) Genetic and physical maps of Saccharomyces cerevisiae. Nature 387: 67–73.

32. Cornman RS, Chen YP, Schatz MC, Street C, Zhao Y, et al. (2009) Genomic analyses of the microsporidian Nosema ceranae, an emergent pathogen of honey bees. PLoS Pathog 5: e1000466.

33. Cornman RS, Schatz MC, Johnston JS, Chen YP, Pettis J, et al. (2010) Genomic survey of the ectoparasitic mite Varroa destructor, a major pest of the honey bee Apis mellifera. BMC Genomics 11: 602.

34. Pruesse E, Quast C, Knittel K, Fuchs BM, Ludwig W, et al. (2007) SILVA: a comprehensive online resource for quality checked and aligned ribosomal RNA sequence data compatible with ARB. Nucleic Acids Res 35: 7188–7196.

35. Fujiyuki T, Takeuchi H, Ono M, Ohka S, Sasaki T, et al. (2005) Kakugo virus from brains of aggressive worker honeybees. Advances in Virus Research 65: 1–27.

36. Ongus JR, Peters D, Bonmatin J-M, Bengsch E, Vlak JM, et al. (2004) Complete sequence of a picorna-like virus of the genus Iflavirus replicating in the mite Varroa destructor. Journal of General Virology 85: 3747–3755.

37. Gilliam M (1997) Identification and roles of non-pathogenic microflora associated with honey bees. FEMS Microbiology Letters 155: 1–10.

38. Forsgren E (2010) European foulbrood in honey bees. J Invertebr Pathol 103 Suppl 1: S5–9.

39. Genersch E (2010) American Foulbrood in honeybees and its causative agent, Paenibacillus larvae. J Invertebr Pathol 103 Suppl 1: S10–19.

40. Novakova E, Hypsa V, Moran NA (2009) Arsenophonus, an emerging clade of intracellular symbionts with a broad host distribution. BMC Microbiol 9: 143.

41. Olivier V, Blanchard P, Chaouch S, Lallemand P, Schurr F, et al. (2008) Molecular characterisation and phylogenetic analysis of Chronic bee paralysis virus, a honey bee virus. Virus Research 132: 59–68.

42. Ghabrial SA, Ochoa WF, Baker TS (2008) Partitiviruses: general features. In: Mahy BWJ, van Regenmortel MHV, editors. Encyclopedia of Virology. 3rd ed. Oxford: Elsevier. 68–75.

43. Bromenshenk JJ, Henderson CB, Wick CH, Stanford MF, Zulich AW, et al. (2010) Iridovirus and microsporidian linked to honey bee colony decline. PLoS One 5: e13181.

44. Gisder S, Hedtke K, Mockel N, Frielitz MC, Linde A, et al. (2010) Five-year cohort study of Nosema spp. in Germany: does climate shape virulence and assertiveness of Nosema ceranae? Appl Environ Microbiol 76: 3032–3038.

45. Berthoud H, Imdorf A, Haueter M, Radloff S, Neumann P (2010) Virus infections and winter losses of honey bee colonies (Apis mellifera). Journal of Apicultural Research 49: 60–65.

46. Dainat B, Evans JD, Chen YP, Gauthier L, Neumann P (2012) Dead or alive: Deformed wing virus and Varroa destructor reduce the life span of winter honeybees. Appl Environ Microbiol 78: 981–987.

47. Maori E, Lavi S, Mozes-Koch R, Gantman Y, Peretz Y, et al. (2007) Isolation and characterization of Israeli acute paralysis virus, a dicistrovirus affecting honeybees in Israel: Evidence for diversity due to intra- and inter-species recombination. Journal of General Virology 88: 3428–3438.

48. Blanchard P, Schurr F, Celle O, Cougoule N, Drajnudel P, et al. (2008) First detection of Israeli acute paralysis virus (IAPV) in France, a dicistrovirus affecting honeybees (Apis mellifera). Journal of Invertebrate Pathology 99: 348–350.

49. Palacios G, Hui J, Quan PL, Kalkstein A, Honkavuori KS, et al. (2008) Genetic analysis of Israel acute paralysis virus: distinct clusters are circulating in the United States. Journal of Virology 82: 6209–6217.

50. Foster L (2011) Interpretation of data underlying the link between Colony Collapse Disorder (CCD) and an invertebrate iridescent virus. Molecular and Cellular Proteomics 10: M110.006387.

51. Ribiere M, Olivier V, Blanchard P (2010) Chronic bee paralysis: a disease and a virus like no other? J Invertebr Pathol 103 Suppl 1: S120–131.

6

A Quantitative Model of Honey Bee Colony Population Dynamics

David S. Khoury[1], **Mary R. Myerscough**[1,2]*, **Andrew B. Barron**[3]

1 School of Mathematics and Statistics, The University of Sydney, Sydney, New South Wales, Australia, **2** Centre for Mathematical Biology, The University of Sydney, Sydney, New South Wales, Australia, **3** Department of Biology, Macquarie University, Sydney, New South Wales, Australia

Abstract

Since 2006 the rate of honey bee colony failure has increased significantly. As an aid to testing hypotheses for the causes of colony failure we have developed a compartment model of honey bee colony population dynamics to explore the impact of different death rates of forager bees on colony growth and development. The model predicts a critical threshold forager death rate beneath which colonies regulate a stable population size. If death rates are sustained higher than this threshold rapid population decline is predicted and colony failure is inevitable. The model also predicts that high forager death rates draw hive bees into the foraging population at much younger ages than normal, which acts to accelerate colony failure. The model suggests that colony failure can be understood in terms of observed principles of honey bee population dynamics, and provides a theoretical framework for experimental investigation of the problem.

Editor: James A. R. Marshall, University of Sheffield, United Kingdom

Funding: This work was supported by The School of Mathematics and Statistics, The University of Sydney. The funders had no role in study design, data collection and analysis, decision to publish, or preparation of the manuscript.

Competing Interests: The authors have declared that no competing interests exist.

* E-mail: mary.myerscough@sydney.edu.au

Introduction

A honey bee colony is a population of related and closely interacting individuals that form a highly complex society. The population dynamics of this group is complicated, because the fates of individuals within it are not independent, and an individual's lifespan is strongly influenced by their role in the colony. To aid exploration of honey bee population dynamics here we describe a simple mathematical representation of how the social regulation of worker division of labour can influence the longevity of individual bees, and colony growth. The model also allows simulation of how demographic disturbances can impact colony growth, or contribute to colony failure.

The life cycle of individual bees in the hive is well understood. Worker bees enter the population from eggs laid by the queen, and the existing population of workers raise a proportion of these eggs to adulthood [1]. It takes three weeks for worker bees to develop from eggs to adults [1], but their lifespan as adults is strongly influenced by their behavioural role in the colony. Survival of bees in the protected hive environment is high, but the survival of forager bees is much lower [1]. The average foraging life of a bee has been estimated as less than seven days, because of the many risks and severe metabolic costs associated with foraging [2]. As a consequence of this it might be expected that a bee's overall lifespan would be strongly influenced by the age at which she commenced foraging.

The division of labour among worker bees in a colony is age dependent: typically young adults work within the hive on colony maintenance tasks and brood care (nursing), but change to foraging tasks when they are older [3,4]. This process of behavioural development is sensitive to social feedback. If there

is a decline in the number of foragers, hive bees accelerate their behavioural development and begin foraging precociously to compensate [5,6]. Similarly, if there is a surfeit of foragers and a lack of nurses, bees can reverse their behavioural development and switch back from foraging to nursing roles [5,7]. The pheromonal mechanism mediating this 'social inhibition' of foraging has been identified [8]. Old forager bees transfer ethyl oleate to young hive bees via trophallaxis, which delays the age at which they begin foraging [8].

As a consequence of this social regulation of division of labour, one would predict an interaction between the composition of the colony workforce, and longevity of individual bees. If social inhibition is reduced and bees initiate foraging when young they would be expected to have an overall reduced lifespan (since foraging is associated with such high mortality), and therefore have less time to contribute to colony growth. Here we present a simple mathematical model that allows a formal exploration of how a loss of foragers and reduced social inhibition might impact colony growth.

This issue is salient because of the current concern over globally declining bee populations. Since 2006 beekeepers worldwide have reported elevated rates of colony losses [9,10,11]. Since 2006 the average overwinter loss of honey bee colonies in the United States has exceeded 30% consistently [9], and elevated colony losses have been reported across Europe, the Middle East and Japan [11]. The impact of the parasitic mite *Varroa destructor* is certainly a major factor behind the global increase in colony failure rates [11,12,13,14], but other stressors include various bee diseases (but especially *Nosema sp.* [15]), changes in bee management practice [16], factors related to climate change and seasonal shifts [17] and pesticide exposure [10,12,18,19,20]. These have all been linked to colony failure.

Extreme cases of mysterious mass colony death where there is no clear causal agent have become known as colony collapse disorder, or CCD [10]. Diagnostic of this syndrome are vacant hives containing dead brood and food stores but few or no adult bees, suggesting very rapid catastrophic depopulation [10]. Surveys of pathogens associated with colony collapse events have identified many disease organisms present [10,21,22,23], and several newly described bee pathogens have been linked with CCD [22,24], but at the time of writing no definite single agent has been identified as the cause of CCD. The current prevailing opinion is that colony collapse is not a result of a single new causal factor [17]. The problem is considered multicausal and may reflect the outcome of an accumulation of stressors on a honey bee colony [11,12].

CCD has focused attention on the problem of colony failure, and the many stressors now impacting colony survival. It is clear that while an enormous amount is know about honey bee sociobiology, comparatively little is know about the social responses of bees to population stresses on a colony. The presented model explores how varying the rate of forager bee mortality might impact colony growth, which may be a useful tool to aid research into the complex problem of colony failure.

Materials and Methods

Constructing a demographic model to explore the process of colony failure: the hypothesis

We hypothesise that colony failure occurs when the death rate of bees in the colony is unsustainable. At this point normal social dynamics break down, it becomes impossible for the colony to maintain a viable population, and the colony will fail.

We hypothesise that any factor that causes an elevated forager death rate will reduce the strength of social inhibition, resulting in a precocious onset of foraging behaviour in young bees [5]. Because foraging is high-risk [2], precocious foraging shortens overall bee lifespan. Precocious foragers are also less effective and weaker than foragers that have made the behavioural transition at the normal age [25,26]. Consequently, as the mean age of the foraging force decreases forager death rates increase further, which accelerates the population decline. A precocious onset of foraging reduces the population of hive bees engaged in brood care. This reduces colony brood rearing capacity, and the population crashes. A similar hypothesis has been proposed to explain the impact of *Nosema ceranae* on colonies [15], but we argue this hypothesis is applicable to any factor that chronically elevates forager bee death rates. We explore this hypothesis using the following simple mathematical model.

The model

A mathematical model allows us to explore the effects of different factors and forces on the population of the hive in a quantitative way. Such a model has the potential to make predictions for the outcome of various manipulations, and to allow a preliminary exploration of the problem before investing in experimental work.

We construct a simple compartment model for the worker bee population of the hive (Fig. 1). Our model only considers the population of female workers since males (drones) do not contribute to colony work. Let H be the number of bees working in the hive and F the number of bees who work outside the hive, referred to here as foragers. We assume that all adult worker bees can be classed either as hive bees or as foragers, and that there is no overlap between these two behavioural classes [1,4]. Hence the total number of adult worker bees in the colony is $N = H + F$.

Our model does not consider the impact of brood diseases on colony failure, however we believe our approach is still useful

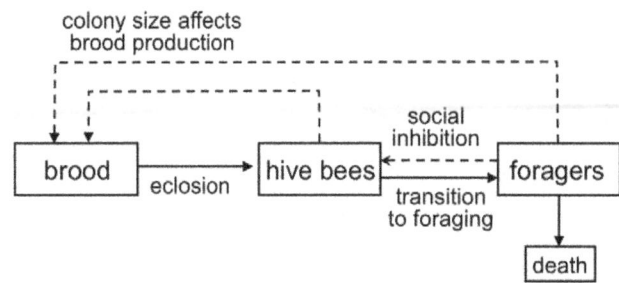

Figure 1. Elements of honey bee social dynamics considered by our model. Eggs laid by the queen are reared as brood that eclose three weeks later as adult bees. Adult bees work in the hive initially before becoming foragers. Our model considers the death rate of adult bees within the hive to be negligible, but forager death rate is a parameter varied in our simulations. We assume the amount of brood reared is influenced by the size of the colony (number of hive and forager bees) and that the rate at which bees transition from hive bees to forager bees is influenced by the number of foragers to represent the effect of social inhibition.

because many cases of colony failure and CCD are not caused by brood diseases [21,22,23]. Hive bees eclose from pupae and mature into foragers. Death rates of adult hive bees in a healthy colony are extremely low as the environment is protected and stable. We assume that the death rate of hive bees is negligible. Workers are recruited to the forager class from the hive bee class and die at a rate m. Let t be the time measured in days. Then we can represent this process as a differential equation model:

Rate of change of hive bee numbers:

$$\frac{dH}{dt} = E(H,F) - HR(H,F) \quad (1)$$

eclosion recruitment to forager class

Rate of change of forager numbers:

$$\frac{dF}{dt} = HR(H,F) - mF. \quad (2)$$

recruitment death

The function $E(H,F)$ describes the way that eclosion depends on the number of hive bees and foragers. The recruitment rate function $R(H,F)$ models the effect of social inhibition on the recruitment rate.

It is known that the number of eggs reared in a colony (and hence the eclosion rate) is related to the number of bees in the hive. Big colonies raise more brood [27,28,29]. The nature of this dependence is not known, however. We assume that the maximum rate of eclosion is equivalent to the queen's laying rate L and that the eclosion rate approaches this maximum as N (the number of workers in the hive) increases. In the absence of other information we use the simplest function that increases from zero for no workers and tends to L as N becomes very large:

$$E(H,F) = L\left(\frac{N}{w+N}\right) = L\left(\frac{H+F}{w+H+F}\right). \quad (3)$$

Here w determines the rate at which $E(H,F)$ approaches L as N gets large. Figure 2 shows $E(H,F)$ as a function of N for a range of values of w.

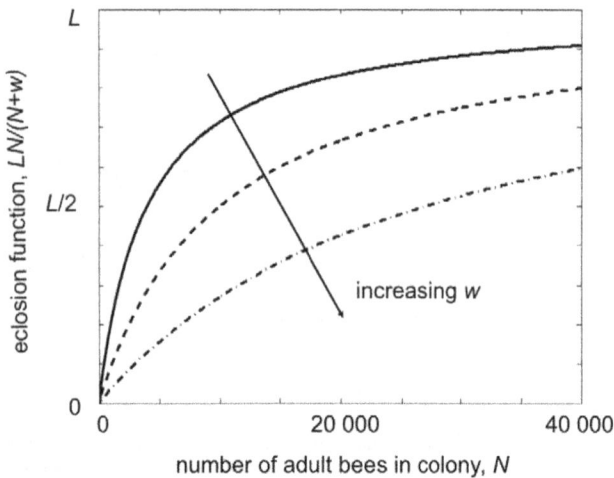

Figure 2. Plot of the eclosion function $E(h,F) = LN/(w+N)$ where $N = H+F$ for different values of w. The solid line has $w = 4000$; the dashed line, $w = 10\,000$ and the dash-dot line, $w = 27\,000$.

We write the recruitment function as

$$R(H,F) = \alpha - \sigma\left(\frac{F}{H+F}\right). \qquad (4)$$

The first term α represents the maximum rate that hive bees will become foragers when there are no foragers present in the colony. The second term $-\sigma F/(F+H)$ represents social inhibition and, in particular, how the presence of foragers reduces the rate of recruitment of hive bees to foragers. We have assumed that social inhibition is directly proportional to the fraction of the total number of adult bees that are foragers, such that a high fraction of foragers in the hive results in low recruitment. In the absence of any foragers new workers will become foragers at a minimum of four days after eclosing [30], so an appropriate choice for the rate of uninhibited transition to foraging is $\alpha = 0.25$. We chose $\sigma = 0.75$ since this factor implies that a reversion of foragers to hive bees would only occur if more than one third of the hive are foragers. We also chose $L = 2000$ as the daily laying rate of the queen [31] and $w = 27,000$.

Analysis of the model

The equations (1) and (2) with the functions (3) and (4) were analysed using standard linear stability analysis and phase plane analysis [32].

The model has a globally stable steady state (H_0, F_0) where

$$F_0 = \frac{L}{m} - w\frac{J}{J+1}, \quad H_0 = \frac{1}{J}F_0 \quad \text{where}$$

$$J = \frac{1}{2}\left[\left(\frac{\alpha}{m} - \frac{\sigma}{m} - 1\right) + \sqrt{\left(\frac{\alpha}{m} - \frac{\sigma}{m} - 1\right)^2 + 4\frac{\alpha}{m}}\right] \qquad (5)$$

when

$$m < \frac{L}{2w}\left(\frac{\alpha + \sigma + \sqrt{(\alpha-\sigma)^2 + 4\frac{L\sigma}{w}}}{\alpha - \frac{L}{w}}\right) \quad \text{and} \quad \alpha - \frac{L}{w} > 0. \qquad (6)$$

Otherwise the state with no adult bees is an attractor and the hive population goes to zero.

Figure 3 shows phase plane solutions for a low death rate, $m = 0.24$, when the populations tend to a positive steady state, and a higher death rate $m = 0.40$, when the population goes extinct. In each case the solution rapidly approaches the line $F = JH$ so that the ratio of hive bee numbers to forager numbers is close to being constant. The population size adjusts more slowly to either a positive steady state or to zero. Figure 4 shows the decline of a doomed population as a function of time (dotted line). If the foragers become less able and more likely to die as they get younger then the decline will be more rapid (solid line).

Figure 5 is a bifurcation diagram, which shows that for low values of the forager death rate m there are large numbers of bees in the colony, but once m passes a critical value the colony population cannot support itself and the colony fails.

Figure 6 shows how the average age at commencement of foraging and the average age at death depend on the forager death rate m. The model predicts that at a higher death rate the forager population will be smaller and also made up of younger bees.

We compared results from the model to experimental observations of Rueppell et al [33]. We used the observed flightspan [the number of days bees were observed foraging 33], to estimate the death rate of foragers since m is the reciprocal of flightspan. With these values of m we used the model to calculate the average age of onset of foraging (AAOF) and the lifespan of worker bees for each colony and compared these model values to observed results. These observed and calculated results are shown in Table 1. Even with the somewhat rough estimates of parameters, the model matches the observational data well for average age at onset of foraging, although it is slightly high for worker lifespan. Nevertheless, given that the model is a very simple representation of honey bee demographics, the results are encouraging.

Results and Discussion

Our model clarifies how forager death rate influences colony population, and suggests that very rapid population decline can result from chronically high forager death rates. The model emphasizes the role social feedback mechanisms within the honey bee colony may play in colony failure, and suggests that colony

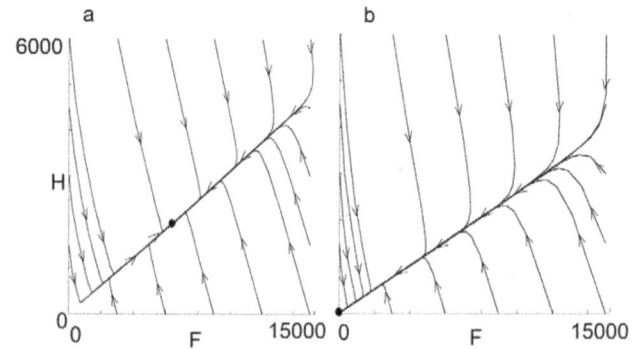

Figure 3. Phase plane diagrams of solutions to the model for different values of m. Each line on the diagrams represents a solution trajectory, giving the number of foragers F and the number of hive bees H. As time t increases the solutions change along the trajectory in the direction of the arrows. In (a) $m = 0.24$ and the populations tend to a stable equilibrium population, marked by a dot. In (b), $m = 0.40$ there is no nonzero equilibrium and the hive populations collapses to zero. Parameter values are $L = 2000$, $\alpha = 0.25$, $\sigma = 0.75$ and $w = 27\,000$.

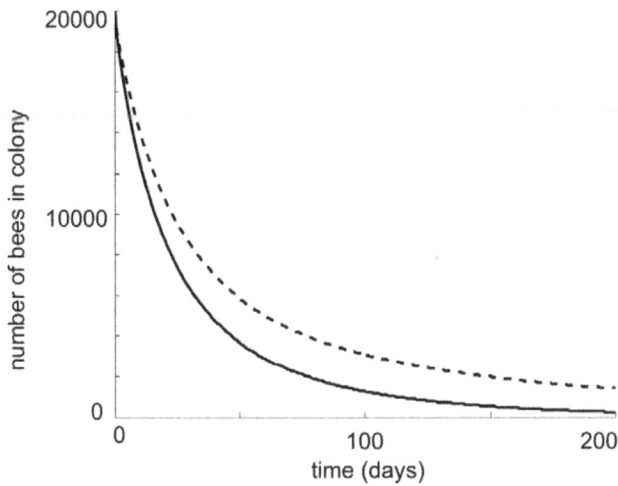

Figure 4. The effect of inefficient precocious foraging on population decline. This plot shows the time course of colony decline when all foragers perform equally well (dashed line) and when precocious foragers die faster than mature foragers (solid line). The effect of precocious foraging is modeled by replacing the death rate m by $m = m_l R^2/(\lambda^2 + R^2)$ whenever $R < 0$ where R is the recruitment rate of foragers given in eqn (4). Parameter values are $L = 2000$, $\alpha = 0.25$, $\sigma = .75$, $w = 27\ 000$, $m_l = 0.6$ and $\lambda^2 = 0.059$.

failure can be explored as both a sociobiological as well as an epidemiological question.

The model proposes a bifurcation point in the death rate parameter such that when death rate is below a critical threshold, colony population reaches an equilibrium point determined by model parameters, but when forager death rate is sustained above the threshold, colony population declines to zero and the colony fails. This bifurcation point represents the point at which the colony cannot maintain brood production at a rate sufficient to replace losses of forager bees in the field. The model suggests that if a high forager death rate is sustained, colony population decline

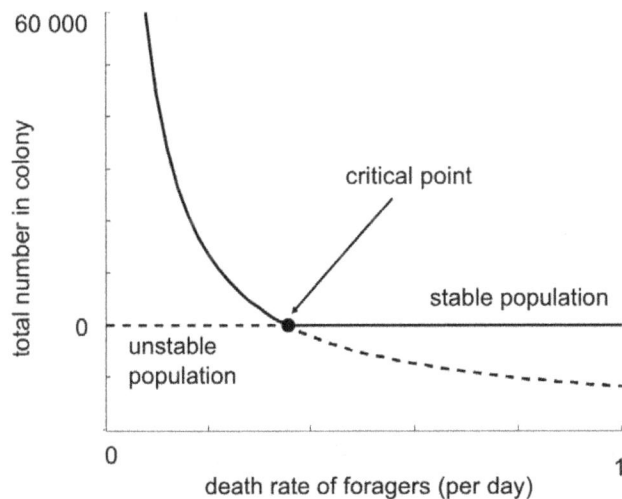

Figure 5. The dependence of the colony population at equilibrium on the death rate of foragers. For this set of parameter values, when the death rate m exceeds 0.355, the only stable equilibrium population is zero. Parameter values are the same as Figure 3.

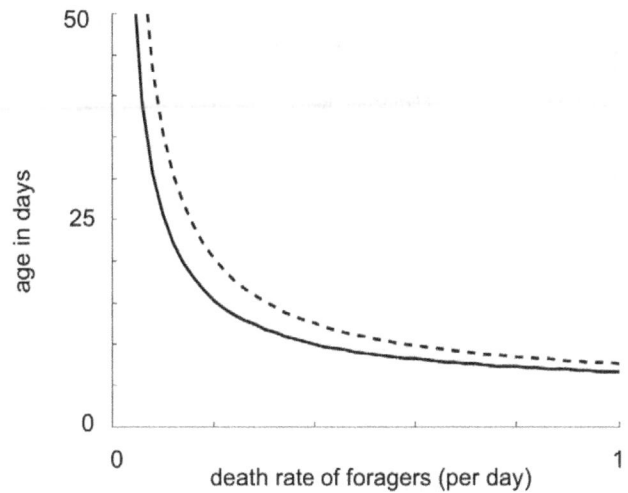

Figure 6. The average age of adult worker bees (dashed line) and the average age of onset of foraging (solid line) as a function of forager death rate. Parameter values are the same as Figure 3.

can be rapid (Fig. 4) since the social consequences of high forager losses accelerate colony failure. When forager death rate is high, nurse bees begin foraging precociously (Fig. 6). While this restores the proportion of foragers in the population, it shortens the overall lifespan of adult bees (Fig. 6) and reduces the time each bee can contribute to colony growth and brood production. This reduces the brood-rearing capacity of the colony. Since precocious foragers are less effective and resilient than normal foragers [25,26] forager death rate increases further, the pressure on colony population is compounded and the rate of colony decline is increased (Fig. 4).

In our simulations the bifurcation point was $m = 0.355$ which would imply that if the average duration of bees' foraging lives is reduced to just 2.8 days of foraging, and if this population stress is sustained colonies are likely to fail. In healthy colonies bees survive about 6.5 days of foraging on average [2], therefore our model predicts that chronic stressors that reduce the forager survival by approximately two thirds will place a colony at risk. Exploration of the model suggested that a high forager death rate in isolation would not cause colony failure, rather colony failure is caused by the social consequences resulting from a high forager death rate driving a decline in brood rearing alongside sustained forager losses.

The importance of forager longevity for equilibrium colony size has also been recognised by earlier modeling approaches [34,35], but the function of these earlier models was to simulate patterns of growth observed in real colonies, whereas the modeling approach that we use here is a more abstract representation of colony population dynamics and its purpose is to explore why forager death rate has such a strong influence on population size.

The model that we present here is very simple and focuses on the effect of varying forager death rate on brood and adult bee population dynamics. We have also constructed and explored more complicated models which include, for example, the effects of stored food in the hive and the effects of the presence of brood on bee behaviour, but we found that this leaner model was the most revealing and conceptually useful. The aim of this model is simply to provide a basic theoretical understanding of colony dynamics in an idealised state. We have not considered seasonal and climatic variation in queen egg laying rate and forager

Table 1. Comparison of experimental data and model results for average age of onset of foraging (AAOF) and lifespan.

Colony	Flightspan (days)	Deathrate, m (days^{-1})	AAOF		Lifespan	
			Observed	Model	Observed	Model
1 (Large)	7.5	0.133	18.6	19.4	22.8	26.9
2 (Large)	6.5	0.154	18.4	17.7	22.3	24.2
3 (Small)	6.7	0.149	23.8	17.6	26.6	24.3
4 (Small)	8.8	0.114	22.2	20.4	26.4	29.2

Experimental data is from Rueppell et al [33] and model results were obtained by running the model for 40 days (approximately the observational period used by Rueppell et al). At the start of each model run $H = 9000$ for large colonies and 4500 for small colonies and $F = 0$. The parameters were $L = 2000$, $w = 27000$, $\alpha = 0.25$ and $\sigma = 0.75$.

mortality rate, but these elements could be incorporated as elaborations of the basic model.

Does the current simplistic model usefully represent colony social dynamics and the process of colony failure? In some ways, simulations from the model effectively mimic the performance of natural colonies. The model predicts that from any initial starting population of hive bees and foragers, colonies move towards an equilibrium point by rapidly establishing a stable and consistent proportion of nurses and foragers (Fig. 3) while the total population size adjusts more slowly until the equilibrium point is reached. These simulations reflect experimental observations [5]. Colonies constructed with either no foragers, or 100% foragers rapidly adjusted the proportions of foragers and hive bees to values closer to those seen in normal hives [5,7,36]. When colonies are experimentally depleted of foragers they rapidly restore the ratio of hive bees to forager bees by accelerating the behavioural development of hive bees [5], but adjustments in colony size occurred more slowly. The model also predicted worker age at onset of foraging and lifespans that were a reasonable match to observed experimental data (Table 1).

While the current model suggests how social processes might contribute to colony failure, in its current form the model does not capture all features associated with the very dramatic colony failure observed in cases of CCD. Rapid population decline is one key characteristic of CCD. The rate of decline is not precisely defined [10] and may vary between cases, but the amount of abandoned brood found in CCD colonies suggests a very large drop in population within a few weeks [10]. The model predicts rapid initial declines in colony population (Fig. 4), but the current model does not effectively represent the absolute colony abandonment, which is also diagnostic of CCD [10]. Our simulations take about 200 days to reach close to zero population (Fig. 4). The current model does not consider factors that might accelerate the terminal decline of a honey bee colony once the population becomes small. Colonies with small populations are not able to thermoregulate effectively, which will weaken or kill developing brood [20,23]. Stressed colonies will cannibalise developing larvae [37], which will further reduce brood production and accelerate colony failure. Stressed colonies will sometimes abscond when the remaining bees and the queen leave the hive box altogether. It seems likely that population decline will

accelerate once colony population becomes small, but this process has not been well studied experimentally.

One of the mysterious aspects of CCD is the abandonment of brood by adult bees [38]. Our model suggests that this may occur because as populations dwindle, bees make the transition from hive bees to become foragers. Whether this extreme failure of division of labour would occur in natural colonies is not known, but experimental evidence has shown that the response of bees to various stressors is to change behaviour from brood care to foraging [25,39]. This suggests that when bees are starving or diseased or face other factors that shorten their individual lifespan, the motivation to forage overrides the motivation to attend to brood. In CCD cases the amount of brood left abandoned would suggest that this total collapse of normal division of labour must occur quite rapidly. Rigorous experimental observation of this process is needed urgently to understand how CCD compares to less dramatic cases of colony failure.

The model that we have presented focuses attention on forager death rate and the social consequences of this as a driver of colony failure. If brood production and the eclosion rate are too low to support a sustained level of forager losses then a colony will fail. One inference from this understanding is that factors that affect the survival of both brood and adult bees could leave colonies particularly vulnerable to collapse. Examples of such factors would be the mite *Varroa destructor*, which affects both brood and forager survival [14,40] and *Nosema* infections [15], both of which are known causes of colony failure [11,12,15]. The model also predicts that treatment strategies to restore failing colonies should focus on preventing precocious foraging to extend the useful lifespan of adult bees in the colony, and boosting brood production to restore the colony to a point at which recruitment into the population is sufficient to sustain ongoing forager losses.

Experimental testing of the model predictions will hopefully yield a better understanding of the process of catastrophic colony failure, and how best to intervene to restore failing colonies.

Author Contributions

Conceived and designed the experiments: ABB MRM DSK. Performed the experiments: DSK. Analyzed the data: DSK. Contributed reagents/materials/analysis tools: MRM. Wrote the paper: ABB MRM.

References

1. Winston ML (1987) The biology of the honey bee. Cambridge: Harvard University Press. 281 p.
2. Visscher PK, Dukas R (1997) Survivorship of foraging honey bees. Insectes Sociaux 44: 1–5.
3. Seeley TD (1982) Adaptive significance of the age polyethism schedule in honeybee colonies. Behavioral Ecology and Sociobiology 11: 287–293.
4. Seeley TD (1995) The Wisdom of the Hive. Cambridge: Harvard University Press. 295 p.
5. Huang Z-Y, Robinson GE (1996) Regulation of honey bee division of labor by colony age demography. Behavioral Ecology and Sociobiology 39: 147–158.
6. Robinson GE, Page RE, Huang Z-Y (1994) Temporal polyethism in social insects is a developmental process. Animal Behaviour 48: 467–469.

7. Robinson GE, Page RE, Strambi C, Strambi A (1992) Colony integration in honey bees: mechanisms of behavioural reversion. Ethology 90: 336–350.

8. Leoncini I, Le Conte Y, Costagliola G, Plettner E, Toth AL, et al. (2004) Regulation of behavioral maturation by a primer pheromone produced by adult worker honey bees. Proceedings of the National Academy of Sciences of the United States of America 101: 17559–17564.

9. VanEngelsdorp D, Hayes Jr. J, Underwood RM, Pettis JS (2010) A survey of honey bee colony losses in the United States, fall 2008 to spring 2009. Journal of Apicultural Research 49: 7–14.

10. VanEngelsdorp D, Evans JD, Saegerman C, Mullin C, Haubruge E, et al. (2009) Colony Collapse Disorder: A Descriptive Study. PLoS ONE 4.

11. Neumann P, Carreck NL (2010) Honey bee colony losses. Journal of Apicultural Research 49: 1–6.

12. Ratnieks FLW, Carreck NL (2010) Carity on honey bee collapse? Science 327: 152–153.

13. Oldroyd BP (1999) Coevolution while you wait: Varroa jacobsoni, a new parasite of western honeybees. Trends in Ecology & Evolution 14: 312–315.

14. Dahle B (2010) The role of *Varroa Destructor* for honey bee colony losses in Norway. Journal of Apicultural Research 49: 124–125.

15. Higes M, Martin-Hernandez R, Botias C, Bailon EG, Gonzalez-Porto AV, et al. (2008) How natural infection by Nosema ceranae causes honeybee colony collapse. Environmental Microbiology 10: 2659–2669.

16. VanEngelsdorp D, Hayes J, Underwood RM, Pettis J (2008) A Survey of Honey Bee Colony Losses in the US, Fall 2007 to Spring 2008. PLoS ONE 3.

17. Watanabe ME (2008) Colony collapse disorder: Many suspects, no smoking gun. Bioscience 58: 384–388.

18. Desneux N, Decourtye A, Delpuech JM (2007) The sublethal effects of pesticides on beneficial arthropods. Annual Review of Entomology 52: 81–106.

19. Chauzat M-P, Martel A-C, Blanchard P, Clément M-C, Schurr F, et al. (2010) A case report of a honey bee colony poisoning incident in France. Journal of Apicultural Research 49: 113–115.

20. Medrzycki P, Sgolastra F, Bortolloti L, Bogo G, Tosi S, et al. (2010) Influence of brood rearing temperature on honey bee development and susceptibility to poisoning by pesticides. Journal of Apicultural Research 49: 52–59.

21. Pettis J, Vanengelsdorp D, Cox-Foster D (2007) Colony collapse disorder working group pathogen sub-group progress report. American Bee Journal 147: 595–597.

22. Cox-Foster DL, Conlan S, Holmes EC, Palacios G, Evans JD, et al. (2007) A metagenomic survey of microbes in honey bee colony collapse disorder. Science 318: 283–287.

23. Oldroyd BP (2007) What's killing American honey bees? PLoS Biology 5: 1195–1199.

24. Bromenshenk JJ, Henderson CB, Wick CH, Stanford MF, Zulich AW, et al. (2010) Iridovirus and microsporidian linked to honey bee colony decline. PLoS ONE 5: e13181. doi:13110.11371/journal.pone.0013181.

25. Woyciechowski M, Moron D (2009) Life expectancy and onset of foraging in the honeybee (*Apis mellifera*). Insectes Sociaux 56: 193–201.

26. Oskay D (2007) Plasticity in flight muscle development and honey bee division of labor. San Juan: University of Puerto RIco.

27. McLellan AR (1978) Growth and Decline of Honeybee Colonies and Inter-Relationships of Adult Bees, Brood, Honey and Pollen. Journal of Applied Ecology 15: 155–161.

28. Allen MD, Jeffree EP (1956) The influence of stored pollen and of colony size on the brood rearing of honeybees. Annals of Applied Biology 44: 649–656.

29. Harbo JR (1986) Effect of population size on brood production, worker survival and honey gain in colonies of honeybees. Journal of Apicultural Research 25: 22–29.

30. Fahrbach SE, Robinson GE (1996) Juvenile hormone, behavioral maturation and brain structure in the honey bee. Developmental Neuroscience 18: 102–114.

31. Cramp D (2008) A Practical Manual of Beekeeping. London: How To Books. 304 p.

32. Edelstein-Keshet L (1988) Mathematical Models in Biology. New York: Random House. 608 p.

33. Rueppell O, Kaftanouglu O, Page RE (2009) Honey bee (Apis mellifera) workers live longer in small than in large colonies. Experimental Gerontology 44: 447–452.

34. DeGrandi-Hoffman G, Curry R (2004) A mathematical model of varroa mite (Varroa destructor Anderson and Trueman) and honeybee (Apis mellifera L.) population dynamics. International Journal of Acarology 30: 259–274.

35. DeGrandi-Hoffman G, Roth SA, Loper GL, Erickson Jr. EH (1989) Beepop: a honeybee population dynamics simulation model. Ecological Modelling 45: 133–150.

36. Schulz DJ, Robinson GE (2001) Octopamine influences division of labor in honey bee colonies. Journal of Comparative Physiology A 187: 53–61.

37. Schmickl T, Crailsheim K (2001) Cannibalism and early capping: strategy of honeybee colonies in times of experimental pollen shortages. Journal of comparative physiology A, Sensory, neural, and behavioral physiology 187: 541–547.

38. Debnam S, Westerverlt D, Bromenshenk J, Oliver R (2009) Colony collapse disorder. Bee Culture 137: 34–35.

39. Schulz DJ, Huang Z-Y, Robinson GE (1998) Effects of colony food shortage on behavioral development in honey bees. Behavioral Ecology and Sociobiology 42: 295–303.

40. Boecking O, Genersch E (2008) Varroosis - the ongoing crisis in bee keeping. Journal of Comsumer Protection and Food Safety 3: 221–228.

7

Metagenomic Detection of Viral Pathogens in Spanish Honeybees: Co-Infection by Aphid Lethal Paralysis, Israel Acute Paralysis and Lake Sinai Viruses

Fredrik Granberg[1,2*,◑], Marina Vicente-Rubiano[3,◑], Consuelo Rubio-Guerri[3], Oskar E. Karlsson[1,2,4], Deborah Kukielka[3], Sándor Belák[1,2,5], José Manuel Sánchez-Vizcaíno[3]

1 Department of Biomedical Sciences and Veterinary Public Health (BVF), Swedish University of Agricultural Sciences (SLU), Uppsala, Sweden, 2 The OIE Collaborating Centre for the Biotechnology-based Diagnosis of Infectious Diseases in Veterinary Medicine, Uppsala, Sweden, 3 Animal Health Department, Faculty of Veterinary, Complutense University of Madrid, Madrid, Spain, 4 SLU Global Bioinformatics Center, Department of Animal Breeding and Genetics (HGEN), SLU, Uppsala, Sweden, 5 Department of Virology, Immunobiology and Parasitology, VIP, National Veterinary Institute (SVA), Uppsala, Sweden

Abstract

The situation in Europe concerning honeybees has in recent years become increasingly aggravated with steady decline in populations and/or catastrophic winter losses. This has largely been attributed to the occurrence of a variety of known and "unknown", emerging novel diseases. Previous studies have demonstrated that colonies often can harbour more than one pathogen, making identification of etiological agents with classical methods difficult. By employing an unbiased metagenomic approach, which allows the detection of both unexpected and previously unknown infectious agents, the detection of three viruses, Aphid Lethal Paralysis Virus (ALPV), Israel Acute Paralysis Virus (IAPV), and Lake Sinai Virus (LSV), in honeybees from Spain is reported in this article. The existence of a subgroup of ALPV with the ability to infect bees was only recently reported and this is the first identification of such a strain in Europe. Similarly, LSV appear to be a still unclassified group of viruses with unclear impact on colony health and these viruses have not previously been identified outside of the United States. Furthermore, our study also reveals that these bees carried a plant virus, Turnip Ringspot Virus (TuRSV), potentially serving as important vector organisms. Taken together, these results demonstrate the new possibilities opened up by high-throughput sequencing and metagenomic analysis to study emerging new diseases in domestic and wild animal populations, including honeybees.

Editor: Amit Kapoor, Columbia University, United States of America

Funding: This work was mainly supported by the Award of Excellence (Excellensbidrag), provided to SB by the Swedish University of Agricultural Sciences (SLU), and an FPU grant from the Ministry of Education, Culture and Sports of Spain, which supported MVR and CRG. Special recognition is given to the Ministry of Education of Spain for the FPU grant of Consuelo Rubio-Guerri and Marina Vicente-Rubiano. The work was also partly supported by/executed in the framework of the EU-project AniBioThreat (Grant Agreement: Home/2009/ISEC/AG/191) with the financial support from the Prevention of and Fight against Crime Programme of the European Union, European Commission - Directorate General Home Affairs. This publication reflects the views only of the authors, and the European Commission cannot be held responsible for any use, which may be made of the information contained therein. The funders had no role in study design, data collection and analysis, decision to publish, or preparation of the manuscript.

Competing Interests: The authors have declared that no competing interests exist.

* E-mail: fredrik.granberg@slu.se

◑ These authors contributed equally to this work.

Introduction

In recent years, the world's population of honeybee (*Apis mellifera*), the main insect pollinator in the USA and Europe, has been decreasing. This loss of bee population has been called Colony Collapse Disorder (CCD), defined as the disappearance of bees from the beehives without any dead bees around the affected hives, presence of abundant breeding cells, pollen, and honey despite a small population of adult bees and without any characteristic symptom of disease [1]. The importance of this phenomenon stems not only from the large direct losses of honey-producing countries, of which Spain has the largest honey production in the European Union (EU) and EU ranks third in the world honey production CCD may also cause indirect losses for a lack of crop pollination, some of them very important in Spain such as citrus crop. To illustrate the serious effects of

honeybee decline, which is the most efficient pollinator of most crops [2], the global value of insect pollination has been estimated at US\$ 212 billion (€153 billion), which represents about 9.5% of the total value of agricultural production. Specifically, the value of insect pollination to agriculture for EU25 is US\$ 19.8 billon (€14.2 billion) [3]. As a result, the CCD phenomenon has become a growing concern for governments and international organizations, which have led to increased investments in terms of research on its origin.

The manifestations of increased honeybee mortality and decline of managed hives due to unclear diseases, such as CCD, have to a great extent been associated with viral infections [4,5]. At least 18 different viruses with the ability to infect honeybees have so far been identified [6,7], and an additional four were recently suggested [8]. Several of them have been demonstrated to have a global spread [9,10] and colonies frequently suffer from multiple viral co-infections [11–14]. In addition, since most of the honeybee viruses tend to persist as covert infections with no obvious symptoms, requiring stress to be activated [15] even seemingly healthy colonies can harbour a variety of potentially harmful viral infections [16,17]. Among the natural activation factors, especially mite infestations have been correlated with outbreak of disease. The parasitism is believed to induce virus proliferation by causing a general decline in immune capacity of the hosts and mites have also been shown to act as vectors of honeybee viruses [18–20].

Regarding the above listed serious problems and losses, there is a high need to investigate the occurrence, emergence and effects of various pathogens in the European honeybee populations, with special regard to known and "unknown", emerging new viruses and to the combination of various agents in complex infections. For the time being, the direct diagnosis of the various disease forms is based on two main approaches: a) traditional diagnostic methods, such as virus isolation and electron microscopy; b) molecular diagnostic methods, such as PCR and microarrays, among others.

The traditional paradigm of detecting and identifying pathogens relies upon diagnostic tests available for the detection of known agents. This makes it difficult, or impossible, to identify unexpected or novel pathogens by using conventional methods.

Virus isolation has a very powerful diagnostic capacity, considering that this method is able to detect a very wide range of viruses directly, in a single system. However, even this excellent method has many weaker sides, e.g., the inability of many viruses, including honeybee viruses, to replicate in the used cell cultures and cause visible signs of virus replication, such as cytopathic effects.

Molecular diagnostic techniques, such as PCR, isothermal amplification, and microarrays have rapidly been replacing the traditional diagnostic approaches and have opened new alternatives for virus detection and identification. However, even these methods have important weak sides, such as the restricted detection range. The detection range can be improved by the application of wide range microarrays, DNA-chips, however, even these system may fail to detect a wide range of pathogens, for example "unknown", emerging new viruses. Considering that many of these viruses may cause severe diseases, malfunctions, synergetic effects with other pathogens, may influence the immune system, and many other effects, it is crucial to improve our diagnostic capacities and to extend our detection capacities.

The introduction of viral metagenomics has opened up a new range of possibilities for the improved detection of both known and unknown viruses. These cell culture and nucleotide sequence independent approaches allow the detection of a very wide range of viruses and other pathogens and they have the capacity to determine the entire infectious flora in different host species. Furthermore, viral metagenomics is able to shed light not only of the presence of various infectious agents, but also on the biodiversity of the detected viruses, bacteria and other infectious agents. This enables us to achieve a better understanding of emerging novel diseases and the complex infection biology of various disease complexes. The comprehensive metagenomic techniques, such as high-throughput nucleotide sequencing, have the potential to detect the full spectrum of emerging new pathogens, including novel viruses and fastidious bacteria, as demonstrated and reviewed [21–24]. The *OIE Collaborating Centre for the Biotechnology-based Diagnosis of Infectious Diseases in Veterinary Medicine* in Uppsala, Sweden has established skills and state-of-the-art facilities for the metagenomic detection of various known and unknown viruses, such as novel bocaviruses, Torque Teno viruses, astroviruses and other infectious agents [25–27].

Given the unclear diagnosis of many honeybee viral diseases, the frequent covert infection of these viruses and the high prevalence of multiple viral co-infections, we hypothesize that a metagenomic approach should be particularly useful to find various potentially causative agents in the bee colonies. This has also been demonstrated by previous studies aimed at characterizing the microflora, or microbiome, of the honeybee in search of microbial agents involved in CCD [4,8]. In this study, honeybees from Spain were investigated using a high-throughput sequencing approach to identify all potential etiological agents.

Materials and Methods

Specimens

The sample of honeybees (*Apis mellifera*) was collected with the owners' permission from one colony belonging to one apiary of 25 commercial hives located in Los Arcos, Navarre, North of Spain. The colony was sampled by the veterinary services due to lack of vitality of adult worker honeybees and unusual depopulation, especially in the brood frames. Furthermore, symptoms compatible with CCD such as drastically reduced adult population in presence of abundant food and breeding were observed. There were no specific symptoms compatible with viral diseases. The sample consisted of approximately 50 adult worker bees from inside and outside the hives to ensure the presence of young and adult bees. The bees were collected in sterile containers and frozen until delivered to the Department of Animal Health at the Complutense University of Madrid for routine testing with standard RT-PCR assays for identification of common bee viruses. Homogenates were manually prepared from 20 whole bees, in a 30 ml Wheaton glass homogenizer containing 6 ml of sterile phosphate-buffered saline (1×PBS).

The sample was analyzed by amplification of virus-specific nucleic acid for the presence of seven honeybee viruses: Deformed Wing Virus (DWV), Black Queen Cell Virus (BQCV), Sacbrood Bee Virus (SBV), Acute Bee Paralysis Virus (ABPV), Chronic Bee Paralysis Virus (CBPV), Kashmir Bee Virus (KBV) and Israeli Acute Paralysis Virus (IAPV). One step real time RT-PCRs based on SYBR-Green dye were carried out, following previously described protocols [13,28–31]. The sample tested positive for IAPV and the viral load was estimated to be 7.5×10^4 genome equivalent copies (GEC) per bee.

Sample Preparation and Nucleic Acid Isolation

The homogenates were centrifuged at 4.000 rpm for 10 min and the collected supernatants were syringe-filtered through disposable 0.45 µm PVDF filters (Millipore). Aliquots of 200 µl supernatant in a final concentration of 1xDNase buffer were nuclease treated with 400 U/ml DNase I (Roche Applied Science) and 8 µg/ml RNase A (Invitrogen) at 37°C for 2 h. DNA was extracted using the QIAamp DNA Mini Kit (Qiagen) according to the manufacturer's spin protocol for blood and body fluid. RNA was isolated using TRIzol LS Reagent (Invitrogen) and further purified using the RNeasy mini kit (Qiagen) according to the manufacturer's instructions.

Tag Labeling and Random Amplification

Extracted DNA and RNA were separately labeled with an identical sequence tag contained in the primer FR26RV-N (GCCGGAGCTCTGCAGATATCNNNNNN) [32]. For the labeling of DNA, 10 µl of template was mixed with 1.5 µl 10× NEBuffer 2, 1.5 µl dNTPs (10 mM of each), and 2 µl FR26RV-N (10 mM). The mixture was denatured at 94°C for 2 min and chilled on ice before the addition of 0.5 µl (2.5 U) 3′–5′ exo-

Klenow DNA polymerase (New England Biolabs). The initial extension at 37°C for 1 h was followed by an identical second cycle, starting with denaturation as above, after which the enzyme was inactivated by heating at 75°C for 10 min. The synthesis of the sequence tagged cDNA was prepared by adding 1.5 µl dNTP (10 mM of each) and 2 µl FR26RV-N (10 mM) to 10 µl of RNA template. The mixture was incubated at 65°C for 5 min and chilled on ice before the addition of 4 µl 5× First-Strand buffer (Invitrogen), 1 µl DTT (100 mM), 1 µl (40 U) RNaseOUT (Invitrogen) and 1 µl (200 U) Superscript III reverse transcriptase (Invitrogen). The RT reaction was incubated at 25°C for 5 min, 50°C for 1 h and 70°C for 15 min, after which it was chilled on ice. Second-strand synthesis was performed by adding 0.5 µl (2.5 U) 3′–5′ exo- Klenow DNA polymerase (New England Biolabs) and incubate at 37°C for 1 h. A final incubation at 75°C for 10 min inactivated the enzyme.

Amplification was performed by PCR using the complementary primer FR20RV (GCCGGAGCTCTGCAGATATC) [32]. Each 50 µl reaction was carried out with 2.5 µl of labeled template and 0.5 µl (2.5 U) Ampli-Taq Gold DNA polymerase (Applied Biosystems) in a final concentration of 1x GeneAmp PCR buffer 2 (Applied Biosystems), 0.2 mM dNTPs, 2.5 mM $MgCl_2$, and 0.8 µM FR20RV. The thermal cycling was initiated with a denaturation step at 95°C for 10 min, followed by 40 cycles of 95°C for 1 min, 58°C for 1 min, 72°C for 1 min, and a final extension at 72°C for 10 min. PCR products were purified with the QIAquick PCR purification kit (Qiagen), both before and after the tag sequence was removed with EcoRV (New England Biolabs), according to the manufacturers' instructions. The final products were checked on agarose gel and quantified using a NanoDrop ND-100 spectrophotometer (NanoDrop Technologies).

Library Preparation and Sequencing

The library preparation and sequencing were performed at the SNP&SEQ Technology Platform in Uppsala. Briefly, the amplification products were pooled and separated into two size fractions of approximately 250–400 bp and 400–550 bp, respectively. Each fraction was labeled, without any further fragmentation, with an indexing sequence using GS FLX Titanium Rapid Library MID Adaptors (Roche) and sequenced on 1/16 of a GS FLX Titanium PicoTiterPlate (Roche/454 Life Sciences) according to the manufacturer's protocol.

Data Handling and Bioinformatics

The sequence data, trimmed of adaptor regions and in Standard Flowgram Format (SFF), were combined into a single FASTA and quality file using the sff_extract python application distributed with the MIRA software package (http://sourceforge.net/projects/mira-assembler). All sequence reads were then assembled using MIRA [33] with the standard settings for *de novo* assembly of 454 data.

Taxonomic classification of unassembled sequence reads was enabled by BLASTN and BLASTX searches against local copies of NCBI's nucleotide and protein databases using NCBI's blastall program [34] with default settings. The resulting outputs were committed into MEGAN 4 [35] with the NCBI taxonomy data for assigning taxa. Each sequence read was placed on a node in the NCBI taxonomy according to the lowest common ancestor (LCA) based on a subset of the best scoring BLAST matches. The parameters for the LCA algorithm were: Min support 5, Min Score 65, Top Percent 10, and Min complexity 0.3. Resulting trees were explored for host genome, bacterial content and viral community.

Evaluating the taxonomic data for potential pathogens, candidate reference genomes were identified and retrieved from GenBank in FASTA format. Alignment of matching contigs from the whole assembled dataset against the nucleotide sequences of the reference genomes were performed using the CodonCode Aligner software (CodonCode Corporation). This allowed analysis of similarities and visualisation of gaps.

Confirmation and Retrieval of Near Full Genome Sequences

Based on the results from the alignments, PCR primers were designed to confirm the presence of viruses in the original material and to close gaps using the Primer3 program [36]. Total RNA was extracted from filtered homogenate as above, but without nuclease treatment, and cDNA was generated using the Superscript III first-strand synthesis system (Invitrogen) with random hexamers according to the manufacturer's instructions. Products with an expected length shorter than 1.500 bp were amplified with an AmpliTaq Gold-based PCR protocol (Applied Biosystems), while the Phusion Hot Start II High Fidelity DNA Polymerase system (Thermo Scientific) was used for longer fragments. The amplified products were size-separated on an agarose gel, purified with the QIAquick Gel Extraction Kit (Qiagen), and sent for sequencing (Macrogen Europe). The obtained sequences were incorporated into the alignments against the reference genomes using the CodonCode Aligner software (CodonCode Corporation). Longer distances were covered by iterative primer walking. The resulting viral sequences reported in this paper have been deposited in GenBank.

Comparison between Sequences and Phylogenetic Analysis

Direct comparisons between two sequences on a nucleotide and amino acid level were performed by using the NCBI's BLAST 2 sequences tool [37]. To enable multiple species and strain comparisons, annotated sequences were retrieved from GenBank in FASTA format. While the complete sequences were directly used for phylogenetic analysis, the partial were aligned against a reference genome using the CodonCode Aligner software (CodonCode Corporation). By using the genomic region with the highest number of overlaps, a maximum number of strains were allowed to be compared. The analysis of the phylogenetic relationships were conducted in MEGA 5 [38] using Clustal W [39] to align the sequences and the Maximum Likelihood method with 1.000 bootstrap replicates to generate the trees [40]. The Kashmir Bee Virus (KBV), which is a *Dicistroviridae* member genetically closed to IAPV [41], was used as outgroup.

Results

Sequence Data and Assembly

GS FLX Titanium sequencing (Roche/454 Life Sciences) of the nuclease treated and amplified sample returned 161.170 reads for the short fraction (250–400 bp) and 80.790 for the long (400–550 bp). For de novo assembly using MIRA, both fractions were combined resulting in a total of 241.960 reads and 54,98 Mbp of sequence data. The assembly generated 6.350 contigs, ranging in length from 40 to 2.945 bp with mean length 393,3 bp and an average GC content of 51,1%.

Taxonomic Classification

The assignment of unassembled sequence reads to taxa based on BLASTN results revealed bacteria to be the largest taxonomic group. Most of the 110.137 reads in this group mapped to Gammaproteobacteria, but the presence of Alpha- and Betaproteobacteria, Bacilli (Firmicutes), Actinobacteria, and Bacteroidetes were also indicated, see Figure 1a. No bacterial pathogens, such as *Paenibacillus larvae*, *Melissococcus plutonius* or *Spiroplasma* were detected. Thus, the observed bacterial diversity is similar and in agreement with results from metagenomic studies aimed at characterizing the normal gut flora of honey bees [42,43]. Among the reads in Eukaryota, 918 corresponded to the host organism *Apis mellifera* and related species, which constitute only 0,38% of the total amount of reads. This indicates that most of the host nucleic acid was removed during the sample preparation step. Furthermore, no matches against mites, such as *Varroa destructor*, or pathogenic fungi, such as certain members of *Nosema*, were detected in the eukaryotic group. A total of 4.310 reads were taxonomically assigned to viruses, the majority belonging to ssRNA viruses. Even though we used a relatively low threshold for allowing assignment, a large number of reads still displayed too short homologies to be uniquely defined and were classified as "Not assigned". In addition, for 2.194 reads, no hits were found in the NCBI database, and 21.961 were disregarded due to low complexity of the reads. Remaking the taxonomic assignment based on BLASTX searches resulted in a similar distribution as described, the main difference being that approximately 10% of the reads were moved from the not assigned group into bacteria. Even so, no bacterial pathogens were identified.

Detected Viral Genomic Sequences

As illustrated in Figure 1b, most of the sequence reads in the virus group were divided between four viruses. While both ALPV and IAPV are members of the *Dicistroviridae* family, Turnip Ringspot Virus (TuRSV) and Turnip Yellow Mosaic Virus (TYMV) are plant-infecting viruses belonging to the families *Secoviridae* and *Tymoviridae*, respectively. Low numbers of reads with similarities to Lake Sinai Virus (LSV), retro-transcribing viruses and bacteriophages were also identified.

For each of the specific viruses, a best matching candidate reference genome was retrieved to enable direct comparisons against all assembled contigs. The resulting alignments revealed regions of high sequence coverage and the total numbers of contigs and reads for each virus are summarized in Table 1. In general, the number of sequence reads for a particular virus is proportional

Table 1. Identified ssRNA viruses with family classifications and numbers of aligned sequences.

Virus	Virus Family/Taxa	Reads	Contigs
Aphid lethal paralysis virus (ALPV)	*Dicistroviridae*	664	16
Israel acute paralysis virus (IAPV)	*Dicistroviridae*	1.048	7
Lake Sinai virus (LSV)	Unclassified	14	1
Turnip ringspot virus (TuRSV)	*Secoviridae*	1.968	14
Turnip yellow mosaic virus (TYMV)	*Tymoviridae*	563	1

to its abundance in the investigated sample. However, the size of the genome and similarity with existing reference genomes can be confounding factors. This makes it hard to judge the lower amount of reads for LSV.

Regarding the indicated presence of retro-transcribing viruses, two contigs were generated and both displayed a high degree of similarity with Moloney murine leukemia virus and Xenotropic Murine Leukemia Virus (XMLV). However, attempts to amplify a larger fragment spanning the gap between the contigs by PCR failed to generate a product (data not shown). Given the dubious findings of XMLV in many different biological samples [44], it is likely that the findings indicate the presence of contaminants [45] rather than a gammaretrovirus.

A Virus Similar to Aphid Lethal Paralysis Virus (ALPV)

The 16 contigs displaying similarity with ALPV were distributed in four non-overlapping clusters, spanning a total of 1.469 nt, as demonstrated by alignment against the first complete ALPV genome that was published (GenBank AF536531). Using the consensus sequences of the clusters as starting points, PCR primers were designed in order to fill the gaps by Sanger sequencing and longer distances were bridged by primer walking. This resulted in a single 9.327 nt sequence (GenBank JX045858) covering approximately 95% of the ALPV genome. While the nucleotide

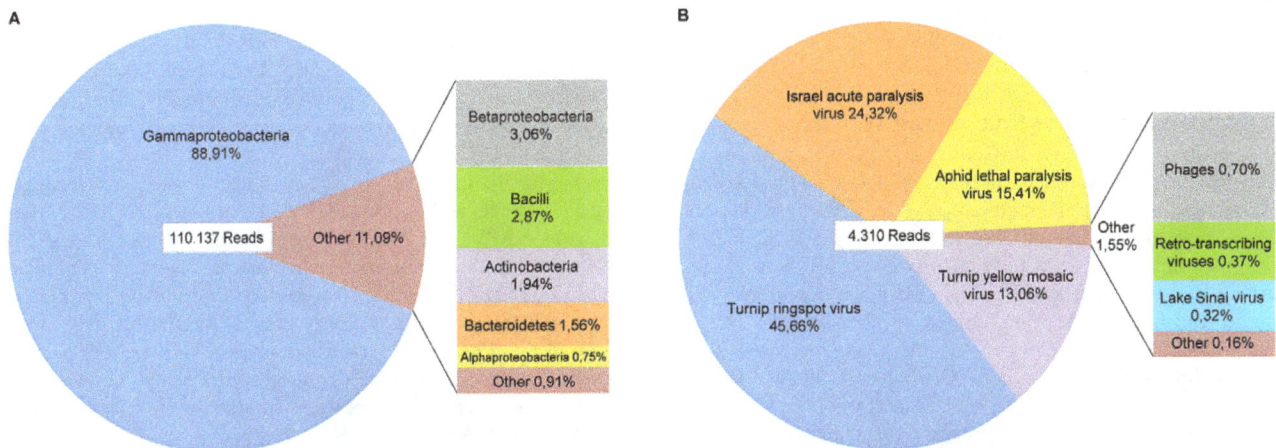

Figure 1. Taxonomic distributions. The distribution of sequence reads within the taxonomic groups of (a) bacteria, and (b) viruses. The taxonomic assignment was performed based on the BLASTN search results using the MEGAN 4 software with the following LCA settings: Min support 5, Min Score 65, Top Percent 10, and Min complexity 0.3.

homology with the reference genome, as shown by BLAST, was 82%, the recovered sequence displayed an even closer resemblance, 96%, to the recently published ALPV strain Brookings (GenBank HQ871932). Unlike the classical ALPV, which has not yet been associated with infections of honeybees, the Brookings strain was first found in diseased honeybees using a next generation sequencing approach [8]. However, due to the relatively high resemblance with the reference sequence, the question whether or not the new strain constitutes a novel species was left unanswered. A comparison at the protein level, translated DNA against protein sequences, demonstrated that the non-structural polyprotein sequence recovered in this study aligned with 91% amino acid identity to the reference genome and 99% to the Brookings strain. It thus appears likely that we have identified the first European member of a subgroup of ALPV being able to infect honeybees and putatively causing disease.

A Novel Variant of Israel Acute Paralysis Virus (IAPV)

Using the closest blast hit as reference sequence (GenBank EU224279), the seven contigs corresponding to IAPV were aligned into three non-overlapping clusters, covering 816 nt in total. A longer continuous sequence was obtained in a similar manner as above, resulting in a stretch of 8.882 nt (GenBank JX045857) representing approximately 93% of the IAPV genome. The sequence was compared against all publicly available complete IAPV genomes deposit in GenBank. As shown in Figure 2a, the IAPV identified in this study had the highest resemblance to the Australia strain and was more similar to the strains from United States than the ones from Israel and China. In an extended phylogenetic analysis, a stretch of approximately 700 nt from the IGR, containing RNA polymerase and structural polyprotein genes, was chosen to compare the obtained sequence to 67 IAPV strains with publicly available nucleotide sequences. The IGR region of the IAPV genome also contains an independent internal ribosome entry site (IRES) and has proven suitable for the inference of phylogenetic relationships in previous studies [4,30]. Alignment and clustering indicated that the IAPV identified in this study shared more similarity with strains from France [46], as illustrated in Figure 2b, than the only strain previously identified in Spain (GenBank FJ821506) [47]. At a nucleotide level, the 700 nt of the obtained sequenced displayed 96% similarity to the Spanish strain and 99% to the French. This indicates that the virus is a variant of IAPV with close resemblance to already sequenced strains. Moreover, the high degree of similarity to the strain previously identified in France is correlating well with the geographic proximity.

A Virus Similar to Lake Sinai Virus (LSV)

A single contig of 631 nt (GenBank JX045859) indicated the presence of a virus similar to LSV, which was only recently discovered [8]. The gene organization of LSV has been described to resemble the genome of Chronic Bee Paralysis Virus (CBPV). However, only a low degree of similarity is retained between LSV and CBPV at the gene level, e.g. the Orf1 genes only display 18% amino acid identity. As demonstrated by blast analysis, the best match for the recovered contig is the Orf1 region of LSV1 (GenBank HQ871931) and LSV2 (GenBank HQ888865), and the nucleotide similarity is 78 and 77%, respectively. Since only one contig was obtained, iterative primer walking could not be applied to obtain a longer sequence. Nevertheless, the presence of the contig sequence in the original sample was verified by PCR and Sanger sequencing. On an amino acid level, a similar comparison revealed that the contig shared 87 and 80% max identity with LSV1 and LSV2, respectively. This indicates that we have made

the first observation of a virus within the same group as LSV outside of the United States.

Turnip Ringspot Virus (TuRSV) & Turnip Yellow Mosaic Virus (TYMV)

Aligning the contigs corresponding to TuRSV against the closest BLASTN match, isolate Toledo segment RNA 1 (GenBank FJ712026), resulted in three non-overlapping clusters with a combined length of 1.169 nt (GenBank JX045854-6). A direct comparison of these clusters with the reference genome revealed that the nucleotide sequence similarity was approximately 90 to 94%. The presence of TuRSV in Spain has previously been established [48], but the published sequence do not overlap with the contigs making a direct comparison impossible (GenBank AJ489259). For TYMV, one contig of 225 bp was generated and it shared 91% nucleotide sequence similarity with its most similar reference genome (GenBank X07441), but only over a stretch of 56 bp in the middle. Since the ends did not show any resemblance with the reference, this could either indicate a new type of TYMV-like virus or an incorrectly assembled contig. As only one short contig of uncertain nature was obtained, we did not proceed to verify this finding. Although likely to be associated with pollen and nectar, honeybees have not explicitly been shown to be a vector for the spread of TuRSV or TYMV.

Discussion

The high occurrence of co-infections in colonies and honeybees has made it desirable to investigate multiple pathogens when attempting to identify the causative etiological agents in the known and in the recently emerging, "unknown" infectious diseases of the various populations. The detection of known and "unknown" infectious agents has recently been greatly facilitated by the application of metagenomic approaches, exploiting the emergence of high-throughput sequencing techniques, which allows the simultaneous detection and characterization of various micro-organisms, including bacteria, viruses, fungi and parasites. Using this type of unbiased metagenomic approach, we here identified and confirmed the presence of three viruses, ALPV, IAPV and LSV, in honeybees from Spain. The existence of a subgroup of ALPV with the ability to infect bees was, together with LSV, only recently reported [8]. Thus, according to our knowledge, this is the first identification of these strains in Europe. Interestingly, this study revealed not only bee viruses in the examined honeybee sample, but also a plant pathogen, TuRSV.

Viruses affecting honeybees have been demonstrated to have a wide spread within the pollinator community. For instance, ABPV has been described to cause covert infections in bumble bee species and KBV has been detected in both bumble bees and wasps [49,50]. In addition, a more in-depth study to investigate the host range and transmission of common honeybee viruses, such as IAPV and BQCV, found that these viruses are disseminating freely among the pollinators via the flower pollen itself [51]. The same study also revealed that non-*Apis* hymenopteran pollinators near honeybee apiaries affected by IAPV were more likely to carry the virus themselves. There are thus several potential reservoirs for these viruses in nature, which motivates the inclusion of other pollinators when conducting prevalence studies of honeybee viruses.

Until recently, no virus similar to ALPV on the sequence level had been reported in association with honeybees. This was changed, soon after this study began, with the finding of the Brookings strain in the United States, where the initial discovery also enabled the virus to be retroactively detected in honeybee

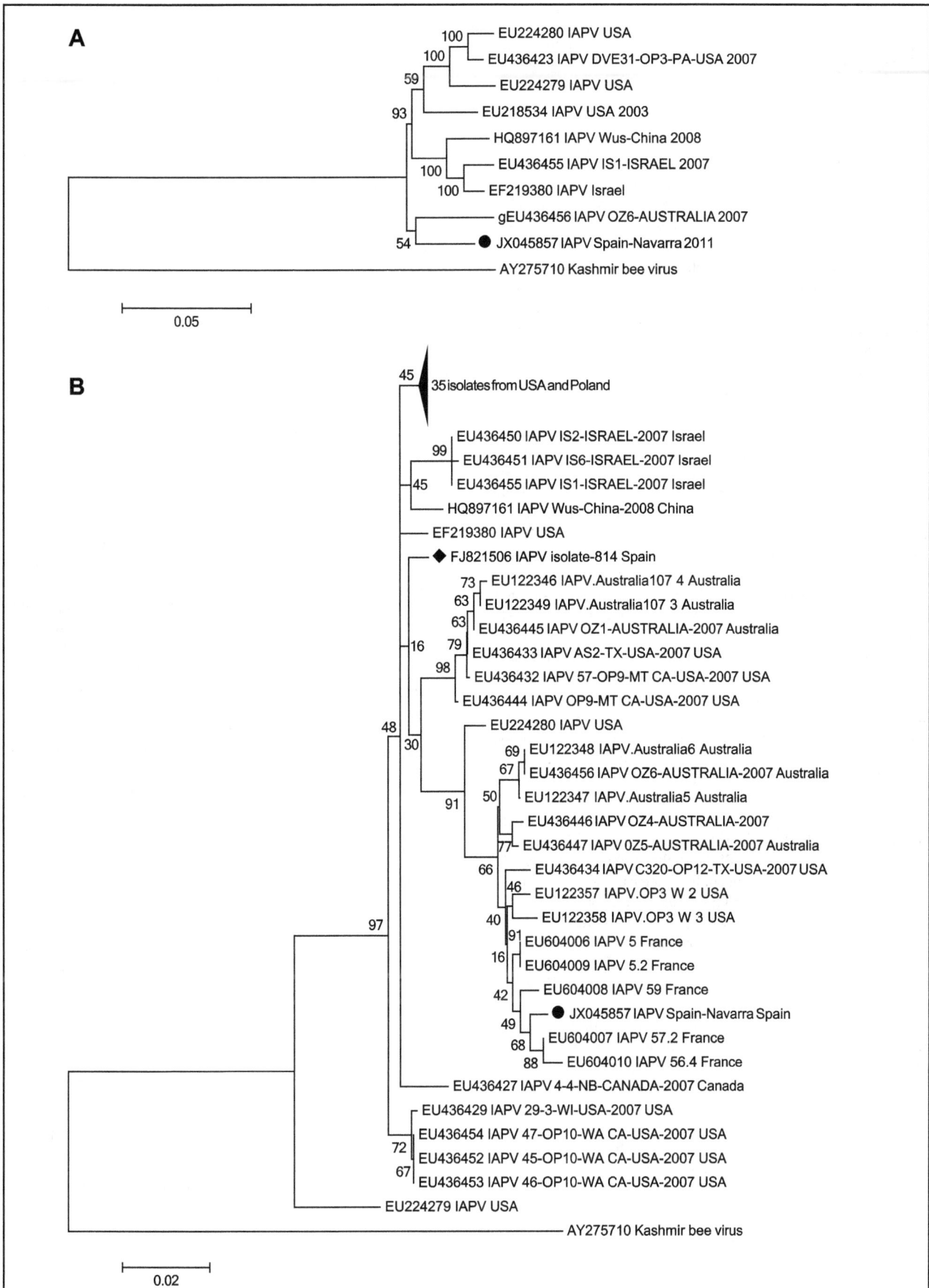

Figure 2. Phylogenetic relationships between strains of IAPV. The trees were based on (a) near full-length genomes, and (b) a 700 nt region (the upper part, containing genetically similar strains from USA and Poland, has been collapsed for clarity). The analyses were conducted in MEGA 5 using Clustal W alignment and the Maximum Likelihood method (bootstrap resampling 1.000 times). The Kashmir bee virus was used as outgroup. The IAPV identified in this study is denoted by (●) and the strain previously identified in Spain with (◆).

samples from multiple geographic locations collected during the course of several months [8]. The frequent detection made the authors conclude that the virus is not just passively being transported together with forage (nectar and pollen) from flowers shared with other insects. However, they could not associate infection by the Brookings strain to any specific symptoms and called for further studies to determine whether the virus is commensal, using honeybees as vector, or a pathogen.

We here report the first finding and description of an ALPV-like virus in honeybees from Europe. Even though the pathogenicity of ALPV in honeybees thus remain unknown [52], the high sequence similarity to the Brookings strain makes it reasonable to assume that the viruses affect their hosts in a similar manner. Given the frequent occurrence of the Brookings strain in the United States and the discovery of an almost identical virus in Europe, further studies of these ALPV-like viruses are highly motivated. A first step would be to conduct further prevalence studies, both in honeybees as well as related pollinators, to determine distribution and host range, as well as assessing the transmissibility and pathogenicity.

In agreement with the initial correlation of IAPV with CCD [4], it has been demonstrated that specific treatment against IAPV can improve the health of affected colonies [53]. However, causative relationship between IAPV presence and CCD has not been established yet and IAPV has also been reported to exist in many hives with no symptoms of CCD around the world [46,54], including Spain [55]. The variant of the virus detected in this study shared most similarity with a strain identified in France during an attempt to correlate IAPV with increased colony mortality [46]. Furthermore, the French strain was phylogenetically demonstrated to belong to a sub-lineage comprised of IAPV isolates from apparently healthy bees. In our extended analysis, a similar division was observed, indicating that this IAPV variant might belong to a group of viruses unable to cause overt infections in affected bees and hence, unable to cause CCD alone.

For its part, Lake Sinai virus has only been described in the USA, and this is the first time that it is described in honeybees from Europe. Related to CBPV and the *Nodaviridae* family, the pathogenic implication and the epidemiological relevance remains unknown, despite the fact that LSV could be one of the honeybee viruses previously described by serology or electron microscopy for which no molecular information is available (Bee virus X and Y, Arkansas bee virus and Berkeley bee virus) [8]. More studies should deepen the knowledge and understanding of this honeybee virus.

The finding of TuRSV through the metagenomic approach indicates the presence of this plant virus in the bee samples, but does not provide sufficient information yet on the reason why this plant virus is present in the homogenised bee organs. It is likely that the bees are passively carrying the plant viruses, e.g., in infected pollen particles attached to their body, and it does not seem to be likely that active plant virus replication would occur in the bodies of the bees. However, further investigations are surely required to clarify exactly the infection biology of this scenario. Even these initial results clearly indicate that the honeybees are

potentially serving as important vector organisms for transmitting infections from plant to plant.

Given the results, it would have been of interest to perform a follow-up of the sanitary status of the studied colony. Unfortunately, this could not be specifically performed as it was a commercial hive and beekeepers often use new brood frames from healthy colonies to avoid the death of the weak colony. However, the apiary did not present any other problem of depopulation and weakening of colonies in the following years, indicating an improvement of the sanitary status of the colonies.

Taken together, these results serve to illustrate the new possibilities offered by metagenomic analysis, including general amplification, high-throughput sequencing and bioinformatic analysis, to allow investigation of whole virome or microbiome in search of unexpected and previously unknown etiological agents. Our investigation of a colony with CCD-like symptoms revealed that it was not only co-infected by two dicistroviruses and by one unclassified virus, but it also harboured a plant virus. This finding can be compared with standard diagnostic methods, i.e. RT-PCR assays. However, due to their principles and limitations, the PCR methods were able to ascertain exclusively the presence of IAPV. The extensive information obtain by a metagenomic approach is thus of great value when trying to understand multifactorial diseases, such as CCD. Considering the importance of honey bees in agriculture, it is also noteworthy that we could further establish the role of honeybees as vectors of pollen-borne viruses of plants by the detection and identification of the TuRSV. Although this initial study managed to identify both a novel subtype of bee-infecting ALPV and LSV not previously reported in Europe, more studies are called for in order to establish the normal genetic and microbiomic background in bees from different geographic regions. Such information would enable comparison between metagenomic profiles from apparent healthy bees and from diseased animals, in order to provide further novel insights into the complex diseases of the honeybee.

Acknowledgments

The authors would like to thank Coordination of Agricultural and Livestock Organizations (COAG), the official veterinary services of Navarre and Navarre beekeepers for the collection of the samples. Acknowledgment is also given to Belen Rivera and Rocio Sanchez, for their technical support, to Ulrika Liljedahl and co-workers at the SNP&SEQ Technology Platform in Uppsala, for performing the GS FLX Titanium sequencing, and to Martin Norling of the SLU Global Bioinformatics Group, for assistance with bioinformatics software applications. Special recognition is given to the Ministry of Education of Spain for the FPU grant of Marina Vicente-Rubiano and Consuelo Rubio-Guerri.

Author Contributions

Conceived and designed the experiments: FG SB JMSV. Performed the experiments: FG MVR CRG. Analyzed the data: FG OEK. Contributed reagents/materials/analysis tools: MVR SB JMSV. Wrote the paper: FG MVR CRG DK SB JMSV.

References

1. van Engelsdorp D, Hayes J, Jr., Underwood RM, Pettis J (2008) A survey of honey bee colony losses in the U.S., fall 2007 to spring 2008. PLoS One 3: e4071.
2. National Research Council (NRC) (2006) Status of Pollinators in North America. Washington, DC: National Academy of Sciencies.
3. Gallai N, Salles J-M, Settele J, Vaissière BE (2009) Economic valuation of the vulnerability of world agriculture confronted with pollinator decline. Ecological Economics 68: 810–821.
4. Cox-Foster DL, Conlan S, Holmes EC, Palacios G, Evans JD, et al. (2007) A metagenomic survey of microbes in honey bee colony collapse disorder. Science 318: 283–287.
5. Bromenshenk JJ, Henderson CB, Wick CH, Stanford MF, Zulich AW, et al. (2010) Iridovirus and microsporidian linked to honey bee colony decline. PLoS One 5: e13181.
6. Chen YP, Siede R (2007) Honey bee viruses. Adv Virus Res 70: 33–80.
7. Genersch E, Aubert M (2010) Emerging and re-emerging viruses of the honey bee (Apis mellifera L.). Vet Res 41: 54.
8. Runckel C, Flenniken ML, Engel JC, Ruby JG, Ganem D, et al. (2011) Temporal analysis of the honey bee microbiome reveals four novel viruses and seasonal prevalence of known viruses, Nosema, and Crithidia. PLoS One 6: e20656.
9. Allen M, Ball B (1996) The incidence and world distribution of honey bee viruses. Bee World 77: 141–162.
10. Ellis JD, Munn PA (2005) The worldwide health status of honey bees. Bee World 86: 88–101.
11. Chen Y, Zhao Y, Hammond J, Hsu HT, Evans J, et al. (2004) Multiple virus infections in the honey bee and genome divergence of honey bee viruses. J Invertebr Pathol 87: 84–93.
12. Berenyi O, Bakonyi T, Derakhshifar I, Koglberger H, Nowotny N (2006) Occurrence of six honeybee viruses in diseased Austrian apiaries. Appl Environ Microbiol 72: 2414–2420.
13. Tentcheva D, Gauthier L, Zappulla N, Dainat B, Cousserans F, et al. (2004) Prevalence and seasonal variations of six bee viruses in Apis mellifera L. and Varroa destructor mite populations in France. Appl Environ Microbiol 70: 7185–7191.
14. Forgach P, Bakonyi T, Tapaszti Z, Nowotny N, Rusvai M (2008) Prevalence of pathogenic bee viruses in Hungarian apiaries: situation before joining the European Union. J Invertebr Pathol 98: 235–238.
15. Di Prisco G, Zhang X, Pennacchio F, Caprio E, Li J, et al. (2011) Dynamics of Persistent and Acute Deformed Wing Virus Infections in Honey Bees, Apis mellifera. Viruses 3: 2425–2441.
16. Dall DJ (1985) Inapparent infection of honey bee pupae by Kashmir and sacbrood bee viruses in Australia. The Annals of applied biology 106: 461–468.
17. Anderson DL, Gibbs AJ (1988) Inapparent Virus Infections and their Interactions in Pupae of the Honey Bee (Apis mellifera Linnaeus) in Australia. Journal of General Virology 69: 1617–1625.
18. Bailey L, Ball BV, Perry JN (1983) Association of viruses with two protozoal pathogens of the honey bee. Annals of Applied Biology 103: 13–20.
19. Ball BV, Allen MF (1988) The prevalence of pathogens in honey bee (Apis mellifera) colonies infested with the parasitic mite Varroa jacobsoni. Annals of Applied Biology 113: 237–244.
20. Brodsgaard CJ, Ritter W, Hansen H, Brodsgaard HF (2000) Interactions among Varroa jacobsoni mites, acute paralysis virus, and Paenibacillus larvae larvae and their influence on mortality of larval honeybees in vitro. Apidologie 31: 543–554.
21. Tang P, Chiu C (2010) Metagenomics for the discovery of novel human viruses. Future Microbiol 5: 177–189.
22. Cheval J, Sauvage V, Frangeul L, Dacheux L, Guigon G, et al. (2011) Evaluation of high-throughput sequencing for identifying known and unknown viruses in biological samples. J Clin Microbiol 49: 3268–3275.
23. Nakamura S, Nakaya T, Iida T (2011) Metagenomic analysis of bacterial infections by means of high-throughput DNA sequencing. Exp Biol Med (Maywood) 236: 968–971.
24. Lipkin WI (2010) Microbe hunting. Microbiol Mol Biol Rev 74: 363–377.
25. Blomström AL, Belák S, Fossum C, Fuxler L, Wallgren P, et al. (2010) Studies of porcine circovirus type 2, porcine boca-like virus and torque teno virus indicate the presence of multiple viral infections in postweaning multisystemic wasting syndrome pigs. Virus Res 152: 59–64.
26. Blomström AL, Belák S, Fossum C, McKillen J, Allan G, et al. (2009) Detection of a novel porcine boca-like virus in the background of porcine circovirus type 2 induced postweaning multisystemic wasting syndrome. Virus Res 146: 125–129.
27. Blomström AL, Widén F, Hammer AS, Belák S, Berg M (2010) Detection of a novel astrovirus in brain tissue of mink suffering from shaking mink syndrome by use of viral metagenomics. J Clin Microbiol 48: 4392–4396.
28. Kukielka D, Esperon F, Higes M, Sanchez-Vizcaino JM (2008) A sensitive one-step real-time RT-PCR method for detection of deformed wing virus and black queen cell virus in honeybee Apis mellifera. J Virol Methods 147: 275–281.
29. Kukielka D, Sanchez-Vizcaino JM (2009) One-step real-time quantitative PCR assays for the detection and field study of Sacbrood honeybee and Acute bee paralysis viruses. J Virol Methods 161: 240–246.
30. Palacios G, Hui J, Quan PL, Kalkstein A, Honkavuori KS, et al. (2008) Genetic analysis of Israel acute paralysis virus: distinct clusters are circulating in the United States. J Virol 82: 6209–6217.
31. Ribière M, Triboulot C, Mathieu L, Aurières C, Faucon J-P, et al. (2002) Molecular diagnosis of chronic bee paralysis virus infection. Apidologie 33: 339–351.
32. Allander T, Tammi MT, Eriksson M, Bjerkner A, Tiveljung-Lindell A, et al. (2005) Cloning of a human parvovirus by molecular screening of respiratory tract samples. Proc Natl Acad Sci U S A 102: 12891–12896.
33. Chevreux B, Pfisterer T, Drescher B, Driesel AJ, Muller WE, et al. (2004) Using the miraEST assembler for reliable and automated mRNA transcript assembly and SNP detection in sequenced ESTs. Genome Res 14: 1147–1159.
34. Altschul SF, Gish W, Miller W, Myers EW, Lipman DJ (1990) Basic local alignment search tool. J Mol Biol 215: 403–410.
35. Huson DH, Auch AF, Qi J, Schuster SC (2007) MEGAN analysis of metagenomic data. Genome Res 17: 377–386.
36. Rozen S, Skaletsky H (2000) Primer3 on the WWW for general users and for biologist programmers. Methods Mol Biol 132: 365–386.
37. Tatusova TA, Madden TL (1999) BLAST 2 Sequences, a new tool for comparing protein and nucleotide sequences. FEMS Microbiol Lett 174: 247–250.
38. Tamura K, Peterson D, Peterson N, Stecher G, Nei M, et al. (2011) MEGA5: Molecular Evolutionary Genetics Analysis Using Maximum Likelihood, Evolutionary Distance, and Maximum Parsimony Methods. Molecular Biology and Evolution 28: 2731–2739.
39. Thompson JD, Higgins DG, Gibson TJ (1994) CLUSTAL W: improving the sensitivity of progressive multiple sequence alignment through sequence weighting, position-specific gap penalties and weight matrix choice. Nucleic Acids Res 22: 4673–4680.
40. Felsenstein J (1985) Confidence Limits on Phylogenies: An Approach Using the Bootstrap. Evolution 39: 783–791.
41. de Miranda JR, Cordoni G, Budge G (2010) The Acute bee paralysis virus-Kashmir bee virus-Israeli acute paralysis virus complex. J Invertebr Pathol 103 Suppl 1: S30–47.
42. Moran NA, Hansen AK, Powell JE, Sabree ZL (2012) Distinctive Gut Microbiota of Honey Bees Assessed Using Deep Sampling from Individual Worker Bees. PLoS One 7.
43. Engel P, Martinson VG, Moran NA (2012) Functional diversity within the simple gut microbiota of the honey bee. Proc Natl Acad Sci U S A 109: 11002–11007.
44. Weiss RA (2010) A cautionary tale of virus and disease. BMC Biol 8: 124.
45. Sato E, Furuta RA, Miyazawa T (2010) An endogenous murine leukemia viral genome contaminant in a commercial RT-PCR kit is amplified using standard primers for XMRV. Retrovirology 7: 110.
46. Blanchard P, Schurr F, Celle O, Cougoule N, Drajnudel P, et al. (2008) First detection of Israeli acute paralysis virus (IAPV) in France, a dicistrovirus affecting honeybees (Apis mellifera). J Invertebr Pathol 99: 348–350.
47. Kukielka D, Sanchez-Vizcaino JM (2010) Short communication. First detection of Israeli Acute Paralysis Virus (IAPV) in Spanish honeybees. Spanish Journal of Agricultural Research 8(2): 308–311.
48. Segundo E, Martín-Bretones G, Ruiz L, Velasco L, Janssen D, et al. (2003) First Report of Turnip mosaic virus in Pisum sativum in Spain. Plant Disease 87: 103–103.
49. Ward L, Waite R, Boonham N, Fisher T, Pescod K, et al. (2007) First detection of Kashmir bee virus in the UK using real-time PCR. Apidologie 38: 181–190.
50. Anderson DL (1991) Kashmir bee virus: a relatively harmless virus of honey bee colonies. American Bee Journal 131 767–770.
51. Singh R, Levitt AL, Rajotte EG, Holmes EC, Ostiguy N, et al. (2010) RNA viruses in hymenopteran pollinators: evidence of inter-Taxa virus transmission via pollen and potential impact on non-Apis hymenopteran species. PLoS One 5: e14357.
52. Van Munster M, Dullemans AM, Verbeek M, Van Den Heuvel JF, Clerivet A, et al. (2002) Sequence analysis and genomic organization of Aphid lethal paralysis virus: a new member of the family Dicistroviridae. J Gen Virol 83: 3131–3138.
53. Hunter W, Ellis J, Vanengelsdorp D, Hayes J, Westervelt D, et al. (2010) Large-scale field application of RNAi technology reducing Israeli acute paralysis virus disease in honey bees (Apis mellifera, Hymenoptera: Apidae). PLoS Pathog 6: e1001160.
54. Reynaldi FJ, Sguazza GH, Tizzano MA, Fuentealba N, Galosi CM, et al. (2011) First report of Israeli acute paralysis virus in asymptomatic hives of Argentina. Rev Argent Microbiol 43: 84–86.
55. Garrido-Bailon E, Martin-Hernandez R, Bernal J, Bernal JL, Martinez-Salvador A, et al. (2010) Short communication. The detection of Israeli Acute Paralysis virus (IAPV), fipronil and imidacloprid in professional apiaries are not related with massive honey bee colony loss in Spain. Spanish Journal of Agricultural Research 8: 658–661.

8

Action versus Result-Oriented Schemes in a Grassland Agroecosystem: A Dynamic Modelling Approach

Rodolphe Sabatier[1,2]*, **Luc Doyen[3]**, **Muriel Tichit[1,2]**

1 INRA, UMR 1048 SADAPT, Paris, France, **2** AgroParisTech, UMR 1048 SADAPT, Paris, France, **3** CNRS, UMR 7204 CERSP, MNHN, Paris, France

Abstract

Effects of agri-environment schemes (AES) on biodiversity remain controversial. While most AES are action-oriented, result-oriented and habitat-oriented schemes have recently been proposed as a solution to improve AES efficiency. The objective of this study was to compare action-oriented, habitat-oriented and result-oriented schemes in terms of ecological and productive performance as well as in terms of management flexibility. We developed a dynamic modelling approach based on the viable control framework to carry out a long term assessment of the three schemes in a grassland agroecosystem. The model explicitly links grazed grassland dynamics to bird population dynamics. It is applied to lapwing conservation in wet grasslands in France. We ran the model to assess the three AES scenarios. The model revealed the grazing strategies respecting ecological and productive constraints specific to each scheme. Grazing strategies were assessed by both their ecological and productive performance. The viable control approach made it possible to obtain the whole set of viable grazing strategies and therefore to quantify the management flexibility of the grassland agroecosystem. Our results showed that habitat and result-oriented scenarios led to much higher ecological performance than the action-oriented one. Differences in both ecological and productive performance between the habitat and result-oriented scenarios were limited. Flexibility of the grassland agroecosystem in the result-oriented scenario was much higher than in that of habitat-oriented scenario. Our model confirms the higher flexibility as well as the better ecological and productive performance of result-oriented schemes. A larger use of result-oriented schemes in conservation may also allow farmers to adapt their management to local conditions and to climatic variations.

Editor: Raphaël Arlettaz, University of Bern, Switzerland

Funding: This work was carried out with the financial support of the "ANR - Agence Nationale de la Recherche - The French National Research Agency" under the "SYSTERRA program - Ecosystems and Sustainable Development," project "ANR-08-STRA-007, FARMBIRD - Coviability models of FARMing and BIRD biodiversity." The funders had no role in study design, data collection and analysis, decision to publish, or preparation of the manuscript.

Competing Interests: The authors have declared that no competing interests exist.

* E-mail: rodolphe.sabatier@forst.uni-goettingen.de

Introduction

After 15 years of implementation, the effectiveness of agri-environment schemes (AES) is still under debate [1]. Result-oriented AES have been proposed to improve the efficiency of conservation policies [2]. They rely on payment for effective biodiversity conservation (e.g. abundance, richness) independently from the management practices implemented by farmers. Such AES have been studied in the case of carnivores [3], grassland flora [4] or grassland birds [5]. If quite a large number of result-oriented schemes already exist, most of them are either experimental or have been run for too short a term and on too small a scale to be properly evaluated [2]. This situation could explain why few comparisons between result-oriented and action-oriented schemes are available and why no clear difference has been found in their effects on population sizes [6].

One of the main advantages of result-oriented schemes is to allow farmers to develop innovative management practices that would be efficient on both productive and ecological performance. By relaxing constraints on management, these schemes make it possible to implement a wider set of management strategies (i.e. sequences of management practices over time). Widening the range of management strategies may offer two advantages. First, out of the new management strategies some of them may be more

efficient either on the ecological or productive performance without decreasing performance on the other dimension. Second, it may give more flexibility to the farming system [2]. Due to the difficulties of implementing and monitoring result-oriented schemes, a third kind of scheme has been created. These schemes aim at producing suitable habitat for biodiversity [7]. Their evaluation is based on indicators of habitat quality and not directly on biodiversity levels [8]. Hereafter, we will call these schemes habitat-oriented schemes. By providing suitable habitats for target species, such schemes are expected to lead to better ecological performance than action-oriented ones. However, the potential of innovation may be limited by the constraints applied on the habitat instead of on biodiversity levels. For example, result-oriented schemes allow inter-annual variability and strategies with successions of ecology-oriented and production-oriented years may appear. Moreover, these schemes may not systematically ensure good ecological performance whereas result-oriented ones should always lead, by definition, to good levels of biodiversity.

The objective of this study was to compare three scenarios corresponding to the different kinds of agri-environment schemes: action-oriented, habitat-oriented and result-oriented schemes. We first assess their differences in productive and ecological performance. A scenario will lead to better performance if it performs better in one dimension without performing worse in the other.

Secondly, we explore the management flexibility linked with each scenario. A scenario will have a higher flexibility if it allows more management option than another. Finally, we illustrate the importance of management flexibility in the face of climate shock. The overall comparison of the three scenarios is based on two hypotheses:

Hypothesis 1: For a given result-oriented scenario, there is no habitat-oriented one that leads to better performances and for a given habitat-oriented scenario, there is no action oriented one that lead to better performances.

Hypothesis 2: For a given result-oriented scenario, there is no habitat oriented one that leads to a higher flexibility.

Formal definitions of these two hypotheses will be given in the core of the text.

As a case study, we focused on the conservation of lapwings *Vanellus vanellus* in wet grasslands of the French Atlantic coast (46°22′N, 1°25′W). Due to their high position in trophic networks and their close connection with wet grasslands, wader species give good information about the health of the ecosystem. The lapwing life cycle is deeply linked to the management of grassland [9] and lapwing was one of the first species to benefit from result-oriented schemes [5,6]. Wet grasslands were the first habitats targeted by agri-environment schemes in France during the early 90's and the conservation of lapwings in these agroecosystem has long been of major concern. To compare different AES in their ability to ensure productive and ecological performance in the long term, we developed a dynamic model linking grazed grassland dynamics and lapwing population dynamics. This model focuses on the effect of AES and is thus limited to the impact of farming practices on bird dynamics. The model is built under the viable control approach [10] which is closely related to the viability theory [11]. This framework enables the satisfaction of production, socio-economic and environmental constraints and is, in this respect, a multi-criteria approach. It makes it possible to find the whole set of viable management strategies that keep a system within some constraints. As it focuses on a set of management strategies and not on a single optimal one, it is of high interest to study management flexibility, i.e. the system ability to adapt to internal or external changes. Viability analysis has been applied to biodiversity management [12], and the sustainability of agricultural systems [13,14].

Methods

Model overview

In line with the model of Sabatier, Doyen & Tichit [15], our model relies on a state-control approach that represents a grassland agroecosystem which is the breeding habitat of a bird species, the lapwing, and the feeding resource for domestic cattle. It is a discrete time model linking grazed grass dynamics to bird population dynamics (Fig. 1). Time step is defined on a monthly basis, which is coherent with farmers' management as most farmers implement middle term grazing sequences (three weeks to several months). In the grazed grass sub-model, biomass is harvested through grazing. The biomass represents a single grassland patch homogeneously managed without any spatial dimension. The grassland patch is one of the feeding resources available for cattle. We assumed that when cattle do not graze the grassland patch, they are fed elsewhere with other resources (either on temporary grasslands or indoor). The bird sub-model simulates population changes over time in response to the direct and indirect

effects of grazing on bird life traits. Even if other factors than grazing may also play a role e.g. field wetness or predation, grazing indisputably remains a major factor driving the life cycle of waders (review in [9]). We therefore focus on the effects of grazing on wader dynamics. Grazing intensity has a direct effect on clutch size through nest trampling by cattle [16]. Grass height (i.e. habitat quality), generated by grazing is a key factor for foraging [17] and impacts juvenile survival. Grass height is also an important predictor of habitat nest selection [9]; however, in the absence of spatial dimension in our model, we did not model this process. The model computes two indicators summarizing the ecological and productive performance of each grazing strategy.

We studied the co-viability of the grassland agroecosystem in three scenarios (action-oriented, habitat-oriented and result-oriented scenario) by looking for viable management strategies that satisfy both ecological and productive constraints. The type of ecological constraints applied to the system differs from one scenario to the other and reflects their specificities. In the action-oriented scenario, constraints correspond to thresholds on minimal and maximal stocking density during the nesting period. Such management requirements aim at limiting the effects of trampling, while ensuring a minimum level of grazing so as to reduce grass

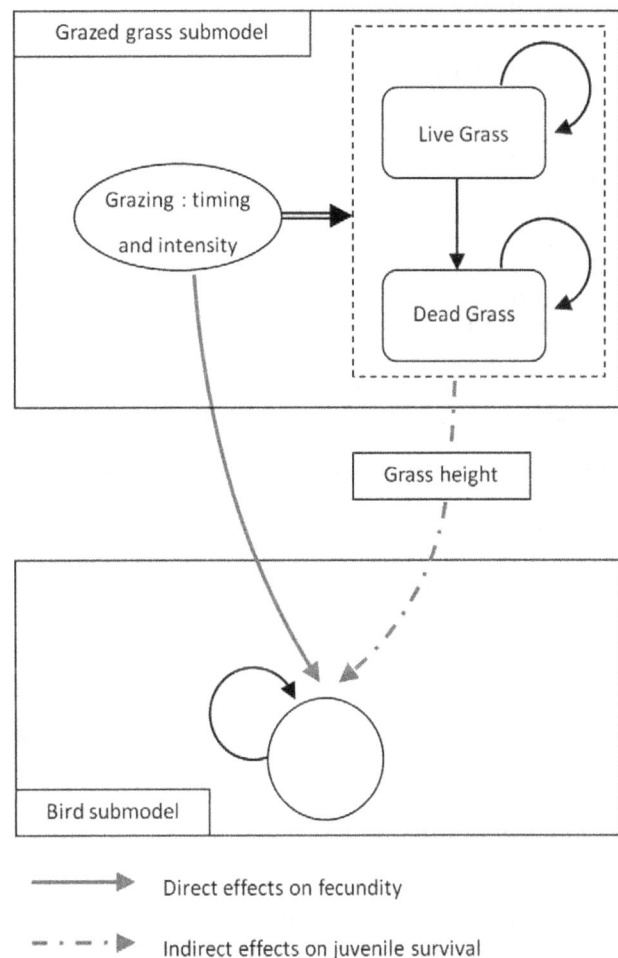

Figure 1. Conceptual model of the direct and indirect effects of grazing on bird population dynamics. Dynamics of grass biomass (black arrows) is controlled through timing and intensity of grazing; double arrow represents cattle consumption of standing live and dead biomass. The bird model is a single stage matrix model.

height, heading toward better habitats. Habitat-oriented scenario combines thresholds on bird fecundity and grass heights during the chick rearing period. It ensures both fecundity and juvenile survival to be maintained to a high level. In the result-oriented scenario, constraints correspond to a threshold on minimal bird population size throughout time. Management is free and any management strategy maintaining the bird population through time is considered as viable. Productive constraints do not differ between the three scenarios. For each scenario, the model computed the viable grazing strategies meeting the constraints. The number of viable management strategies is used as an indicator of the flexibility of the system. Due to their extremely high number, viable strategies could not be counted directly and the number of states reached by the viable strategies (or size of the viability tube) was used as an index of the system flexibility.

Grazed grass state and dynamics

The first state of the system represents a grass biomass vector $B(t)$ considered monthly and partitioned into live and standing dead grass $(B_L(t), B_D(t))$ both expressed in organic matter (g OM ha^{-1}). Grass dynamics is controlled by the timing and intensity of grazing $u(t)$, expressed in livestock unit per ha (LU ha^{-1}). The dynamics of the grass biomass $B(t)$ controlled by grazing intensity $u(t)$ is summarized as follows:

$$B(t+1) = A(t,B(t)B(t)) - G(u(t),B(t)) \quad for \ t = 0, 1, \ldots, T \quad (1)$$

where matrix A is a time dependent matrix that encompasses the transition rates from t to $t+1$. It includes grass growth, senescence and decay rates that are time dependent on a monthly basis. G is a vector representing biomass harvest through grazing. The state of the biomass is linked to grass height through a linear function $h(B)$. Databases from the Ouest-du-Lay marsh were used to parameterize the grazed grass dynamics ([15]; appendix S3). For further details on biomass dynamics and parameter values, see Appendix S1 and Table S1.

Bird state and dynamics

The second state of the system describes the lapwing life cycle. By contrast with Sabatier, Doyen & Tichit [15], the bird model is deterministic and represents the female portion of a single class population. During the nesting period, cattle trampling impacts clutch size and during the chick rearing period grass height is a variation factor of juvenile survival. Assuming a pre-breeding census, the monthly dynamics of birds $N(t)$ from t to $t+1$ reads as follows:

$$N(t+1) = M(t,u(t),B(t),N(t))N(t) \quad (2)$$

where $N(t)$ is the population size and $M(t,u,B,N)$ the population growth function:

$$M(t,u,B,N) = 1 \qquad \qquad if \ t \neq t* \quad (3)$$

$$M(t,u,B,N) = s_2 + \alpha.f(u).\sigma.s_1(h(F(B,u)))/(1+c.N) \quad if \ t = t* \quad (4)$$

with

$$F(B,u) = A(t*,B)B - G(u,B) \quad (5)$$

where $t*$ is the nesting month, s_2 the annual adult survival, α the proportion of breeding females, $f(u)$ the clutch size depending on

cattle density $u(t)$, σ the primary sex ratio and $s_1(h(F(B,u)))$ the chick survival that depends on grass height $h(B)$ at time $t*+1$. Grass height depends on grass biomass $B(t*+1)$ and therefore on $F(B(t*),u(t*))$. We consider that breeding success is affected by an intra-specific competition. We use a Beverton-Holt-like density dependence function to model this competition in which c measures the strength of competition. A full description of the bird model along with parameter values are given in Appendix S2 and Table S2.

Viability constraints

Three types of viability constraints formalize the multiple roles played by the grazed grassland. Constraints applied to the three scenarios are listed in Table 1.

Cattle feeding requirement constraint. Given a monthly biomass demand per livestock unit q, the feeding requirement constraint is defined as follows:

$$(\mathbf{a}) \quad qu(t) \leq B*(t) \qquad for \ t = 0,1,..,T. \quad (6)$$

This feeding requirement constraint limits stocking density which cannot exceed the available biomass $B*(t)$. It assumes that cattle cannot graze below a minimal biomass threshold and situations where insufficient grass availability could lead to a poorer body condition of livestock are not considered.

Productive constraint. A second constraint defines a minimal level of productive performance necessary for the farmer. Productive performance $P(u,T)$ corresponds to the number of grazing days (simplifying to 30 days per month) associated with a grazing strategy $u = [u(0),...,u(T)]$. The model does not incorporate explicitly any spatial scale but the quantification of the productive performance is given for one hectare. The productive constraint corresponds to a lower threshold on the number of grazing days on the whole time period studied. It does not imply any minimum time period or upper threshold for grazing. It reads as follows:

$$(\mathbf{b}) \quad P(u,T) = \sum_{t=0}^{T} 30 \cdot u(t) \geq P^b \quad (7)$$

where P^b is the threshold of minimal productive performance. Its value was defined by the 10% lower quantile of a dataset of 344 real grazing strategies recorded on our study site [18].

Ecological constraints. Ecological constraints are defined in three different ways so as to capture the three kinds of scenarios.

In the action-oriented scenario (AO), the model includes two ecological constraints. The first one is related to trampling mechanisms. An upper threshold $u^{\#}$ is imposed on cattle density during the nesting month $t*$:

$$(\mathbf{c}) \quad u(t*) \leq u^{\#} \quad (8)$$

The second constraint is related to grass height during the first month following chick birth. It is represented by a lower threshold on cattle density during nesting month to induce a minimum level of grazing:

$$(\mathbf{d}) \quad u(t*) \geq u^b \quad (9)$$

Table 1. Constraint sets of the three scenarios.

		Scenarios	
Constraints	Action-oriented	Habitat-oriented	Result-oriented
Productive performance $P(u,T) > P^b$	X	X	X
Cattle feeding requirements $q.u(t) < B^*(t)$	X	X	X
Trampling $u(t) \leq u^\#$	X		
Fecundity $f(t) \leq f^b$		X	
Grazing $u(t) \geq u^b$	X		
Habitat quality $h^b \leq h(t) \leq h^\#$		X	
Population size $N(t) \geq N^b$			X

Productive performance constraint imposes that productive performance $P(u,T)$ stays over a minimal productive performance P^b (the minimal annual number of grazing days per hectare associated with a grazing strategy u). Cattle feeding requirement constraint imposes that cattle demand $q.u(t)$ is always lower than the available biomass $B^*(t)$. Cattle density constraint is an upper threshold $u^\#$ on cattle density $u(t)$ during the nesting month. A habitat quality constraint imposes grass height to remain within a minimal h^b and maximal $h^\#$ grass heights during chick rearing. Population size constraint imposes that populations size $N(t)$ stays over a minimum population size N^b throughout time.

In the habitat-oriented scenario (HO), the model still includes two ecological constraints. The first one is related to clutch size in relation with trampling mechanisms. During the nesting month t*, a lower threshold, $f^\#$ is imposed on clutch size $f(u)$. As f is a decreasing function this constraint is similar to *eqn 8*:

$$(\mathbf{e}) \quad f(u(t*)) \geq f^b \qquad (10)$$

In addition to the previous constraint (eqn 10), the model also includes a constraint on habitat quality. It is imposed on grass height during the first month following chick birth (t^*+1) in order to ensure suitable habitat for chicks. It is bounded by minimal and maximal grass heights as follows:

$$(\mathbf{f}) \quad h^b \leq h(t*+1) \leq h^\# \qquad (11)$$

In the result-oriented scenario (RO) the model involves a single ecological constraint that imposes a minimum population size N^b throughout time:

$$(\mathbf{g}) \quad N(t) \geq N^b \quad for \ t = 12, \ldots, T \qquad (12)$$

In the action-oriented scenario, the ecological constraints bound the control variable. In habitat-oriented scenario, ecological constraints combine both control and state constraints. It still limits cattle density to ensure a good clutch size and also focuses on an intermediate management objective linked with grass height to achieve a good juvenile survival. In the result-oriented scenario, no constraint is set either on cattle density or grass height and the only ecological constraint corresponds to a state constraint on the management goal which is the maintenance of the bird population size above a minimal threshold at any time step. Using such a state constraint relaxes all management restriction on farmer's decision.

Co-viability analysis

The viability framework is used to identify combinations of biomass $B(.)$, population size $N(.)$ and cattle density $u(.)$ that satisfy viability constraints throughout time. It relies on the computation of the so called viability kernel [11]. In the present case, this viability

kernel depends on time and we prefer to speak of a viability corridor *Viab(t)*. In this section we will refer to three concepts: the viability corridor, the viable grazing strategy and the viability tube.

Viability corridor. The viability corridor *Viab (t)* is the set of grass biomass conditions and bird population sizes (states, $B_L(t)$, $B_D(t)$ and $N(t)$ from which at least one grazing strategy is viable. At $t = t_0$, the Viability corridor *Viab(t_0)* is thus defined differently in each scenario.

In the action-oriented scenario (AO), it is defined as follows:

$Viab_{AO}(t_0) = \{(B(t_0), N(t_0)) | there \ exists \ grazing \ u(t) \ and$

$a \ sequence \ of \ states \ (B(t), N(t)) starting \ from$

$(B(t_0), N(t_0)) satisfying \ constraint \ (a) \ for \ any \ time \qquad (13)$

$t = t_0, \ldots, T, satisfying \ constraints \ (c) \ and \ (d) \ for \ any \ time$

$t > 12 \ and \ satisfying \ constraints \ (b) \ at \ time \ T\}$

In the habitat-oriented scenario (HO), it is defined as follows:

$Viab_{HO}(t_0) = \{(B(t_0), N(t_0)) | there \ exists \ grazing \ u(t) \ and$

$a \ sequence \ of \ states \ (B(t), N(t)) starting \ from \ (B(t_0), N(t_0))$

$satisfying \ constraint \ (a) \ for \ any \ time \ t = t_0, \ldots, T, \qquad (14)$

$satisfying \ constraints \ (e) \ and \ (f) \ for \ any \ time \ t > 12$

$and \ satisfying \ constraints \ (b) \ at \ time \ T\}$

In the result-oriented scenario (RO), it is defined as follows:

$Viab_{RO}(t_0) = \{(B(t_0), N(t_0)) | there \ exists \ grazing \ u(t) \ and$

$a \ sequence \ of \ states \ (B(t), N(t)) starting \ from (B(t_0), N(t_0))$

$satisfying \ constraints \ (a) \ for \ any \ time \ t = t_0, \ldots, T, \qquad (15)$

$satisfying \ constraint \ (g) \ for \ any \ time \ t > 12 \ and$

$satisfying \ constraints \ (b) \ at \ time \ T\}$

For the three scenarios, constraints (c) to (g) were not taken into account the first year (t<12) so as to enable a transition of the grazed system toward AES. This choice reflects a conventionally driven system in which AES would be introduced at the end of the first year.

Viable grazing strategies. Once the viability corridor has been found, we compute the viable grazing strategies that verify the different constraints over the period of time involved. Such U exist as long as the state $(B(t), N(t))$ lies within the viability corridor $Viab(t)$. We thus consider the set of the viable grazing strategy at time t for a given viable state $(B(t), N(t))$. A viable grazing strategy is a temporal sequence of grazing intensities that keeps the whole system within the constraint set. To each viable grazing strategy corresponds a viable state trajectory defined in terms of grass biomass and population size. These viable grazing strategies U (t, B, N) are defined through a dynamic programming structure.

In the action-oriented scenario (AO), it is defined as follows:

$$U_{AO}(t,B,N)=\left\{u(t)\Big|\begin{array}{l} \text{u(t) satisfies (a), (b), (c) and (d)}\\ (B(t+1),N(t+1))\in Viab_{AO}(t+1)\end{array}\right\}\quad(16)$$

In the habitat-oriented scenario (HO), it is defined as follows:

$$U_{HO}(t,B,N)=\left\{u(t)\Big|\begin{array}{l} u(t)\ satisfies\ (a),\ (b),\ (e)\ and\ (f)\\ (B(t+1),N(t+1))\in Viab_{HO}(t+1)\end{array}\right\}\quad(17)$$

In the result-oriented scenario (RO), it is defined as follows:

$$U_{RO}(t,B,N)=\left\{u\Big|\begin{array}{l} u(t)\ satisfies\ (a),\ (b)\ and\ (g)\\ (B(t+1),N(t+1))\in Viab_{RO}(t+1)\end{array}\right\}\quad(18)$$

Viability tube. Finally, we identify the Viability tube VT (t). It is the temporal succession of biomass conditions that are reachable by viable grazing strategies. It takes into account the fact that not every viable state can be reached by a viable grazing strategy. Some states are viable (i.e. starting from them, there is at least one viable grazing strategy) but they can only be reached by grazing strategies that are not viable. The viability tube is defined as follows:

$$VT(0)=Viab(0)\quad(19)$$

$$VT(t+1)=\left\{(B,N)\left|\begin{array}{l}\exists(\tilde{B},\tilde{N},\tilde{u})\\ \tilde{u}\in U(t,\tilde{B},\tilde{N})\\ (\tilde{B}(t+1),\tilde{N}(t+1))=(B,N)\end{array}\right.\begin{array}{l}(\tilde{B},\tilde{N})\in Viab(t)\\ \\ \end{array}\right\}\quad(20)$$

As they differ among scenarios, we distinguished VT_{HO}, VT_{RO} and VT_{AO}. We characterized the Viability tubes by their volumes $\Theta(VT)$.

$$\Theta(VT)=\sum_{t=t_0}^{T}\iint\limits_{VT(t)}dB_D dB_L\quad for\ a\ given\ N(t_0)\quad(21)$$

$\Theta(VT)$ (expressed in $g^2.s.ha^{-2}$) is a viability metric and an indicator of the quantity of viable state trajectories. Our system includes three state dimensions (B_L, B_D and N). So as to be able to

plot the viability tubes, we limited the tubes to two states (B_L and B_D). The tubes therefore corresponded to projections of the 4 dimensional tubes on the three dimensional spaces defined by B_L, B_D and t for a given initial abundance $N(t_0)$.

Hypotheses

The two hypotheses can be formalized as follows:

Hypothesis 1: For a given result-oriented scenario, there is no habitat-oriented one that leads to better performances and for a given habitat-oriented scenario, there is no action oriented one that lead to better performances.

$$\left\{\begin{array}{c} E(N_{AO}(T))<E(N_{HO}(T))<E(N_{RO}(T))\\ and\\ E(P_{AO}(T))<E(P_{HO}(T))<E(P_{RO}(T))\end{array}\right.\quad(22)$$

whith $E(N(T))$ the average value of $N(T)$ over a set of 10 000 random viable grazing strategies.

$$E[N(T)]=\frac{1}{10000}\sum_{i=1}^{10000}N_i(T)\quad(23)$$

where $(N_i(T),B_i(T))\in VT(T)$

Similarly, $E(P(T))$ is the average value of $P(T)$ over the same set of 10 000 random viable grazing strategies.

Hypothesis 2: For a given result-oriented scenario, there is no habitat oriented one that leads to a higher flexibility.

$$\Theta(VT_{HO})<\Theta(VT_{RO})\quad(24)$$

The volume of the viability tube is used as an index of flexibility. A scenario leading to a bigger viability tube will allow more management strategies, and is considered being more flexible.

Simulations

To test hypotheses 1 and 2, we followed a two step approach. First we tested them for a given set of ecological constraints and initial conditions ($u^b = 0.5$; $u^{\#} = 2$; $f^b = 2.5$; $h^b = 0$; $h^{\#} = 14$; $N^b = 30$; $N(t_0) = 30$). Then we performed a sensitivity analysis to verify the generality of our results under a wider range of ecological constraints and initial conditions ($u^b = [0, 0.5, 1, 1.5, 2]$; $u^{\#} = [1, 1.5, 2, 3, 4, 5]$; $f^b = [3.2, 2.5, 1.9, 1.5, 1.1]$; $h^b = [0, 5, 7, 10, 12, 13, 14]$; $h^{\#} = [10, 12, 14, 17, 20, 30]$; $N(t_0) = [25, 30, 35]$). Constraint values were chosen to explore the range of possible states and controls observed in our study area on lapwing nesting fields ($0 \le h \le 30$ and $0 \le u \le 5$; [18]). As $f(u)$ is a monotonous function of u, values of f^b were thus chosen to correspond to the different thresholds on $u^{\#}$.

A dynamic programming algorithm [10] was used to identify viable initial conditions $(B(t_0), N(t_0))$, viable grazing strategy $U(t,B,N)$, grass state trajectories $B(t)$ and bird population state trajectories $N(t)$ respecting the different constraints at each time step over a period of $T = 96$ months. The numerical computations were performed with Scilab 4.1.2 software (http://www.scilab.org/; Scilab Consortium 2007). Once viable grazing strategies and state trajectories were found, their ecological and productive performances $N(T)$ and $P(T)$ were assessed. The performance of the agroecosystem under the three scenarios AO, HO and RO was compared with a permutation tests using Python 2.6 (http://www.python.org/) so as to test Hypothesis 1. For a given performance (ecological or productive one) and for a given pair of scenarios, the test calculates a criterion (the difference of the average performances) and compares it to the distribution of this criterion for n = 10000 random permutations within the two sets of trajectories tested. The p value of the test is the proportion of permuted situations for which the criterion is larger (in absolute value) than the criterion of the not permuted situation. More details on permutation tests can, for example, be found in [19]. In order to investigate the advantage of the improved flexibility of the result-oriented scenario in facing climatic variations, we tested the effect on the viability tubes of a shock in climatic conditions represented by an increased grass growth in year 5. Parameters of matrix A were modified so as to simulate an earlier grass onset in the season (i.e. one month earlier) and a stronger grass growth (i.e. +25%).

Results

Hypothesis 1: scenarios differ in performance

Fig. 2 shows the ecological and productive performance of a sample of 10 000 grazing strategies for each of the three scenarios. Comparison of both average ecological and productive performance of the three scenarios showed significant differences (permutation test, p-value = 0). However, differences between the habitat and result-oriented scenarios were much lower than differences between the action-oriented scenario and the other two scenarios (Table 2). The result-oriented scenario led to better performances than the habitat-oriented one and the latter scenario led to much better ecological performance than the action-oriented one and slightly better productive ones. However, it should be kept in mind that the habitat and result-oriented scenarios were very similar for both performance criteria.

Hypothesis 2: the result-oriented scenario improves management flexibility

We restricted the comparison of flexibility to the other two scenarios since the action-oriented scenario did not maintain bird populations. The inclusion of the tubes, their shape and their volumes showed that more states and controls were viable in the result-oriented scenario than in the habitat-oriented one. Numerical computations showed that the habitat-oriented tube was included in the result-oriented one:

$$\forall t, VT_{HO}(t) \subset VT_{RO}(t) \qquad (25)$$

The inclusion of the two tubes means that the flexibility of the result-oriented scenario at least as high as the flexibility of the habitat-oriented one. For these two scenarios, ensuring similar levels of performance (Table 2), tubes were bigger in the result-oriented than in the habitat-oriented scenario. Indeed, the calculation of $\Theta(VT)$ showed 1.5 more viable grass states in the result-oriented scenario than in the habitat-oriented one ($\Theta(VT_{RO}) = 6842$ versus $\Theta(VT_{HO}) = 4997$ g^2.s.ha^{-2}). A larger range of grass biomass conditions was thus available for farmers throughout time. The shape of the *Viability tube* for both habitat and result-oriented scenarios illustrates the couples of possible viable states (B_L, B_D) throughout time and the higher flexibility of the result-oriented scheme (Fig. 3.a and 3.b).

These results illustrate that more flexibility was given to the grazing strategies in the result-oriented scenario. We have therefore validated Hypothesis 2. In terms of management this means that the farmer could implement a wider range of grazing strategies in the result-oriented scenario than in the habitat-oriented one (appendix S4). Especially, higher cattle densities can be implemented in spring with the result-oriented scheme.

Sensitivity analysis

Results of the sensitivity analysis are presented in Appendix S5. Sensitivity analysis showed one limit case ($h^{\#} = 30$ cm) for which Hypothesis 1 was falsified. In this situation both performances of the action-oriented scenario were higher than those of the habitat-oriented one. Apart from this case, when scenarios could be ranked, action-oriented scenario always led to worse performances than habitat-oriented one and both action and habitat-oriented scenarios led to worse performances than result-oriented scenario. Hypothesis 1 was therefore acceptable for most constraint values. Whatever the parameter settings, Hypothesis 2 was always true.

Illustrating the importance of flexibility

We examined the interest of the improved flexibility of the result-oriented scenario in facing environmental variations. It turned out that the state of the system still lied within the result-oriented viability tube VT_{RO} despite the disturbance, while it left the habitat-oriented viability tube VT_{HO}. In other words, no couple of control strategy and state trajectory respected all productive and habitat-oriented ecological constraints. Thus it was not possible for the farmer to produce a suitable grass height for birds every year with low trampling while ensuring good productive performance and satisfying cattle feeding requirements. However, the result-oriented tube was not empty and it was possible to find viable state trajectories and control strategies. As illustrated with one simulation (Fig. 4), a viable result-oriented grazing strategy did not respect habitat-oriented constraints every year but it did, however, maintain bird populations throughout time due to inter-annual compensations. In this example, grazing intensity in spring was low in 2009 and 2010 (Fig. 4.a). It implied low levels of trampling and an increase in bird population sizes (Fig. 4.c). In 2011, spring grazing intensity was stronger and bird population decreased but still remained above the population threshold. This result shows how, in the result-oriented scenario, the farmer can adapt his management to climatic shocks by implementing an inter-annual variation of management strategies. Such inter-annual variation in management was not available in the habitat-oriented scenario. This result again emphasized the advantages of the increased flexibility provided by the result-oriented scenario.

Discussion

First, our results showed that in most cases the habitat and result-oriented scenarios led to much better ecological performance than the action-oriented scenario. Productive performance was quite similar among the scenarios. Secondly, our results showed that the result-oriented scenario had a higher flexibility than the habitat-oriented one. This difference in flexibility was even greater when the grazed grassland agroecosystem was exposed to climatic variation.

A modelling approach to compare management schemes

Using a modelling approach gave us the opportunity to compare situations all other things being equal, as we would have done in a controlled trial. We therefore did not include mechanisms such as environmental stochasticity or landscape source/sink mechanisms. These mechanisms are of high importance in the real world but management through grazing has low (if any) impact on them and including them in the model would only have blurred the simulation results. These considerations have to be kept in mind when considering the results. As an example, one of the main differences between the model and reality is the absence of migration. Here, we considered a closed population of birds to assess the effects of management practices. Using population size as an indicator of ecological performances was therefore possible as well as very convenient and illustrative. In the field, such an indicator would raise questions. In the one hand it does not only reflect mechanisms occurring at field scale but in the other hand, this indicator is much closer to the final objective of a conservation policy than a bird productivity index would be.

Our results showed several undetermined situations. They could occur for two reasons: either the three scenarios could not be ranked or it was impossible to find any viable grazing strategies. Changing the values of the constraints oriented the set of viable strategies to either better ecological or better productive performance, illustrating the trade-off between production and conservation in such agroecosystems [15,20]. It could lead to extreme situations with very high performances on one dimension and very low on the other dimension. These situations could not be put in a hierarchy. In other cases, the constraint values tested pushed the system too far and no viable grazing strategy could be found. Consequently, nothing could be said on Hypothesis 1 since no performance could be assessed.

Result-oriented schemes aim at protecting the whole agroecosystem by targeting umbrella species. We could here focus on management strategies that impact the whole agroecosystem and offer advantages to other species with similar ecological requirements and similar sensitivity to management. However, in the field, farmers may implement very specific measures only benefiting the target species. For example, in the result-oriented scheme implemented in the Netherlands, it happened that farmers only build an electric fence around the nest [5]. If this management leads to better hatching success for the target species, it is of minor interest for other species in the agroecosystem. This measure has been strongly criticized for its lack of cost effectiveness [21] and was cancelled in the new scheme. To avoid it, the evaluation of management must be done on an indicator as close to the final objective of conservation as possible. Considering several species [8,22] could be a powerful solution. Best effects are expected with management options having broad effects on the agroecosystem. In this respect, management options at field scale include grazing sequences, amount and timing of fertilization as well as mowing techniques and dates. At upper scale, the proportion of land uses [23,24] as well as their spatial arrangement [25,26,27] could also be efficient management options that would impact the whole agroecosystem.

Improving management flexibility

Multi-criterion analysis mainly looks for optimal performance but do not take into account the issue of flexibility in decision making. Optimality is well adapted to static situations or stable environments but flexibility is of major concern for systems exposed to uncertainties [28]. The viable control approach makes it possible to go beyond the search of optimum and to look for a diversity of management strategies. Although management strategies were quite similar in terms of performance, the number of viable management strategies gave a strong advantage to result-oriented schemes. Greater flexibility of management is one of the major arguments in favour of result-oriented schemes [2]. First, it is expected to improve the resilience of the agroecosystem as farmers may choose alternative management strategies to adapt to inter-annual climatic variability. The agroecosystems we studied are low input, extensively grazed grasslands. Such systems are highly dependent on climatic conditions and flexibility in grassland use is a major component in coping with unexpected events [29]. In comparison with habitat-oriented schemes that impose constraints on habitat and fecundity every year and forces periodic management strategies, the result-oriented schemes allow for inter-annual variability. It gives the possibility of segregating ecological and productive objectives among years (e.g. to adapt grazing strategies to climatic conditions). These new strategies are the basis of the higher flexibility of the result oriented schemes. Our study focussed on temporal flexibility of grazing strategies but we conjecture that in the same way, spatial flexibility would allow farmers to adapt their management to variations in external conditions among several fields. Further development of the model will take these spatial variations into account. A second advantage of this greater flexibility would be to allow farmers to look for innovative management strategies. Our results suggest that loosening the ecological constraints of the agroecosystem gives farmers a higher degree of freedom. Matzdorf & Lorenz [30] indicate that this potential of innovation is very well used by farmers involved in result-oriented schemes. It also leads the farmers to become more involved in conservation and increases their willingness to improve ecological performance of their fields [2]. In this study, we focussed on a well known species. However, such detailed knowledge is not often available. In the absence of stabilized knowledge on the effects of farming activities on biodiversity, the high potential of innovation, associated to the willingness to improve ecological performance that result-oriented schemes provide may help finding ecological sound management strategies. In such a context, biodiversity becomes a joined-production that could be considered as a new "crop" and the capacity of farmers (in link with local environmental managers and/or researchers) to produce the empirical knowledge needed should not be underestimated. In this transition phase, the modalities of the compensation payments may however be reconsidered and a form of payment for knowledge production could replace the payment for results. In the model, such an imperfect knowledge could be integrated by adding uncertainty on the key parameters in the form of stochasticity. Using algorithms of stochastic viability [10] would make it possible to maintain the viability approach in such a context.

Result-oriented schemes have many advantages. They seem moreover to be very well accepted by farmers since they do not necessarily imply extra-costs and allow for more room for manoeuvre in the management of their farm [8]. The set up of such schemes in the field seems to be more limited by legal issue than by acceptance by local stakeholders. Indeed, the Rural Development Regulation, based on a strict interpretation of the World Trade Organisation rules, restricts payments for farmers to compensations of income losses or additional costs due to a change of management practice. This rule fits well to Action-Oriented Schemes but result-oriented ones are seen as distorting measures and public stakeholders are often reluctant to implement them. This legal problem is one of the reasons for the abandonment of the Dutch result-oriented scheme [31]. According to Schwartz et al. [2], a window of negotiation seems however to be available in the WTO rules but would imply high level negotiations.

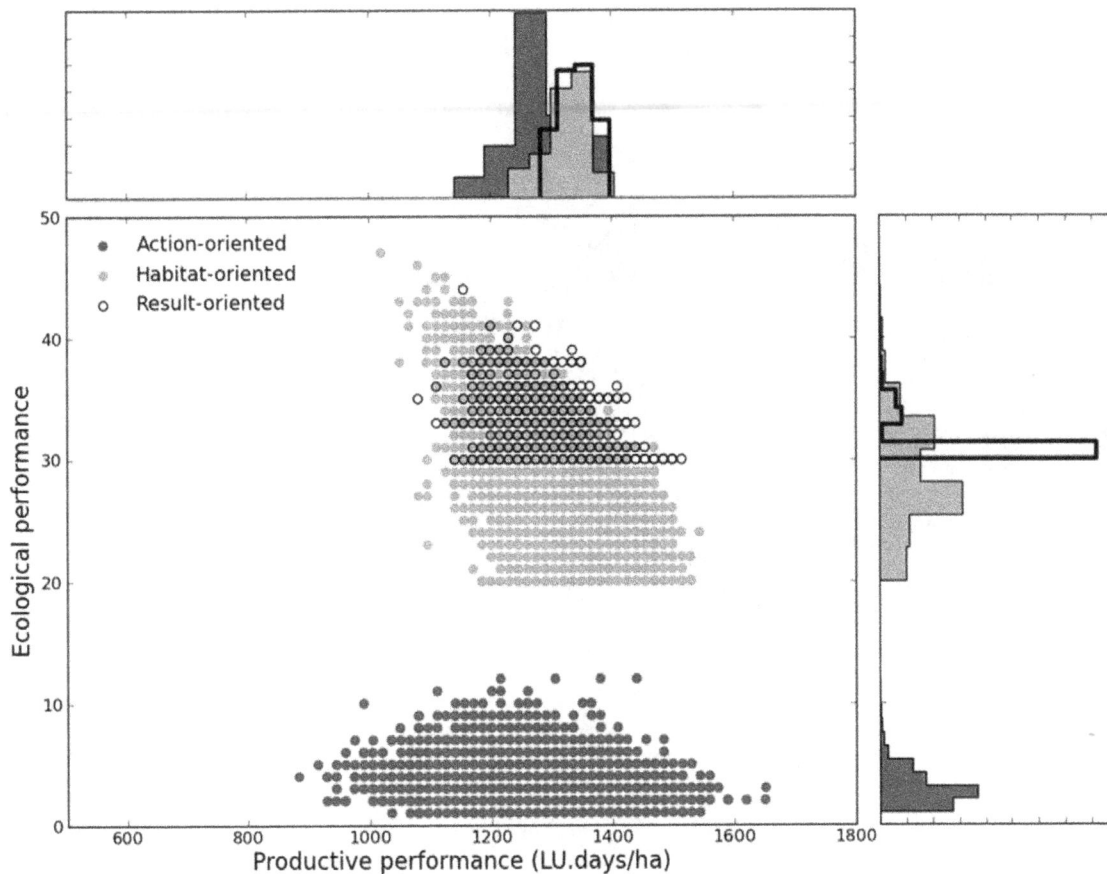

Figure 2. Ecological performance $N(T)$ **and productive performance** $P(u,T)$ **and histograms of distributions of action-oriented, habitat-oriented and result-oriented.** For each scenario, results are plotted for a sample of 10 000 couples of viable state trajectories and viable grazing strategies. The action-oriented scenario (dark gray) is run with cattle density constraint ($u^b = 0.5$; $u^\# = 2$ livestock units per hectare); the habitat-oriented scenario (light gray) is run with fecundity and habitat quality constraints ($f^b = 2.5$, $h^b = 0$ cm and $h^\# = 14$ cm); the result-oriented scenario (empty black) is run with minimum population size ($N^b = 30$); all, scenario involve constraints on productive performance and cattle feeding requirement; all scenarii are run with initial population size $N(t_0) = 30$.

Toward increased spatial scales

Other mechanisms may improve the effectiveness of result-oriented schemes. For instance, farmers frequently allocate schemes to fields with the lowest productivity so as to limit the impact on the overall performance of the farm [2]. Therefore, the localisation of AES fields is often defined regardless of its expected

Table 2. Ecological and productive performance of action, habitat and result-oriented scenarios.

	Action oriented	Habitat oriented	Result oriented
Productive performance (LU.days/ha)	1313 (95)	1321 (74)	1339 (58)
Ecological performance (Population size)	4 (1)	29 (4)	31 (2)

Means and standard deviation () are given for three random samples of 10 000 viable state trajectories and viable grazing strategies. Productive performance $P(u,T)$ is the number of livestock unit.days ha^{-1} (LU.days/ha) characterizing a grazing strategy. The ecological performance $N(T)$ is the bird population size at time horizon (starting with $N(t_0) = 30$).

ecological outcome. With result-oriented schemes both productive and ecological performance would have to be taken into account as the ecological outcome would be of major concern to farmers. Such schemes could thus be expected to reach better levels of effectiveness. The level of payment would however need to be addressed with caution for the scheme to remain attractive. Our model does not include economic incentives yet and development in this direction should help defining these levels of payment.

Beyond the legal issues mentioned at the end of the former section, other limits of result-oriented schemes arise from the possible difficulties to assess the ecological outputs. Schwarz *et al.* [2] recommend focusing in a first step on plant communities as ecological and agricultural processes fit into the same scale: the field. Methods that prove to be fair to the farmer have been developed in Germany [22] and in France [8] to provide assessments in the case of grassland flora. However, concerning mobile species, such as birds, with larger home ranges, assessment at field scale is more difficult. First, birds are not present in the field all the time and accurate surveys imply heavy monitoring protocols. A solution to this first problem was to focus on local indicators such as breeding success but results were mitigated. [5,6]. The second difficulty, which is linked to the latter point, is that bird population trends not only depend on processes

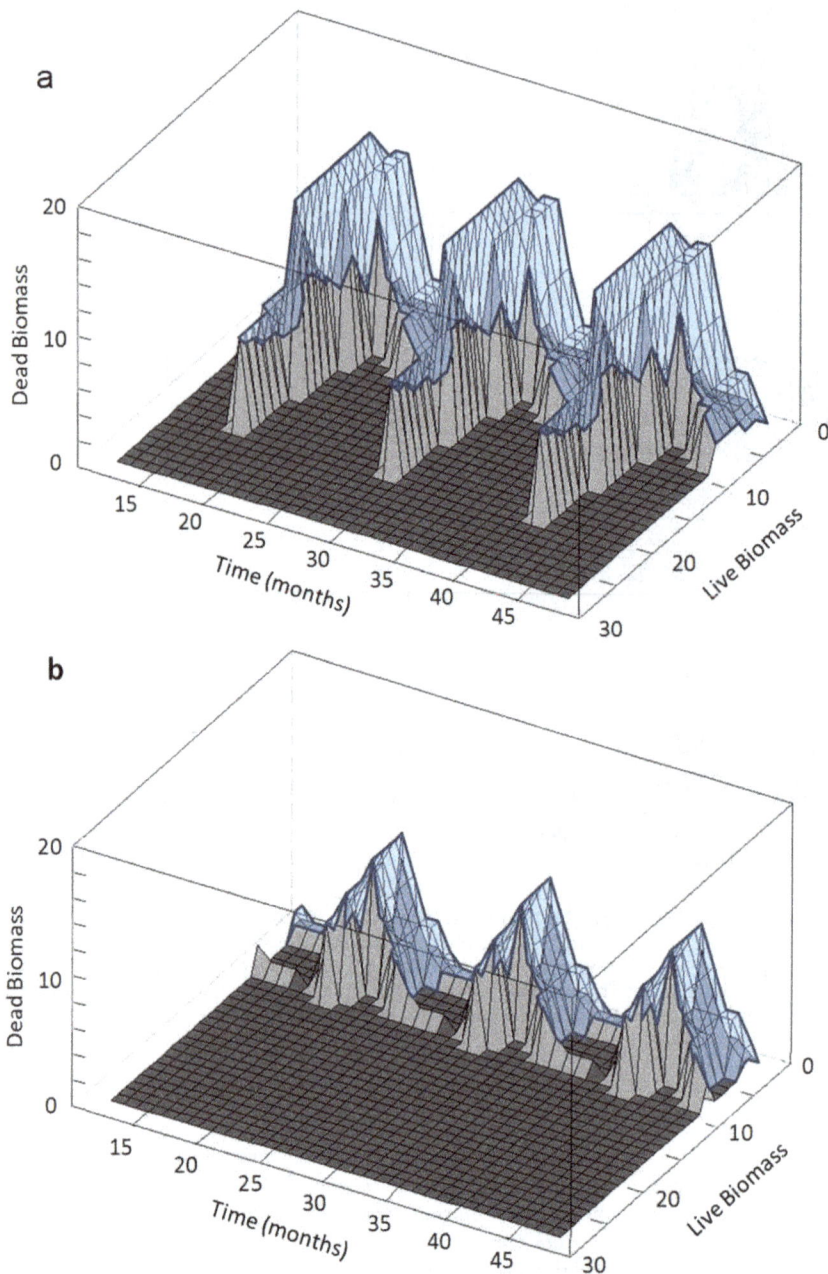

Figure 3. Zoom on three years of the viability tubes (VT_{RO} and VT_{HO}) for the result-oriented (fig a) and habitat-oriented (fig b) scenarios. The tubes show the set of viable states throughout time (in months). The two state dimensions are the live biomass and the dead biomass both expressed in organic matter (10^{-4} g OM ha^{-1}). The viability tube corresponds to the volume (in blue) between the light gray surface and the wireframe. Dark gray areas are the ones for which no viable state exists.

occurring at the field scale but also on processes occurring at a larger scale (i.e. a set of neighbouring fields). A solution to this problem could be to develop schemes at a scale matching the home range of species under concern. However, management at larger scales involving several land owners may lead to situations where some land owners behave as free-riders and compromise the success of the scheme. This issue has been taken into consideration in Sweden in the case of carnivores with very large home ranges [3]. In this case, payments by results were not given directly to individuals but to the communities. The efficiency of the conservation policy thus relied on collective action. Result-oriented schemes at the landscape scale based on collective action

would have another major advantage. Groups of farmers could both adapt their management practices at the field scale and modify the spatial allocation of management practices at the landscape scale in order to create habitat heterogeneity. Increased landscape heterogeneity could improve ecological performance as it makes spatial complementarities among habitats possible [26]. Improvement of the model presented here to account for these spatial effects (nest site selection, landscape heterogeneity,...) is another major perspective of this work that we are currently handling [27,32].

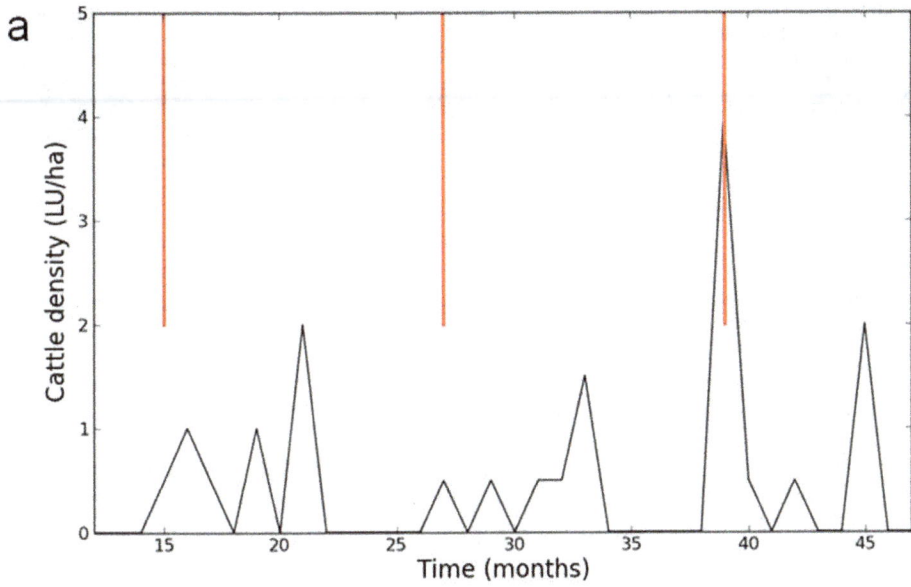

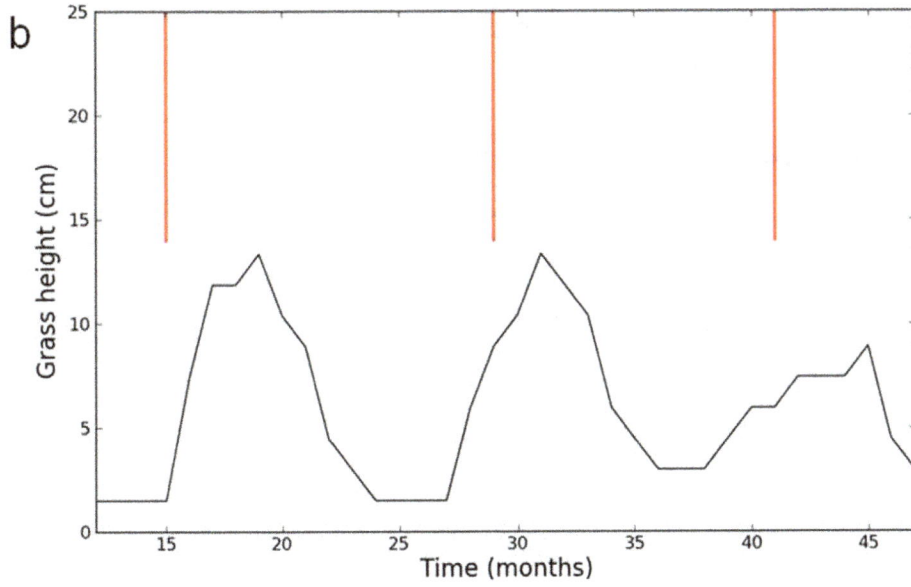

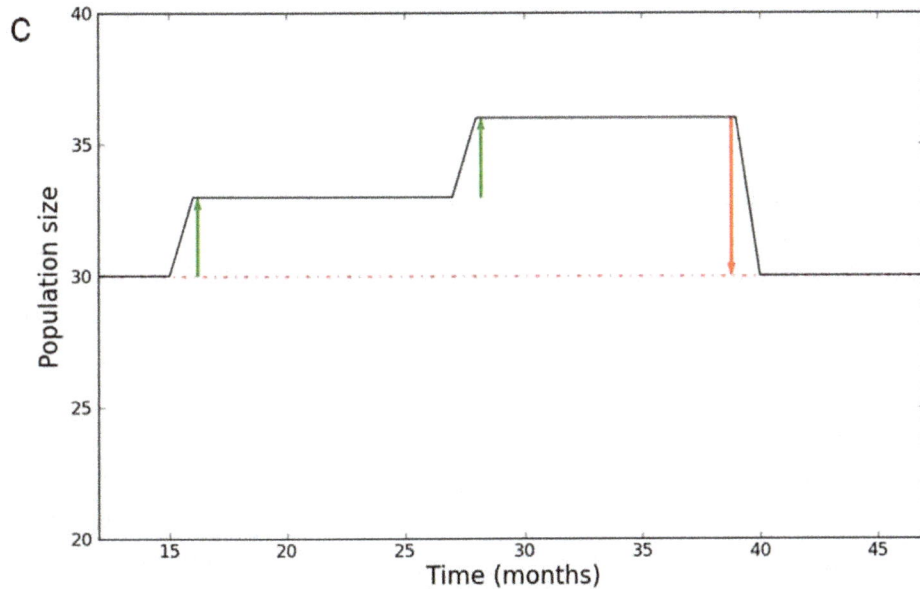

Figure 4. Example of one viable grazing strategies and state trajectories in the result-oriented scenario with a climatic perturbation (zoom on the three years around the climatic perturbation). The different constraints are plotted to illustrate the fact that viable result-oriented strategies would not respect action and habitat oriented constraints. Figure a, viable grazing strategy in the result oriented scheme. Red bars represent the cattle density constraint $u^{\#}$ applied in the habitat and action oriented schemes. Figure b, viable grass height trajectory h in the action and result oriented schemes. Red bars represent the grass height constraint $h^{\#}$ applied in the habitat and action oriented schemes. Figure c, viable bird population trajectory N. The red dotted line stands for the population size constraint N^{b}. The green and red arrows highlight the mechanisms of inter-annual compensation.

Conclusion

Compared with action-oriented schemes, our study shows that improvement of ecological performance is high when schemes are habitat or result-oriented. Differences in performances between habitat and result-oriented schemes remained limited. The main advantage of result-oriented schemes is to increase the overall management flexibility of the grassland agroecosystem. Such improved flexibility may also allow farmers to adapt their management to climatic variations. Further model developments will focus on both the spatial and temporal dimensions of farming flexibility. This next step will make it possible to better match management and ecological processes.

Supporting Information

Appendix S1 Discrete time dynamics of the grazed grassland.

Appendix S2 Discrete time dynamics of the wader population.

Appendix S3 Model calibration.

Appendix S4 Different degrees of freedom in grazing sequences.

Appendix S5 Sensitivity analysis.

Table S1 Parameters used in the grazed grass model.

Table S2 Parameters used in the bird model.

Acknowledgments

We thank Donald White for proofreading the manuscript. We also thank Dr Melman for fruitful discussion on Dutch agri-environment schemes.

Author Contributions

Conceived and designed the experiments: RS MT LD. Performed the experiments: RS. Analyzed the data: RS. Wrote the paper: RS LD MT.

References

1. Kleijn D, Baquero RA, Clough Y, Diaz M, De Esteban J, et al. (2006) Mixed biodiversity benefits of agri-environment schemes in five European countries. Ecology Letters 9: 243–254.

2. Schwarz G, Moxey A, McCracken DI, Huband S, Cummins R (2008) An analysis of the potential effectiveness of a Payment-by-Results approach to the delivery of environmental public goods and services supplied by Agri-Environment Schemes. 108 p. Available: http://www.lupg.org.uk/pdf/LUPG_Payment_by_Results_Feb09.pdf.

3. Zabel A, Holm-Muller K (2008) Conservation performance payments for carnivore conservation in Sweden. Conservation Biology 22: 247–251.

4. Klimek S, Kemmermann AR, Steinmann HH, Freese J, Isselstein J (2008) Rewarding farmers for delivering vascular plant diversity in managed grasslands: A transdisciplinary case-study approach. Biological Conservation 141: 2888–2897.

5. Verhulst J, Kleijn D, Berendse F (2007) Direct and indirect effects of the most widely implemented Dutch agri-environment schemes on breeding waders. Journal of Applied Ecology 44: 70–80.

6. Musters CJM, Kruk M, De Graaf HJ, Ter Keurs WJ (2001) Breeding birds as a farm product. Conservation Biology 15: 363–369.

7. SNH (2005) East Scottland Grassland Management Scheme. 12 p. Available: http://www.snh.org.uk/pdfs/NatCare/GrasslandScheme.pdf.

8. Mestelan P, Agreil C, Marie CdS, Meuret M, Mailland-Rosset S (2007) Implementing agro-environmental measures based on ecological results. The case of meadows and rangelands in the massif des Bauges regional Park. Rencontres Recherche Ruminants. Paris. Available: http://www.journees3r.fr/IMG/pdf/2007_04_pastoralisme_06_Mestelan.pdf.

9. Durant D, Tichit M, Kerneis E, Fritz H (2008) Management of agricultural wet grasslands for breeding waders: integrating ecological and livestock system perspectives - a review. Biodiversity and Conservation 17: 2275–2295.

10. De Lara M, Doyen L (2008) Sustainable management of natural resources; Allan RUF, W. Salomons, eds. Berlin: Springer. 266 p.

11. Aubin J-P, ed (1991) Viability theory. Boston. 542 p.

12. Tichit M, Doyen L, Lemel JY, Renault O, Durant D (2007) A co-viability model of grazing and bird community management in farmland. Ecological Modelling 206: 277–293.

13. Tichit M, Hubert B, Doyen L, Genin D (2004) A viability model to assess the sustainability of mixed herds under climatic uncertainty. Animal Research 53: 405–417.

14. Baumgartner S, Quaas MF (2009) Ecological-economic viability as a criterion of strong sustainability under uncertainty. Ecological Economics 68: 2008–2020.

15. Sabatier R, Doyen L, Tichit M (2010) Modelling trade-offs between livestock grazing and wader conservation in a grassland agroecosystem. Ecological Modelling 221: 1292–1300.

16. Beintema AJ, Muskens GJDM (1987) Nesting success of birds breeding in Dutch agricultural grassland. Journal of Applied Ecology 24: 743–758.

17. Devereux CL, McKeever CU, Benton TG, Whittingham MJ (2004) The effect of sward height and drainage on Common Starlings Sturnus vulgaris and Northern Lapwings Vanellus vanellus foraging in grassland habitats. Ibis 146: 115–122.

18. Durant D, Tichit M, Fritz H, Kerneis E (2008) Field occupancy by breeding lapwings Vanellus vanellus and redshanks Tringa totanus in agricultural wet grasslands. Agriculture Ecosystems & Environment 128: 146–150.

19. Sokal R, Rohlf F (1995) Biometry (3rd edn). WH Freeman and company: New York.

20. Groot JCJ, Rossing WAH, Jellema A, Stobbelaar DJ, Renting H, et al. (2007) Exploring multi-scale trade-offs between nature conservation, agricultural profits and landscape quality–A methodology to support discussions on land-use perspectives. Agriculture, Ecosystems & Environment 120: 58–69.

21. Brunner A, Huyton H (2005) Agri-environment schemes and biodiversity: lessons learnt and examples from across Europe Bird Life International. 14 p. Available: http://www.birdlife.org/eu/pdfs/Agrienvironment_schemes_lesson_learnt.pdf.

22. Wittig B, Kemmermann ARG, Zacharias D (2006) An indicator species approach for result-orientated subsidies of ecological services in grasslands - A study in Northwestern Germany. Biological Conservation 133: 186–197.

23. Sabatier R, Doyen L, Tichit M (2008) Assessing the effect of stocking density thresholds on productive and ecological performances of livestock systems based on grasslands with high biodiversity stakes Institut National de la Recherche Agronomique (INRA). pp 213–216. Available: http://www.journees3r.fr/IMG/pdf/2008_07_environnement_01_Sabatier.pdf.

24. Tichit M, Puillet L, Sabatier R, Teillard F (2011) Multicriteria performance and sustainability in livestock farming systems: Functional diversity matters. Livestock Science 139: 161–171.

25. Melman TCP (2010) A web-based tool for tailor made management for meadow birds. In: biologists Aoa, editor. Leicester.

26. Sabatier R (2010) Multiscale trade-offs between agricultural production and biodiversity in a grassland agroecosystem. Paris: Agroparistech. 226 p. Available: https://www.versailles-grignon.inra.fr/sadapt/content/download/4266/40211/version/1/file/SABATIER_Manuscrit_THESE_+XIV.pdf.

27. Sabatier R, Doyen L, Tichit M (2010) Reconciling production and conservation in agrolandscape, does landscape heterogeneity help? Montpellier, France. 10p

p. Available: http://hal.archives-ouvertes.fr/docs/00/51/05/52/PDF/Sabatier_Reconciling_production.pdf.

28. Gunderson L (1999) Resilience, flexibility and adaptive management - antidotes for spurious certitude? Conservation Ecology 3.

29. Martin G, Cruz P, Theau JP, Jouany C, Fleury P, et al. (2009) A multi-site study to classify semi-natural grassland types. Agriculture Ecosystems & Environment 129: 508–515.

30. Matzdorf B, Lorenz J (2010) How cost-effective are result-oriented agri-environmental measures? An empirical analysis in Germany. Land Use Policy 27: 535–544.

31. NEAS (2007) Executive summary: Ecological Evaluation of Nature Conservation Schemes run under the Stewardship Programme and the Dutch National Forest Service 2000–2006 Bilthoven. 46 p. Available: http://www.mnp.nl/images/500410004%20Ecological%20Evaluation_tcm61-35640.pdf.

32. Sabatier R, Tichit M (2011) Does landscape heterogeneity modulate the trade-off between production and biodiversity? Proceedings of the 3rd Farming system design conference: Resilient Food systems for a Changing World, Brisbane. pp 54–55. Available: http://aciar.gov.au/files/node/13992/does_landscape_heterogeneity_modulate_the_trade_of_55592.pdf.

9

Varroa-Virus Interaction in Collapsing Honey Bee Colonies

Roy M. Francis*, Steen L. Nielsen, Per Kryger

Department of Agroecology, Science and Technology, Aarhus University, Slagelse, Denmark

Abstract

Varroa mites and viruses are the currently the high-profile suspects in collapsing bee colonies. Therefore, seasonal variation in varroa load and viruses (Acute-Kashmir-Israeli complex (AKI) and Deformed Wing Virus (DWV)) were monitored in a year-long study. We investigated the viral titres in honey bees and varroa mites from 23 colonies (15 apiaries) under three treatment conditions: Organic acids (11 colonies), pyrethroid (9 colonies) and untreated (3 colonies). Approximately 200 bees were sampled every month from April 2011 to October 2011, and April 2012. The 200 bees were split to 10 subsamples of 20 bees and analysed separately, which allows us to determine the prevalence of virus-infected bees. The treatment efficacy was often low for both treatments. In colonies where varroa treatment reduced the mite load, colonies overwintered successfully, allowing the mites and viruses to be carried over with the bees into the next season. In general, AKI and DWV titres did not show any notable response to the treatment and steadily increased over the season from April to October. In the untreated control group, titres increased most dramatically. Viral copies were correlated to number of varroa mites. Most colonies that collapsed over the winter had significantly higher AKI and DWV titres in October compared to survivors. Only treated colonies survived the winter. We discuss our results in relation to the varroa-virus model developed by Stephen Martin.

Editor: Stephen J. Martin, Sheffield University, United States of America

Funding: The authors wish to thank Promilleafgiftsfonden for funding this study. The funders had no role in study design, data collection and analysis, decision to publish, or preparation of the manuscript.

Competing Interests: The authors have declared that no competing interests exist.

* E-mail: royfrancis.mathew@agrsci.dk

Introduction

Honey bees are important insects economically and ecologically. Large scale losses of managed honey bees in the recent years have perplexed beekeepers and bee researchers. Honey bee colonies in temperate climates that do not survive the winter are referred to as winter losses. These losses may be a result of natural causes or varroa infestation and viruses [1,2,3]. DWV in combination with varroa mites have been shown to reduce lifespan of winter bees [4]. Colony Collapse Disorder (CCD) is a syndrome describing the large-scale loss of managed honey bees worldwide first reported in 2006–2007 [5]. Rise in colony losses were quickly reported from several locations worldwide [6]. Several studies have investigated and reported various causes for this sudden decline is bees such as viruses [7,8], varroa mites [9], microsporidean *Nosema spp.* [10,11,12], and pesticides [13,14,15]. Due to the lack of consensus on a possible causal agent of colony losses, it is being investigated extensively [16] and it is becoming clear that a single causal agent is difficult to identify and that causes are possibly multiple and complex [8,17,18,19,20]. A combination of varroa and viruses are now frequently implicated in collapsing colonies [9,21,22,23].

The single greatest threat to honey bee populations worldwide is the recently introduced invasive mite *Varroa destructor* Andersen & Trueman. The life cycle of the varroa mites is tightly adapted to the development of the honey bees. Varroa mites are serious and devastating ectoparasites of the honey bee. During the phoretic phase, the varroa mites live on the bodies of honey bees and feed on their haemolymph. The symptoms arising out of heavy mite infestation is referred to as varoosis [24,25]. The reproductive phase of varroa mites happens exclusively in the capped cells of developing bee pupae [26]. Several studies have documented the ill effects of varroa infestation on honey bees including reduced lifespan [27], decreased survivorship [28] and weight loss in drones [29].

Varroa mites are also efficient vectors for transmission of viral diseases [30]. The rise of viral infection in bees after the introduction of varroa mites into Europe has been reported quite extensively [31]. Several honey bee viruses have been reported to be transmitted by varroa mites including DWV [32], KBV [33,34], SBV [35], ABPV [30] and IAPV [36]. Several extensive survey studies have been carried out to monitor virus and mite levels over an entire season [37,38,39]. Thus we know that mite population increases from spring to autumn during peak brood production in the colony [40]. Drone brood in particular accelerates the growth of the mite population. Hence, mite levels and subsequently viral prevalence are highest towards autumn. During winter, the small cluster of bees ceases brood production which restricts mite reproduction. However, the phoretic mites may shift host thereby transmitting viruses within the colony.

Given the complex interactions between honey bees, varroa mites and viruses, modelling approaches have been applied to understand dynamics in the hive [41,42,43]. The model proposed by Stephen Martin [44], predicts that the less virulent DWV would be highly prevalent in varroa-infested colonies whereas the more virulent ABPV should disappear as it rapidly kills its host. Putting this model to test requires improved sampling strategies for estimating varroa and virus prevalence. A few methods have been put forward to estimate varroa load at colony and apiary level [40,45,46] while no guidelines have been established for optimal sampling for virus prevalence. So far, most studies have tried to determine the viral titre in a sample of workers from colonies, however, this cannot be used to estimate the prevalence of virus-infected individuals in the colony. In particular, there exists a huge variation in viral titres amongst bees, ranging from uninfected individuals to carriers of billions of viral particles, rendering the interpretation of data from pooled samples cumbersome. A recent study of 293 colonies from 35 apiaries across three years in Hawaii [23], reported a combination of DWV and varroa as the destructive force behind collapsing colonies. DWV infection was studied in varroa-free colonies through the phase of varroa infestation. The DWV titre in varroa infested colonies were 10^{10} copies per bee compared to 10^4 copies in varroa-free colonies. As the DWV titres increased, the genetic diversity of DWV decreased ultimately leading to a single high-virulent species.

Beekeepers use various acaricides to keep mite levels low and to avoid outbreak of viral diseases. Acaricides control mites through different pathways and their side-effect on the colony may vary [47,48]. In Denmark, most popular chemical methods include organic acids (formic acid, lactic acid, oxalic acid) and pyrethroid flumethrin. Now that, there is considerable evidence to show that a combination of viruses along with varroa is playing an important role in colony health, monitoring the changing viral load in relation to mite number over a whole season is vital to understand the nature of this interaction. Viral analysis was carried out on subsamples to estimate the prevalence of virus infected bees in the colonies. This study investigated viral load in bees and varroa mites across 23 colonies from 16 apiaries under three treatment conditions over a year.

Materials and Methods

Sample collection and processing

Honey bee worker samples were procured from our own experimental hives and from the following Danish beekeepers: Aksel Jørgensen, Arne Jensen, Bent Larsen, Christian Petersen, Ditlev Bluhme, Flemming Thorsen, Gunner Borg, Jørgen Jørgensen, Karen Poulsen, Leif Johanssen, Orla Overby and Willy Svendsen. Samples were collected from 23 colonies in 15 apiaries (Fig. S5). The colonies were categorised based on the method of treatment used to control mite population. Category A included 11 colonies (A01 to A11) from seven apiaries which were treated using organic acids mostly formic acid and always oxalic acid. Formic acid is used in early autumn followed by oxalic acid in early winter. Category B included nine colonies (B01 to B09) from seven apiaries which were treated using the pyrethroid Flumethrin. The beekeepers use non-standardised treatment methods which result in diverse methods of application and treatment times. Category C included three experimental colonies (C01, C02 and C03) from one apairy which were not treated in any manner. These colonies were also not treated the year before. One colony C02 was removed in September, due to American foulbrood and hence, not sampled in October 2011. Seven colonies (A04, A05, A09, B09, C01, C02, C03) died over winter and were not sampled in April 2012.

Approximately 200 bees were collected from every colony in the first week of every month from April 2011 to October 2011 and again in April 2012. All bee samples were received alive after being transported by post in 400 ml plastic containers with breathing holes and sugar candy food in 1.5 ml vials. Live bees were used to ensure high quality RNA. The bees were anaesthetised with carbon dioxide gas for 15 min and shaken in a polythene bag to release mites. The bee numbers and mite counts were recorded and approximately 200 bees were split into 10 separate subsamples of maximum 20 bees each. The mites from each colony were pooled together as a separate sub-sample. If samples received were less than 200 bees, then subsamples were split to 10 bees or 15 bees. Samples with more than 200 bees were stored as extras. From these extra bees, six individuals from eight samples (four colonies from September and four colonies from October) were analysed individually to estimate prevalence of diseased bees. Samples were freeze-dried on a Heto LyoPro 6000 apparatus for 72 hours at pressure 0.05 hPa and temperature $-80°C$. After lyophilisation, all samples were stored at $-80°C$ in 50 ml plastic bottles with tight fitting screw-caps.

RNA extraction and cDNA conversion

Metal beads were added to the sample bottles and the samples were homogenized on a genogrinder 2000 for 1 min at 1500 rpm. A pinch (approximately 10–15 mg) of the crushed material was used to extract RNA. In a pool of 20 bees, a pinch would be approximately 0.40 bee. Total RNA was extracted using NucleoMag® 96 RNA Kit (Macherey-Nagel) on a Kingfisher Magnetic Extractor using a custom program following the manufacturer's guidelines. The extracted RNA (100 µl) was stored in 96 well plates at $-80°C$ for further use. The RNA was transcribed to cDNA using High Capacity cDNA Reverse-Transcription Kit (Applied Biosystems (AB)). 10 µl of RNA was added to 10 µl cDNA master mix yielding a 20 µl cDNA solution. The incubation conditions were as recommended by the manufacturer: 10 min at 25°C, 120 min at 37°C and 5 min at 85°C. The cDNA solution was then diluted 10-fold in water and stored at $-80°C$.

Real-time PCR

The three assays tested in this study are AKI, DWV and beta-actin. The ABPV-KBV-IAPV viruses were detected in a single assay using a single pair of primers referred to as 'AKI' primers [49]. Two sets of primers were used for each virus (AKI and DWV), referred to as outer primers and inner primers [39]. Primers used in this study are listed in Table S1. Standard PCR was carried out on previously known positive samples using outer primers to generate stock of positive controls. Ten ten-fold dilutions, prepared from these stocks and were used on every plate to estimate plate to plate variation. All samples and controls were tested using inner primers. RNase-free water was used as template for negative controls (NTC). The real-time polymerase chain reaction (qPCR) assays were carried out on an ABI PRISM 7600HT (AB) using SYBR Green DNA binding dye (AB). The volume for qPCR reactions was 12 µl with a final primer concentration of 0.4 µM. All reactions were loaded on optical 384 well PCR plates (ABgene) and run in replicates of two.

Data processing and analysis

Eight dilutions for AKI, beta-actin4 and DWV3 were selected for standard curves and subsequent regression analysis based on their linearity within the dynamic range. The baseline was automatically set and a manual threshold of 0.19 was used for all control runs and test runs. Dissociation profiles for all reactions were visually examined and flagged. Data from the qPCR runs were analysed in R 2.15.1 [50] and Microsoft Excel^{TM}. Replicates showing coefficient of variation (CV) greater than 10% were flagged and replicates were examined and manually corrected if required. Samples which did not cross the threshold before cycle 40 were given a C_t value of zero or no virus. Samples with incorrect melting curve profile were given a C_t value of zero.

The slope and intercepts calculated from the standard curves were used to estimate DNA copies from known C_t values (Fig. S1). DNA concentration was converted to copies using equation [copies $= (c \times N)/M$] where where c = concentration in g, N = avogrado's constant and M = molecular mass of the amplicon in Daltons. While converting to copies, copies near 1 were rounded to 0 or 1 as it is assumed there has to be one copy of the virus/beta-actin or none. Based on the standard curves, a C_t Value of 34 was chosen as a cut-off because the standard curves were no longer linear after cycle 34. Based on the regression, cycle 34 corresponds to 13, 10 and 243 copies of AKI, beta-actin and DWV respectively.

Statistical analyses

Statistical analysis and data handling was carried out in Microsoft Excel and R [50]. All figures in this paper were generated in R. All statistical tests were non-parametric as the data was non-normal and heteroscedastic due to zero-clumping. All tests were carried out on $\log_{10}$ viral copies. Significance testing between two independent groups was done using Mann-Whitney-U Test (Wilcoxon Rank Sum Test) (W) and tests for more than two groups were done using Kruskal-Wallis test (KW). Viral prevalence in the colonies were calculated online using EpiTools pooled prevalence calculator (http://epitools.ausvet.com.au) [51] using method 2 [52,53]. For a few colonies, all 10 monthly sub-samples were positive or negative rendering them unsuitable for prevalence calculation. For five of these colonies, viral titres in six extra bees were analysed individually.

Results

Survival and varroa levels

Of the 23 hives, all except seven hives survived to spring 2012. One untreated hive (C02) was lost in September 2011 due to American foulbrood and therefore not an overwintering loss. Three hives from the organic group (A04, A05, A09), one hive from the pyrethroid group (B09) and two hives from the untreated group (C01, C03) died over winter. The mean bee count across 176 samples from eight months (excluding dead hives) was 231 ± 68 bees ranging from 112 to 484 bees. The mean varroa mite count across 176 samples from eight months (excluding dead hives) was 8 ± 23 mites ranging from 0 to 242 mites.

The mite indices for three treatment groups are shown in Figure 1. The mite index was significantly different between treatment groups for the months of April 2011 (KW P = 0.023), May 2011 (KW P = 0.0098) and July 2011 (KW P = 0.02). The curves for treated organic and pyrethoid groups showed similar trend except in September where the organic group showed a sharp rise. This drastic rise in mites for September was mainly based on five colonies from one organic apiary. Mite indices in treated colonies were significantly lower than untreated colonies for all months except Aug and Sep (Figure 1). The mite load in the untreated group continued to rise until they collapsed. The untreated group was not treated the year before, hence, they started the season with more mites than the two treated groups.

Viral titres

DWV titre was generally higher than AKI titre in all three treatment groups (Figure 2). For AKI, the viral titre rose from April to July in all three treatment groups. In organic and pyrethroid groups, AKI titres decreased in connection with the

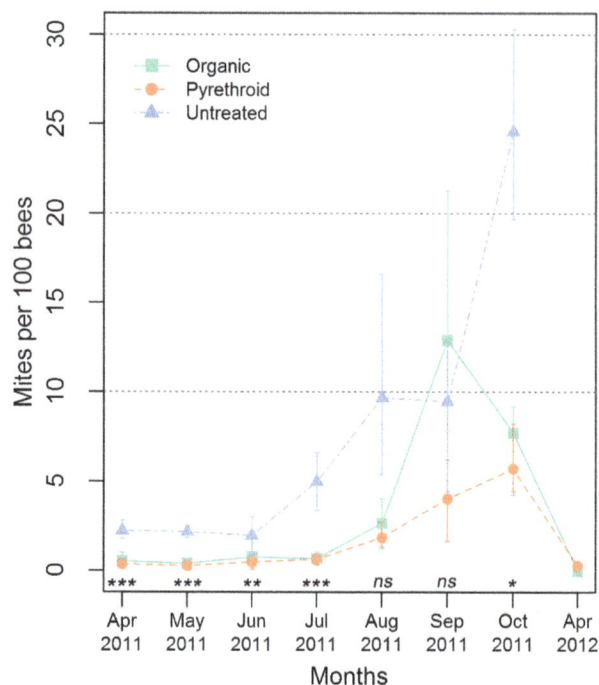

Figure 1. Varroa mite indices for the three treatment groups over eight months. Mite indices were computed using sum of bees and sum of mites in each treatment group. These are not mean of mite indices of samples. Error bars show standard error. Error bars were computed using n as follows: Organic group (n = 11 colonies except Apr 2012 where n = 8), Pyrethoid group (n = 9 except Apr 2012 where n = 8), Untreated group (n = 3 except Oct 2011 where n = 2 and Apr 2012 where n = 0). Asterix indicate significance (Wilcoxon test) between treated and untreated colonies (* P<0.05, ** P<0.01, *** P<0.001).

varroa treatment during August, but then increased again. In the untreated group, AKI titre peaked in September. For DWV, all groups displayed a steady rise in viral titre throughout the season. Viral titres in mites also showed a rising trend over the season (Figure S2). AKI titres in treated colonies were significantly different (Wilcoxon) from untreated colonies only in May and July. DWV titres in treated colonies were significantly different (Wilcoxon) from untreated colonies only June and Sep.

The sum of viral titres in the sub-samples of AKI and DWV copies per hive was correlated to mite index (mites per 100 bees) per hive (Figure 3). In all three treatment groups, viral copies were significantly (P<0.01) correlated to mite index.

Prevalence of diseased bees

Splitting each sample into 10 subsamples allowed us to determine the prevalence of virus-infected bees using pooled prevalence calculations. Depending on the time of the year, we find huge variation in the number of infected bees. For AKI, we had zero of 230 (0%) sub-samples infected in May to 130 of 220 (59%) sub-samples infected in October. For DWV, we had 65 of 230 (28%) sub-samples infected in June up to 216 of 220 (98%) infected sub-samples in October. Of the dying colonies, all (100%) sub-samples were infected with DWV while 56 of 60 (93%) sub-samples were infected with AKI in October.

In order to determine the fraction of workers with detrimental infection, our solution was to include only sub-samples with titres above a defined threshold. We considered sub-samples of 20 bees with viral titres above 10^7 copies an indication of at least one

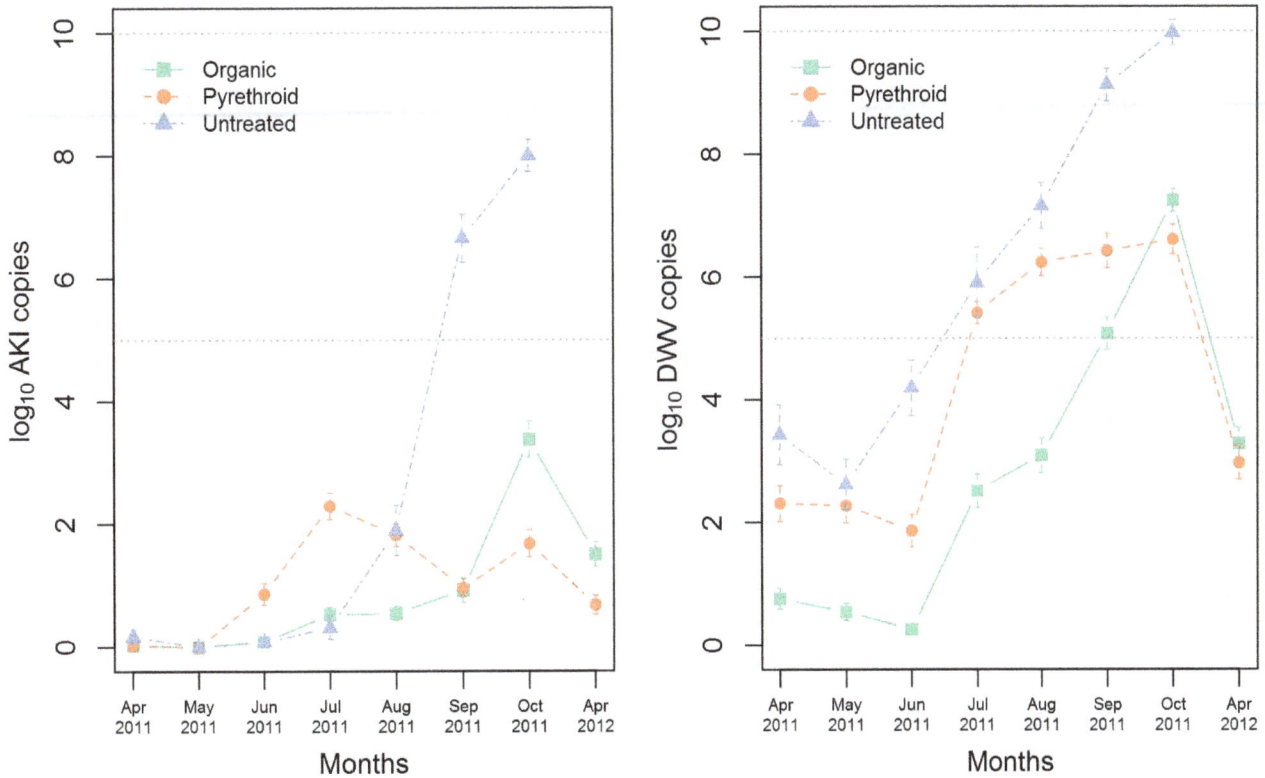

Figure 2. Mean viral titres in bees (subsamples) for three treatment groups across eight months for AKI (left) and DWV (right).
Number of sub-samples in the organic group (n = 110 except in Apr 2012 where n = 80), pyrethroid group (n = 90 except in Apr 2012 where n = 80), untreated group (n = 30 except in Oct 2011 where n = 20). Error bars show standard error. Untreated colonies showed significantly higher AKI titres in May and July and significantly higher DWV titres in June and Sep compared to treated colonies.

individual bee within the sub-sample being diseased. We compared the prevalence of diseased bees for every month in colonies that died to the prevalence of diseased bees in colonies that survived the following winter. Monthly prevalence was calculated based on the number of sub-samples showing greater than 10^7 copies by combining all sub-samples from surviving

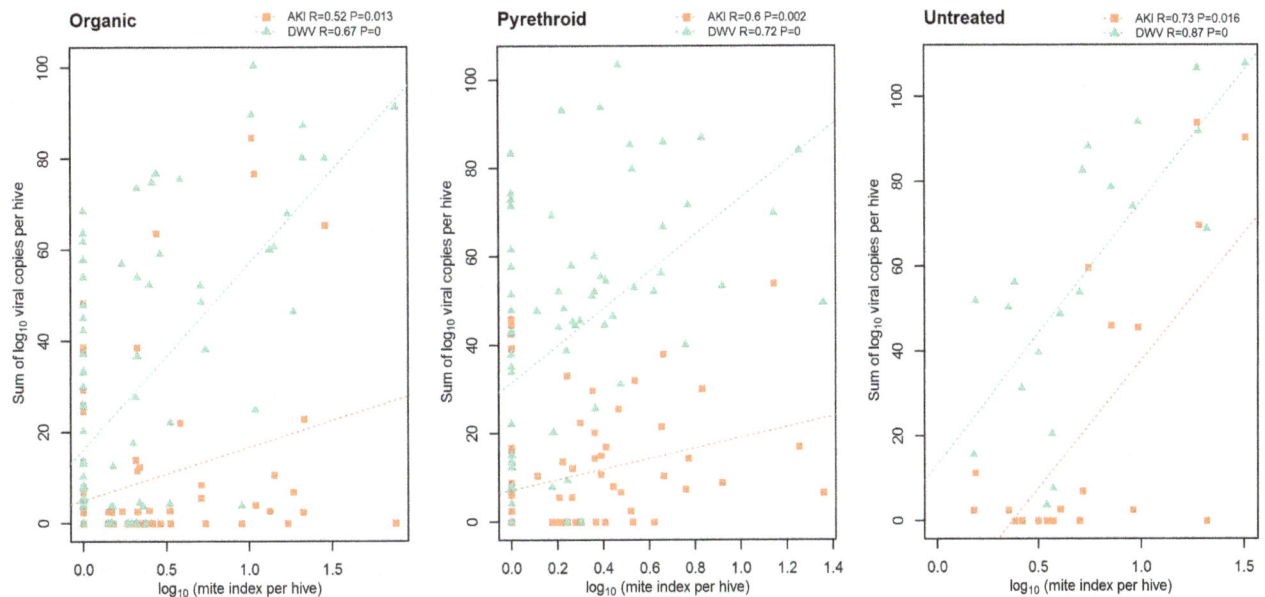

Figure 3. Correlation of mite index versus viral copies for the three treatment methods. Organic (n = 85), Pyrethroid (n = 71), Untreated (n = 20). For all three treatment groups and for both viruses, the correlations are highly significant.

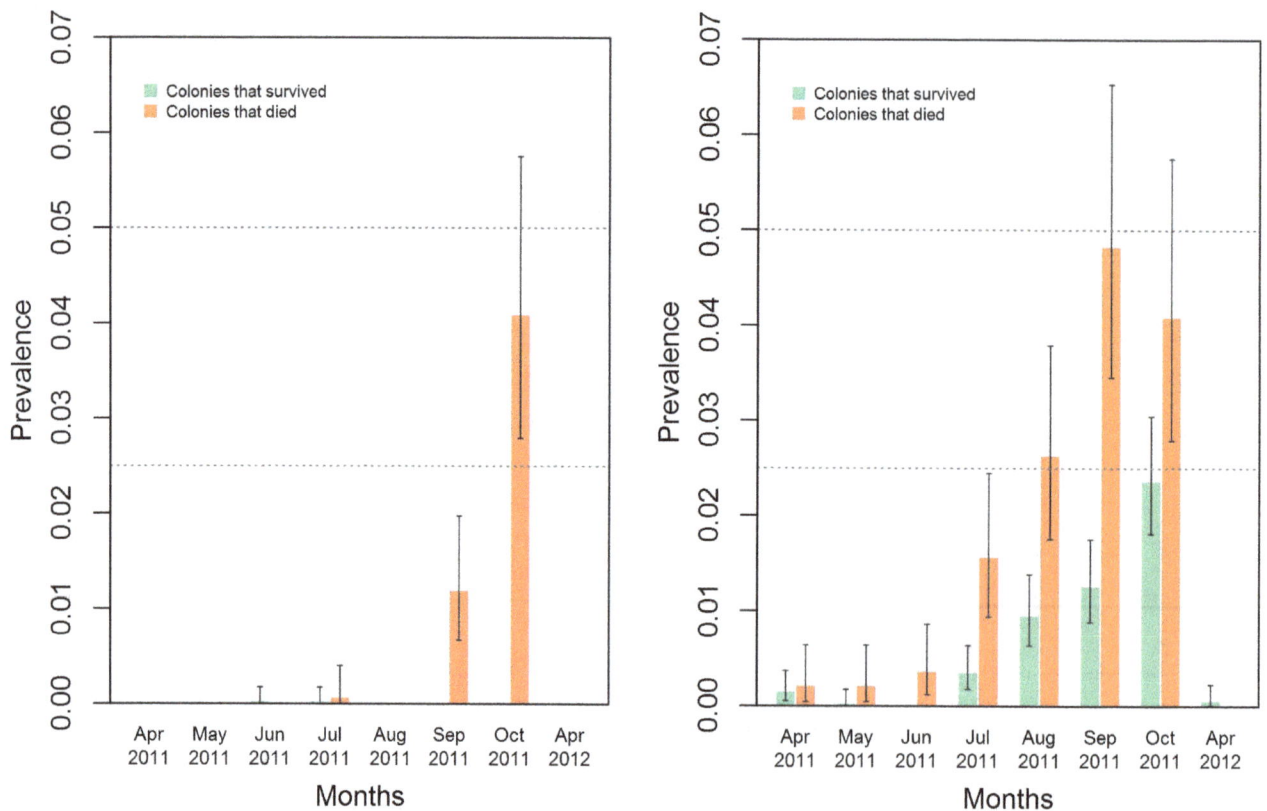

Figure 4. Prevalence of workers infected with greater than 10^7 copies of AKI and DWV viruses across eight months in subsamples (n = 160) from colonies that survived (n = 16) and subsamples (n = 70 except Oct 2011 where n = 60 and Apr 2012 where n = 0) from colonies that died (n = 7 except Oct 2011 where n = 6 and Apr 2012 where n = 0) during the experiment. Error bars show 2.5% and 97.5% confidence intervals.

(n = 160) and dying (n = 70 except October n = 60) colonies, respectively. Colonies that died showed significantly higher prevalence of AKI and DWV infected bees during the months of September and October (Figure 4).

The pooled prevalence calculator is unable to determine the prevalence when all 10 sub-samples of a colony are positive. But in order to directly verify the prevalence of bees with viral titres above 10^7 copies, we tested individual bees from four colonies in September (A02, A03, A04, A09) and four colonies in October (A02, A03, A04, A05) (Figure S8). This included two colonies that survived (A02, A03) and three colonies that died (A04, A05, A09). For AKI, we found considerable variation amongst the tested individuals. All individual bees from the two colonies that survived (A02, A03) had viral titres below 3500 copies. In contrast, the individuals from the three colonies that died had AKI titres ranging from zero to 6.9×10^9 copies. For DWV, the picture is quite different. In the two colonies that survived, none of the 12 bees had viral titres above 10^7 copies in September. However, by October, one of the surviving colonies had two of six bees with titres above 10^7 copies. In the dying colonies, 12 of 24 bees were above 10^7 copies. The viral copies observed in individual bees cannot be directly compared to pooled sub-samples due to differences in methodology.

Case study

We investigated the mite infestation and viral titres of seven colonies that died compared to 16 colonies that survived winter. The overall mite index of the colonies that died was significantly

higher (W P = 5.8×10^{-5}) than those colonies remaining alive. The mite index is shown in Table 1. The September and October AKI titres of the colonies that collapsed were significantly higher than those that survived (W P = 7.7×10^{-8}). DWV titres for colonies that collapsed compared to those that survived were significantly higher (W P<0.05) during all months except July. From Figure 5, it is clear that during the months of September and October, the AKI titres of the dying colonies were substantially higher with the exception of colony A09, which however showed a sudden rise in AKI titre in October. For DWV, a similar scenario exists. However, colony A09 has one of the lowest titres. So it seems unlikely that DWV was the cause of mortality in this colony, but the sudden rise of AKI in October could indicate that the colony had contact with an infected source. AKI titre in colony A05 remained low or zero throughout the season until September and then showed a drastic rise to 10^8 copies in October. This also coincided with a huge rise in mite index in September possibly due to influx of mites from the surroundings. Colony C02 showed rapidly rising titres until September, when we were forced to destroy it due to an outbreak of American foulbrood.

Comparing the viral titres in the varroa mites to titres in bees, we find considerable discrepancy. There are several examples of cases where the mite sample was positive while the accompanying bees were negative. Similarly, we have also found cases where mite sample was negative while the bee subsamples were positive. Viral titres in the mite samples across months are shown in Figure S2.

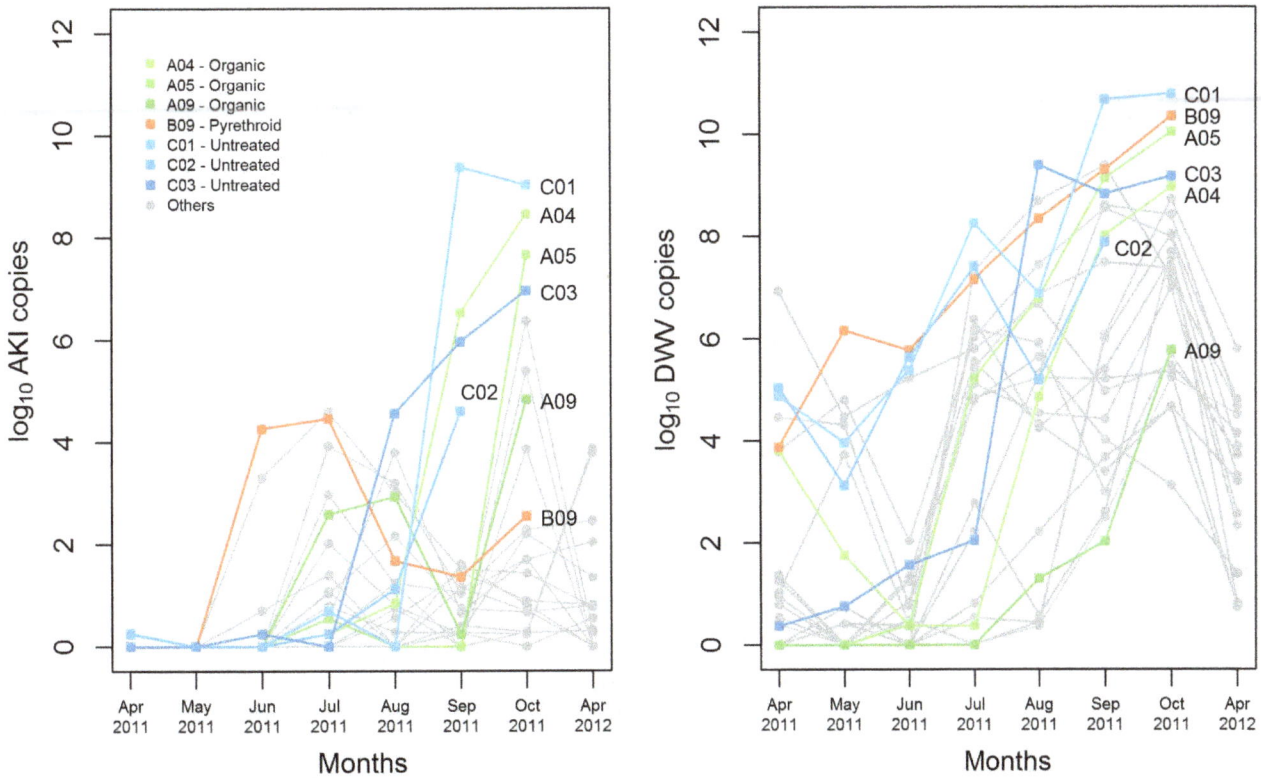

Figure 5. AKI and DWV titres (sum of subsample titres) of colonies that collapsed (coloured, ■) in comparison to surviving colonies (gray, ●) for each month. Green colour denotes organic group, red colour denotes pyrethroid group, blue colour denotes untreated group and gray colour denotes surviving colonies regardless of treatment. Hive C02 was destroyed in September due to American foulbrood. Colonies that died had significantly higher AKI titres in Sep and Oct compared to colonies that survived. Colonies that collapsed had significantly higher DWV titres for all months except July compared to colonies that collapsed.

Discussion

The drastic loss of honey bee colonies in recent years has seriously impacted beekeeping worldwide. Although causes remain indecisive, varroa mites and associated viruses are often mentioned

Table 1. Varroa mite index (mites per 100 bees) from surviving and dead colonies.

Month	Surviving (Treated)	Dead (Treated)	Dead (Untreated)	Signif[¤]
APR	0.2	1.6	2.3	**
MAY	0.3	0.7	2.2	*
JUN	0.2	2.3	2.0	*
JUL	0.6	1.2	5.0	ns
AUG	1.5	5.3	11.0	ns
SEP	4.4	29.2	8.6	ns
OCT	6.0	8.9	25.0[#]	ns
n	16	4	3[#]	

All surviving colonies (n = 16) were treated. Four colonies that died over the winter were treated. All untreated colonies died over winter.
[#]n = 2 in October. Despite the treatment, mite infestation level increased, especially in the succumbing colonies. This could be due to ineffective treatment or subsequent mite reinvasion.
[¤]Significant difference in mite index between surviving and dead colonies (* P<0.05, ** P<0.01, *** P<0.001).

as likely culprits. Therefore, we decided to investigate the relationship between colony survival, varroa treatment and viral load. Our results suggest that the methods used do not always yield satisfactory results. Mite numbers and viral titres in September and October continued to rise in several apiaries despite the treatment in August. This had major consequences for colony health as seven colonies out of 23 were lost over the winter period. Of the seven colonies that died during the experiment, one was destroyed due to AFB, but the death of all other colonies coincided with the sudden increase in viral titres in autumn. We consider the remaining six colonies to have died as a result of varroa-virus interaction.

For beekeepers and scientists alike, estimating varroa load in colonies is difficult and cumbersome [46]. The mite index defined as the number of mites per 100 bees is a commonly used measurement of mite load in the colony. The rise in mite load from spring to summer is slow due to increased brood production. The steep rise in varroa mites from July till September and the decreasing bee population in the same period results in considerable variation in mite index over time. Using mite index to determine the mite load demands several samplings over time and remains impractical for commercial beekeeping. Therefore, many beekeepers treat all hives on a set schedule irrespective of mite index. A large German bee monitoring project [2] reported mite index for October in surviving colonies (3.2–3.6, n = 2906) to be significantly different from collapsing colonies (14.6–16.5, n = 368). In our study, the mite index in September for the six colonies that we presume died of varroa/viruses were 0, 3, 5, 18, 28, and 74 which progressed to 0, 15, 18, 31, 9 and 10 in October respectively (Figure S7). The reduction in mite index for the last two colonies from September to October is possibly due to treatment. We see a more uniform distribution of the mites in October as compared to September, indicating flow of mites between colonies in the vicinity of each apiary.

According to the established varroa-virus model [40,44], increasing varroa population in the colony leads to higher transmission of viruses amongst the bees. Build-up of DWV leads to high proportion of bees with reduced survivorship impacting colony survivorship. Colonies may recover from a large DWV mite population if mites are removed early enough, allowing successful overwintering of bees. As a consequence of the beekeepers' annual treatment, the bees, the mites and DWV co-exist from one season to the next. In stark contrast, ABPV rapidly kills infected pupae preventing the developing bee to emerge, therefore ABPV vectoring mites are unlikely to reproduce successfully. Collapse of a colony due to ABPV occurs only if a large mite population is present when the virus is introduced, according to the varroa-virus model [44].

The German bee monitoring project [2] further considered the impact of viral infections on colony survivorship. In particular, they noticed that the presence of DWV was a good indicator for the mortality of colonies. A weaker link was noticed for ABPV and no significant relationship was noted for KBV. An American study [7] reported IAPV to be a strong predictor for CCD colonies. A study in Hawaii also showed strong increase in titre and prevalence of DWV in colonies as a result of varroa introduction, but no such relationship was noted for ABPV, KBV or IAPV viruses (Martin, 2012). In our study, we find a link between varroa infestation as well as viral titres for both AKI and DWV, to colony mortality. In September and October, most dying colonies showed higher AKI and DWV titres than surviving colonies (Figure 5). The sum of viral titres from all subsamples is plotted in Figure 5, and it can be noted that there is considerable variation between the subsamples (Figure S6). Hence, examining samples of less than 200 bees could lead to an uncertain estimate of the titre. For the survival of the bee colony, the number of diseased individual bees is a better indicator than the overall viral titre. With the help of the 10 subsamples of 20 individual bees, it is possible to infer the prevalence of diseased bees. The observed prevalence of AKI infection in colonies during early spring was extremely low, indicated by only four positive sub-samples out of 460 in April and May 2011. However, later in the season, the prevalence of AKI-positive subsamples increased steadily (Figure S3). It is more informative only to consider sub-samples that exceed our threshold value of 10^7 copies, representing diseased bees (Figure S4). Since all 10 subsamples were positive for several of the colonies, it was not possible to calculate the prevalence for each colony. Therefore, we considered the subsamples of all surviving colonies and the dying colonies as two separate pools, from which we could calculate the month-wise prevalence of diseased individuals (Figure 4). The prevalence of AKI-diseased bees peaked in October reaching 4% of the workers in the dying colonies compared to zero percent in surviving colonies. For DWV, the difference in October is less pronounced at 4% in dying colonies compared to 2.5% in survivors. This makes DWV less predictive of colony death than AKI. The 48 bees that were analysed individual from surviving and dying colonies (Figure S8) confirms that considerable variation exists among individuals. The viral copies between individual bees and pooled subsamples are not directly comparable due to difference in methodology. The samples from non-surviving colonies indicate that only 4% of the workers exhibit detrimental viral titres. This suggests a threshold level of viral prevalence beyond which a domino effect

sets in, probably mediated by varroa mites shifting host [44,54] leading to colony death.

Our results support the notion that high varroa load leads to build-up of high viral titres. As an exception, no varroa mites were detected in the bee samples for the entire year in colony A09. Yet, in October 2011, this colony suddenly exhibited increased level of virus prevalence for both AKI and DWV before collapsing in winter. This suggests that viral infections can spread between colonies even at low mite loads. The observed absence of varroa mites may be due to inadequate sampling. This is a typical problem when one has to balance between optimal sampling size and influencing survival of the studied colonies due to bee and mite removal. Too large samples or too frequent sampling will remove bees and varroa mites from the studied colonies. Beyond the influence of sampling on the final result, there is also the aspect of available time and money.

Based on the experiments done here, it is clear that a uniform sampling size across the season is not ideal. Much more can be learned from analysing biological replicates. We used a standard subsample size of 20 bees, based on the assumption that few bees have high viral load in spring and expected a build up over the season. In May, none of the subsamples were positive. In September and October, we observed all 10 subsamples to be positive for several colonies. In other words, subsamples of 20 bees were too large. Ideally, about half the pools are to be positive and half negative to achieve a reliable estimation of prevalence. Since the number of diseased bees was unknown, pools of 20 individuals were used throughout. In retrospect, it might have been wiser to split the analysis into several phases. An initial screening of two subsamples of 20 bees each followed by increased or decreased pool size depending on the initial results.

This study shows a clear trend of rising viral titres in bee populations over the course of a season from spring to autumn. Despite the application of acaricide, varroa mites and viral titres continued to rise while remaining below the levels observed in the untreated group. In many cases, the treatment only helped to control varroa loads and viral titres to a limited extend, and led to no significant reduction in the prevalence of virus-sick bees. In part, this may be explained by the recent findings [47,48], that show negative impact of acaricides on honey bee immunity. Adding to this effect of acaricides, the observed failure of the treatment to reduce varroa numbers in several colonies might have led to immuno-suppression syndrome as observed in other studies [18,55]. Our findings differ from a previous study [56], where an immediate drop of DWV titre was observed following varroa treatment. It has been noted in several studies that DWV reduces the lifespan of overwintering workers often in connection with varroa mites [1,3]. DWV has also been suggested to lead to colony losses independent of varroa infestation possibly in synergy with an uncharacterised stress factor [4]. Thus, we hypothesise that the effect of treatment may constitute this stress factor.

AKI and DWV viruses play a major role in declining colonies in Denmark. Varroa mites are the driving factor that has led to an upsurge in viral titres. Beekeepers implement various treatment regimes against varroa mites to prevent colony losses, which in turn lead to persistence viral infection in the colonies. The immune-suppression resulting from pesticide usage in the hive linked with the higher prevalence of viral infections may result in sudden colony loss. Beekeepers often experience low efficacy after years of recurring treatments. This subsequently leads to more aggressive treatment methods resulting in an unsustainable warfare on varroa mites, also detrimental to bee health. One practical solution to this malady would be selective treatment of colonies exhibiting high mite load. Knowledge of mite and virus levels in bee colonies is essential to establish a threshold beyond which colony loss is unavoidable. Ideally this threshold can be use in selection programs to identify and multiply colonies for which yearly treatment is not required. In Germany, the 'Arbeitsge-meinschaft Toleranzzucht', a co-operation of queen breeders in collaboration with researchers are working towards achieving this goal (www.toleranzzucht.de) [57].

Supporting Information

Figure S1 Dynamic range of quantification for AKI, beta-actin and DWV primers. Correlation (R^2) and reaction efficiency (E) for each primer pair are shown. Error bars show standard deviation based on two replicates each on eight plates (n = 16 for each point except * where n = 10).

Figure S2 Viral titres in varroa mites in three treatment groups across eight months (Left: AKI, Right: DWV). Error bars show standard error.

Figure S3 Proportion of bee sub-samples showing presence of viral infection in three treatment groups across eight months. Left: Proportion of bee sub-samples showing AKI infection. The organic and pyrethroid groups are reduced by treatment while untreated group continues to rise over the season. Right: Proportion of bee sub-samples showing DWV infection. All groups start at a higher infection proportion. All groups show rising proportion of infection. Treatment does not seem to suppress the spread of infection.

Figure S4 Proportion of bee sub-samples showing greater than 10^7 viral copies in three treatment groups across eight months. Left: AKI and Right: DWV.

Figure S5 Geographical locations of colonies used in this study. Inset bottom: Bornholm is an island located about 140 km east of Denmark.

Figure S6 AKI and DWV titre data is shown in $\log_{10}$ copies after C_t 34 cut-off. Sampling months are shown row-wise which includes 10 subsamples, varroa-free subsample as 0 (if sampled) and varroa as VAR (if present). Colonies are shown in columns. Contiguous background fill colour for the colony names represent apiaries. The colour of the colony text denotes treatment category. Green - Organic, Red - Flumethrin, Blue - Untreated. Viral titres are colour-coded into four groups. Zero or null virus is not coloured. 10–10^4 copies or low-level infection is coloured green, 10^4–10^7 copies or medium-level infection is coloured yellow and greater than 10^7 copies is coloured red showing serious damaging infection.

Figure S7 Raw mite counts, bee counts and mite indices are shown month-wise for all colonies. Colony names and colours are as explained for figure S6. The colour scheme is applied independently in the three tables. Colonies that died are marked as 'x'.

Figure S8 $\log_{10}$ AKI and DWV titres for 48 bees that were individually analysed. 'S' denotes colonies that survived while 'D' denotes colonies that died.

Table S1 List of primers used in this study. Outer primers were used to prepare standard curves.

Acknowledgments

We wish to thank beekeepers Aksel Jørgensen, Arne Jensen, Bent Larsen, Christian Petersen, Ditlev Bluhme, Flemming Thorsen, Gunner Borg, Jørgen Jørgensen, Karen Poulsen, Leif Johanssen, Orla Overby and Willy Svendsen for providing us with samples on time. We wish to thank lab technicians Hanne-Birgitte Christiansen, Henriette Nyskjold and Tina Tønnersen for help with sample handling and lab work.

Author Contributions

Conceived and designed the experiments: PK SLN. Performed the experiments: RMF. Analyzed the data: PK RMF. Contributed reagents/materials/analysis tools: RMF. Wrote the paper: RMF PK.

References

1. Dainat B, Evans JD, Chen YP, Gauthier L, Neumann P (2012) Dead or alive: deformed wing virus and Varroa destructor reduce the life span of winter honeybees. Appl Environ Microbiol 78: 981–987.

2. Generisch E, von der Ohe W, Kaatz H, Schroeder A, Otten C, et al. (2010) The German bee monitoring project: a long term study to understand periodically high winter losses of honey bee colonies. Apidologie 41: 332–352.

3. Berthoud H, Imdorf A, Haueter M, Radloff S, Neumann P (2010) Virus infections and winter losses of honey bee colonies (Apis mellifera). Journal of Apicultural Research 49: 60–65.

4. Highfield AC, El Nagar A, Mackinder LC, Noel LM, Hall MJ, et al. (2009) Deformed wing virus implicated in overwintering honeybee colony losses. Appl Environ Microbiol 75: 7212–7220.

5. Vanengelsdorp D, Evans JD, Saegerman C, Mullin C, Haubruge E, et al. (2009) Colony collapse disorder: a descriptive study. Plos One 4: e6481.

6. Carreck N, Neumann P (2010) Honey bee colony losses. Journal of Apicultural Research 49: 1–6.

7. Cox-Foster DL, Conlan S, Holmes EC, Palacios G, Evans JD, et al. (2007) A metagenomic survey of microbes in honey bee colony collapse disorder. Science 318: 283–287.

8. Dainat B, Evans JD, Chen YP, Gauthier L, Neumann P (2012) Predictive markers of honey bee colony collapse. Plos One 7: e32151.

9. Le Conte Y, Ellis M, Ritter W (2010) Varroa mites and honey bee health: can varroa explain part of the colony losses? Apidologie 41: 353–363.

10. Higes M, Martin-Hernandez R, Botias C, Bailon EG, Gonzalez-Porto AV, et al. (2008) How natural infection by Nosema ceranae causes honeybee colony collapse. Environmental Microbiology 10: 2659–2669.

11. Higes M, Martin-Hernandez R, Garrido-Bailon E, Gonzalez-Porto AV, Garcia-Palencia P, et al. (2009) Honeybee colony collapse due to Nosema ceranae in professional apiaries. Environmental Microbiology Reports 1: 110–113.

12. Paxton RJ (2010) Does infection by Nosema ceranae cause "Colony Collapse Disorder" in honey bees (Apis mellifera)? Journal of Apicultural Research 49: 80–84.

13. Cresswell JE, Desneux N, Vanengelsdorp D (2012) Dietary traces of neonicotinoid pesticides as a cause of population declines in honey bees: an evaluation by Hill's epidemiological criteria. Pest Manag Sci 68: 819–827.

14. Gross M (2008) Pesticides linked to bee deaths. Current Biology 18: R684–R684.

15. Johnson RM, Ellis MD, Mullin CA, Frazier M (2010) Pesticides and honey bee toxicity - USA. Apidologie 41: 312–331.

16. VanEngelsdorp D, Speybroeck N, Evans JD, Nguyen BK, Mullin C, et al. (2010) Weighing risk factors associated with bee colony collapse disorder by classification and regression tree analysis. Journal of Economic Entomology 103: 1517–1523.

17. Runckel C, Flenniken ML, Engel JC, Ruby JG, Ganem D, et al. (2011) Temporal analysis of the honey bee microbiome reveals four novel viruses and seasonal prevalence of known viruses, Nosema, and Crithidia. Plos One 6: e20656.

18. Nazzi F, Brown SP, Annoscia D, Del Piccolo F, Di Prisco G, et al. (2012) Synergistic Parasite-Pathogen Interactions Mediated by Host Immunity Can Drive the Collapse of Honeybee Colonies. PLoS Pathogens 8: e1002735.

19. Evans JD, Schwarz RS (2011) Bees brought to their knees: microbes affecting honey bee health. Trends in Microbiology 19: 614–620.

20. Vejsnæs F, Nielsen SL, Kryger P (2010) Factors involved in the recent increase in colony losses in Denmark. Journal of Apicultural Research 49: 109–110.

21. Carreck NL, Ball BV, Martin S (2010) Honey bee colony collapse and changes in viral prevalence associated with Varroa destructor. Journal of Apicultural Research 49: 93–94.

22. EU (2011) Guidelines for a pilot surveillance project on honey bee colony losses. 1–34 p.

23. Martin SJ, Highfield AC, Brettell L, Villalobos EM, Budge GE, et al. (2012) Global honey bee viral landscape altered by a parasitic mite. Science 336: 1304–1306.

24. Piotrowski F (1982) Varroosis-the correct term for varroatosis. Angew Parasitol 23: 49.

25. Boecking O, Genersch E (2008) Varroosis – the Ongoing Crisis in Bee Keeping. Journal für Verbraucherschutz und Lebensmittelsicherheit 3: 221–228.

26. Ifantidis MD (1983) Ontogenesis of the Mite Varroa-Jacobsoni in Worker and Drone Honeybee Brood Cells. Journal of Apicultural Research 22: 200–206.

27. Kralj J, Brockmann A, Fuchs S, Tautz J (2007) The parasitic mite Varroa destructor affects non-associative learning in honey bee foragers, Apis mellifera L. Journal of Comparative Physiology A: Neuroethology, Sensory, Neural, and Behavioral Physiology 193: 363–370.

28. Yang X, Cox-Foster D (2007) Effects of parasitization by Varroa destructor on survivorship and physiological traits of Apis mellifera in correlation with viral incidence and microbial challenge. Parasitology 134: 405–412.

29. Duay P, De Jong D, Engels W (2002) Decreased flight performance and sperm production in drones of the honey bee (Apis mellifera) slightly infested by Varroa destructor mites during pupal development. Genetics and Molecular Research 1: 227–232.

30. Ball BV (1983) The association of Varroa jacobsoni with virus diseases of honey bees. Experimental and Applied Acarology 19: 607–613.

31. Ball BV, Allen MF (1988) The prevalence of pathogens in honey bee (Apis mellifera) colonies infested with the parasitic mite Varroa jacobsoni. Annals of Applied Biology 113: 237–244.

32. Bowen-Walker PL, Martin SJ, Gunn A (1999) The transmission of deformed wing virus between honeybees (Apis mellifera L.) by the ectoparasitic mite Varroa jacobsoni Oud. Journal of Invertebrate Pathology 73: 101–106.

33. Chen YP, Pettis JS, Evans JD, Kramer M, Feldlaufer MF (2004) Transmission of Kashmir bee virus by the ectoparasitic mite Varroa destructor. Apidologie 35: 441–448.

34. Shen M, Yang X, Cox-Foster D, Cui L (2005) The role of varroa mites in infections of Kashmir bee virus (KBV) and deformed wing virus (DWV) in honey bees. Virology 342: 141–149.

35. Shen MQ, Cui LW, Ostiguy N, Cox-Foster D (2005) Intricate transmission routes and interactions between picorna-like viruses (Kashmir bee virus and sacbrood virus) with the honeybee host and the parasitic varroa mite. Journal of General Virology 86: 2281–2289.

36. Di Prisco G, Pennacchio F, Caprio E, Boncristiani HF Jr, Evans JD, et al. (2011) Varroa destructor is an effective vector of Israeli acute paralysis virus in the honeybee, Apis mellifera. J Gen Virol 92: 151–155.

37. Nordstrom S, Fries I, Aarhus A, Hansen H, Korpela S (1999) Virus infections in Nordic honey bee colonies with no, low or severe Varroa jacobsoni infestations. Apidologie 30: 475–484.

38. Tentcheva D, Gauthier L, Zappulla N, Dainat B, Cousserans F, et al. (2004) Prevalence and seasonal variations of six bee viruses in *Apis mellifera* L. and Varroa destructor mite populations in France. Applied and Environmental Microbiology 70: 7185–7191.

39. Gauthier L, Tentcheva D, Tournaire M, Dainat B, Cousserans F, et al. (2007) Viral load estimation in asymptomatic honey bee colonies using the quantitative RT-PCR technique. Apidologie 38: 426–U427.

40. Martin S (1998) A population model for the ectoparasitic mite Varroa jacobsoni in honey bee (Apis mellifera) colonies. Ecological Modelling 109: 267–281.

41. Vetharaniam I (2012) Predicting reproduction rate of varroa. Ecological Modelling 224: 11–17.

42. Martin SJ (2001) Varroa destructor reproduction during the winter in Apis mellifera colonies in UK. Experimental and Applied Acarology 25: 321–325.

43. Sumpter DJT, Martin SJ (2004) The dynamics of virus epidemics in Varroa-infested honey bee colonies. Journal of Animal Ecology 73: 51–63.

44. Martin SJ (2001) The role of Varroa and viral pathogens in the collapse of honeybee colonies: a modelling approach. Journal of Applied Ecology 38: 1082–1093.

45. Branco MR, Kidd NAC, Pickard RS (2006) A comparative evaluation of sampling methods for Varroa destructor (Acari: Varroidae) population estimation. Apidologie 37: 452–461.

46. Lee KV, Moon RD, Burkness EC, Hutchison WD, Spivak M (2010) Practical sampling plans for Varroa destructor (Acari: Varroidae) in Apis mellifera (Hymenoptera: Apidae) colonies and apiaries. Journal of Economic Entomology 103: 1039–1050.

47. Locke B, Forsgren E, Fries I, de Miranda JR (2012) Acaricide Treatment Affects Viral Dynamics in Varroa destructor-Infested Honey Bee Colonies via both Host Physiology and Mite Control. Applied and Environmental Microbiology 78: 227–235.

48. Boncristiani H, Underwood R, Schwarz R, Evans JD, Pettis J, et al. (2012) Direct effect of acaricides on pathogen loads and gene expression levels in honey bees Apis mellifera. Journal of Insect Physiology 58: 613–620.

49. Francis RM, Kryger P (2012) Single Assay Detection of Acute Bee Paralysis Virus, Kashmir Bee Virus and Israeli Acute Paralysis Virus. Journal of Apicultural Science 56: 137–146.

50. R Development Core Team (2011) R: A language and environment for statistical computing. R Foundation for Statistical Computing. Vienna.

51. Sergeant E (2012) Epitools epidemiological calculators. AusVet Animal Health Services and Australian Biosecurity Cooperative Research Centre for Emerging Infectious Disease.

52. Hauck WW (1991) Confidence intervals for seroprevalence determined from pooled sera. Ann Epidemiol 1: 277–281.

53. Cowling DW, Gardner IA, Johnson WO (1999) Comparison of methods for estimation of individual-level prevalence based on pooled samples. Preventive veterinary medicine 39: 211–225.

54. Le Conte Y, Arnold G (1987) Influence de L'age des abeilles (Apis mellifera L.) et de la chaleur sur le comportement de Varroa jacobsoni Oud. Apidologie 18: 305–320.

55. Yang XL, Cox-Foster DL (2005) Impact of an ectoparasite on the immunity and pathology of an invertebrate: Evidence for host immunosuppression and viral amplification. Proceedings of the National Academy of Sciences of the United States of America 102: 7470–7475.

56. Martin SJ, Ball BV, Carreck NL (2010) Prevalence and persistence of deformed wing virus (DWV) in untreated or acaricide-treated Varroa destructor infested honey bee (Apis mellifera) colonies. Journal of Apicultural Research 49: 72–79.

57. Buchler R, Berg S, Le Conte Y (2010) Breeding for resistance to Varroa destructor in Europe. Apidologie 41: 393–408.

10

Killing Them with Kindness? In-Hive Medications May Inhibit Xenobiotic Efflux Transporters and Endanger Honey Bees

David J. Hawthorne*, Galen P. Dively

Department of Entomology, University of Maryland, College Park, Maryland, United States of America

Abstract

Background: Honey bees (*Apis mellifera*) have recently experienced higher than normal overwintering colony losses. Many factors have been evoked to explain the losses, among which are the presence of residues of pesticides and veterinary products in hives. Multiple residues are present at the same time, though most often in low concentrations so that no single product has yet been associated with losses. Involvement of a combination of residues to losses may however not be excluded. To understand the impact of an exposure to combined residues on honey bees, we propose a mechanism-based strategy, focusing here on Multi-Drug Resistance (MDR) transporters as mediators of those interactions.

Methodology/Principal Findings: Using whole-animal bioassays, we demonstrate through inhibition by verapamil that the widely used organophosphate and pyrethroid acaricides coumaphos and τ-fluvalinate, and three neonicotinoid insecticides: imidacloprid, acetamiprid and thiacloprid are substrates of one or more MDR transporters. Among the candidate inhibitors of honey bee MDR transporters is the in-hive antibiotic oxytetracycline. Bees prefed oxytetracycline were significantly sensitized to the acaricides coumaphos and τ-fluvalinate, suggesting that the antibiotic may interfere with the normal excretion or metabolism of these pesticides.

Conclusions/Significance: Many bee hives receive regular treatments of oxytetracycline and acaricides for prevention and treatment of disease and parasites. Our results suggest that seasonal co-application of these medicines to bee hives could increase the adverse effects of these and perhaps other pesticides. Our results also demonstrate the utility of a mechanism-based strategy. By identifying pesticides and apicultural medicines that are substrates and inhibitors of xenobiotic transporters we prioritize the testing of those chemical combinations most likely to result in adverse interactions.

Editor: Guy Smagghe, Ghent University, Belgium

Funding: This work was supported by the United States Department of Agriculture. The funders had no role in study design, data collection and analysis, decision to publish, or preparation of the manuscript.

Competing Interests: The authors have declared that no competing interests exist.

* E-mail: djh@umd.edu

Introduction

Honey bees are in trouble. Widespread depopulation of colonies often characterized by high overwintering losses has occurred since at least 2006 in the United States, threatening the sustainability of North American apiculture. Despite considerable effort, no single cause of the phenomenon called colony collapse disorder (CCD) has been identified, though associations of several pathogens and parasites appear to increase the risk of colony collapse [1,2]. Pesticides are also among the suspected contributing factors of colony collapse both because bees encounter a diverse array of pesticides when foraging and because more than 120 different pesticides have been found within bee hives [2,3,4,5,6]. Some pesticides have received extra scrutiny, notably the acaricides coumaphos and τ-fluvalinate, applied to bee hives for control of parasitic varroa mites, and the widely used neonicotinoid insecticides. These acaricides are applied directly to bee hives, accumulate in wax and were found in nearly all hives recently tested in both N. American and France [5,6]. The neonicotinoids (especially imidacloprid) are of concern because

they are toxic to honey bees, used on many crops and ornamental plants, and they tend to be systemically distributed within treated plants, potentially contaminating nectar and pollen of treated and rotational crops not initially treated with these products [7,8,9,10].

Although pesticide drift and overdosing cause accidental bee kills no single pesticide has been directly implicated with widespread overwintering losses or CCD [2,5]. It remains possible however, that combinations of toxins may cause adverse additive or synergistic effects that would be difficult to detect through surveys of beekeepers or analysis of their apiaries without dedicated multifactorial analysis. It has been shown, for example, that the toxicity to bees of some pyrethroid and neonicotinoid insecticides increases significantly when combined with certain fungicides [11,12]. Similarly, Johnson et al. [13] found that coumaphos and τ-fluvalinate each synergize the other's toxicity to honey bees, perhaps through competitive inhibition of the metabolic enzymes that detoxify those pesticides. Given the many pesticides that bees encounter there may be adverse combinations of them eroding hive health in both subtle and dramatic ways.

The problem, of course, is the large number of potentially adverse pesticide combinations which prevents evaluation of all, or even most, combinations of them. This problem challenges our ability to anticipate the risks associated with bee's exposure to a novel pesticide or to identify combinations of toxins contributing to a colony collapse. If we could identify mechanisms of the honey bee xenobiotic metabolism and excretion systems that systematically mediate multiple-toxin interactions, we could reduce the overwhelming number of candidate pesticide interactions to a smaller set of compounds that are substrates or inhibitors of the most predictive mechanisms.

The membrane-bound transporter proteins from the ABC transporter family of proteins are found in all phyla [14,15]. The xenobiotic transporters in this family actively shuttle toxins across cell membranes to reduce the intracellular toxin and metabolite concentrations. Working in concert with metabolic enzymes, these transporters mediate a baseline tolerance to a diverse array of toxins including numerous drugs, pesticides and phytochemicals [16,17]. Several of these transporters, especially members of the ABCB, ABCC, and ABCG subfamilies of transporters (referred to here as Multiple Drug Resistance, or MDR transporters), are of medical importance, playing a role in resistance to multiple cancer and anti-parasite drugs [17,18,19].

MDR transporters are relatively unstudied in insects, and completely neglected in honey bee toxicology. These transporters act in several insect tissues, including the cuticle [20], malpighian tubules [21,22], midgut [23] and at the blood-brain barrier [24,25] to transport toxins, including pesticides, towards excretion [17]. The honey bee genome contains genes coding for orthologues of these proteins, which presumably protect bees from toxins as they do in *Drosophila melanogaster* [24,26,27], chironomid flies [28], mosquitoes [29], *Heliothis virescens* (tobacco budworm) [20] and *Manduca sexta* (tomato hornworm) [21,25]. It seems reasonable therefore to consider the role that these proteins play in honey bee tolerance of pesticides and to begin an analysis of potentially inhibitory compounds that bees commonly encounter.

The most well studied MDR transporter, p-glycoprotein (p-gp), has both a diverse range of substrates and is inhibited by an array of drugs, pesticides and plant compounds [17]. This inhibition is a mechanism by which MDR transporters would cause adverse interactions among many chemicals; one compound inhibits the transporters thereby increasing sensitivity to other toxic substrates. The drug verapamil is a potent inhibitor of p-gp and possibly other MDR transporters [30,31]. It is frequently used as the standard inhibitor of p-gp where it increases the sensitivity of treated cells, tissues or organisms to toxic transporter substrates [17,18,26]. Here we use verapamil inhibition to determine if 5 pesticides are substrates of MDR transporters and therefore potentially synergized by other inhibitors more likely to be encountered by honey bees. Remarkably, three widely used in-hive pesticides and medications (the previously mentioned acaricides coumaphos and τ-fluvalinate and the antibiotic oxytetracycline) are known substrates and/or inhibitors of mammalian p-gp [31,32,33]. We suspect that these in-hive medications and pesticides may be interacting with bee's MDR transporters, increasing their sensitivity to these and perhaps other pesticides and toxins. The frequent contamination of hive wax with these acaricides [6] and routine treatment of hives with oxytetracycline [34,35,36,37] undoubtedly increases the exposure of bees to these compounds, with potentially significant consequences if they are indeed substrates or inhibitors of honey bee MDR transporters.

Interaction of neonicotinoid insecticides with insect MDR transporters has not yet been reported. Because of the likelihood

of exposure of bees to these insecticides we ask if the neonicotinoid insecticides imidacloprid, acetamiprid and thiacloprid are substrates of honey bee MDR transporters. Evidence of neonicotinoid processing by MDR transporters would be significant because inhibition of those transporters could cause mortality at lower doses than normally expected for individual compounds.

Results

When fed to bees verapamil significantly increased the toxicity of all 5 acaricides/insecticides. Mean mortality of young worker bees topically treated with the acaricides coumaphos or τ-fluvalinate was significantly higher when bees were pretreated with verapamil (Fig. 1, Table 1). Control mortality following topical application of acetone was 0% for both sucrose and sucrose+verapamil fed bees. Acute oral toxicity was also significantly higher for all three neonicotinoids (acetamiprid, thiacloprid, imidacloprid) when bees were pretreated with verapamil (Fig. 1, Table 2). Increased mortality at higher concentrations and at the later end point (48 h) was observed for thiacloprid, and at 48 h for imidacloprid. The effect of verapamil pretreatment did not differ among concentrations of these insecticides (Table 2). Control mortality of sucrose only and sucrose+verapamil cohorts averaged 2–3%.

Oxytetracycline significantly increased the mortality of bees exposed to coumaphos and τ-fluvalinate (Fig. 2). For comparison with the verapamil synergism reported above, mean mortality of bees treated with 2 ug/ul coumaphos increased from 7% (n = 4 cages) to 51% (n = 4 cages) following feeding of OTC (1.4 mM), a significant but smaller increase than that caused by verapamil (Fig. 2A, Table 1). OTC feeding increased the mortality of bees treated with 3 ug/ul τ-fluvalinate from 5.6% (n = 10 cages) to 39% (n = 8 cages) (Fig. 2B, $p = 0.002$). Mean mortality of cohorts fed OTC alone were below 10% and were not significantly different from those fed sucrose alone (Fig. 2).

Discussion

Here we provide the first evidence that the MDR transporter(s) inhibited by verapamil play a role in protecting honey bees from pesticides, and that the acaricides coumaphos and τ-fluvalinate, and 3 neonicotinoid insecticides are substrates of these transporters in insects. The observation that coumaphos and τ-fluvalinate are substrates of honey bee p-gp or another MDR transporter was anticipated from previous study of mouse cells, and suggests that insect and mammalian MDR transporters share substrates. Clearly, the abundance of these pesticides found in the wax and pollen of bee hives [6] coupled with evidence that their toxicity to bees is increased through inhibition of MDR transporters implicates them as toxins of interest in any multifactorial explanation of high overwintering colony losses.

This is the first report that neonicotinoid insecticides are substrates of insect MDR transporters. In efforts to protect honey bees, energetic opposition to the neonicotinoids has arisen in North America and Europe, but direct implication of them in overwintering losses has not been sustained by recent research [2,6]. Estimates of the environmental exposure of bees to imidacloprid are typically low relative to the LD_{50} [6], and studies have not demonstrated hive-level consequences of imidacloprid contamination [38]. Our results suggest that inhibition of MDR transporters may reduce the LD_{50} of neonicotinoids possibly amplifying acute and chronic effects to bees at lower concentrations.

The large increases in sensitivity to pesticides by inhibition of MDR transporters and the chemical diversity of the synergized

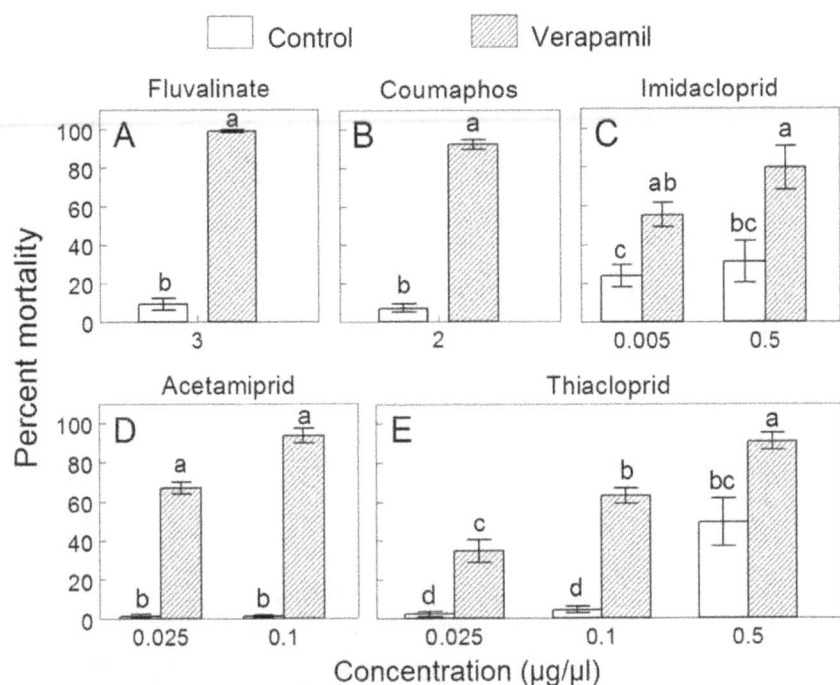

Figure 1. Verapamil synergizes honey bee mortality by five acaricides/insecticides. Mean mortality (±SE) of honey bees (average of 24 and 48 h) following topical (A, B) and oral (C, D, E) exposure to pesticides. Bees were pre-fed sucrose or sucrose+verapamil (1 mM) solution. For each pesticide, different letters indicate significant differences between treatments ($p<0.05$).

pesticides suggest that these transporters may mediate adverse synergisms of diverse toxins in bees. Because of its clinical importance in human health, proven and candidate p-gp substrates and inhibitors of many types have been identified [16,17]. Knowledge of these compounds may help us identify chemicals likely to interact with honey bee MDR transporters. In the first application of this mechanism-based strategy to honey bees, we uncover a significant negative interaction among three medications routinely applied to bee hives [35,36,37]. OTC, coumaphos and τ-fluvalinate are all known to interact with mammalian p-gp [31,32,33]. If honey bee transporters behave similarly, we would expect increased toxicity following co-application of a toxic transporter substrate and an inhibitor. As anticipated, concentrations of OTC similar to those applied to bee hives increased bee's sensitivity to both coumaphos and τ-fluvalinate. OTC is applied to bee hives in the late fall and/or early spring, often in tandem with one of the acaricides [36]. Our results suggest that co-application of these compounds could increase the likelihood of intoxication by the acaricides and other

pesticides contaminating beeswax and food stores. These results raise the possibility that adverse interactions of medications (such as OTC) and pesticides (coumaphos and τ-fluvalinate) contribute to the loss of honey bee colonies during the winter or early spring, a common feature of CCD. Although the per-bee concentration of OTC used here is similar to field application rates, the pesticide concentrations are higher than those found in bee hives (see [6]). Therefore, although we have demonstrated that verapamil and OTC increase bee's sensitivity to these pesticides in acute laboratory bioassays, additional testing of lower pesticide concentrations over longer time periods is necessary to fully understand the field relevance of these interactions. Additional work is also required to directly demonstrate that OTC inhibits p-gp or other efflux transporters in honey bees. Nevertheless, we show here using OTC and the acaricides as an example, how identification of MDR transporter substrates and inhibitors can highlight potentially dangerous chemical combinations and improve the assessment and management of toxicological risks faced by honey bees.

Table 1. Repeated-measures analysis of variance of honey bee mortality.

	Pesticide treatment (Pretreatment)								
	Coumaphos (Verapamil)			τ-Fluvalinate (Verapamil)			Coumaphos (OTC)		
	df	F	p	df	F	p	df	F	p
Pretreatment	1,14.5	61.89	<0.0001	1,10	57.77	<0.0001	1,11	10.83	0.0072
Time	1,11.2	3.91	0.07	1,10	1.46	0.26	1,9.8	10.64	0.0088
Pretreatment×Time	1,11.2	3.20	0.10	1,10	1.46	0.26	1,9.8	1.66	0.2277

Bees were pretreated with verapamil, oxytetracycline (OTC), or sucrose syrup then treated with the acaricides coumaphos or τ-fluvalinate.

Table 2. Repeated-measures analysis of variance of honey bee mortality.

	Imidacloprid			Acetamiprid			Thiacloprid		
	df	F	p	df	F	p	df	F	p
Pretreatment	1,28	17.78	0.0002	1,12	128.54	<0.0001	1,24	65.53	<0.0001
Concentration	1,28	2.75	0.11	1,12	0.26	0.62	1,24	27.93	<0.0001
Time	1,28	43.12	<0.0001	1,12	1.24	0.29	1,24	94.97	<0.0001
Pretreatment×Concentration	1,28	0.80	0.38	1,12	0.27	0.61	1,24	2.39	0.11
Pretreatment×Time	1,28	1.72	0.2	1,12	0.63	0.44	1,24	53.31	<0.0001
Concentration×Time	1,28	0.66	0.42	1,12	1.02	0.33	1,24	58.17	<0.0001
Pre×Conce×Time	1,28	3.51	0.07	1,12	0.80	0.39	1,24	69.75	<0.0001

Bees were pretreated with verapamil or sucrose syrup and then fed one of three neonicotinoid insecticides.

Materials and Methods

Insects

Bees were collected for laboratory bioassays from newly established colonies reared on new frames and freshly drawn comb. Colonies were not treated with apicultural medications or pesticides. Frames with emerging workers were taken from hives and placed into dark growth chambers maintained at $33\pm2°C$ and (70–80%) RH. Newly-emerged bees were collected from the frames daily and maintained in groups of 20–30 in 80×100 mm metal mesh cages capped at each end by standard polystyrene petri dishes. Bees were fed sucrose solution (50%; w:v) through 1 mm holes from a 2.0 ml microfuge tube.

Chemicals

Terramycin (oxytetracycline, 5.5% soluble powder, Pfizer) was purchased from Dadant and Sons (Hamilton, Illinois). Coumaphos, τ-fluvalinate (both technical grade) and verapamil were purchased from Sigma-Aldrich Inc. (St. Louis, MO). Commercial formulations of imidacloprid (Admire Pro) and thiacloprid (Calypso) were provided by Bayer CropScience (Durham, NC), and acetamiprid (Assail) was provided by United Phosphorous Inc. (King of Prussia, PA).

Drug pretreatments

Verapamil (1 mM) and oxytetracycline (OTC, 1.4 mM) were incorporated into 50% sucrose solutions for oral dosing of 1–3 day

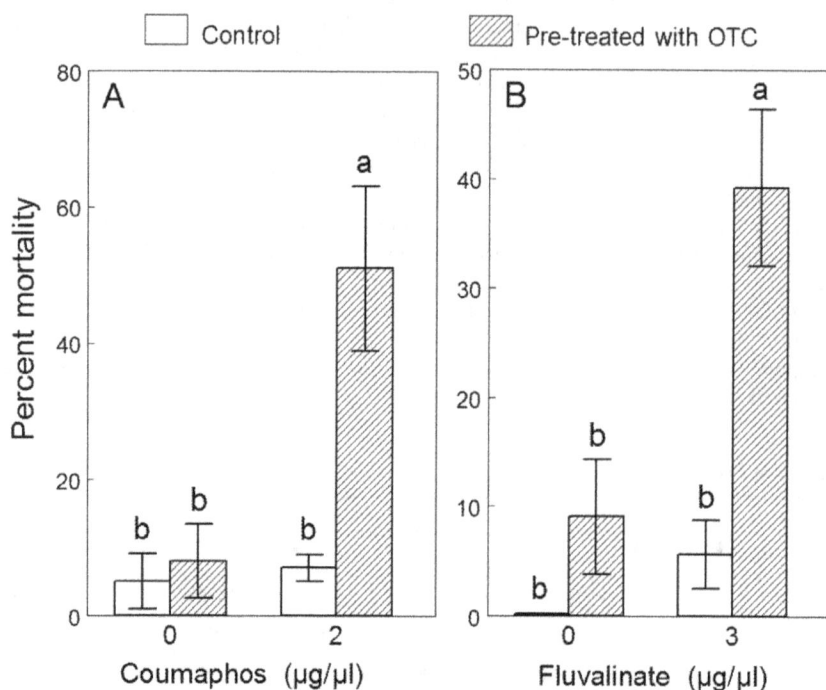

Figure 2. Oxytetracycline (OTC) synergizes honey bee mortality by in-hive acaricides. Mean mortality ($\pm$SE) of honey bees pre-fed sucrose solution (50%) or sucrose+oxytetracycline (1.4 mM) and topical application of (A) coumaphos (average of 24 and 48 h) and (B) τ-fluvalinate (24 h). For each pesticide, different letters indicate significant differences between treatments ($p<0.05$).

old workers. Preliminary feeding trials of 1 mM solutions of oral verapamil revealed no toxicity. The 1.4 mM concentration of OTC provides a per-bee exposure comparable to that of the label-recommended dosage of 600 mg applied to a hive containing 12,000 bees—a typical colony size entering winter [39]. Sucrose-drug solutions were made fresh every 3 days and the vials supplying each cage were replaced as needed.

Topical bioassays of insecticides/acaricides

Cohorts of 3–6 day old workers pretreated by feeding with the two sucrose-drug solutions were anesthetized with CO_2, and 1 ul of coumaphos (2 ug/ul)or τ-fluvalinate (3 ug/ul) in acetone (or acetone alone for controls) was applied to the dorsal thorax of each bee using an ISCO microapplicator driving a 1/4 cc tuberculin syringe. After application, bees were returned to cages containing the sucrose-drug or sucrose-only solution. Mortality of bees in each cage was recorded at 24 and 48 h. 5–10 replicate cohorts of 25 bees were tested for each acaricide - pretreatment combination.

Oral bioassays of insecticides

Pre-fed cohorts of 3–6 day old workers were fed sucrose syrup containing one of the neonicotinoids. Mortality of each cage was recorded at 24 and 48 hours. Imidacloprid was tested at 5 and 50 ng/ul, acetamiprid at 25 and 100 ng/ul, and thiacloprid at 25, 100 and 500 ng/ul. These concentrations caused low-intermediate

mortality of bees fed sucrose-only solution in preliminary range-finding experiments. 2–13 replicate cohorts of 25 bees were tested for each toxin concentration - pretreatment combination.

Analysis. The effects of verapamil or OTC pretreatment on insecticide/acaricide mortality were tested using a repeated measures analysis of variance (Proc Mixed, SAS). Following transformation (arcsine square-root), mortality was analyzed with a model that included pretreatment, insecticide concentration if multiple levels were used, and time endpoints (24 and 48 h) as fixed factors to assess the main effects and their interactions. Because only mortality at 24 h was available, analysis of τ-fluvalinate combined with OTC, was performed using a simple t-test, comparing the τ-fluvalinate and the τ-fluvalinate+OTC treatments.

Acknowledgments

We thank Jeffrey Pettis, Mike Embrey, Peter Swaan, Jeff Scott and David Onstad for thoughtful discussions and suggestions.

Author Contributions

Conceived and designed the experiments: DJH. Performed the experiments: DJH. Analyzed the data: DJH GPD. Contributed reagents/materials/analysis tools: DJH GPD. Wrote the paper: DJH.

References

1. Cox-Foster DL, Conlan S, Holmes EC, Palacios G, Evans JD, et al. (2007) A metagenomic survey of microbes in honey bee colony collapse disorder. Science 318: 283–287.
2. vanEngelsdorp D, Evans JD, Saegerman C, Mullin C, Haubruge E, et al. (2009) Colony collapse disorder: a descriptive study. PLoS ONE 4: e6481.
3. Ratnieks FLW, Carreck NL (2010) Clarity on honey bee collapse? Science 327: 152–153.
4. Frazier M, Mullin C, Frazier J, Ashcraft S (2008) What have pesticides got to do with it? Am Bee J 148: 521–523.
5. Chauzat MP, Carpentier P, Martel AC, Bougeard S, Cougoule N, et al. (2009) Influence of pesticide residues on honey bee (Hymenoptera: Apidae) colony health in France. Environ Entomol 38: 514–523.
6. Mullin CA, Frazier M, Frazier JL, Ashcraft S, Simonds R, et al. (2010) High levels of miticides and agrochemicals in North American apiaries: implications for honey bee health. PLoS ONE 5: e9754.
7. Maus C, Cure G, Schmuck R (2003) Safety of imidacloprid seed dressings to honey bees: a comprehensive overview and compilation of the current state of knowledge. Bull Insectol 56: 51–57.
8. Rortais A, Arnold G, Halm MP, Touffet-Briens F (2005) Modes of honeybees exposure to systemic insecticides: estimated amounts of contaminated pollen and nectar consumed by different categories of bees. Apidologie 36: 71–83.
9. Laurent FM, Rathahao E (2003) Distribution of [14C]imidacloprid in sunflowers (Helianthus annuus L.) following seed treatment. J Agric Food Chem 51: 8005–8010.
10. Elbert A, Haas M, Springer B, Thielert W, Nauen R (2008) Applied aspects of neonicotinoid uses in crop protection. Pest Manage Sci 64: 1099–1105.
11. Iwasa T, Motoyama N, Ambrose JT, Roe RM (2004) Mechanism for the differential toxicity of neonicotinoid insecticides in the honey bee, Apis mellifera. Crop Protect 23: 371–378.
12. Pilling ED, Jepson PC (1993) Synergism between EBI fungicides and a pyrethroid insecticide in the honeybee (Apis mellifera). Pestic Sci 39: 293–297.
13. Johnson RM, Pollock HS, Berenbaum MR (2009) Synergistic interactions between in-hive miticides in Apis mellifera. J Econ Entomol 102: 474–479.
14. Gottesman MM, Fojo T, Bates SE (2002) Multidrug resistance in cancer: role of ATP-dependent transporters. Nat Rev Cancer 2: 48–58.
15. Sturm A, Cunningham P, Dean M (2009) The ABC transporter gene family of Daphnia pulex. BMC Genomics 10.
16. Didziapetris R, Japertas P, Avdeef A, Petrauskas A (2003) Classification analysis of P-glycoprotein substrate specificity. J Drug Targeting 11: 391–406.
17. Buss DS, Callaghan A (2008) Interaction of pesticides with p-glycoprotein and other ABC proteins: A survey of the possible importance to insecticide, herbicide and fungicide resistance. Pestic Biochem Physiol 90: 141–153.
18. Leslie EM, Deeley RG, Cole SPC (2005) Multidrug resistance proteins: role of P-glycoprotein, MRP1, MRP2, and BCRP (ABCG2) in tissue defense. Toxicol Appl Pharmacol 204: 216–237.
19. Kerboeuf D, Guegnard F (2011) Anthelmintics are substrates and activators of nematode p-glycoprotein. Antimicrob Agents Chemother 55: 2224–2232.
20. Lanning CL, Ayad HM, AbouDonia MB (1996) P-glycoprotein involvement in cuticular penetration of C-14 thiodicarb in resistant tobacco budworms. Toxicol Lett 85: 127–133.
21. Gaertner LS, Murray CL, Morris CE (1998) Transepithelial transport of nicotine and vinblastine in isolated malpighian tubules of the tobacco hornworm (Manduca sexta) suggests a P-glycoprotein-like mechanism. J Exp Biol 201: 2637–2645.
22. Leader JP, O'Donnell MJ (2005) Transepithelial transport of fluorescent p-glycoprotein and MRP2 substrates by insect Malpighian tubules: confocal microscopic analysis of secreted fluid droplets. J Exp Biol 208: 4363–4376.
23. Tapadia MG, Lakhotia SC (2005) Expression of mdr49 and mdr65 multidrug resistance genes in larval tissues of Drosophila melanogaster under normal and stress conditions. Cell Stress Chaperones 10: 7–11.
24. Mayer F, Mayer N, Chinn L, Pinsonneault RL, Kroetz D, et al. (2009) Evolutionary conservation of vertebrate blood-brain barrier chemoprotective mechanisms in drosophila. J Neurosci 29: 3538–3550.
25. Murray CL, Quaglia M, Arnason JT, Morris CE (1994) A putative nicotine pump at the metabolic blood-brain-barrier of the tobacco hornworm. J Neurobiol 25: 23–34.
26. Callaghan A, Denny N (2002) Evidence for an interaction between p-glycoprotein and cadmium toxicity in cadmium-resistant and -susceptible strains of Drosophila melanogaster. Ecotoxicol Environ Saf 52: 211–213.
27. Vache C, Camares O, Cardoso-Ferreira MC, Dastugue B, Creveaux I, et al. (2007) A potential genomic biomarker for the detection of polycyclic aromatic hydrocarbon pollutants: Multidrug resistance gene 49 in Drosophila melanogaster. Environ Toxicol Chem 26: 1418–1424.
28. Podsiadlowski L, Matha V, Vilcinskas A (1998) Detection of a P-glycoprotein related pump in Chironomus larvae and its inhibition by verapamil and cyclosporin A. Comparative Biochemistry and Physiology B-Biochemistry & Molecular Biology 121: 443–450.
29. Porretta D, Gargani M, Bellini R, Medici A, Punelli F, et al. (2008) Defence mechanisms against insecticides temephos and diflubenzuron in the mosquito Aedes caspius: the P-glycoprotein efflux pumps. Med Vet Entomol 22: 48–54.
30. Tsuruo T, Iida H, Tsukagoshi S, Sakurai Y (1981) Overcoming of vincristine resistance in P388 leukemia invivo and invitro through enhanced cyto-toxicity of vincristine and vinblastine by verapamil. Cancer Res 41: 1967–1972.
31. Bain LJ, McLachlan JB, LeBlanc GA (1997) Structure-activity relationships for xenobiotic transport substrates and inhibitory ligands of P-glycoprotein. Environ Health Perspect 105: 812–818.
32. Bain LJ, LeBlanc GA (1996) Interaction of structurally diverse pesticides with the human MDR1 gene product P-glycoprotein. Toxicol Appl Pharmacol 141: 288–298.
33. Schrickx J, Fink-Gremmels J (2007) P-glycoprotein-mediated transport of oxytetracycline in the Caco-2 cell model. J Vet Pharmacol Ther 30: 25–31.
34. Spivak M, Reuter GS (2001) Resistance to American foulbrood disease by honeybee colonies Apis mellifera bred for hygienic behavior. Apidologie 32: 555–565.

35. Sammataro D (1998) The beekeeper's handbook. IthacaNY: Comstock Pub. pp 190.
36. Sanford MT, Bonney RE (2010) Storey's Guide to Keeping Honey Bees. North Adams, MA: Storey Publishing.
37. Delaplane KS, Lozano LF (1994) Using Terramycin(R) in honey-bee colonies. Am Bee J 134: 259–261.

38. Faucon JP, Aurieres C, Drajnudel P, Mathieu L, Ribiere M, et al. (2005) Experimental study on the toxicity of imidacloprid given in syrup to honey bee (Apis mellifera) colonies. Pest Manage Sci 61: 111–125.
39. Winston ML (1987) The biology of the honey bee. CambridgeMass.: Harvard University Press. 281 p.

11

Sub-Lethal Effects of Pesticide Residues in Brood Comb on Worker Honey Bee (*Apis mellifera*) Development and Longevity

Judy Y. Wu, Carol M. Anelli, Walter S. Sheppard*

Department of Entomology, Washington State University, Pullman, Washington, United States of America

Abstract

Background: Numerous surveys reveal high levels of pesticide residue contamination in honey bee comb. We conducted studies to examine possible direct and indirect effects of pesticide exposure from contaminated brood comb on developing worker bees and adult worker lifespan.

Methodology/Principal Findings: Worker bees were reared in brood comb containing high levels of known pesticide residues (treatment) or in relatively uncontaminated brood comb (control). Delayed development was observed in bees reared in treatment combs containing high levels of pesticides particularly in the early stages (day 4 and 8) of worker bee development. Adult longevity was reduced by 4 days in bees exposed to pesticide residues in contaminated brood comb during development. Pesticide residue migration from comb containing high pesticide residues caused contamination of control comb after multiple brood cycles and provided insight on how quickly residues move through wax. Higher brood mortality and delayed adult emergence occurred after multiple brood cycles in contaminated control combs. In contrast, survivability increased in bees reared in treatment comb after multiple brood cycles when pesticide residues had been reduced in treatment combs due to residue migration into uncontaminated control combs, supporting comb replacement efforts. Chemical analysis after the experiment confirmed the migration of pesticide residues from treatment combs into previously uncontaminated control comb.

Conclusions/Significance: This study is the first to demonstrate sub-lethal effects on worker honey bees from pesticide residue exposure from contaminated brood comb. Sub-lethal effects, including delayed larval development and adult emergence or shortened adult longevity, can have indirect effects on the colony such as premature shifts in hive roles and foraging activity. In addition, longer development time for bees may provide a reproductive advantage for parasitic *Varroa destructor* mites. The impact of delayed development in bees on *Varroa* mite fecundity should be examined further.

Editor: Frederic Marion-Poll, INRA - Paris 6 - AgroParisTech, France

Funding: Sources of funding that supported this work included State Beekeeping Associations from WA, OR, ID, CA and MT. The funders had no role in study design, data collection and analysis, decision to publish, or preparation of the manuscript.

Competing Interests: The authors have declared that no competing interests exist.

* E-mail: shepp@wsu.edu

Introduction

Losses associated with colony collapse disorder (CCD) represent a continuation in sudden and often catastrophic population crashes in honey bee (*Apis mellifera*) colonies that have become commonplace since the mid 1980s, when two species of parasitic mites were discovered in the United States [1]. Over 60 contributing factors of CCD have been identified, including *Varroa destructor* mites, poor nutrition, exposure to both agrochemicals and beekeeper-applied pesticides, and various other pests and pathogens [2]. Honey bee health decline and colony losses have not been limited to the U.S. Many studies in Europe have examined potential correlations between major recent bee losses and pesticide exposure, particularly, the class of neonicotinoid insecticides [3,4,5]. Studies from Spain have focused mainly on the effects of *Nosema ceranae*, a microsporidian pathogen that targets the honey bee midgut and deprives infected bees of nutrients [6]. There is a lack of agreement about which factors are more important in colony collapse and

some researchers have focused on interaction effects of combined factors. For example, pesticide exposure increases honey bee susceptibility to *Nosema ceranae* spore infection and *vice versa* [7,8].

Honey bee colony health can be affected by many factors including hygienic behavior, innate immunity, pesticide sensitivity, nutrition, adult age, and temperature. As social insects, honey bees have evolved various traits, such as grooming or other hygienic behaviors (including removal of mites and dead or diseased brood) that protects the colony against pests and pathogens. Social immunity provides significant protection for honey bee colonies and it has been suggested that this may explain why, compared to non-social insects, honey bees are relatively immunologically deficient (i.e., express fewer immune response proteins) [9]. Honey bees have about half as many detoxifying enzymes as pesticide resistant insects [10]. This deficiency increases the sensitivity of honey bees to pesticide exposure and can further reduce their ability to fight bacterial or viral infections. Pesticide sensitivity and physiological condition may also vary due to bee age and nutritional status, which can affect overall colony health [11]. Older bees (foragers) are more susceptible to pesticide exposure due to foraging activity than younger bees that remain in the hive [12,13]. Honey bees fed high quality pollen are less susceptible to pesticide exposure than bees fed protein-deficient pollen or pollen substitutes [12]. Migratory commercial beekeepers typically provide pollen substitute to colonies during transport and seasonal dearth to maximize brood production prior to and during pollination services. Adult honey bees are also more susceptible to pesticides when reared at lower temperatures (33°C) [14], a potential added stress factor associated with the commonly employed transportation of honey bee colonies.

In this study we examined the sub-lethal effects of developmental exposure to pesticide residues on worker bees. Worker bees were reared in brood comb containing high levels of known pesticide residues or in brood comb relatively free of pesticide residues. We discuss implications of sub-lethal and indirect effects of pesticide residues in brood comb on colony health and structure.

Materials and Methods

Experimental combs

Frames of treatment brood comb originated from migratory Pacific Northwest beekeeping operations that used miticides and from colonies provided by the USDA-ARS honey bee laboratory, Beltsville, MD that were suspected to have died from Colony Collapse Disorder. Pesticide residue analyses were performed on brood comb samples and thirteen frames of brood combs positive for high levels of pesticide residues were cut into treatment blocks (11×11-cm), each containing roughly 450 cells. Control brood combs were newly drawn out from a single colony or sampled from feral colonies that tested negative for pesticide residue contamination.

Experimental design

Standard Langstroth frames, with the center area (22×11-cm) of the frame removed, were used as frame supports for a pair of comb blocks, i.e., one low pesticide residue control comb placed next to a treatment comb block containing high pesticide residue levels (n = 17). Three colonies of similar strength were used from May through August of 2008 and 2009 to host experimental frames supporting paired comb blocks. Placing control and treatment combs within the same colony during larval development equalized possible effects of colony activity and quality of resources fed to brood. Laying sister queens from each colony were caged for 24 hours over experimental frames, allowing access to both control and treatment comb blocks. Queens were released the following day and excluded to the bottom box for the duration of the experiment. Frames containing a patch of 224 eggs on control and treatment blocks were photographed and frames with insufficient number of eggs were removed from the experiment. Egg patches were monitored for larval mortality on days 4, 8,12, and 19 of development, and photographs taken of larvae developing in control and treatment comb were mapped using Microsoft Paint 2007. On day 19, experimental frames containing pupae reared in control and treatment comb were incubated at 33±1°C. Push-in cages were used to isolate treatment and control blocks. Emergence of adult bees was recorded daily and bees were counted, tagged with Testor's enamel, and placed in a 3.2 mm mesh metal cage (11×9×5-cm). Bees reared in treatment blocks were placed in the same cage with bees reared in corresponding control blocks from the same frame. Worker bees were fed water, 50% sucrose syrup, and pollen supplement *ad libitum* and mortality was recorded daily. Some experimental frames (n = 9) containing a

pair of control and treatment comb blocks were reused up to three times during the experiment. Experimental frame supports containing comb blocks that had not yet been used in the experiment (Rep 1) were introduced to host colonies at the same time as other frames that had gone through multiple brood cycles (Rep 2 & 3) to minimize seasonal foraging and thermal effects on larval survival. A total of twenty-eight replicates were completed in this study between May and August 2008 and 2009.

Chemical analysis

Brood comb samples were sent to Roger Simonds USDA-AMS-National Science Laboratory, Gastonia, NC to be analyzed using QuEChERS method. Pesticide residue extraction and analysis was accomplished using liquid chromatography combined with tandem mass spectrometry (LC/MS/MS - Agilent 1100 LC equipped with a Thermo Quantum Discovery Max Triple Quadrupole Mass Spectrometer or equivalent), gas chromatography coupled with mass selective detection in electron impact mode (GC/MS-EI - Agilent 6890 GC equipped with a Agilent 5975 Mass Selective Detector in EI mode or equivalent), and gas chromatography coupled with mass selective detection in negative chemical ionization mode (GC/MS-NCI - Agilent 6890 GC equipped with a Agilent 5975 Mass Selective Detector in NCI mode or equivalent). Pesticide residues extracted from comb samples were quantified using matrix matched calibration standards of known concentrations prepared from neat standard reference material. Measurements were reported in nanograms of active ingredient per gram of wax (ng/g) or parts per billion (ppb). Identification of extracted residues was achieved through mass spectral comparison of ion ratios with standards, 171 of the most commonly used pesticides and their metabolites, of known identity. Limits of detection were in the low parts per billion (ppb).

Measurements

To assess the sub-lethal effects of exposure to pesticide residues, biologically meaningful parameters were measured throughout the main stages of the honey bee life cycle. Egg eclosion, or successful hatching was measured at day 4; larval mortality and development time from egg to pupa were recorded at day 8; pupation was recorded at day 12 and 19; adult emergence rate was recorded on day 20 and continued daily until emergence was no longer observed; and adult longevity was recorded daily until all caged bees were dead. Observations of abnormal larval development and signs of disease or pest infection were also recorded. Taken together, these life cycle parameters enabled assessment of the health effects of exposure to sub-lethal pesticide residues in brood comb.

Statistical analysis

Pairwise comparisons with repeated measures were performed on larval mortality, adult longevity, and adult emergence rate of worker bees reared in relatively uncontaminated brood comb and brood comb containing high levels of pesticide residues. Comparisons of both treatments were made by sample day (4, 8, 12 and 19) and by the number of brood cycles (Rep 1, 2, 3). Differences in pesticide analyses, specifically the number of pesticide residues and the levels detected in control and treatment comb used multiple times, were compared before and after the experiment. Normality assumptions were accepted for bee mortality on day 4, 9, 12, and 19 in both control and treatment combs (Shapiro-Wilk W = 0.844 and 0.929, respectively). Statistical differences were detected by one-way analysis of variance (ANOVA) followed by paired two-tailed t-tests on control and treatment combs with significance determined at $p \leq 0.025$.

Results

Chemical analysis of brood combs

The number of different pesticide residues detected in treatment combs ranged from 4 to 17, averaging 10. The total number of pesticides detected in all treatments was 39 including 7 fungicides, 2 herbicides, 23 insecticides (miticides included) and 7 metabolites (Table 1). The three most frequently detected pesticide residues in treatment combs were the beekeeper applied miticides fluvalinate, coumaphos, and coumaphos oxon metabolite. Fluvalinate, a pyrethroid pesticide, was detected in treatment combs at levels as high as 24,340 ppb and averaged 6,712 ppb. Coumaphos and its oxon metabolite were detected at levels as high as 22,100 ppb and 3,140 ppb, averaging 8,079 ppb and 596 ppb, respectively.

Table 1. Pesticide residues detected in treatment combs (n = 13) used to rear worker bees in experiments.

Active ingredient	Chemical Family	Purpose of use	Toxicity honey bee	Average (ng/g)	% detected	min	max	LOD
2,4 Dimethylphenyl formamide (DMPF)		metabolite		145	15	142	147	4
3-hydroxycarbofuran		metabolite		23	8	*	23	4
Aldicarb	Carbamate	INSECT	High	20	8	*	20	4
Azoxystrobin	Strobilurin	FUNG		19	38	5	29	2
Boscalid	Carboxamide	FUNG		35	15	35	64	4
Carbendazim (MBC)		metabolite		21	31	4	48	5
Carbofuran	Carbamate	INSECT	High	32	8	*	32	5
Chlorothalonil	Chloronitrile	FUNG		17	62	4	66	1
Chlorpyrifos	Ogranophosphate	INSECT	High	8	62	3	15	1
Clothianidin	Neonicotinoid	INSECT	High	35	8	*	35	20
Coumaphos	Ogranophosphate	INSECT	Mod	8079	100	281	22100	1
Coumaphos oxon		metabolite		596	100	10	3140	1
Cyfluthrin	Pyrethroid	INSECT	Low	43	17	8	79	2
Cypermethrin	Pyrethroid	INSECT	High	2	8	*	2	2
Cyprodinil	Anilinopyrimidine	FUNG		27	31	13	61	16
Diazinon	Ogranophosphate	INSECT	High	1	15	1	2	1
Dicofol	Organochlorine	INSECT	Low	6	23	4	8	1
Dinotefuran	Neonicotinoid	INSECT	High	97	8	*	97	30
Diphenylamine	Amine	INSECT		151	23	20	281	1
Endosulfan I	Organochlorine	INSECT	Mod	2	54	1	4	1
Endosulfan II	Organochlorine	INSECT	Mod	2	38	1	5	1
Endosulfan sulfate		metabolite		1	31	1	2	1
Esfenvalerate	Pyrethroid	INSECT	High	5	46	1	12	1
Fenhexamid	Hydroxyanilide	FUNG		46	8	*	46	6
Fenpropathrin	Pyrethroid	INSECT	High	7	8	*	7	1
Fluvalinate	Pyrethroid	INSECT	High	6712	100	164	24340	1
Imidacloprid	Neonicotinoid	INSECT	High	45	8	*	45	20
Iprodione	Dicarboximde	FUNG		283	8	*	283	20
Malathion oxon		metabolite		22	8	*	22	4
Norflurazon	Fluorinated pyridazinone	HERB		5	8	*	5	6
Oxamyl	Carbamate	INSECT	High	22	8	*	22	5
Oxyfluorfen	Diphenyl ether	HERB		2	23	1	2	1
Permethrin total	Pyrethroid	INSECT	High	103	8	*	103	10
Phosalone	Ogranophosphate	INSECT	Mod	32	8	*	32	10
Pyrethrins	Pyrethroid	INSECT	High	229	8	*	229	50
Thiacloprid	Neonicotinoid	INSECT	Low	113	8	*	113	8
Thiamethoxam	Neonicotinoid	INSECT	High	38	8	*	38	20
THPI		metabolite		96	15	93	99	50
Vinclozolin	Dicarboximde	FUNG		1	8	*	1	1

Toxicity category for honey bee: High; LD50 ≤2 µg/bee = highly toxic; Mod; LD50 2–11 µg/bee = moderately toxic; minimum and maximum ranges of pesticides detected, LOD; limit of detection.

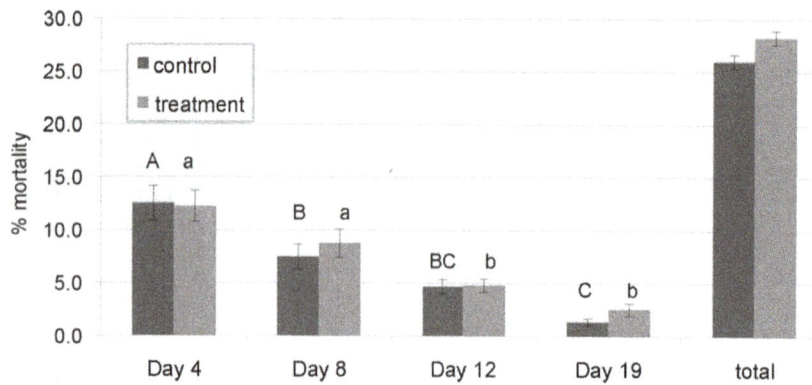

Figure 1. Percent larval mortality for bees reared in control and treatment comb at each sample date and overall total. Significance denoted with different letters.

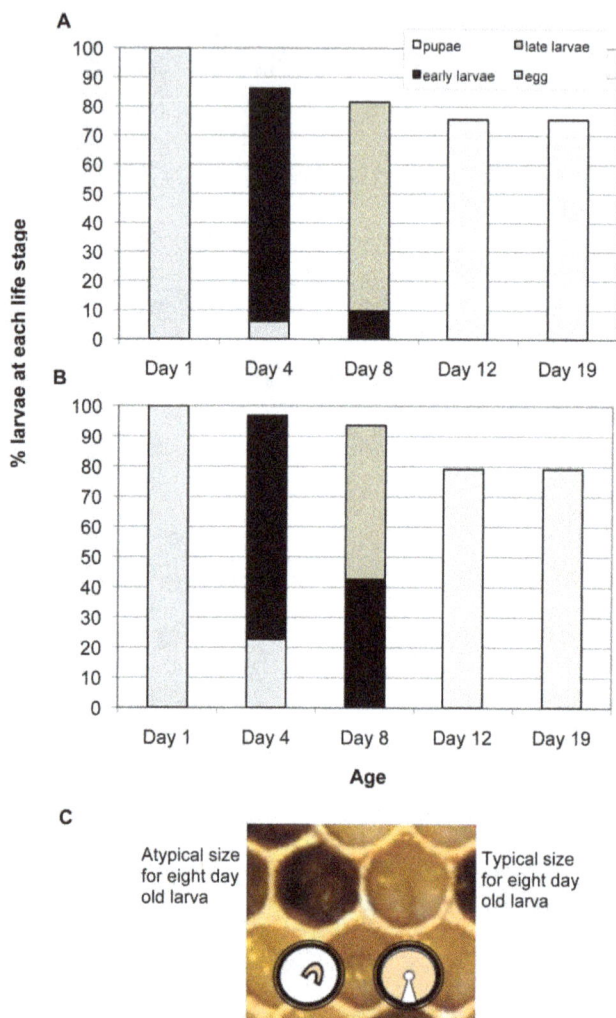

Figure 2. Larval development of worker bees from day 1 (egg stage) through day 19 (late pupal stage). (A) Normal larval development of bees reared in relatively uncontaminated control brood comb. (B) Larval development of bees reared in brood comb containing 17 different pesticides, expressing delayed development at day 4 and day 8. (C) Worker brood reared in brood comb containing 17 different pesticides at day 8 of development. Left: delayed growth. Right: normal development.

Coumaphos was the only residue detected in newly drawn out control combs (21 ppb).

Brood effects

There was no statistical difference in total larval mortality between bees reared in control and treatment combs (26 and 33%, respectively; p = 0.059) (Fig. 1). Delayed development at day 4 and 8 was observed in bees reared from four different combs with high levels of pesticide residues originating from colonies suspected to have CCD (Fig. 2A–C).

Brood mortality in bees reared from control comb was significantly greater on day 4 of development than on days 8, 12, and 19 (p = 0.0243; p = 0.0005; p<0.0001, respectively). In contrast, brood mortality in bees reared in treatment combs was not significantly different between days 4 and 8, although mortality was significantly higher on days 4 and 8 than on days 12 and 19 (p≤0.017 and p = 0.0001, respectively). The repeated use of experimental frames over several replicates may have allowed the migration of pesticide residues from treatment to control blocks, reducing the difference in residue levels between treatment and control combs and treatment effect differences (Table 2). Mortality was significantly higher in control bees reared from frames that were used in the experiment more than once and had experienced multiple brood cycles (Fig. 3). Total larval mortality increased with the repeated use of experimental frames in control combs from 13% through the first brood cycle (Rep 1) to 28% and 39% through the second (Rep 2) and third (Rep 3) brood cycles, respectively (Fig. 3). Brood mortality in bees reared through the third brood cycle in control comb was significantly higher than in the first and second brood cycles (p = 0.023; p = 0.048, respectively). In treatment comb blocks containing high levels of pesticide residues, overall mean larval mortality increased from 17% to 37% then decreased to 22% for the first, second, and third brood cycles, respectively (Fig. 3). Brood mortality in treatment combs was significant only between the first and second brood cycles (p = 0.013).

Chemical analysis of comb

Comparisons of chemical analyses, performed on five paired control and treatment combs before and after the experiment (n = 10), confirmed pesticide residue transfer and contamination of control combs over a 3-month period. Four additional new pesticide residues were detected in control comb, on average, compared to a reduction of 3 pesticide residues in treatment combs after the experiment (Fig. 4). The quantity or concentration of active ingredients also increased in control combs and decreased in

Table 2. Total amount of pesticide residues detected in five pairs of control and treatment combs before & after experiments.

Chemical	Class	F1 Control before	F1 Control After	F1 Treatment before	F1 Treatment after	F2 Control before	F2 Control after	F2 Treatment before	F2 Treatment after	F3 Control before	F3 Control after	F3 Treatment before	F3 Treatment after	F4 Control before	F4 Control after	F4 Treatment before	F4 Treatment after	F5 Control before	F5 Control after	F5 Treatment before	F5 Treatment after
Boscalid	Carboxamide (F)											35									
Chlorothalonil	Chloronitrile (F)			4				66							2						
Cyprodinil	Anilinopyrimidine (F)											61	72				58				
Fenhexamid	Hydroxyanilide (F)											46									
Iprodione	Dicarboximide (F)											432				283	1030				463
Myclobutanil	Dithiocarbamate (F)																				31
THPI	Phthalimide (F)															93				99	
Oxyfluorfen	Diphenyl ether (H)											1				2					
Diphenylamine	Amine (I)			281																	
Aldicarb	Carbamate (I)			20																	
Carbofuran	Carbamate (I)			32																	
Oxamyl	Carbamate (I)			22																	
Clothianidin	Neonicotinoid (I)			35																	
Dinotefuran	Neonicotinoid (I)			97																	
Imidacloprid	Neonicotinoid (I)			45																	
Thiacloprid	Neonicotinoid (I)			113																	
Thiamethoxam	Neonicotinoid (I)			38								1									
Endosulfan 1	Organochlorine (I)			1																2	
Endosulfan II	Organochlorine (I)																			2	
Endosulfan sulfate	Organochlorine (I)		3																		
Chlorpyrifos	Organophosphate (I)											8	9			5				9	13
Coumaphos	Organophosphate (I)		703	22100	9920	21	4550	281	859		451	3140	1580	21	2830	8200	14300	21	669	7230	7090
Phosalone	Organophosphate (I)																			32	
Cyfluthrin	Pyrethroid (I)			79																	
Esfenvalerate	Pyrethroid (I)																			12	6
Fluvalinate	Pyrethroid (I)		43	164	159		1400	11280	2330		998	24340	14500	21	1420	9850	7130		2250	6800	3980
Permethrin total	Pyrethroid (I)							103													
Pyrethrins	Pyrethroid (I)							229													
Paradichlorobenzene	Halogenated organic (I)		104	104			310				54	109			188		62		184		174
2,4 Dimethylphenyl formamide (DMPF)	Amidine (m)							147					39							142	
3-hydroxycarbofuran	Carbamate (m)			23																	
Chlorferone	Organophosphate (m)			944			511				602				255		2160				785

Table 2. Cont.

Chemical	Class		Frame 1				Frame 2				Frame 3				Frame 4				Frame 5			
			Control		Treatment		Control		Treatment		Control		Treatment		Control		Treatment		Control		Treatment	
			before	After	before	after	before	after	before	after	before	after	before	after	before	after	before	after	before	after	before	after
Coumaphos oxon	Organophosphate	m		99	1850	617		276	10	101	64		3140	112	335		474	438	92		231	246
Malathion oxon	Organophosphate	m			22																	
Total # compounds			0	5	17	5	1	5	7	7	0	4	9	9	1	6	7	7	1	4	10	9

Results reported in ng/g or parts per billion. (F = fungicide; H = herbicide; I = insecticide; m = metabolite).

treatment combs after the experiment, further supporting the transfer of pesticide residue from areas of comb contaminated with high levels of pesticide residues to uncontaminated areas with low levels of residues. Insecticides, including the 3 most frequently detected compounds (coumaphos, coumaphos oxon, and fluvalinate) initially in treatment combs, increased in concentration in control combs and decreased in treatment combs after the experiment. Concentrations for coumaphos oxon, fluvalinate and combined insecticides were significantly higher in control comb after the experiment than before (p<0.025; p<0.01; p<0.025; respectively). High levels of metabolites were also detected in control combs after the experiment suggesting possible metabolism of active compounds as a result of pesticide residue migration. Fluvalinate residue levels were significantly lower in treatment combs after the experiment than before (p<0.025). The majority of new compounds found in control combs after the experiments were compounds previously detected in treatment combs at higher levels before the experiment than after (Table 2). Fungicides were the only pesticide group that was detected at higher concentrations in treatment combs after the experiment than before the experiment, an increase that was not statistically significant (averaging 280 ppb). These results illustrate that pesticide residues quickly diffuse through wax or across comb surface in an active honey bee colony.

Adult emergence and longevity

Worker bees reared in relatively uncontaminated brood comb lived an average of 4 days longer than bees reared in comb containing high levels of pesticide residues (Fig. 5, p = 0.005). Emergence time was also affected by contamination of control comb after multiple brood cycles, resulting in a shift in the proportion of worker bees that emerged on days 20, 21 and 22 (Fig. 6A–C). During the first brood cycle (Rep 1), a significantly higher proportion of bees emerged from control combs on days 20 and 21 of development compared to emergence on day 22. Of worker brood reared in control combs, 42% and 53% emerged as adults on days 20 and 21, respectively, while only 5% emerged on day 22 (p<0.0007). In contrast, by the third brood cycle (Rep 3) adult emergence from control comb on day 22 was much higher (18%) than emergence on day 22 during the first brood cycle (Rep 1) (5%). In addition, only 2% of worker brood reared in control comb on the third replicate emerged as adults on day 20 compared to 42% of brood that emerged on day 20 during the first brood cycle. The majority (80%) of brood from replicate 3 emerged on day 21 of development (Fig. 6A–C). These data suggest a shift in the occurrence of adult emergence from day 20–21 to 21–22 and delayed emergence for developing worker bees as a result of pesticide residue exposure to contaminated brood comb.

Discussion

Honey bees of all ages and castes are susceptible to effects from pesticide exposure [13]. Older adult bees may be exposed to pesticides during flight and foraging, while younger adults remain in the hive but may be exposed to incoming contaminated pollen and nectar. They may also be exposed to beekeeper-applied pesticides commonly used in-hive to control *Varroa destructor* mites, serious external parasites that infect honey bee brood. Prior to adult emergence, eggs and developing bees may be exposed to pesticide residues through contaminated comb cell walls or food sources. Queen bees can be exposed to pesticides by contact with contaminated bees, wax, and food. Sub-lethal pesticide exposure through wax can have adverse reproductive consequences such as reduced egg laying, early supercedure, increased queen cell rejection, and reduced ovarian weight in queen bees [15,16].

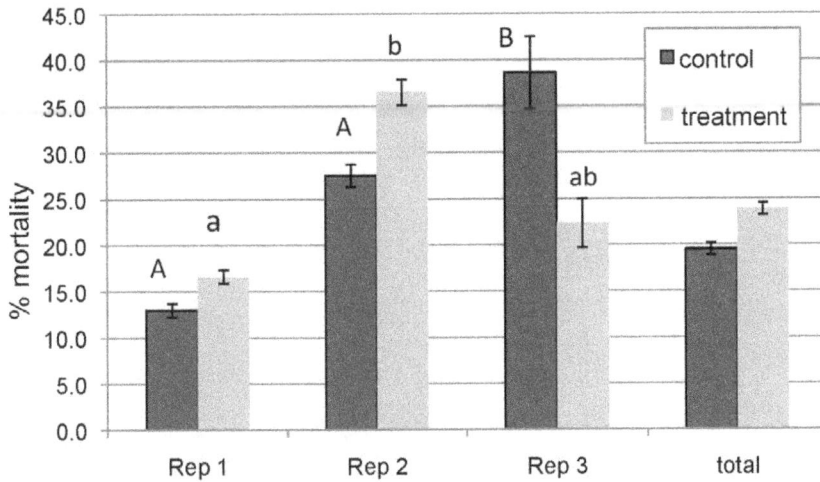

Figure 3. Percent mortality in larvae reared in control and treatment comb over multiple replications (Rep 1, 2, and 3; n = 28). Significance denoted with different letters.

In this study, worker bees reared in comb containing high levels of pesticide residues had lower survivorship than bees reared in relatively uncontaminated comb. Comb age may have been a factor as well, given that brood mortality was higher in newly drawn control comb than in older control comb sampled from feral colonies. Newly drawn comb lacks exuviae (molted larval cuticles), which contain brood pheromone cues that indicate brood presence to nurse bees and increase larval survivorship [14]. However, while initial larval survivability can be lower in bees reared in new comb, overall colony health in hives using old brood comb is compromised by higher incidences of pests and pathogens [17]. The paired-block setup allowed pests or pathogens from older treatment combs to migrate or transfer (via nurse bees) over

to larvae reared in new control comb. While this design was intended to help reduce differences due to pathogen loads, we cannot exclude the possibility that some pathogens, exuviae and brood pheromones embedded in the cell walls of the comb would not have been transferrable.

For economic reasons, beekeepers typically reuse wax foundation, but pesticide residues accumulate in wax and may persist for years [18–20]. Contamination of reused control brood combs in this experiment illustrated how quickly pesticide residues could penetrate and migrate through or across brood comb wax. The presence of additional pesticide residues in control combs detected after the experiment confirmed pesticide residue transfer and contamination of control combs. Incoming pesticides brought back

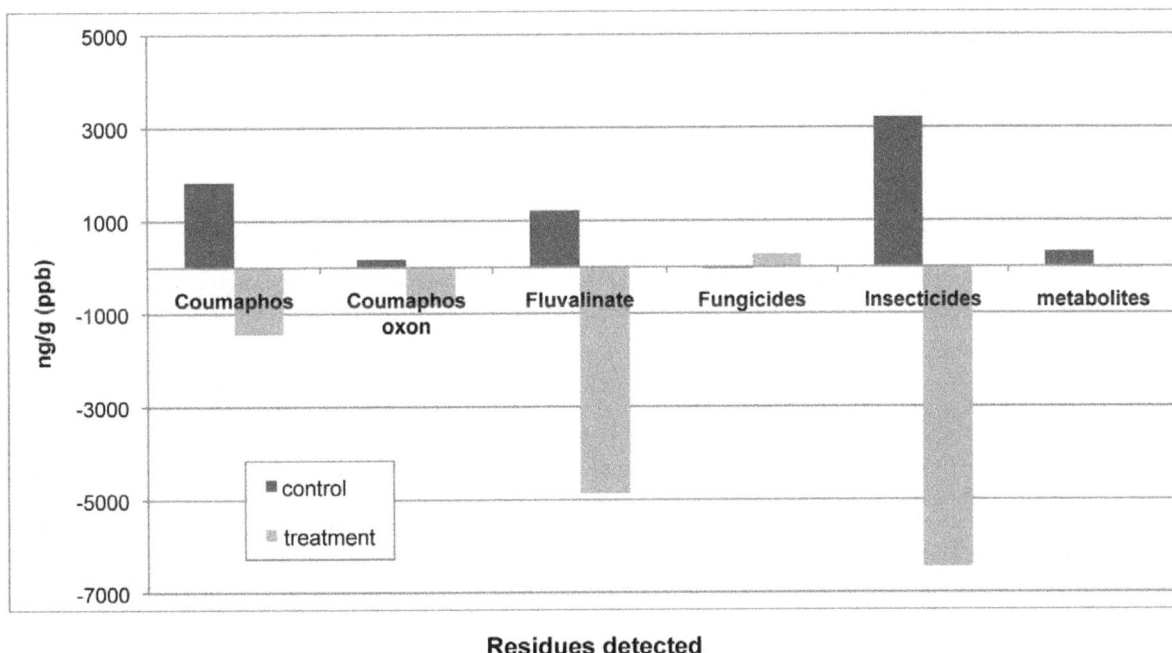

Residues detected

Figure 4. Average difference in quantity (ppb) of pesticide residues detected between pre- and post-experimental analyses for control and treatment brood combs after 2 or 3 replicates.

A

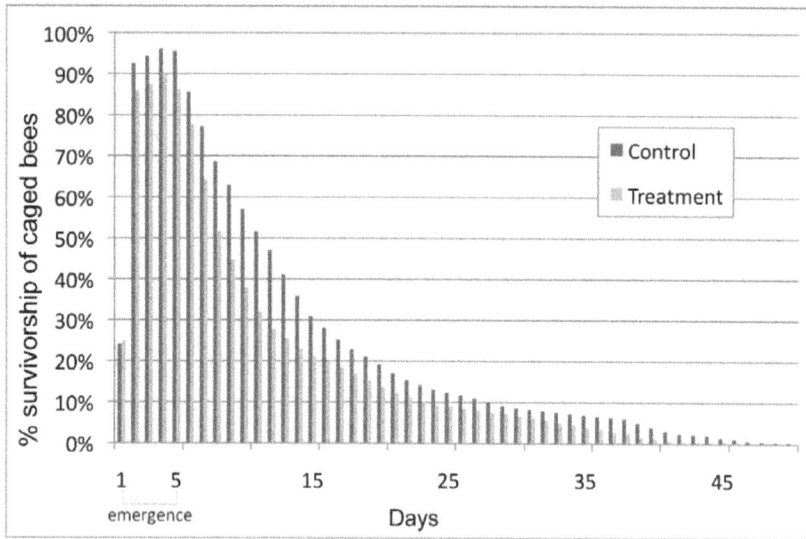

B

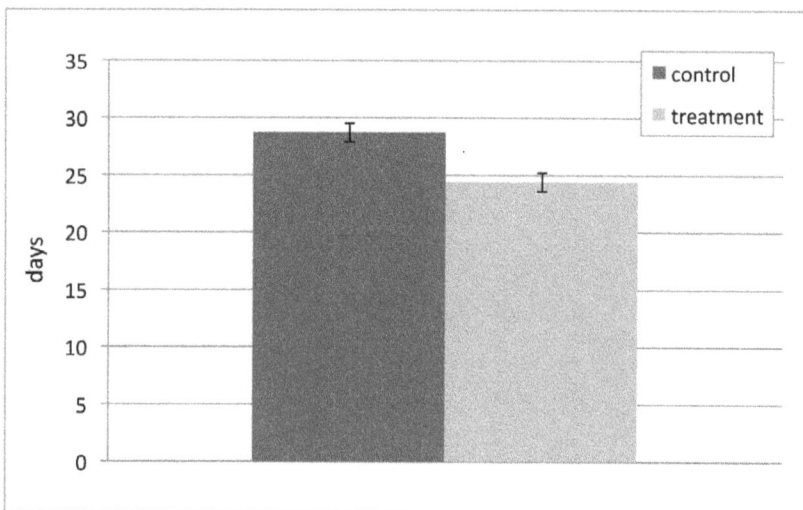

Figure 5. Adult emergence and longevity of bees reared in control and treatment brood comb over first brood cycle. (A) Percent emergence and survivorship of caged control and treatment bees over 50 days. (B) Average adult longevity of caged control and treatment bees. Adult bees reared in control combs lived an average of 4 days longer than adult bees reared in combs containing high levels of pesticides (p = 0.005).

by foragers from external sources would have been detected in both control and treatment combs, because the experimental frames were housed in the same colony. High levels of pesticide metabolites detected in control combs after the experiment also suggest possible metabolism of active compounds during migration. Metabolites can be more harmful to organisms than parent compounds and can have delayed effects [11,21]. In the paired comb blocks, detection of increasing mortality for bees reared in control blocks and decreasing mortality for treatment blocks, over time, confirms toxicological consequences from pesticide residue migration.

Brood effects of pesticide exposure

Sub-lethal effects of pesticides on bees, including delayed adult emergence, may seem inconsequential but may provide a reproductive advantage for *Varroa* mites. A gravid foundress mite

can invade a cell occupied by a developing bee larva and lay four eggs in 30 hour intervals. The first egg results in a male, with subsequent eggs developing into multiple daughter mites [22–24]. The most injurious effects of *Varroa* mites occur when the foundress and her multiple offspring feed on the hemolymph of a pupating bee, causing reductions in emergence weight and metabolic reserves and physical deformities in host bees [25,26]. Normally, the third daughter mite only has a 13% chance of reaching maturity before the pupating bee emerges from the cell after 20 to 21 days of development [27]. However, with delayed adult bee emergence the likelihood that the third daughter mite will successfully reach maturity increases. In this study, delayed development occurred in bees reared in treatment comb containing 17 different pesticides, including 9 systemic compounds and 5 neonicotinoid insecticides (Table 3). As the queen in these experiments laid eggs in

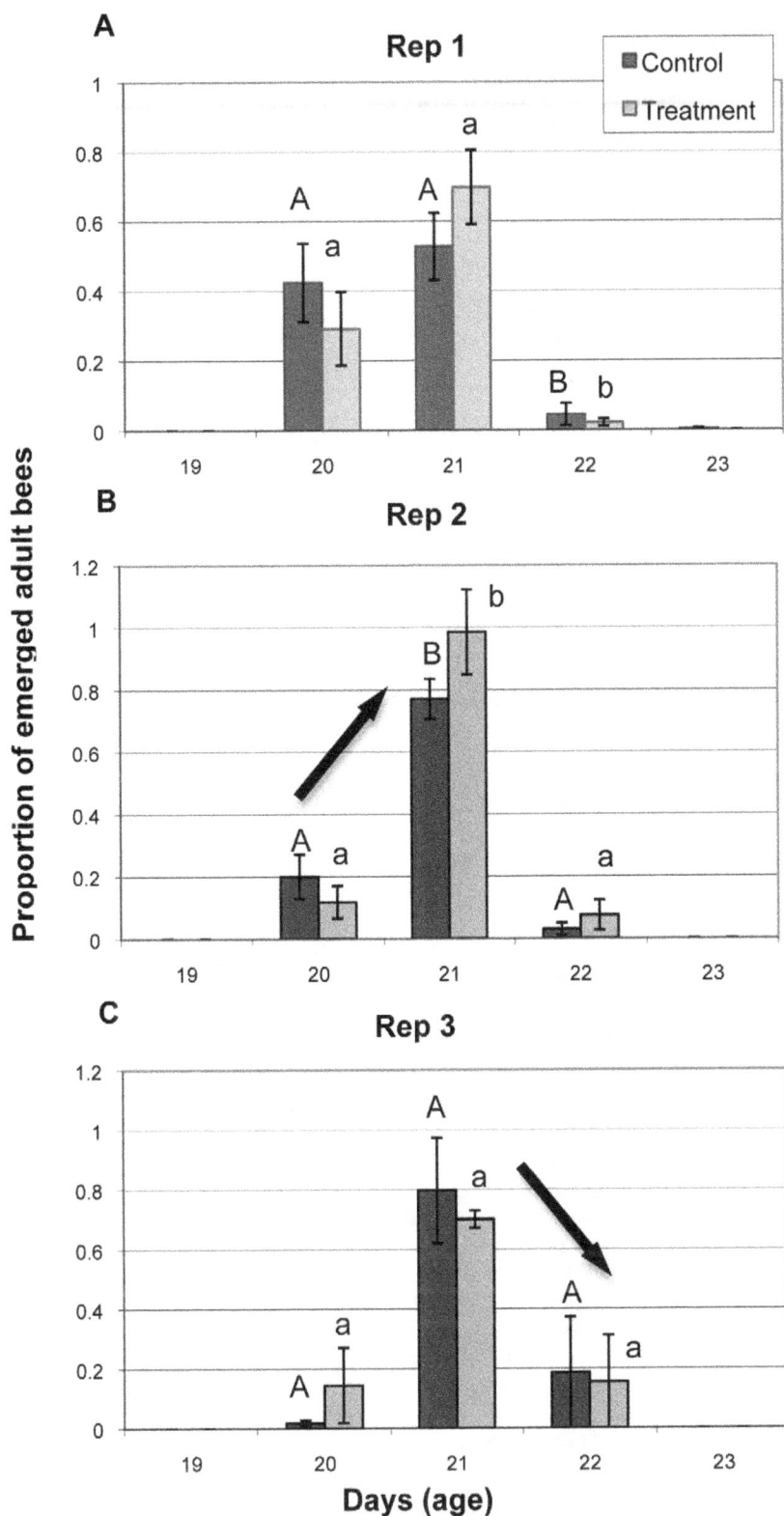

Figure 6. Adult emergence of bees reared in control and treatment brood comb over multiple replications. (A) First brood cycle emergence (Rep 1). (B) Second brood cycle emergence (Rep 2). (C) Third brood cycle emergence (Rep 3). Different capital letters denote significant differences in emergence of control bees on different days; different lower case letters denote significant differences in emergence of treatment bees on different days.

Table 3. Pesticide residues contained in treatment brood comb with observed delayed development of worker honey bees.

Pesticides	Chemical family	Systemic	Toxicity honey bee	(ng/g) ppb	LOD
3-hydroxy-carbofuran	metabolite	Systemic		23	4
Aldicarb	Carbamate	Systemic	High	20	4
Carbofuran	Carbamate	Systemic	High	32	5
Chlorothalonil	Fungicide		---	4	1
Clothianidin	Neonicotinoid	Systemic	High	35	20
Coumaphos	Organophosphate		Moderate	22100	1
Coumaphos oxon	metabolite			1850	5
Cyfluthrin	Pyrethroid		High	7.9	2
Dinotefuran	Neonicotinoid	Systemic	High	97	30
Diphenylamine	Amine		---	281	1
Endosulfan 1	Organochlorine		Moderate	1	1
Fluvalinate	Pyrethroid		High	164	1
Imidacloprid	Neonicotinoid	Systemic	High	45	20
Malathion Oxon	metabolite			22	4
Oxamyl	Carbamate	Systemic	High	22	5
Thiacloprid	Neonicotinoid	Systemic	High	113	8
Thiamethoxam	Neonicotinoid	Systemic	High	38	20

Toxicity category for honey bee: High; LD50 ≤2 µg/bee = highly toxic; Mod; LD50 2–11 µg/bee = moderately toxic; LOD; limit of detection.

both control and treatment comb within a 24 hour period, the normal growth pattern was expected to be uniform. However, by day 4, 23% of eggs were unhatched in the treatment comb and by day 8, over 46% of remaining larvae reared in the contaminated treatment comb were small and their development visually stunted or delayed (Figs. 6A–C). Another three treatment combs, sampled from colonies suspected to have colony collapse disorder (CCD), had similar patterns of egg hatch and development. An average of 19% of eggs laid in comb sampled from CCD colonies containing high levels of pesticides remained unhatched on day 4, and 60%–90% of unhatched eggs were removed by bees before the next sampling date. Inefficiencies in brood production place energetic stresses on honey bee colonies. Nurse bees must remove unhatched eggs rather than tend to developing brood, and high brood mortality increases the demand for egg-laying by the queen. Egg-laying efficiency is further reduced when queen bees are unable to deposit eggs in a general area but, instead, must seek empty cells scattered throughout the brood nest [28].

Adult longevity

Worker bees reared in treatment comb containing high levels of pesticide residues lived an average of 4 days less than bees reared in relatively uncontaminated control combs in cage trials (Fig. 5). To place this in context, the mean lifespan of honey bees after entering the ranks of foragers is less than 8 days [29]. Reduced adult longevity within the ranks of foraging bees can lead to precocious foraging by "under-aged" replacement bees. Over the long term, this activity could affect an entire cascade of hive activities including brood care, food processing and storage, queen care, hygienic behavior and foraging efficiency, disrupting age-based polyethism and its role in colony homeostasis. Precocious foraging has been reported to have a major impact on colony size and viability, through reduction of the "younger" nurse bee population from which replacement foragers are derived [30]. In fact, in their model of induced precocious foraging, Thompson and co-workers [30] found that to simulate

the "sublethal" effect caused by the reduction in nurse bee capabilities (the mean number of larvae that can be reared per nurse bee), the mortality rate of foragers would have to be increased by 500%.

Conclusion

Combined effects from honey bee exposure to pesticide residue in brood comb, such as reduced adult longevity, increased brood mortality, higher fecundity of *Varroa* mites (due to delayed development and emergence of adult bees) and increased susceptibility to pathogens, may contribute to reduced honey bee colony health, as affected queens and worker bees are unable to meet the demand for brood production and resources needed to sustain large colony populations.

Honey bees are biological indicators, picking up chemicals and other pollutants from their environment both external and internal to their hives. Our findings suggest that one of the underlying commonalities in the worldwide reports of a decline in honey bee health and observations of Colony Collapse Disorder (CCD) may be exposure of honey bees and bee products to pesticides. Developmental exposure of honey bees to pesticide contaminated brood comb may appear subtle and indirect, but can lead to sublethal effects that actually have serious consequences.

Acknowledgments

We thank E. Olson and numerous other beekeepers, and J. Pettis (USDA ARS) for providing comb samples, R. Simonds from the USDA-AMS-National Science Laboratory in Gastonia, NC for chemical residues analysis, M. Evans (WSU) for assistance with statistics, and J. Stark (WSU) for comments that improved the manuscript.

Author Contributions

Conceived and designed the experiments: JYW CMA WSS. Performed the experiments: JYW. Analyzed the data: JYW. Contributed reagents/materials/analysis tools: WSS. Wrote the paper: JYW CMA WSS.

References

1. Van Engelsdorp D, Underwood RM, Caron D, Hayes J, Jr. (2007) An estimate of managed colony losses in the winter of 2006–2007: a report commissioned by the Apiary Inspectors of America. Am Bee J 147: 599–603.

2. Van Engelsdorp D, Evans J, Saegerman C, Mullin C, Haubruge E, et al. (2009) Colony collapse disorder: a descriptive study. PLoS ONE 4: e6481.

3. Bonmatin JM, Moineau I, Charvet R, Colin ME, Fleche C, et al. (2005) Behaviour of imidacloprid in fields. Toxicity for honey bees. In: Lichtfouse E, Schwarzbauer J, Didier R, eds. Environmental Chemistry: green chemistry and pollutants in ecosystems. Berlin: Springer. pp 483–494.

4. Ramirez-Romero R, Chaufaux J, Pham-Delègue M (2005) Effects of Cry1Ab protoxin, deltamethrin and imidacloprid on the foraging activity and the learning performances of the honeybee Apis mellifera, a comparative approach. Apidologie 36: 601–11.

5. Girolami V, Mazzon L, Squartini A, Mori N, Marzaro M, et al. (2009) Translocation of neonicotinoid insecticides from coated seed to seedling guttation drops: a novel way of intoxication for bees. J Econ Entomol 102(5): 1808–1815.

6. Higes M, Martín-Hernández R, Garrido-Bailón E, González-Porto AV, García-Palencia P, et al. (2009) Honeybee colony collapse due to Nosema ceranae in professional apiaries. Environ Microbiol Reports 1: 110–113.

7. Ladas A (1972) The influence of some internal and external factors upon the insecticide resistance of honeybees. Apidologie 3: 55–78.

8. Alaux C, Brunet J, Dussaubat C, Mondet F, Tchamitchan S, et al. (2009) Interactions between Nosema microspores and a neonicotinoid weaken honeybees (Apis mellifera). Environ Microbiol 12: 774–782.

9. The honeybee genome sequencing consortium (2006) Insights into social insects from the genome of the honeybee Apis mellifera. Nature 443: 931–949.

10. Claudianos C, Ranson H, Johnson RM, Biswas S, Schuler MA, et al. (2006) A deficit of detoxification enzymes: pesticide sensitivity and environmental response in the honeybee. Insect Mol Biol 15: 615–636.

11. Suchail S, Guez D, Belzunces LP (2001) Discrepancy between acute and chronic toxicity induced by imidacloprid and its metabolites in Apis mellifera. Environ Toxicol Chem 20: 2482–2486.

12. Wahl O, Ulm K (1983) Influence of pollen feeding and physiological condition on pesticide sensitivity of the honey bee Apis mellifera carnica. Oecologia 59: 106–128.

13. Rortais A, Arnold G, Halm M, Touffet-Briens F (2005) Modes of honeybee exposure to systemic insecticides: estimated amounts of contaminated pollen and nectar consumed by different categories of bees. Apidologie 36: 71–83.

14. Medrzycki P, Sgolastra F, Bortolotti L, Bogo G, Tosi S, et al. (2009) Influence of brood rearing temperature on honey bee development and susceptibility to poisoning by pesticides. J Apic Res 49: s52–s60.

15. Haarmann T, Spivak M, Weaver D, Weaver B, Glenn T (2002) Effects of fluvalinate and coumaphos on queen honey bees (Hymenoptera: Apidae) in two commercial queen rearing operations. J Econ Entomol 95: 28–35.

16. Pettis JS, Collins AM, Wilbanks R, Feldlaufer MF (2004) Effects of coumaphos on queen rearing in the honey bee, Apis mellifera. Apidologie 35: 605–610.

17. Berry JA, Delaplane KS (2001) Effects of comb age on honey bee colony growth and brood survivorship. J Apic Res 40: 3–8.

18. Bogdanov S, Kilchenmann V, Imdorf A (1996) Acaricide residues in beeswax and honey. In: Mizrahi A, Lensky Y, eds. Bee products: properties, applications and apitherapy. New York: Plenum Press. pp 247–252.

19. Chauzat MP, Faucon JP (2007) Pesticide residues in beeswax samples collected from honey bee colonies (Apis mellifera L.) in France. Pest Manag Sci 63: 1100–1106.

20. Mullin CA, Frazier M, Frazier JL, Ashcraft S, Simonds R, et al. (2010) High levels of miticides and agrochemicals in North American apiaries: implications for honey bee health. PLoS ONE 5: e9754.

21. Sparlings DW, Fellers G (2007) Comparative toxicity of chlorpyrifos, diazinon, malathion and their oxon derivatives to larval Rana boylii. Environ Pollut 147: 535–539.

22. Ifantidis MD (1983) Ontogenesis of the mite Varroa jacobsoni in worker and drone honeybee brood cells. J Apic Res 22: 200–6.

23. Donzé G, Guerin PM (1997) Time-activity budgets and space structuring by the different life stages of Varroa jacobsoni in capped brood of the honey bee, Apis mellifera. J Insect Behav 10: 371–93.

24. De Jong D (1997) Mites: varroa and other parasites of brood, 3rd ed. In: Morse RM, Flottum PK, eds. Honey bee pests, predators, and diseases. Medina: Root. pp 279–327.

25. Bowen-Walker PL, Gunn A (2001) The effect of the ectoparasitic mite, Varroa destructor on adult worker honeybee (Apis mellifera) emergence weights, water, protein, carbohydrate, and lipid levels. Entomol Exp Appl 101: 207–217.

26. Amdam GV, Hartfelder K, Norberg K, Hagen A, Omholt SW (2004) Altered physiology in worker honey bees (Hymenoptera: Apidae) infested with the mite Varroa destructor (Acari: Varroidae): a factor in colony loss during overwintering? J Econ Entomol 97: 741–747.

27. Varroa jacobsoni reproduction: Varroa mites reproduce in capped brood cells. USDA ARS. Baton Rouge, LA. Available: http://www.ars.usda.gov/services/docs.htm?docid = 2744&page = 14. Accessed 2011 Jan 28.

28. Mackensen O (1951) Viability and sex determination in the honey bee (Apis mellifera L.). Genetics 36: 500–509.

29. Vosscher PK, Dukas R (1997) Survivorship of foraging honey bees. Insectes Sociaux 44: 1–5.

30. Thompson HM, Wilkins S, Batterby AH, Waite RJ, Wilkinson D (2007) Modelling long-term effects of IGRs on honey bee colonies. Pest Manag Sci 63: 1081–1084.

12

Modelling Food and Population Dynamics in Honey Bee Colonies

David S. Khoury[1¤], **Andrew B. Barron**[2], **Mary R. Myerscough**[1,3]*

1 School of Mathematics and Statistics, The University of Sydney, Sydney, New South Wales, Australia, **2** Department of Biological Sciences, Macquarie University, Sydney, New South Wales, Australia, **3** Centre for Mathematical Biology, The University of Sydney, Sydney, New South Wales, Australia

Abstract

Honey bees (*Apis mellifera*) are increasingly in demand as pollinators for various key agricultural food crops, but globally honey bee populations are in decline, and honey bee colony failure rates have increased. This scenario highlights a need to understand the conditions in which colonies flourish and in which colonies fail. To aid this investigation we present a compartment model of bee population dynamics to explore how food availability and bee death rates interact to determine colony growth and development. Our model uses simple differential equations to represent the transitions of eggs laid by the queen to brood, then hive bees and finally forager bees, and the process of social inhibition that regulates the rate at which hive bees begin to forage. We assume that food availability can influence both the number of brood successfully reared to adulthood and the rate at which bees transition from hive duties to foraging. The model predicts complex interactions between food availability and forager death rates in shaping colony fate. Low death rates and high food availability results in stable bee populations at equilibrium (with population size strongly determined by forager death rate) but consistently increasing food reserves. At higher death rates food stores in a colony settle at a finite equilibrium reflecting the balance of food collection and food use. When forager death rates exceed a critical threshold the colony fails but residual food remains. Our model presents a simple mathematical framework for exploring the interactions of food and forager mortality on colony fate, and provides the mathematical basis for more involved simulation models of hive performance.

Editor: Adrian G. Dyer, Monash University, Australia

Funding: This work was supported by a Hermon Slade Foundation grant to ABB. The funders had no role in study design, data collection and analysis, decision to publish, or preparation of the manuscript.

Competing Interests: The authors have declared that no competing interests exist.

* E-mail: mary.myerscough@sydney.edu.au

¤ Current address: The Centre for Vascular Research, Faculty of Medicine, The University of New South Wales, Sydney, New South Wales, Australia

Introduction

A honey bee colony gathers dispersed floral resources (pollen and nectar) from the environment to a central place, and processes them to provide food to support the current population and rearing of the next cycles of brood. Previously we proposed a simple mathematical model of honey bee population dynamics to explore the impact of varying forager death rate on colony growth and development [1]. This model was a deliberate simplification to consider how interactions between adult foragers and hive bees and brood might influence colony growth. However in natural colonies food availability may impose limits on colony development. Here we present a new model to explore how changes in food availability might interact with behavioural and social processes in the colony to influence colony growth.

This issue is pertinent because the amount of honey that can be extracted from commercial bee hives for human use depends on bees collecting nectar in excess of what is needed to support their population, and storing the excess as honey. The honey industry is therefore reliant on manipulating the flux of food through a colony to maximize the excess, and understanding the relationship between food availability and colony growth may improve colony management practice. Further, recent concerns about the sustainability of bee populations [2] have highlighted a need to

better understand how healthy colonies function, and why they may sometimes fail.

All the nutritional demands of a honey bee colony are met by supplies of pollen and nectar gathered by foragers [3]. Nectar is entirely carbohydrate in the form of simple sugars (with sometimes some trace minerals and allelochemicals) [4,5]. Pollen provides bees with lipids, protein, and vitamin and mineral nutrients [3,5]. Nectar is transferred from foragers to non-foraging hive bees who deposit the nectar in cells and, over time, process and concentrate it to form honey [5]. Pollen is deposited directly in cells by foragers, but mixed with a small amount of nectar and packed by hive bees for storage [5]. Honey, nectar and pollen are consumed by hive bees and used to produce a protein rich brood food, which is fed to the queen and developing larvae [3,5]. In seasonal climates during winter the colony relies on stored pollen and nectar collected over the summer [5].

Honey bees have a very typical pattern of age polyethism performing various functions within the hive for the first two to three weeks of their adult life before transitioning to foraging [5], but bee behaviour is very sensitive to changes in levels of stored food or food influx. A shortage of food within the colony stimulates a precocious onset of foraging in adult bees truncating the amount of time they spend as hive bees [6,7]. Brood rearing is especially sensitive to food shortages [3]. Typically a colony does not maintain a large store of pollen, and interruptions in pollen inflow to a colony can trigger cannibalism of developing larvae by worker bees [8,9,10]. This is interpreted as an adaptive response by workers to reduce the size of the brood population to that most likely able to be successfully reared when food is limited [3,9].

The interactions between food and population dynamics in a bee colony are therefore quite complex. Food collection is influenced by the size of the forager population, and in turn food flux through the colony can influence the size of the forager population by altering the rate at which hive bees become foragers and the size of the brood population, which will eventually become the next generation of foragers. The model we present here offers a simple theoretical framework with which to explore how the dynamics of food flow through a colony might interact with population dynamics to determine colony growth.

Methods

Constructing a demographic model which includes brood and food dynamics

In Khoury et al., [1] we constructed a model for the population of a hive of honey bees which only included the adult bees. These were divided into two classes; hive bees and foragers. We assumed that the rate that adults emerged from pupation was a function of hive size only and that food was not a limiting factor in hive population dynamics. We also assumed that the hive had sufficient available food so that food scarcity did not affect the population dynamics.

Here we extend the model of Khoury et al., to include both food and brood explicitly (Figure 1). As before, we only consider the population of female worker bees since it is only females that contribute to foraging and colony maintenance. Let B be the number of uncapped brood in the hive, H be the number of hive bees and F the number of foragers. Let f be a measure of the amount of food that is stored in the hive and available for the colony to use. We do not distinguish between pollen and nectar (protein and carbohydrates) here. Our aim is to keep the model simple so that we can perform comprehensive analyses and model gross effects transparently. We assume that the survival of uncapped brood (eggs and larvae) is dependent on the number of hive bees available to tend and feed brood, on food availability and on the laying rate L of the queen. Larvae become pupae inside cells that are capped by worker bees and we assume that pupation occurs at a constant rate proportional to the amount of brood present. This is a simplifying assumption that allows us to continue to use a compartment model rather than a more complicated model with explicit age structure. Adult bees emerge 12 days after pupation and we assume that mortality of capped brood is negligible. We also assume that the death rate of hive bees is negligible. Foragers are recruited from the hive bee class and die at a rate m. Let t be the time in days. Then we can represent the model illustrated in Figure 1 as four differential equations:

Rate of change of brood numbers:

$$\frac{dB}{dt} = LS(H,f) - \phi B \qquad (1)$$

The first term represents laying and survival of brood where L is the laying rate of the queen and $S(H,f)$ is a function of food and hive bee numbers. We assume that $S(H,f)$ becomes constant as f and H become large and that the dependence of food and hive bee numbers is independent of one another. With these assumptions,

we can model $S(H,f)$ as

$$S(H,f) = \frac{f^2}{f^2 + b^2} \frac{H}{H + v} \qquad (2)$$

where b and v are parameters that determine how rapidly $S(H,f)$ tends to one as f and H increase respectively. The first term in $S(H,f)$ models the way that brood survival declines when food stores are low. This decline in brood survival has two causes: brood die because there is not enough food to feed them as they develop; and because workers cannibalise the eggs and young larvae when food is scarce to recycle protein in the colony and so increase the likelihood of older larvae surviving to pupation. We assume that when food levels are very low there is almost no survival, but that survival rates climb rapidly when food reaches a viable level. Consequently we use a sigmoid form for this term. The second term models the impact of hive bee numbers on brood survival. When there are few hive bees there may be insufficient workers to supply all the larvae with the food that they need, even if the hive has large food stores. Also, low hive bee numbers will impact on the colony's ability to keep the brood warm so that they develop properly [11]. We assume that when hive bee numbers are low the amount of brood that is raised is close to a linear function of hive bee numbers and so we choose a Michaelis-Menten type of function for this term.

Rate of change of hive bees:

$$\frac{dH}{dt} = \phi B(t - \tau) - HR(H,F,f) \qquad (3)$$

where $\phi B(t - \tau)$ is the rate that adult bees emerge from pupation. The bees that are emerging at time t are the same bees that entered pupation at $t - \tau$. The function $R(H,F,f)$ gives the proportional rate that hive bees make the transition into foragers. This rate is a function of hive bee and forager numbers, but also depends on stored food supply.

Rate of change of foragers:

$$\frac{dF}{dt} = HR(H,F,f) - mF \qquad (4)$$

where the first term is the rate that hive bees become foragers and the last term is the rate that foragers die.

We assume that the transition from hive bee to forager has an underlying component that is increased by the absence of stored food and reduced by social inhibition due to the presence of foragers in the hive. We write the recruitment function as

$$R(H,F,f) = \alpha_{min} + \alpha_{max}\left(\frac{b^2}{b^2 + f^2}\right) - \sigma\left(\frac{F}{F + H}\right) \qquad (5)$$

where α_{min} is the rate that hive bees become foragers when there is plenty of stored food but no foragers in the hive, α_{max} governs the strength of the effect that low food stores have on the transition to foragers and b controls the rate that the food-dependent terms decrease as food stores increase. Social inhibition depends on the proportion of foragers in the adult bee population, and the strength of this inhibition is governed by σ. The forager-to-hive-bee transition depends on food stores in a similar way to how brood survival depends on food stores as both share the same parameter b. This implicitly assumes that shortage of food for the larvae is one of the stimuli that drive increased forager recruitment.

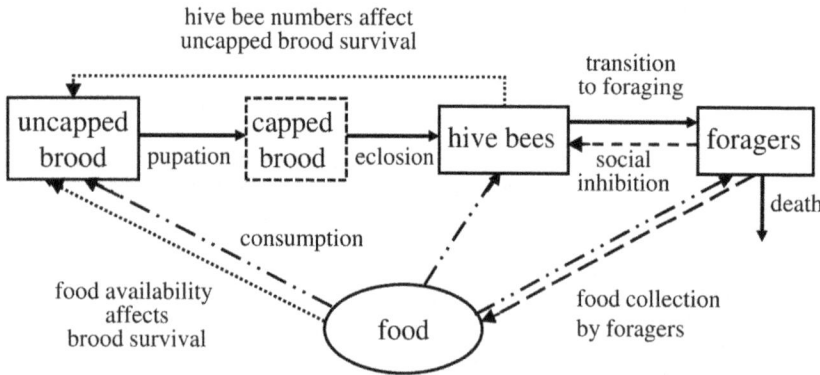

Figure 1. Honey bee social dynamics represented in the model. The dash-dot lines represent consumption of food. The class "capped brood" appears in a box with a dashed border because it is not explicitly modeled although the period that brood spends as capped pupae is accounted for by the delay τ in the equations.

Rate of change in food stores:

$$\frac{df}{dt} = cF - \gamma_B B - \gamma_H H - \gamma_F F \qquad (6)$$

where c is the average amount of food collected per forager per day. In reality, this will vary seasonally, daily or even hourly, but for the current purposes we will assume that it is constant. The consumption of stored food by brood, hive bees and foragers is given by γ_B, γ_H and γ_F respectively. Because the ratio of foragers to hive bees in the hive quickly equilibrates we will assume that we can model food consumption by adult bees using the average consumption γ_A so that equation (6) can be written as

$$\frac{df}{dt} = cF - \gamma_A(F + H) - \gamma_B B. \qquad (7)$$

Solution and analysis of the model. Differential equations (1), (3), (4) and (7), using functions (2) and (5), were solved using MATLAB to get typical plots of brood, hive bees, foragers and food against time. When there are very large stores of food (that is as f→∞) then, to a good approximation

$$S(H,f) = S(H) = \frac{H}{H+v} \text{ and } R(H,F,f) =$$
$$R(H,F) = \alpha_{min} - \sigma \frac{F}{F+H}. \qquad (8)$$

In other words, brood survival and transition to foraging do not depend on the size of the food stores in the hive; there is easily enough food to provide for all the hive's needs. We solved for the steady state of the model, both when food is present in abundance and when food is limited, and we obtained a solution for the value of m where the hive changes from having abundant food to being food-limited. We also found the value of m^*, the critical death rate, where the hive goes extinct. These are all given in Appendix S1.

Choosing parameters for the model. Following Khoury et al., [1], we set the queen's daily egg laying rate at $L = 2000$, the transition rate in the absence of foragers but in the presence of food to $\alpha_{min} = 0.25$ and the parameter governing social inhibition of forager recruitment to $\sigma = 0.75$. These values assume that the youngest age at which a worker can become a forager is 4 days old

[12] and that reversion of foragers to hive bees can only begin if more than one third of the worker bees are foraging [13]. We set the parameter which controls the effect of the hive bees on brood survival at $v = 5000$. This assumes that when there are 5000 hive bees, roughly half the eggs that are laid fail to survive to pupation because of lack of attention by worker bees, which is congruent with our observations of experimental colonies, and above the minimum population size at which colonies can effectively rear brood [14,15]. We chose $\alpha_{max} = 0.25$ so that the recruitment of hive bees in the absence of foragers was doubled when food was absent as well [7]. We set the pupation rate of brood to $\phi = 1/9$, which assumes that it takes 9 days from the time that an egg is laid until the larvae pupates and we set the duration of pupation τ to 12 days.

We measured food in units of grams. Using data from Harbo [16,17], Russell et al [18] estimate that foragers collect approximately 0.1 grams of food per forager per day, when food is plentiful in the environment. Other data from Harbo [17] suggests that it requires about 0.163 g of honey to rear a worker bee to the point of pupation, so if we average this over the nine days that brood is uncapped we get that the consumption rate per brood item $\gamma_B = 0.018$ g/day. In reality food is consumed only by larvae, but we have not separated eggs from larvae in this model and so take an average amount over the whole period before pupation. The rate of honey consumption for adult bees is given as $\gamma_A = 0.007$ [17]. If we assume that the effects of low food stores are not evident when there is a kilogram or more stored food then we can estimate that $b = 500$ so that $b^2/(b^2+f^2) = f^2/(b^2+f^2) = \frac{1}{2}$ when $f = 500$, that is when there is 500 g of stored food.

Results

Figure 2 shows brood, hive bees, forager and food stores as a function of time for increasing death rate m with all other parameters held constant. When death rates are low ($m = 0.1$, Figure 2a), food stores rise rapidly while the bee populations tend to a steady state that is essentially determined by the balance between the death rate of foragers and the laying rate of the queen. At a higher death rate ($m = 0.3$, Figure 2b), the rate of food accumulation is decreased, and the equilibrium population size is reduced. At even higher death rates (illustrated here by $m = 0.42$, Figure 2c) food becomes a limiting factor as the hive does not have enough foragers to collect more food than it consumes, so that the amount of stored food does not continually increase but settles at a steady state. Bee populations are also much lower. For death rates above the critical death rate, $m*$ the bee population goes extinct but stored food remains in the hive, even after all bees have died (Figure 2d). This is probably because the hive bee population and hence the brood rearing effort are so compromised that the population declines faster than residual food stores can be consumed.

These behaviours are summarized in Figure 3a, which is a plot of steady state populations against death rate where all other parameters are fixed, and Figure 3b which plots populations and food against c the rate of collection of food by foragers.

Figure 4 shows results from the model when the death rate of foragers is changed suddenly from a value where food stores are increasing to a death rate where food collection will limit hive growth. In Figure 4a, the initial death rate is low, food is accumulating rapidly and the population of adult bees is large and growing. Increasing the death rate leads to a rapid decline in adult bee numbers while food stores cease to grow and start to decline slightly. If the death rate is already high, but not so high that food is limiting (Figure 4b), increasing death rate to a point at which it limits growth results in a much smaller decline in the adult bee population of the hive, and food stores go from growing slowly to declining slowly. In both cases the amount of uncapped brood does not change significantly.

The model gives reasonable agreement with experimental data from Harbo [16], especially for large colonies (Table 1). However, according to Harbo's [16] observations small colonies have much lower food stores (expressed as honey gain per bee per day) than are predicted from our model. There could be many reasons for this. Small colonies may expend more energy per bee for thermoregulation, or may be less efficient at collecting food since a small forage force will be less efficient at identifying rich forage sources in the environment.

Discussion

The model that we have presented here is not an attempt to simulate reality; rather the intention of this model is to provide a framework with which to consider the factors that influence colony growth and development, and how they might interact. We have focused on food availability and how that interacts with the intrinsic demographic processes within the colony to affect colony health and growth.

The model suggests that both food availability and forager death rate have very strong influences on colony growth and development. When forager death rates are low, low food availability limits both the amount of food accumulated by the colony, and colony population size (Figure 2b). However as food availability increases, the amount of stored food and total hive population both increase (Figure 3b). Colony population eventually stabilizes at an equilibrium size determined by the forager death rate m, and is no longer affected by increasing food availability, whereas food stores continued to increase (Figure 3b). Under conditions of low mortality and high food availability our model predicts an infinite amount of stored food in a colony. Obviously this does not capture the reality of an operating bee colony, but it does reflect to a degree a beekeeper's ideal situation where a colony has the capacity to accumulate a large surplus of honey that can be harvested without compromising colony function.

Varying the death rate has more complex effects on hive dynamics, and as we increase death rates from low to high levels our model produces different categories of equilibrium colony conditions (Figure 3a). Increasing forager death rates from low to moderate levels reduces the equilibrium population size (Figure 2a and 2b). Colonies can still accumulate a surplus of food, although the rate of food accumulation is reduced in smaller colonies (Figure 2a and 2b). At higher death rates, while a colony can

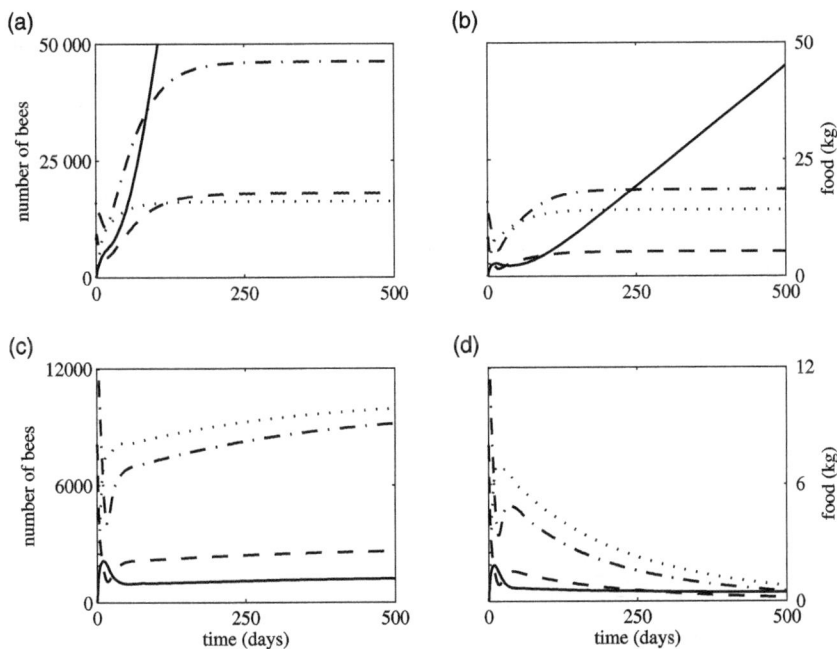

Figure 2. Population and food behaviour over time for different rates of forager mortality. In all plots, the solid line represents food, the dashed line foragers, the dash-dot line hive bees and the dotted line brood. Parameter values are $L = 2000$, $\phi = 1/9$, $v = 5000$, $\sigma = 0.75$, $\alpha_{min} = 0.25$, $\alpha_{max} = 0.25$, $b = 500$, $c = 0.1$, $\gamma_A = 0.007$, $\gamma_B = 0.018$, and $\tau = 12$. In (a) $m = 0.1$; (b) $m = 0.3$;(c) $m = 0.42$; (d) $m = 0.5$. In all plots the hive starts with 16000 hive bees, 8000 foragers and no brood or food at $t = 0$. Note that (c) and (d) have a different vertical scale to (a) and (b).

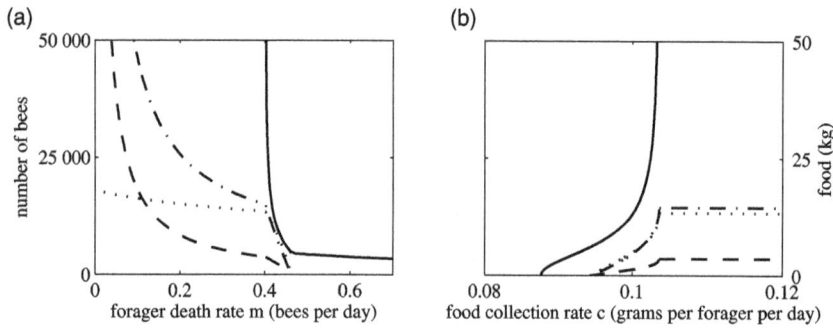

Figure 3. Steady state population as a function of (a) death rate and (b) food collection rate. Parameter values are $L = 2000$, $\phi = 1/9$, $v = 5000$, $\sigma = 0.75$, $\alpha_{min} = 0.25$, $\alpha_{max} = 0.25$, $b = 500$, $c = 0.1$, $\gamma_A = 0.007$, $\gamma_B = 0.018$, and $\tau = 12$. In (a) $c = 0.1$ and in (b) $m = 0.42$. The line styles are the same as Figure 2.

maintain a stable, if small, population it is incapable of accumulating a surplus of food. Rather it maintains a small equilibrium food store reflecting a fine balance of food collection and food consumption by the colony (Figure 2c). The population dwindles to zero at death rates exceeding the threshold at which a colony can maintain a stable population, but food stores remain because colonies fail before they have completely consumed their food reserves (Figure 2d). Therefore the model suggests that different forager death rates result in qualitatively and quantitatively different colony outcomes, which range from a stable population with an excess of food stores, to a stable population with limited food stores, to zero population with residual food stores.

This model suggests a hypothesis for the puzzling observation of colonies dying and completely depopulating but leaving a residual food store. Such a situation has been observed with increasing frequency, and has been considered one of the more perplexing features of colony collapse disorder [19,20]. Our model suggests that depopulation with a small amount of residual food would be expected if colonies suffer a sustained high level of forager mortality (Figure 2d), which is entirely consistent with the rapid declines seen in colony collapse disorder and the lack of dead bees found in the vicinity of a hive [19,20,21]. It is unlikely our simple model accurately captures the dynamics of the terminal phase of a colony. When colony populations get small a host of factors could come into play that would accelerate the colony's death, including inability to incubate brood or maintain nest temperature [15], inability to control nest parasites and compromised food collection

and processing. These would suggest that colonies would fail far more quickly than is indicated by the model.

The model also indicates the ideal parameter space for maximal honey harvest. In the model low forager mortality coupled with high food availability results in a colony that can continue accumulating a food surplus indefinitely. In reality food stores cannot be infinite but the model does, in some way, represent the fact that healthy bee colonies with abundant forage will continue to accumulate honey reserves daily until they become limited by storage space, seasonal changes or a dearth in forage. It is precisely in this state that beekeepers try to maintain their colonies to produce the greatest possible honey harvest. Our model suggests that to achieve this state, bees should be managed to both maximize forager longevity and capitalize on situations of high food availability. Unfortunately this is easier said than done as the two conditions rarely align for long. Floral resources are transient, and in attempting to maximize foraging opportunities for bees many beekeepers in North America and Europe move their colonies to follow flowering periods of different agricultural crops. But moving a colony imposes a cost of increased forager mortality since experienced foragers are unable to successfully navigate the new environment and become lost [22]. It can take a colony weeks to restore normal forager performance after a move [22]. Therefore the twin ideals of low forager mortality and high food availability are hard to achieve, and some trade off must be made in attempting to optimize colony management. Our model provides a simplified framework for exploring consequences of different approaches to this trade off for colony productivity.

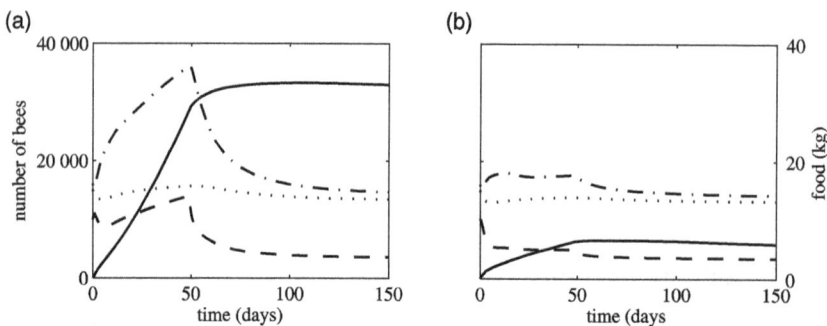

Figure 4. The effect of suddenly increased death rates. Parameter values are as for Figure 2. The solid lines represents food, the dotted lines brood, the dash-dot lines hive bees and the dashed lines foragers. When $t = 50$, the death rate m is reset from its initial value to 0.42. In (a) the initial death rate is $m = 0.1$ and in (b) the initial death rate is $m = 0.3$. In both plots the hive starts with 16000 hive bees, 10000 foragers, 15000 uncapped brood items and no food at $t = 0$.

Table 1. Comparison of model results with observations.

Initial adult bee number		Final brood cell numbers		Brood cells per adult bee		Honey gain per bee per day (mg)		Percentage survival at 22 days	
observed	model	observed	model	observed	model	observed	model	observed	model
2316	2400	4325	4005	2.41	2.13	2.6	23.9	56	56
4515	4500	11162	9154	3.04	2.55	1.6	17.7	64	60
9352	9000	16275	17542	2.21	2.40	10.1	14.7	58	62
17099	18000	22875	26493	1.67	1.79	11.9	15.4	63	64
37061	36000	27875	33599	0.97	1.13	17.7	17.4	55	65

Experimental data is from Harbo, (1986) for hives set up in April. The model results were obtained by running the model for 21 days. The experimental hive was set up for 22 days but the queen was only free to lay after day 2. At the start of each model run, there was no brood or food in the hive and one third of the adult bees were foragers. The parameters used were $L = 2000$, $\phi = 1/9$, $v = 5000$, $\sigma = 1.3$, $\alpha_{min} = 0.25$, $\alpha_{max} = 0.25$, $b = 500$, $c = 0.09$, $\gamma_A = 0.007$, $\gamma_B = 0.018$, $m = 0.06$ and $\tau = 21$ (which prevented any adult bees emerging during the simulations to match the experimental set-up).

We emphasise that monitoring colony productivity demands attention to food availability, forager longevity and adult bee population. Previously most attention has been paid to food availability, and bee management decisions have been focused on moving colonies to exploit good foraging opportunities, but regardless of food availability colonies will not flourish if forager mortality rates are high. There is increasing interest in development of sensors that could be deployed to aid monitoring of colony performance, and numerous scales, thermometers and humidity sensors have been developed for beekeepers to use in colonies in the field. However, our model indicates that colony weight on its own may not be a good index of colony condition. The lag in the model in the change of overall food stores and brood weight in response to increased forager mortality suggests that colony weight would be a poor early indicator of a dwindling forager population (Figure 4). Examination of the brood nest for the amount of brood may also give a poor indicator of a colony under stress as the model suggests brood responds quite slowly to a change in forager mortality (Figure 4a), and this has been reported for colony deaths related to colony collapse disorder [21].

In our model the nurse bee population declines fastest, and as nurse bees transition to forager bees this partially buffers the decline in forager population (Figure 4). As a consequence simply tracking hive traffic could underestimate the rate at which a colony is depopulating. We suggest that the best strategy for monitoring both level and rate of change of stored food and colony population in bee colonies would require monitoring both change in mass, and the rate of loss of field bees.

There are many obvious ways in which this model could be extended. It does not, for example, take into account seasonal variations in food supply or the queen's egg laying rate, nor does the model distinguish between nectar and pollen collection and consumption. There is always a balance between keeping models simple so that the influences of the most important factors on major outcome can be easily explored and formulating complicated models that include more, but where the overall picture can be harder to grasp. In this paper we have chosen to use a simple model to look at major features of colony demographics in a straightforward way but it would be easy to adapt the ideas of this model into a much more complicated computer simulation model. Such a simulation model would be heavily reliant on accurate parameterization if it were to yield meaningful predictions of how environmental changes might alter colony growth and develop-

ment. This would demand far more extensive measurement of how colony parameters vary with availability of pollen and nectar in the environment than are currently available.

Several simulation models for honey bee colonies already exist [23,24,25]. Most of these have been written on particular computational platforms for particular purposes. For example Makela et al [24] created a model in the LISP programming language to explore the factors that gave Africanised honey bee reproductive superiority over pure-bred European honey bees. Schmikl and Crailsheim [25] used Mathematica to formulate a very complicated model that extends the model of DeGrandi-Hoffman et al [23], in part, by including the effect of division of labour in the hive, modeled using ideas from the Foraging-for-Work theory [26]. This theory of task allocation is less relevant in bees than in ants because bees have a strong age-based component in task allocation [27,28] and social inhibition is a very important driver of task specialization [27,29,30,31]. In any case, these simulation models are complicated to understand and to construct and most are tailored to particular situations, as reflected in either their intrinsic structure or their parameterization or both. They are very useful when a specific question needs to be addressed and the necessary input data is available, but it is harder to get a large scale, general picture from complex simulation models than from much simpler, less tightly specific differential equation models such as the one that we present here. The type of model that is most useful will depend, to a very large extent, on the purpose of the modeling and on the scale of the relevant dynamics under investigation; that is, whether, for example, details of individual-to-individual interactions are important to the outcome of the model or whether averages can be taken across the populations of classes of bees as in the model presented here.

Author Contributions

Conceived and designed the experiments: DSK ABB MRM. Performed the experiments: DSK ABB MRM. Analyzed the data: DSK ABB MRM. Wrote the paper: DSK ABB MRM.

References

1. Khoury DS, Myerscough MR, Barron AB (2011) A quantitative model of honey bee colony population dynamics. PLoS ONE 6: e18491.
2. Neumann P, Carreck NL (2010) Honey bee colony losses. Journal of Apicultural Research 49: 1–6.
3. Brodschneider R, Crailsheim K (2010) Nutrition and health in honey bees. Apidologie 41: 278–294.
4. Kearns CA, Inouye DW (1993) Techniques for Pollination Biologists. Boulder CO: University Press of Colorado.
5. Seeley TD (1995) The Wisdom of the Hive. Cambridge: Harvard University Press.
6. Toth AL, Robinson GE (2005) Worker nutrition and division of labour in honeybees. Animal Behaviour 69: 427–435.
7. Schulz DJ, Huang Z-Y, Robinson GE (1998) Effects of colony food shortage on behavioral development in honey bees. Behavioral Ecology and Sociobiology 42: 295–303.
8. Blaschon B, Guttenberger H, Hrassnigg N, Crailsheim K (1999) Impact of bad weather on the development of the broodnest and pollen stores in a honeybee colony (Hymenoptera : Apidae). Entomologia Generalis 24: 49–60.
9. Schmickl T, Crailsheim K (2001) Cannibalism and early capping: strategy of honeybee colonies in times of experimental pollen shortages. Journal of comparative physiology A, Sensory, neural, and behavioral physiology 187: 541–547.
10. Schmickl T, Crailsheim K (2002) How honeybees (Apis mellifera L.) change their broodcare behaviour in response to non-foraging conditions and poor pollen conditions. Behavioral Ecology and Sociobiology 51: 415–425.
11. Jones J, Helliwell P, Beekman M, Maleszka R, Oldroyd B (2005) The effects of rearing temperature on developmental stability and learning and memory in the honey bee, Apis mellifera. Journal of Comparative Physiology A 191: 1121–1129.
12. Fahrbach SE, Robinson GE (1996) Juvenile hormone, behavioral maturation and brain structure in the honey bee. Developmental Neuroscience 18: 102–114.
13. Robinson GE, Page RE, Strambi C, Strambi A (1992) Colony integration in honey bees: mechanisms of behavioural reversion. Ethology 90: 336–350.
14. Becher M, Hildenbrandt H, Hemelrijk C, Moritz R (2010) Brood temperature, task division and colony survival in honeybees: a model. Ecological Modelling 221: 769–776.
15. Rosenkranz P (2008) Report of the Landesanstalt für Bienenkunde der Universität Hohenheim for the year 2007. Stuttgart-Hohenheim, Germany.
16. Harbo JR (1986) Effect of population size on brood production, worker survival and honey gain in colonies of honeybees. Journal of Apicultural Research 25: 22–29.
17. Harbo JR (1993) Effect of brood rearing on honey consumption and the survival of worker honey bees. Journal of Apicultural Research 32: 11–17.
18. Russell R, Barron AB, Harris D (2013) Dynamic modelling of honey bee (Apis mellifera) colony growth and failure. Ecological Modelling. In press.
19. VanEngelsdorp D, Evans JD, Saegerman C, Mullin C, Haubruge E, et al. (2009) Colony Collapse Disorder: A Descriptive Study. PLoS ONE 4.
20. Watanabe ME (2008) Colony collapse disorder: Many suspects, no smoking gun. Bioscience 58: 384–388.
21. Oldroyd BP (2007) What's killing American honey bees? PLoS Biology 5: 1195–1199.
22. Lindauer M (1961) Communication Among Social Bees; Mayr E, Thimann KV, Griffin DR, editors. Cambridge Massachusetts: Harvard University Press.
23. DeGrandi-Hoffman G, Roth SA, Loper GL, Erickson EH Jr (1989) Beepop: a honeybee population dynamics simulation model. Ecological Modelling 45: 133–150.
24. Makela ME, Rowell GA, Sames WJ IV, Wilson LT (1993) An object-oriented intracolonial and population level model of honey bees based on behaviors of European and Africanized subspecies. Ecological Modelling 67: 259–284
25. Schmickl T, Crailsheim K (2007) HoPoMo: A model of honeybee intracolonial population dynamics and resource management. Ecological Modelling 204: 219–245.
26. Franks NR, Tofts C (1994) Foraging for work - how tasks allocate workers. Animal Behaviour 48: 470–472.
27. Beshers SN, Fewell JH (2001) Models of division of labor in social insect colonies. Annual Review of Entomology 46: 413–440.
28. Huang Z-Y, Robinson GE (1996) Regulation of honey bee division of labor by colony age demography. Behavioral Ecology and Sociobiology 39: 147–158.
29. Robson SK, Beshers SN (1997) Division of labour and 'foraging for work': simulating reality versus the reality of simulations. Animal Behaviour 53: 214–218.
30. Leoncini I, Le Conte Y, Costagliola G, Plettner E, Toth AL, et al. (2004) Regulation of behavioral maturation by a primer pheromone produced by adult worker honey bees. Proceedings of the National Academy of Sciences of the United States of America 101: 17559–17564.
31. Huang Z-Y, Robinson GE (1999) Social control of division of labor in honey bee colonies. In: Detrain C, Deneubourg JL, Pasteels JM, editors. Information Processing in Social Insects. Basel: Birkhauser. pp. 165–187.

13

A New Threat to Honey Bees, the Parasitic Phorid Fly *Apocephalus borealis*

Andrew Core[1], Charles Runckel[2], Jonathan Ivers[1], Christopher Quock[1], Travis Siapno[1], Seraphina DeNault[1], Brian Brown[3], Joseph DeRisi[2], Christopher D. Smith[1], John Hafernik[1]*

1 Department of Biology, San Francisco State University, San Francisco, California, United States of America, 2 Department of Biochemistry and Biophysics, University of California, San Francisco, San Francisco, California, United States of America, 3 Entomology Section, Natural History Museum of Los Angeles County, Los Angeles, California, United States of America

Abstract

Honey bee colonies are subject to numerous pathogens and parasites. Interaction among multiple pathogens and parasites is the proposed cause for Colony Collapse Disorder (CCD), a syndrome characterized by worker bees abandoning their hive. Here we provide the first documentation that the phorid fly *Apocephalus borealis*, previously known to parasitize bumble bees, also infects and eventually kills honey bees and may pose an emerging threat to North American apiculture. Parasitized honey bees show hive abandonment behavior, leaving their hives at night and dying shortly thereafter. On average, seven days later up to 13 phorid larvae emerge from each dead bee and pupate away from the bee. Using DNA barcoding, we confirmed that phorids that emerged from honey bees and bumble bees were the same species. Microarray analyses of honey bees from infected hives revealed that these bees are often infected with deformed wing virus and *Nosema ceranae*. Larvae and adult phorids also tested positive for these pathogens, implicating the fly as a potential vector or reservoir of these honey bee pathogens. Phorid parasitism may affect hive viability since 77% of sites sampled in the San Francisco Bay Area were infected by the fly and microarray analyses detected phorids in commercial hives in South Dakota and California's Central Valley. Understanding details of phorid infection may shed light on similar hive abandonment behaviors seen in CCD.

Editor: Nigel E. Raine, Royal Holloway University of London, United Kingdom

Funding: United States National Science Foundation grant DEB-1025922 supported BB. JD was supported by the Howard Hughes Medical Institute. CR was supported by a Genetech Graduate Student Fellowship and Project Apis m. JH and CS were supported by a California State University Program for Education and Research in Biotechnology Faculty-Student Seed Research grant. The funders had no role in study design, data collection and analysis, decision to publish, or preparation of the manuscript.

Competing Interests: The authors have declared that no competing interests exist.

* E-mail: acore13@yahoo.com

Introduction

The honey bee *Apis mellifera* has experienced recent unexplained die-offs around the world [1]. In the United States, Colony Collapse Disorder (CCD), a syndrome characterized by loss of hives and the behavior of hive abandonment, threatens honey bee colonies and has received considerable scientific and media attention. While the United States is the only country for which CCD *sensu stricto* has been documented, there also has been an increase in unexplained colony losses for some regions of Europe and other parts of the world [1–4]. At the same time, some regions of Europe and Asia have reported only normal colony losses. Although catastrophic losses of honey bee colonies have occurred in the past, the magnitude and speed of recent hive losses appear unprecedented [1]. So far, the main causal suspects have been parasitic mites, fungal parasites, viral diseases and interactions amongst them [1–5]. While viral and microsporidian infections have been linked to increased mortality and declining health in honey bee colonies [5], [6], studies have not directly addressed behavioral changes involved in abandonment of hives.

Honey bees suffer from numerous parasites and pathogens including viruses, bacteria, parasitic fungi and ectoparasitic mites

[7]. Infections from agents within any of these pathogen and parasite groups can be fatal to honey bees, but the parasitic *Varroa destructor* mite appears to be the most harmful to colonies overall. *Varroa destructor* is widespread in honey bee hives, affecting every life stage of honey bees from larva to adult [8]. Probably because of this, beekeepers in the United States rank parasites as a bigger threat to their honey bee colonies than CCD [1]. Controlling for parasitic mites is time consuming and costly with damage control estimated in the billions of dollars worldwide [9]. Further, *V. destructor* has been implicated as a vector of many pathogens that can compromise colony health [10–12]. Understanding parasitic infections in honey bees is crucial in predicting the long-term health of honey bee hives.

Here we report that *Apocephalus borealis*, a phorid fly native to North America, previously known to parasitize bumble bees and paper wasps [13], [14], also attacks the non-native honey bee. The genus *Apocephalus* is best known for the "decapitating flies" that parasitize a variety of ant species [15]. *Apocephalus borealis* belongs to the subgenus *Mesophora*, which is a group that contains species that attack hosts other than ants. Although the hosts of most species in the *Mesophora* group are unknown, previously discovered hosts include a variety of arthropods including bees, wasps, beetles and spiders, but not honey bees [14].

In this paper, we show that *A. borealis* has a profound effect on parasitized honey bees, leading them to abandon their hives at night. We use an Arthropod Pathogen Microarray (APM) [16] to detect pathogens that have been implicated in CCD that are associated with adult flies and larvae and to detect the presence of phorids in commercial hives in South Dakota and California's Central Valley. Understanding causes of the hive abandonment behavior we document could explain symptoms associated with CCD. Further, knowledge of this parasite could help prevent its spread into regions of the world where naïve hosts may be easily susceptible to attack.

Results

We found widespread parasitism by *A. borealis* amongst 7,417 honey bees and 195 bumble bees (177 *Bombus vosnesenskii*, 18 *Bombus melanopygus*) sampled from San Francisco Bay Area localities (Figure 1 and Table S1). In all, 77% of our sample sites (24 of 31) yielded honey bees parasitized by *A. borealis*. We reared phorids from 26 *B. vosnesenskii* workers, one *B. vosnesenskii* queen and one *B. melanopygus* worker.

Using DNA barcoding, we confirmed that the phorids that emerged from *Apis* and *Bombus* had no more than 0.2% (1 bp) divergence among samples (Figures S1, S2). The slight variation we found was among those phorids reared from honey bees, not between flies reared from honey bees and those reared from bumble bees. We further confirmed the identity of the phorids using morphological criteria and sequencing of 18S rRNA genes

used on the APM. In addition, our lab infections of honey bees (see below) used phorids that had emerged from both honeybees and bumblebees. Flies from both hosts responded in the same way to the presence of honey bee workers. Taken together these data confirm that the phorids that attack honey bees are the same species as those attacking bumble bees.

Foraging *B. vosnesenskii* showed a higher rate of phorid parasitism than *A. mellifera* foragers (Table S3). Although our individual sample sizes for bumble bees are small due to their relative rarity in summer 2010, we observed parasitism rates as high as 80% (8/10) in one sample of foraging bumble bees from September.

In laboratory infections, female flies attacked honey bees soon after they were placed in an arena with them. Female flies pursued a bee, landed on its abdomen and inserted their ovipositors into it for two to four seconds (Figure 2A, 2B). We observed the same behavior towards honey bees from phorids reared from bumble bees or from honey bees. This interaction is similar to that of other species of phorids that parasitize ants [17] and bees [18]. Mature phorid larvae emerged from the junction between a bee's head and thorax (Figure 2C), on average, seven days after collection (n = 636, Range = 1–14, SD = 1.68) (Figure S3A) and moved away from the bee to pupate. All larvae that emerged from worker bees successfully pupated under laboratory conditions (see methods). Production rates from field-collected bees ranged from one to 13 mature larvae per infected bee (n = 961, Mean = 4.8, SD = 2.4) (Figure. S3B), giving flies the potential to multiply rapidly. In the laboratory, we observed even higher maximal larval production

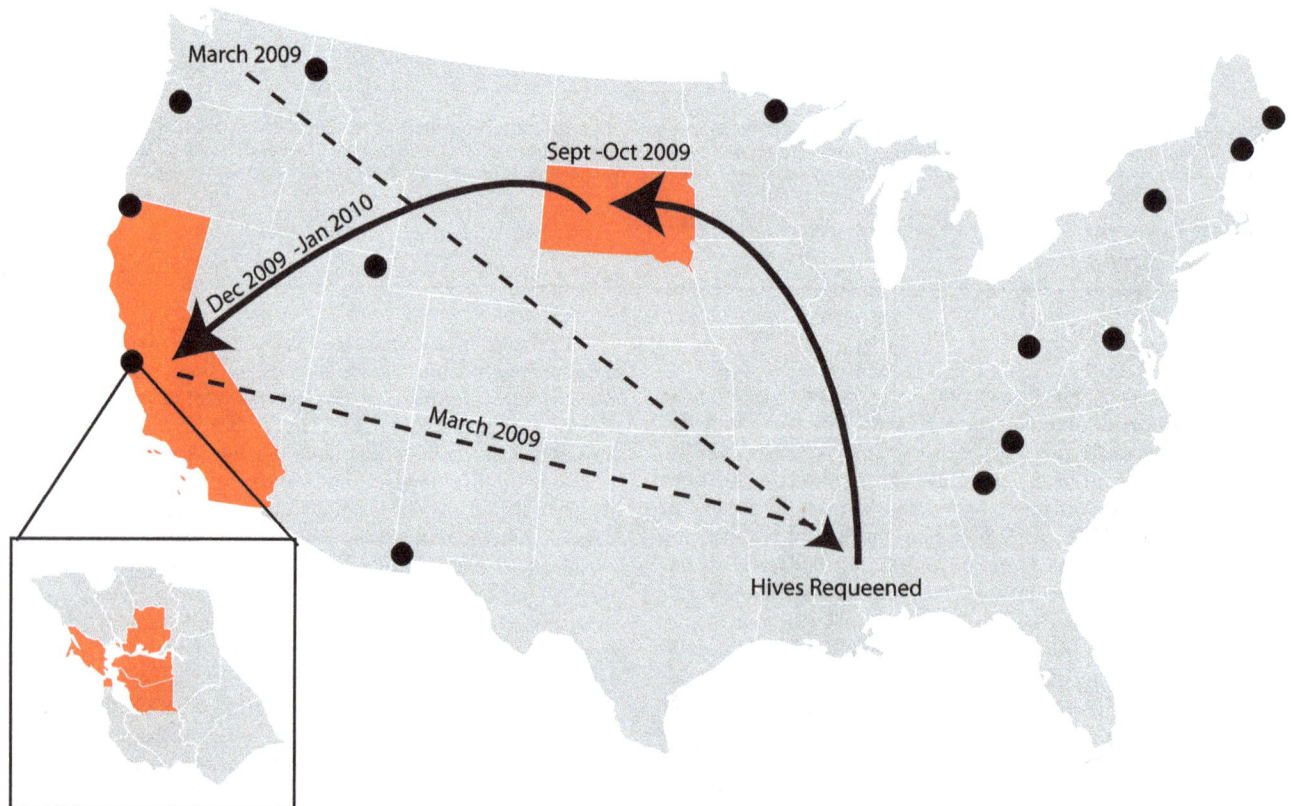

Figure 1. Distribution of phorid-infected honey bees sampled in this study (red). Inset shows the San Francisco Bay Area counties where we found phorid-parasitized honey bees. The routes of commercial hives tested are indicated (arrows), where dotted lines represent states the hives crossed before viral microarray testing and solid lines represent the route of hives during the period of microarray testing. Sites where *A. borealis* was previously known [7] are indicated by black dots.

Figure 2. Images of *Apocephalus borealis* and honey bees. (A) Adult female *A. borealis*. (B) Female *A. borealis* ovipositing into the abdomen of a worker honey bee. (C) Two final instar larvae of *A. borealis* exiting a honey bee worker at the junction of the head and thorax (red arrows).

with one bee producing 25 pupae. Adult flies emerged on average in 28 days (n = 94, Range = 22–36, SD = 1.9) after pupation (Figure S3C).

To investigate internal hive behavior and possible infections within a hive, we kept an observation hive in a laboratory near our primary study hive. Samples taken from the observation hive in June 2010 confirmed infection with *A. borealis*. Rates of infection varied between June 2010 and December 2010 (Mean = 25% Range = 12%–38%) peaking over the sample period in November at 38%. In September, the number of bees in the hive declined and we observed phorid pupae and empty pupal casings among dead bees at the bottom of the hive, indicating emergence of adult phorids within the hive and the potential for *A. borealis* to multiply *within* a hive and infect a queen.

Using an Arthropod Pathogen Microarray (APM) [16], we detected four phorid-positive samples which also shared 99.8% identity over a 432 nt fragment of the 18S rRNA gene (Figure S2) from bees in traveling commercial hives: two from South Dakota during September and October of 2009 and two near Bakersfield, California in January and February of 2010 (Figure 1) [16]. Notably, the APM also detected a higher rate of apparent phorid infection in samples from San Francisco State University on dates when larval emergence assays measured lower levels of parasitism. In this regard, array samples collected between April 23 and June 18, 2010 from various locations on campus (Table S2) detected phorids in 10 of 31 bees (32%) versus only 17 of 244 (7%) detected by our emergence assays (Fishers Exact Test p<0.0002). This difference suggests that the APM is the more sensitive tool to measure infection rates and that our emergence assay data provide a conservative estimate of the abundance of phorids.

We screened phorid adults, larvae and parasitized bees for honey bee pathogens with the APM [16], [19]. Phorid adults and phorid larvae tested positive for infection by *Nosema ceranae* (4/8 adults and 7/8 larvae) and deformed wing virus (DWV) (2/8 adults and 6/8 larvae) (Table S2). Bees from monitored hives and stranded bees sampled from a variety of locations were commonly infected with *N. ceranae* (26/36 bees), and DWV (16/36 bees). Presence of nucleic acid from these pathogens indicates that particles are present, not that they are replicating or are in an infectious form.

While there are previous reports of night activity in honey bees [20], we are the first to link night activity to hive abandonment. We first found stranded worker honey bees beneath lights and within light fixtures on the campus of San Francisco State University (37°43′24.9″N×122°28′31.93″W) (Figure S4A–C) under a variety of weather conditions including cold rainy nights when virtually no other insects were seen around lights. Stranded bees showed symptoms such as disorientation (walking in circles) and loss of equilibrium (unable to stand on legs). Unlike most insects attracted to light, stranded bees remained mostly inactive the next day until they died. Honey bees that left their hives at night had a much higher rate of parasitism by *A. borealis* than bees foraging during the daytime ($\chi^2 = 133$, d.f. 1, p<0.0001) (Figure 3A). From October 2009 to January 2010 parasitism rates were as high as 91% in one sample of nocturnally active bees with a mean parasitism rate of 63% for that period (SD = 18.5, Range = 32%–91%, n = 266 bees) (Figure 3A). During the same period, foraging bees collected at the hive had a mean parasitism rate of only 6% (SD = 8.2, Range = 0%–17.4%, n = 162 bees) (Figure 3A). Phorid parasitism declined from February through spring 2010 before climbing in May and peaking again in autumn 2010 (Figure 3A and Figure S5). During this second recorded peak of parasitism (July 2010–November 2010), stranded bees again had a significantly

A.

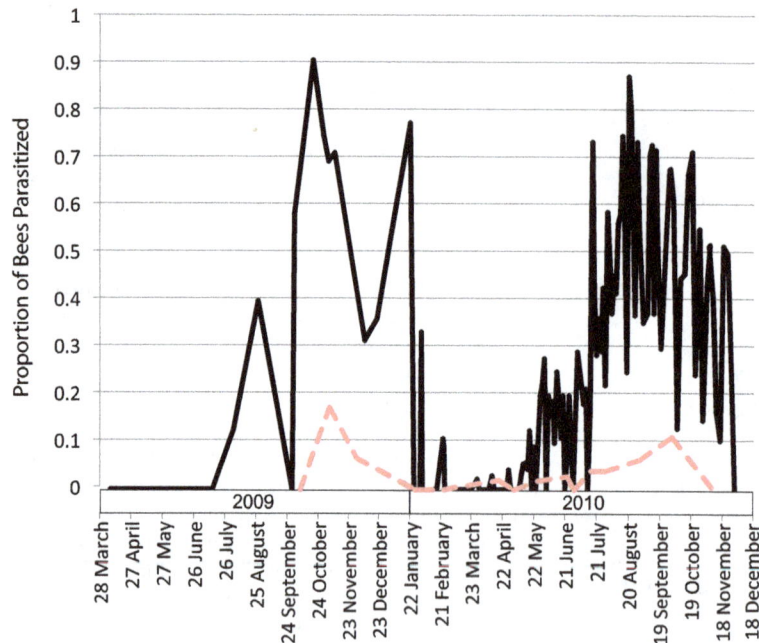

B.

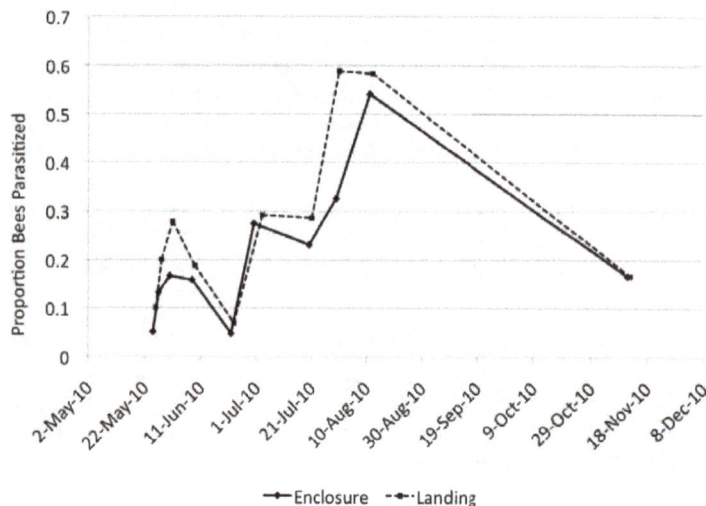

Figure 3. Rates of phorid parasitism in honey bees. (A) Rates of parasitism for bees sampled from April 2009 through November 2010. Black solid line shows rates in stranded bees from under lights on the San Francisco State University campus, while the pink dashed line shows rates in foraging bees. Stranded bees found under lights were sampled at irregular intervals during 2009 and sampled every two days in 2010. Foragers were sampled monthly from our main study hive. A rate of zero indicates that samples from that period contained no parasitized bees. We compared rates of parasitism in stranded and active foraging bees collected at San Francisco State University from October 2009 through January 2010 and from July 2010 to December 2010 (when parasitism rates peaked). 2009–2010 peak rates of parasitism in samples of stranded bees (Mean = 60%, n = 276) were significantly higher than peak rates of parasitism in active foragers from our main study hive (Mean = 6%, n = 164) (χ^2 = 126.7, d.f. 1, p<0.0001). This pattern repeated in 2010 when peak rates of parasitism in samples of stranded bees (Mean = 50%, n = 860) were again significantly higher than rates of parasitism in active foragers (Mean = 4%, n = 422) (χ^2 = 255.3, d.f. 1, p<0.0001). (B) Proportion of honey bees parasitized by phorids in samples from stranded bees collected from the Hensill Hall landing under lights (dotted line) and from samples of bees collected from overnight hive enclosures on adjacent nights (solid line). Parasitism rates of bees trapped in the enclosures closely track rates in stranded bees found under lights during the same period and the number of bees found under lights significantly declined when the enclosure was in place (Welch's t-test p<0.0001) indicating that stranded bees came from our main study hive and were parasitized prior to abandoning the hive.

higher rate of parasitism than foragers (χ^2 = 255.3, d.f. 1, p<0.0001). Parasitism rates in stranded bees again peaked at nearly 90% (Mean = 50%, SD = 19, Range = 11%–88%, n = 860 bees) while foragers had a much lower rate of parasitism (Mean = 4%, Range = 0%–11%, n = 422 bees). These peaks in infection occurred just prior to or during the time of year when losses of honey bee colonies from CCD and other causes peak in the San Francisco Bay Area.

We periodically placed an enclosure over our primary study hive and assessed rates of parasitism of bees that left their hive at night (Figure S4D). Samples of bees trapped in the enclosure (n = 10 samples) ranged from 24–62 bees per night (Mean = 43.5, SD = 15.4). These samples closely tracked the rates of parasitism of stranded bees under nearby lights sampled the day after the enclosure was in place (Figure 3B). Moreover, the number of stranded bees under lights each night significantly declined when the enclosure was in place (Mean = 0.8, SD = 1.14, Range = 0–3, n = 8) compared to a mean of 15.7 (SD = 7.26, Range = 6–29, n = 157) stranded bees for non-enclosure nights (Welch's t-test, p<0.001). This indicates that stranded bees primarily came from our main study hive. The few bees we found stranded on nights when the enclosure was in place probably came from our nearby observation hive. These data confirm that nocturnally active bees were parasitized before leaving their hive and were drawn to the nearby light.

Discussion

The behavior we observed in honey bees is similar to that reported for imported fire ants, Solenopsis invicta parasitized by the phorid, Pseudacteon tricuspis [21], and suggests that A. borealis is manipulating the behavior of its host bees. Such host manipulation has been proposed as an adaptive evolutionary strategy for a number of interactions between a variety of parasites and their hosts [22]. Recent work on gypsy moth larvae infected with nucleopolyhedrovirus identifies the genetic mechanism of host manipulation. The virus manipulates larval behavior inducing larvae to climb to the tops of trees where they die, liquefy and rain virus on the foliage below to infect new hosts [23]. This study provides a clear example of modifications to the expression of a key gene in a host and supports the extended phenotype theory proposed by Richard Dawkins [24], [25]. In the case at hand, perhaps A. borealis manipulates the behavior of honey bees by changing a bee's circadian rhythm, its sensitivity to light or other aspects of its physiology. In order to show that the changes in bee behavior that we document are adaptive for the fly, future studies will need to document that the change in behavior leads to an increase in the fitness of the parasite [22]. Alternatively, phorid infection may be one of several stressors resulting in aberrant nighttime activity (Figure S5). If true, sick bees may altruistically leave their hives to reduce risk to hive mates [26]. A similar response has been proposed for bumble bees parasitized by conopid flies [27] and ants infected by a fungal pathogen [28]. If this explanation is correct, bees might also leave their hive in response to infections such as those that we detected using the APM. Hive mates might also detect parasitized bees due to behavioral or physiological changes associated with parasitism and eject them from the hive. For example, Richard et al. [29] showed that bees intentionally infected with bacterial lipopolysaccarides expressed significantly different cuticular hydrocarbon profiles compared to healthy bees and that coating healthy bees with the hydrocarbon profile of infected bees aroused significant aggression towards those bees by hive mates. If parasitism by A. borealis alters a bee's chemical signature, this could provide a means for workers to detect phorid-infected hive mates.

Our data clearly show that phorid-parasitized bees demonstrate the unusual behavior of abandoning their hives at night. However, we can't exclude the possibility that some parasitized bees also abandon their hive during normal foraging times and die at some distance from the hive. Future experimental studies comparing the daily activity patterns of parasitized versus unparasitized workers are needed to test this possibility.

Until now, North American honey bees have appeared relatively free of parasitoid insects [30], [31]. In South and Central America, honey bees are attacked by numerous species of phorid flies, almost none of which occur in North America [32], [33]. Our study establishes A. borealis as a novel parasite of honey bees and documents hive abandonment behavior consistent with a symptom of CCD. This is a cause for concern because other species of phorid flies can dramatically affect social insect behavior and are used as biocontrol agents of introduced fire ants [21], [34–36]. So far, our main study hive has persisted despite losses to phorid parasitism and infection from a variety of pathogens. Seasonal variation seen in the rates of parasitism in our main study hive is consistent with other honey bee diseases [16], but the relationship, if any, is not fully understood. Seasonal variation could be associated with the life cycle of the fly in which rates of parasitism of honey bees fluctuate as A. borealis populations seasonally increase and decline. The fact that we did not find fly adults within hives may indicate that phorids do not survive in large numbers during the late winter when foraging bees are inactive. A detailed study of a larger sample of hives is needed to measure effects of various densities of phorid parasitism on hive health.

It is possible that A. borealis expanded its host range to include the non-native honey bee many years ago and has gone unnoticed because infected bees abandon their hive and flies occurred undetected in low densities. We believe it is more likely that the phenomenon we report represents a recent host shift and an emerging problem for honey bees. Honey bees are among the most studied insects in North America due to their importance to agriculture. The meticulous attention given to honey bees by humans suggests that phorids would have been detected sooner had the host shift occurred long ago, especially since detection of the parasite does not require sophisticated techniques. Observation of dead bees over as little time as five days should detect phorid presence. Furthermore, honey bees have inhabited areas adjacent to electric lights for at least a century, yet we know of no reports of large numbers of honey bees aggregating around lights until recently. This latter point suggests that, even if the flies were present in low numbers in honey bee colonies in the past, something has happened recently that has increased densities making phorids an emerging threat. To test for the presence of phorids in honey bees at earlier times, the APM could be used to analyze preserved honey bees from previous decades. Additional studies of the distribution and frequency of phorid parasitism of honey bees in North America are needed to assess the scope of this phenomenon and to detect if it is expanding to other areas or is already widespread. The easiest way to monitor nocturnal abandonment of hives is to place light traps nearby and then monitor trapped bees for emergence of phorid larvae. We hope that our study and methods will enable professional and amateur beekeepers to collect vital samples of bees that leave the hive at night, in order to determine if these bees are parasitized by phorids.

The host shift from bumble bees to honey bees has potentially major implications for the population dynamics of A. borealis. Bumble bees live in relatively small colonies that last only a single season with only queens overwintering. Honey bees, on the other hand, live in much larger colonies with tens of thousands of individuals living in hives that are warm even in winter. If these flies have or can gain the ability to reproduce within hives they could greatly increase their population size and levels of virulence. Moreover, hundreds and sometimes thousands of commercial honey bee colonies are often found in close proximity to one another in agricultural areas. Such high host density might lead to population explosions of the fly and major impacts on the hives they parasitize. Further, *A. borealis* is already widely distributed across North America [14] (Figure 1).

Although we did not sample hive bees such as nurses to determine if these workers are being parasitized within the nest, infection rates in foragers alone may still have a strong affect on overall hive health. Koury *et al.* [37] modeled colony population dynamics and predict that significant loss of foragers (beyond a certain threshold) could cause rapid population decline and colony collapse. Their model also predicts that significant loss of foragers leads to hive bees moving into the foraging population at younger ages than normal accelerating colony failure. While our emergence data indicated relatively low infection rates by the fly, our APM data suggest infection rates that are considerably higher. If parasitized bees are numerous or co-occur with other infections, a hive could reach a tipping point leading to its collapse. The detection of *A. borealis* in bees from South Dakota and Bakersfield, CA underlines the danger that could threaten honey bee colonies throughout North America. Movement of commercial hives could quickly spread phorid infection; especially given the number of states that commercial hives cross and are deployed in.

Detection of DWV and *N. ceranae* in adult *A. borealis* raises a number of questions. Do these pathogens have a negative influence on the vitality of the flies or affect their behavior? In this regard, microsporidian infections reduce viability in some insect parasitoids [38] but not in phorid parasitoids of the fire ant *S. invicta* [39]. Are phorids involved in transmission of these and perhaps other diseases among honey bees in a colony? Are phorids involved in transmission of pathogens between the non-native honey bees and native bees? Alternately, are phorids a dead end for pathogens since as parasitoids they might kill their host before the pathogens can multiply? Answering these questions will require more detailed study. However, just because an infectious agent ultimately proves fatal does not mean it cannot be a vector for other pathogens. This is especially true if the development time of phorid larvae is long. Our results document that phorid-infected foragers spend time in their hive before abandoning it. This period of infection (before abandonment) could extend for a week or more providing time for the pathogens to multiply.

In the case of DWV, the virus has been isolated from the feces and intestines of queen honey bees [40]. If this is true of workers, it provides a potential means to transmit the virus in fluids exchanged by honey bees or by close contact. Vectoring of microsporidian infections during oviposition occurs in some parasitic hymenopteran parasitoids [41], [42]. This mode of transmission has been documented under laboratory conditions for at least three different pathogen-parasitoid-host complexes [42]. Similar to *A. borealis*, *Pseudacteon* phorids have tested positive for microsporidian pathogens of fire ants and have been suggested as a possible vector via oviposition [39]. As yet, it is unclear what proportion of *A. borealis* attacks in the wild result in successful parasitism; however, it is conceivable that unsuccessful attacks could still puncture the abdomen and expose the target bee to any pathogens infecting or carried by the phorid. Considering other honey bee parasites, such as the *Varroa destructor* mite, have been implicated as a vector of DWV, Kashmir bee virus, slow paralysis virus, and Israeli acute paralysis virus, [10]–[][12], phorid flies may also act as vectors for DWV or *N. ceranae*. Finally, *N. ceranae* and DWV have been isolated from bumble bees suggesting that exchange of pathogens between honey bees and bumble bees has occurred [43].

Apocephalus borealis may also be a threat to native pollinators since it parasitizes a number of bumble bee species and paper wasps (*Vespula* spp) [13], [14]. Wild bumble bees are experiencing substantial declines in North America [44], [45]. So far, attention has focused on emerging pathogens such as *Crithidia bombi and Nosema bombi*. In the laboratory, bumble bees parasitized by *A. borealis* show a dramatic reduction in life span compared to unparasitized bees [13]. The high rate of parasitism in some of our samples of foraging bumble bees and previous high parasitism rates from Canada [13], suggest that parasitism by *A. borealis*, especially in combination with infection by emerging pathogens, could place significant stress on bumble bee populations. If so, phorid parasitism or pathogen transmission to bumble bees might contribute to a cascade of effects in plant and agricultural communities that rely on bumble bees as pollinators. Furthermore, the domestic honey bee is potentially *A. borealis'* ticket to global invasion. Establishment of *A. borealis* on other continents, where its lineage does not occur, where host bees are particularly naïve, and where further host shifts could take place, could have negative implications for worldwide agriculture and for biodiversity of non-North American wasps and bees.

Methods

Ethics statement

Samples of San Francisco Bay Area honey bees and bumble bees were obtained with appropriate permissions from beekeepers, landowners and the San Francisco Recreation and Parks Department.

Sampling procedures

We sampled honey bees from a variety of circumstances. Our main samples consisted of the following: 1) Bees found stranded under lights near the main entrance to Hensill Hall on the San Francisco State University campus (Figure S4A–C). From April 2009 until January 2010, a portion of bees found stranded under lights was sampled at irregular intervals (Range = 2–112 bees per sample). From February to November 2010, stranded bees were sampled at two-day intervals (Range = 2–56 bees per sample) (Figure 3A). All bees were cleared from beneath the lights prior to sundown to ensure that only bees from one night's flight were included in each sample the next morning. Samples consisted of all bees found stranded under the lights. 2) We collected active, foraging bees monthly from our main study hive on the San Francisco State University campus. Samples consisted of 50 incoming foragers collected in individual *Drosophila* vials and samples of 50 or more outgoing foragers collected by placing a standard aerial insect net in front of the hive entrance for 30–60 seconds. We compared the rate of infection in samples of outgoing foragers and incoming foragers. We found no significant difference between these groups (Fishers Exact Test p = 0.32). Therefore, both groups are used to determine long-term trends in rates of infection in active, foraging bees (Figure 3A). This allowed us to compare infection rates of foraging vs. stranded bees. 3) We periodically placed a 1.83 m×1.83 m×1.83 m enclosure (Nica-maka Pop-Up Beach Shade/Tent) over the hive after sunset and removed it before dawn (Fig S4A). We collected all bees captured in the enclosure. Prior to setting up the enclosure, we removed all bees from the area under nearby lights. This allowed us to compare the number of bees stranded under lights during enclosure experiments to the number of bees stranded the day after enclosure experiments. 4) In April 2010, we established an observation hive that allowed us to observe in-hive activities and check for presence of phorids within the hive.

In order to survey prevalence of parasitism in nearby areas, we collected stranded and foraging bees from a variety of locations in the San Francisco Bay Area and from the hives of local beekeepers who agreed to participate in our study. In two of these locations (Table S1 and S2) bees came from areas near feral hives. The feral hive on the San Francisco State University campus has been in place for a number of years and was present before our main study hive appeared on campus. Bees collected near this feral hive were found stranded under a light that is immediately adjacent to the tree containing the colony. The second feral hive was in a tree near the California Academy of Sciences and was discovered during our study. Its history is unknown. We collected stranded bees from beneath the tree that it occupied. In addition, we collected samples of two bumble bee species from the San Francisco Bay Area, *Bombus vosnesenskii* and *B. melanopygus* (Table S1).

Assessment of parasitism rates

In order to assess parasitism rates, bees from all samples were brought into the laboratory and confined at room temperature (19–20°C) in individual glassine envelopes or *Drosophila* rearing containers from April 2009 to November 2010. We checked confined bees daily for a period of two weeks and recorded the number of phorid larvae that emerged. Additionally, we recorded date of larval emergence for a subset of 636 parasitized bees and duration of the pupal instar for a subset of 94 pupae.

Laboratory phorid-honey bee infections

In order to observe interactions between phorids and honey bees, adult flies were obtained from a hatching chamber provisioned with a feeder (a 2.54 cm plastic straw filled with cotton saturated in sugar water) and allowed to sit for at least one day in a container provisioned with dishes containing cotton soaked in sugar water and honey solutions. Adult flies were then placed into a clear plastic enclosure approximately 24 cm×12 cm×13.5 cm, and individual honeybees were introduced to them. With each introduction, we recorded whether phorids approached the bee and demonstrated oviposition behavior. After exposure, bees were kept alive in containers provisioned with sugar water and honey solutions.

Barcode sequencing and phylogenetic comparison

We used DNA barcoding to confirm that the morphologically similar phorids from bumble bees and honey bees were conspecific (Figure S1). High genetic similarity between the two also would support the view that the native *A. borealis* has expanded its host range to include non-native honey bees. We used Qiagen Blood & Tissue DNA extraction kits (Qiagen, Valencia CA) to extract all cellular DNA from collected honey bees, bumble bees, and phorid pupae. We used standard CO1 primers [46] (IDT, Coralville IA)(FWD, 5′ TAAACTTCAGGGTGACCAAAAAATCA.... REV, 5′ GGTCAACAAATCATAAAGATATTG) and the following PCR conditions (1 cycle of 95°C 1 min; 5 cycles of 95°C 1 min, 45°C 1.5 min, 72°C 1.5 min; 35 cycles of 95°C 1 min, 50°C 1.5 min, 72°C 1.5 min; 1 cycle 72°C 5 min) and visualized products on 1% agarose gels. PCR reactions were purified using QiaQuick columns (Qiagen, Valence CA) and sent to Elim Biopharmaceuticals Inc (Hayward, CA) for standard Sanger dideoxy sequencing in both the forward and reverse direction using the CO1 primers. Reads from each orientation were manually contiged using Sequencher (v4.8 Gene Codes Corporation Ann Arbor, MI), and DNA mismatches were visually compared to the DNA chromatogram to correct miscalled bases. Corrected, contigs were aligned using CLUSTALX [47] known phorid barcode sequences and a neighbor-joining tree was generated using 1000 bootstrap replicates.

Microarray analysis

An Arthropod Pathogen Microarray (APM) [16], [19] including all known honey bee viruses, fungal and bacterial pathogens of honey bees, and mite-specific oligos was augmented with products specific to the phorid 18S rRNA gene. Using phorid larvae, total RNA spiked into unparasitized honey bee total RNA, the PCR assay was capable of detecting one part phorid in 10,000 parts honey bee from 5 ng of cDNA, suggesting that relatively early infections could be detected. In total, 378 samples collected from 2008–2010 were screened, including a 20-hive time-course study sampled approximately biweekly as commercial hives migrated from Mississippi to South Dakota and finally to California (Figure 1). Here, five pooled workers each were screened by PCR and Sanger sequencing of the phorid 18S rRNA gene.

Whole insects were homogenized in 1 mL of 1:1 Trizol:PBS with a 5 mm steel ball in a TissueLyzer II at 30 Hz for 4 min. Total nucleic acid was extracted by the addition of 100 µL chloroform and centrifugation, followed by isopropanol precipitation. For each sample, one quarter of the total nucleic acid (1–5 µg) was randomly primed with Superscript II (Invitrogen) with primer RdA (5′GTTTCCCACTGGAGGATANNNNNNNNN). Second-strand synthesis was performed twice with the same primer and Sequenase DNA polymerase (USB). One quarter of this reaction was amplified with Taq polymerase and a single adapter primer RdB (5′GTTTCCCACTGGAGGATA). This randomly amplified material was used for screening for Phorid rRNA with primer pair Phorid-rRNA-1F (GTACACCTATA-CATTGGGTTCGTACATTAC) and -1R (GAGRGCCA-TAAAAGTAGCTACACC) in a Taq polymerase PCR with an annealing temperature of 57°C.

For pathogen detection by microarray, the randomly amplified material was further amplified and labeled with a dye-linked primer RdC (5′Cy3-GTTTCCCACTGGAGGATA), column purified and hybridized to a 70-mer DNA microarray in 3× SSC, 50 mM HEPES and 0.5% SDS at 65°C overnight. Microarrays were scanned on an Axon 4000A scanner and analyzed visually or with the Cluster analysis package [48]. All microarray spots that indicated the presence of pathogens were further confirmed by PCR and Sanger sequencing with primers *Nosema ceranae* F-4186 (5′-CGGATAAAAGAGTCCGTTACC) and R-4435 (5′-TGAGCAGGGTTCTAGGGAT) [49] and DWV-F-1165 (5′-CTTACTCTGCCGTCGCCCA) -R-1338 (5′-CCGTTAGGAACTCATTATCGCG) [50].

Data availability and compliance with standards

The *A. borealis* mitochondrial barcode sequence (ID# JF798506) and18S rRNA gene sequence (ID# JF808447) have been deposited in Genbank. APM design and results have been submitted to GEO (design accession GPL11490 and array data accession GSE28235) and are MIAME compliant.

Supporting Information

Figure S1 CLUSTALX alignment of 450 bp of cyto-chrome oxidase I DNA barcodes obtained from infected honey bees (samples 19–24,26–31,34,35) and bumble bees (samples 33,36). Bidirectional Sanger sequence indicates that only two positions varied (88, 288) in a single sample each. All samples had less than 0.22% divergence (i.e. 1 bp).

Figure S2 *A. borealis* 18S rRNA and mitochondrial cytochrome oxidase I (COI) DNA sequence used for barcoding and APM.

Figure S3   Timing of life history events in parasitism of honey bees by *A. borealis*. (A) Length of time after sample collection until phorid larvae emerged from their honey bee hosts (Mean = 7.14 days, SD = 1.68, n = 636). (B) Number of phorid larvae per infected bee for samples from various locations (Mean = 4.8, SD = 2.45, n = 961). (C) Length of pupal period (Mean = 27.9 days, SD = 1.9, n = 94).

Figure S4   San Francisco State University Hensill Hall study site. (A) Primary study hive, blue arrow indicates direction that honey bees fly to reach the nearby light. (B) Landing above the hive where stranded bees were collected and the light (C) immediately above the landing showing honey bees attracted to it from the previous night. (D) A typical enclosure setup.

Figure S5   The number of parasitized bees (red) compared to all bees (black) collected at the San Francisco State University Hensill Hall collection site. Notably, numerous bees were collected from the lights and landing in months even when parasitism rate was low. Our direct rearing method may have underestimated the rate of parasitism during spring 2010 since the Arthropod Pathogen Array (APM) indicated a higher rate of parasitism during April and early May than we observed in our rearings. The APM also detected a high level of infection with *Nosema ceranae* and deformed wing virus during that period.

Table S1   Honey bee and bumble bee collection sites in the San Francisco Bay Area. Locations of hives which did not yield parasitism in the San Francisco Bay Area are shaded light grey. Locations where stranded and foraging honey bees and bumble bees were collected are shaded dark grey.

Table S2   Arthropod Pathogen Microarray results. Location codes are main study hive (HHH), stranded on landing near main hive (HHL), main hive enclosure (HHC), observation hive (OH), near feral hive on San Francisco State University campus (GYMA), feral hive near California Academy of Sciences (CAS), X's indicate whether infected by phorids, *Nosema ceranae*, or deformed wing virus.

Table S3   Rate of parasitism for *Bombus vosnesenskii* sampled from San Francisco, California locations from May to November 2010.

Acknowledgments

We thank Gretchen LeBuhn for access to hives on the San Francisco State University Campus; Eric Mussen for advice and for suggesting the connection between stranding and attraction to light; Ilma Abbas, Cory Robinson, Chaundra Cox for assistance in DNA barcoding; Jessica Van Den Berg for use of her automontage images of adult phorids; Erika Bueno and Caitlin Papathakis for help in sampling bees; Stan Williams for advice on beekeeping and the following beekeepers for allowing us to sample their hives: R. MacKimmie, L. Guay, T. Williams, T. Trang, P. Gerrie, J. Sanphillippo, R. Bowen, B. and A. Berger, K. Peteros, L. Gartland, M. Andre, T. Brumleve, K. Bairey, L. McCloy, C. Giaioma, L. Lasar, G. Lawrence, J. Chan, D. and S. Goemmel, S. Willis, P. Wickware, J. Levison, R. Beckett, M. McMillan, A. Henninger, and B. Reese. We thank Bret Adee and the beekeepers at Adee Honey Farms for collections in SD. Andy Zink, Neil Tsutsui, Gene Robinson, Giar-Ann Kung and three anonymous reviewers provided helpful comments on the manuscript.

Author Contributions

Conceived and designed the experiments: JH AC JI CQ SD CR JD CS. Performed the experiments: JH AC JI CQ SD CR TS. Analyzed the data: JH AC CS CR JD TS BB. Contributed reagents/materials/analysis tools: JD CR. Wrote the paper: JH AC CS CR BB. Discovered phorid/bee phenomenon: JH. Photographed phorids and bees: CQ JH.

References

1. Williams GR, Tarpy DR, vanEngelsdorp D, Chauzat M, Cox-Foster DL, et al. (2010) Colony collapse disorder in context. Bioessays 32: 845–846. doi: 10.1002/bies.201000075.

2. Oldroyd BP (2007) What's killing American honey bees? PLoS Biol 5: e168. doi:10.1371/journal.pbio.0050168.

3. vanEngelsdorp D, Evans JD, Saegerman C, Mullin C, Haubruge E, et al. (2009) Colony collapse disorder: a descriptive study. PLoS ONE 4(8): e6481. doi:10.1371/journal.pone.0006481.

4. Ratnieks LWF, Carreck NL (2010) Clarity on honey bee collapse? Science 327: 152–153.

5. Johnson RM, Evans J, Robinson GE, Berenbaum MR (2009) Changes in transcript abundance relating to colony collapse disorder in honey bees (*Apis mellifera*). Proc Natl Acad Sci USA 106: 14790–14795.

6. Higes M, Martín-Hernández R, Garrido-Bailón E, González-Porto AV, García-Palencia P, et al. (2009) Honeybee colony collapse due to *Nosema ceranae* in professional apiaries. Environ Microbiol Rep 1(2): 110–113.

7. Generisch E (2010) Honey bee pathology: current threats to honey bees and beekeeping. Appl Microbiol Biotech 87: 87–97.

8. Anderson DL, Trueman JWH (2000) *Varroa jacobsoni* (Acari: Varroidae) is more than one species. Exper and App Acar 24: 165–189.

9. Cook DC, Thomas MB, Cunningham SA, et al. (2007) Predicting the economic impact of an invasive species on an ecosystem service. Eco Appl 17: 1832–1840.

10. Chen YP, Pettis JS, Evans JD, Kramer M, Feldlaufer MF (2004) Transmission of Kashmir bee virus by the ectoparasitic mite *Varroa destructor*. Apidologie 35: 441–448.

11. Di Prisco G, Pennacchio F, Caprio E, Boncristiani Jr. HF, Evans JD, et al. (2011) *Varroa destructor* is an effective vector of Israeli Acute Paralysis Virus in the honey bee, *Apis mellifera*. J Gen Virol 92: 151–155.

12. Santillan-Galicia MT, Bail BV, Clark SJ, Alderson PG (2010) Transmission of deformed wing virus and slow paralysis virus to adult bees (*Apis mellifera* L.) by *Varroa destructor*. J of Apicult Res and Bee World 49: 141–148.

13. Otterstatter MC, Whidden TL, Owen RE (2002) Contrasting frequencies of parasitism and host mortality among phorid and conopid parasitoids of bumble-bees. Ecol Entom 27(2): 229–237.

14. Brown BV (1993) Taxonomy and preliminary phylogeny of the parasitic genus *Apocephalus*, subgenus *Mesophora* (Diptera: Phoridae). Sys Entom 18: 191–230.

15. Brown BV (1997) Revision of the *Apocephalus attophilus*-group of ant-decapitating flies (Diptera: Phoridae). Contrib in Sci 468: 1–60.

16. Runckel C, Flenniken ML, Engel J, Ganem D, Andino R, et al. (2011) Temporal analysis of the honey bee microbiome reveals four novel viruses and seasonal prevalence of known viruses, *Nosema*, and *Crithidia*. PLoS ONE 6(6): e20656. doi:10.1371/journal.pone.0020656.

17. Morrison LW, Dall'aglio-Holvorcem CG, Gilbert LE (1997) Oviposition behavior and development of *Pseudacteon* flies (Diptera: Phoridae), parasitoids of *Solenopsis* fire ants (Hymenoptera: Formicidae). Environ Entom 26(3): 716–724.

18. Brown BV, Kung G (2006) Revision of the *Melaloncha ungulata*-group of bee-killing flies (Diptera: Phoridae). Contributions in Science 507: 1–31.

19. Wang D, Coscoy L, Zylberg M, Avila PC, Boushey HA, et al. (2002) Microarray-based detection and genotyping of viral pathogens. Proc Natl Acad Sci USA 99(24): 15687–15692.

20. Robinson GE, Morse RA (1982) Number of honey bees that stay out all night. Bee World 63: 173–174.

21. Henne DC, Johnson SJ (2007) Zombie fire ant workers: behavior controlled by decapitating fly parasitoids. Insectes Sociaux 54(2): 150–153.

22. Poulin R (2010) Parasite manipulation of host behavior: an update and frequently asked questions. Adv in the Study of Behav 41: 151–186.

23. Hoover K, Grove M, Gardner M, Hughes DP, McNeil J, Slavicek J (2011) A gene for an extended phenotype. Science 333: 1401.

24. Dawkins R (1982) The Extended Phenotype: The gene as unit of selection. New York: Oxford University Press. 307 p.

25. Lambrechts L, Fellous S, Koella JC (2006) Coevolutionary interactions between host and parasite genotypes. Trends in Parasitology 22: 12–16.

26. Rueppell O, Hayworth MK, Ross NP (2010) Altruistic self-removal of health-compromised honey bee workers from their hive. J Evol Biol 23: 1538–1546.

27. Müller CB, Schmid-Hempel P (1993) Exploitation of cold temperature as defense against parasitoids in bumblebees. Nature 363(6424): 65–67.

28. Heinze J, Bartosz B (2010) Moribund ants leave their nests to die in social isolation. Curr Biol 20(3): 249–252.

29. Richard FJ, Aubert A, Grozinger CM (2008) Modulation of social interactions by immune stimulation in honey bee, *Apis mellifera*, workers. BMC Biology 6: 50. doi:10.1186/1741-7007-6-50.

30. Feener DH, Brown BV (1997) Diptera as parasitoids. Annu Rev Entom 42: 73–97.

31. Schmid-Hempel P (1998) Parasites in social insects. Monographs in Behavior and Ecology. New Jersey: Princeton University Press. 392 p.

32. Brown, BV (2004) Revision of the subgenus *Udamochiras* of *Melaloncha* bee-killing flies (Diptera: Phoridae). Zoological Journal of the Linnean Society 140: 1–42.

33. Gonzalez L, Brown BV (2004) New species and records of *Melaloncha* (*Udamochiras*) bee-killing flies (Diptera: Phoridae). Zootaxa 730: 1–14.

34. Orr MR, Seike SH, Benson WW, Gilbert LE (1995) Flies suppress fire ants. Nature 373(6512): 292–293.

35. Porter SD (2000) Host specificity and risk assessment of releasing the decapitating fly *Pseudacteon curvatus* as a classical biocontrol agent for imported fire ants. Biol Control 19(1): 35–47.

36. Morrison LW, Porter SD (2005) Testing for population-level impacts of introduced *Pseudacteon tricuspis* flies, phorid parasitoids of *Solenopsis invicta* fire ants. Biol Control 33(1): 9–19.

37. Koury DS, Myerscough MR, Barron AB (2010) A quantitative model of honey bee colony population dynamics. PLoS ONE 6(4): e18491. doi:10.1371/journal.pone.0018491.

38. Futerman PH, Layen SJ, Kotzen ML, Franzen C, Kraaijeveld AR, et al. (2006) Fitness effects and transmission routes of a microsporidian parasite infecting Drosophila and its parasitoids. Parasitology 132: 479–492.

39. Oi DH, Porter SD, Valles SM, Briano JA, Calcaterra LA (2009) *Pseudacteon* decapitating flies (Diptera: Phoridae): Are they potential vectors of the fire ant pathogens *Kneallhazia* (= *Thelohania*) *solenopsae* (Microsporidia:Thelohaniidae) and *Vairimorpha invictae* (Microsporidia: Burenellidae)? Biol Control 48: 310–315.

40. Chen YP, Pettis JS, Collins A, Feldlaufer MF (2006) Prevalence and transmission of honeybee viruses. Appl and Exper Microbiol 72(1): 606–611.

41. Brooks WM (1993) Host–parasitoid–pathogen interactions. In: Beckage NE, Thompson SN, Federici BA, eds. Parasites and Pathogens of Insects Vol. 2: Pathogens, Academic Press, San Diego, CA. pp 231–272.

42. Becnel JJ, Andreadis TG (1999) Microsporidia in insects. In: Wittner M, Weiss LM, eds. The Microsporidia and Microsporidiosis, American Society of Microbiology Press, Washington, DC. pp 447–501.

43. Plischuk S, Martín-Hernández R, Prieto L, Lucía M, Botías C, et al. (2009) South American native bumblebees (Hymenoptera:Apidae) infected by *Nosema ceranae* (*Microsporidia*),an emerging pathogen of honeybees (*Apis mellifera*). Environmental Microbiology Reports 1(2): 131–135.

44. Otterstatter MC, Thomson JD (2008) Does pathogen spillover from commercially reared bumble bees threaten wild pollinators? PLoS ONE 3: e2771. doi:10.1371/journal.pone.0002771.

45. Cameron SA, Lozier JD, Strange JP, Koch JB, Cordes N, et al. (2011) Patterns of widespread decline in North American bumble bees. Proc Natl Acad Sci USA 108: 662–667.

46. Folmer O, Black M, Hoeh W, Lutz R, Vrijenhoek R (1994) DNA primers for amplification of mitochondrial cytochrome c oxidase subunit I from diverse metazoan invertebrates. Mol Mar Biol Biotechnol 3: 294–297.

47. Larkin MA, Blackshields G, Brown NP, Chenna R, McGettigan PA, et al. (1997) Clustal W and Clustal X version 2.0. Bioinformatics 23: 2947–2948.

48. Eisen MB, Spellman PT, Brown PO, Botstein D (1998) Cluster analysis and display of genome-wide expression patterns. Proc Natl Acad Sci USA 95(25): 14863–14868.

49. Chen Y, Evans JD, Smith IB, Pettis JS (2007) *Nosema ceranae* is a long-present and wide-spread microsporidian infection of the European honey bee (*Apis mellifera*) in the United States. J Invert Path 97(2): 186–188.

50. Chen YP, Higins JA, Feldlaufer MF (2004) Quantitative real-time reverse transcription-PCR analysis of deformed wing virus infection in the honeybee (*Apis mellifera* L.). App Environ Microbiol 71(1): 436–441.

14

Positive Interspecific Relationship between Temporal Occurrence and Abundance in Insects

Ricardo A. Scrosati*, Ruth D. Patten, Randolph F. Lauff

Department of Biology, Saint Francis Xavier University, Antigonish, Nova Scotia, Canada

Abstract

One of the most studied macroecological patterns is the interspecific abundance–occupancy relationship, which relates species distribution and abundance across space. Interspecific relationships between temporal distribution and abundance, however, remain largely unexplored. Using data for a natural assemblage of tabanid flies measured daily during spring and summer in Nova Scotia, we found that temporal occurrence (proportion of sampling dates in which a species occurred in an experimental trap) was positively related to temporal mean abundance (number of individuals collected for a species during the study period divided by the total number of sampling dates). Moreover, two models that often describe spatial abundance–occupancy relationships well, the He–Gaston and negative binomial models, explained a high amount of the variation in our temporal data. As for the spatial abundance–occupancy relationship, the (temporal) aggregation parameter, k, emerged as an important component of the hereby named interspecific temporal abundance–occurrence relationship. This may be another case in which a macroecological pattern shows similarities across space and time, and it deserves further research because it may improve our ability to forecast colonization dynamics and biological impacts.

Editor: Mark S. Boyce, University of Alberta, Canada

Funding: Funding was provided through grants to RAS from the Natural Sciences and Engineering Research Council of Canada (NSERC, www.nserc-crsng.gc.ca), the Canada Foundation for Innovation (CFI, www.innovation.ca), and the Canada Research Chairs program (CRC, www.chairs-chaires.gc.ca). The funders had no role in study design, data collection and analysis, decision to publish, or preparation of the manuscript.

Competing Interests: The authors have declared that no competing interests exist.

* E-mail: rscrosat@stfx.ca

Introduction

Macroecology investigates the distribution and abundance of organisms at broad scales [1]. One of the most studied macroecological patterns is the interspecific abundance–occupancy relationship. For species assemblages in a region, there is a relationship between the mean local abundance of each species and the proportion of local sites that each species occupies. Studies done on plants and animals have shown that such a relationship is generally positive [2]. The factors that shape the spatial abundance–occupancy relationship are not entirely clear, although likely ones are niche breadth, habitat selection, vital rates, range overlap, body size, and dispersal [2–6]. What is certain is that the search for such factors has stimulated research on the links between species distribution and abundance.

Both species distribution and mean abundance can be viewed across space as well as time [1]. Thus, it is also relevant to investigate whether the temporal distribution and mean abundance of species over time may be related. For this purpose, we define temporal occurrence as the proportion of sampling dates in which a species occurs in a given place, and temporal mean abundance as the number of individuals counted in that place throughout the study period divided by the total number of sampling dates. Finding a link between both traits could have important implications for basic and applied ecology. For example, the frequency of visits of mobile species to an area of interest (e.g., a new agriculture field or a restored habitat) could be predicted from knowledge on the abundance of species in similar neighboring environments, which could help to forecast ecological impact or colonization patterns. Conversely, collecting data from simple counts of species occurrence in traps or species sightings in an area might allow one to estimate species abundances in the region, a useful option when traditional methods of abundance estimation are difficult to implement. Interestingly, however, the possible existence of an interspecific temporal abundance–occurrence relationship remains largely unexplored. To investigate this issue, we used an insect assemblage.

Specifically, we asked whether temporal occurrence and temporal mean abundance are related and, if so, whether the relationship is positive. A number of mathematical models have been proposed to describe the spatial abundance–occupancy relationship, among which the He–Gaston model and the negative binomial model usually yield the greatest fit [7]. These models (see Materials and Methods) are based on the degree of spatial aggregation of species, which is a common feature in plant and animal populations [8]. In fact, this predominant trait of natural systems makes the He–Gaston and negative binomial models theoretically more realistic alternatives than the other proposed models, since the latter are based either only on random distribution patterns or on restricted conditions of aggregation that exclude natural variation [7]. Since aggregated patterns of species occurrence may also happen over time, the He–Gaston and negative binomial models emerged as potentially useful tools to describe temporal relationships. Therefore, we tested their utility for our data.

Materials and Methods

We used data for a species assemblage composed of horse and deer flies (Diptera, Tabanidae) from South Side Harbour, Nova Scotia, Canada (45°40′N, 61°53′W). The daily occurrence and abundance of 31 species (Table 1) were recorded on 37 consecutive days in June–July 2000 (Data S1) using a trap box located in a hay field surrounded by forest vegetation near freshwater marshes. The box design has been described by French and Hagan [9]; it is particularly suitable to detect the abundance of tabanid fly species in the environment. Because most tabanid flies are diurnal [10], the trap was emptied daily at dusk for measurements. The daily samples were frozen within 30 minutes of being collected. Every collected specimen was analyzed under a stereomicroscope and identified using the taxonomic key developed by Teskey [11].

Table 1. Number of days in which each species of tabanid fly occurred in the trap (total n = 37 days) and total number of individuals found for each species during the study period.

Species	Number of days	Total number of individuals
Chrysops aestuans	3	3
Chrysops calvus	6	7
Chrysops carbonarius	1	1
Chrysops cincticornis	4	4
Chrysops cuclux	1	3
Chrysops excitans	31	500
Chrysops frigidus	3	6
Chrysops lateralis	1	1
Chrysops mitis	1	1
Chrysops niger	17	38
Chrysops sordidus	1	4
Chrysops vittatus	2	4
Hybomitra affinis	6	16
Hybomitra arpadi	2	3
Hybomitra epistates	32	531
Hybomitra frontalis	5	9
Hybomitra illota	1	3
Hybomitra lasiopthalma	31	3348
Hybomitra liorinha	2	3
Hybomitra longliglossa	1	3
Hybomitra lurida	2	5
Hybomitra microcephala	1	3
Hybomitra nitidifrons nuda	16	221
Hybomitra pechumani	9	15
Hybomitra trepida	24	91
Hybomitra typhus	12	20
Hybomitra zonalis	3	4
Tabanus marginalis	1	1
Tabanus nigrovittatus	5	19
Tabanus reinwardtii	1	1
Tabanus similis	27	181

For each species, we determined temporal occurrence as the proportion of sampling dates in which individuals were found in the trap. We determined the temporal mean abundance of each species as the total number of individuals found in the trap during the study period divided by the total number of sampling dates (37). We investigated the relationship between temporal occurrence and temporal mean abundance by evaluating the fit of the data to the He–Gaston and negative binomial models. The He–Gaston equation is:

$$p = 1 - (1 + \frac{\alpha \mu^{\beta}}{k})^{-k},$$

where p was originally defined as the spatial occupancy of a species (proportion of surveyed sites occupied by the species), μ as the mean local abundance across the surveyed sites, k as a spatial aggregation parameter, and α and β as generic parameters empirically determined on a case-by-case basis [7]. For our study, we considered p to be the temporal occurrence of a species, μ as the temporal mean abundance of that species, and k as a temporal aggregation parameter. The negative binomial model derives from the He–Gaston model simply by considering that both empirical parameters are 1 [7]:

$$p = 1 - (1 + \frac{\mu}{k})^{-k}.$$

We parameterized both models using nonlinear least-squares regression [12]. We evaluated the degree of model fit by calculating the Pearson correlation coefficient (r) between the observed values and model-predicted values of temporal occurrence. In general terms, a perfect fit of a data set to a nonlinear model should produce a perfect correlation ($r = 1$) using all predicted–observed data pairs. We did the analyses using SYSTAT 5.2 for Macintosh.

A number of studies on the spatial abundance–occupancy relationship have calculated the mean local abundance of each species by dividing the total number of organisms found in the region of interest by the number of sampling units in which the species occurred (not by the total number of surveyed units). However, such an alternative measure of mean local abundance produces a number of undesirable artefacts on the abundance–occupancy relationship [13]. Therefore, for our study, we did not consider the equivalent form of temporal mean abundance (that is, temporal mean abundance calculated for each species using only the sampling dates in which individuals occurred in the trap).

As part of our descriptive statistics, we calculated Simpson's evenness index applied to the abundance values for our insect species [14]:

$$E = (S \sum p_i^2)^{-1},$$

where S was the species richness (31) and p_i was the proportional abundance of each species relative to the total number of individuals counted during the study period. Simpson's evenness index ranges between 0 and 1 [14].

Results

We found 5049 individuals of tabanid flies during the study period (Table 1; Data S1). There was a wide range in the occurrence of species over time, with some appearing only in one date and others almost throughout the entire study period

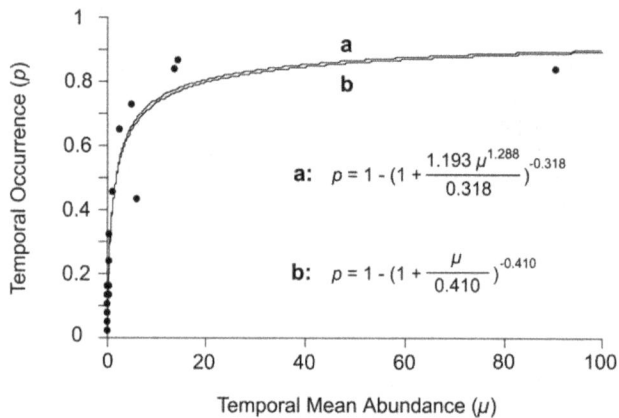

Figure 1. Relationship between the temporal occurrence (p) and temporal mean abundance (μ) of species of an insect assemblage from Nova Scotia, Canada. The temporal mean abundance was calculated for each species based on all sampling dates (n = 37). The upper line (a) is the He–Gaston model and the lower line (b) is the negative binomial model as parameterized for this data set.

(Table 1). Most species, however, occurred in a limited amount of dates: 8.1 ± 1.8 dates (mean ± SE, n = 31 species). Likewise, species abundance also showed a wide range, with some species contributing with only one individual during the study period and others contributing with many (Table 1). Most individuals, however (96%), belonged to just six species, which yielded a low evenness index for this assemblage ($E = 0.07$).

Temporal occurrence and temporal mean abundance were positively related for our insect assemblage. The data showed a high degree of fit to the He–Gaston and negative binomial models. Even by having two empirical parameters (α and β) in addition to the temporal aggregation parameter (k), the He–Gaston model showed only a marginally higher fit ($r = 0.971$, $P < 0.001$) than the negative binomial model ($r = 0.970$, $P < 0.001$; Fig. 1).

Discussion

Our study has revealed a positive relationship between temporal occurrence and temporal mean abundance using tabanid flies as a model species assemblage. In addition, two equations that often describe spatial abundance–occupancy relationships well, the He–Gaston and negative binomial models [7], were also found to describe the temporal abundance–occurrence relationship successfully. Since the He–Gaston model yielded only a marginally

higher fit than the negative binomial model, the temporal aggregation parameter (k) emerges as a key element of the temporal abundance–occurrence relationship. Thus, investigating what determines the timing of occurrence of different species in communities should lead to building functions with appropriate k values to predict outcomes under different scenarios.

It is worth noting that a previous study had found a positive correlation between the number of years in which annual plants occurred in an Arizona desert and their overall abundance over time [15]. However, no attempt was made in that study to test the ability of equations developed for the spatial abundance–occupancy relationship to model temporal data, as done here.

The high degree of model fit found for our data set calls for studies on other species assemblages to test for the generality of the temporal abundance–occurrence relationship. It may also be interesting to examine possible links between temporal and spatial patterns and possible effects of community traits such as species richness and evenness or habitat traits such as environmental suitability [16]. The existence of similar patterns across space and time is not infrequent in ecology, although the factors affecting spatial vs. temporal relationships may differ to some extent (for example, biomass–density patterns in crowded plant stands [17,18]). The aggregation parameter (k) appears to be a key component for both the spatial and temporal relationship between species distribution and abundance. Thus, it may also be pertinent to investigate what processes may affect species aggregation in space and time in comparable ways. Overall, we hope that the present study opens the door to long-term research on the fundamental and applied aspects of the interspecific temporal abundance–occurrence relationship.

Acknowledgments

We thank Sebastián P. Luque and two reviewers for their constructive comments on our manuscript.

Author Contributions

Conceived and designed the experiments: RDP RFL. Performed the experiments: RDP RFL. Contributed reagents/materials/analysis tools: RAS RDP RFL. Wrote the paper: RAS. Generated the idea that motivated the paper, analyzed and interpreted the data in the context of the research goals, and wrote the first version of the article: RAS. Designed the field study, collected the organisms, identified the species, generated the data, and contributed towards improving the manuscript: RDP RFL.

References

1. Blackburn TM, Gaston KJ, (eds) (2003). Macroecology: concepts and consequences. Oxford: Blackwell Science. 464 p.
2. Blackburn TM, Cassey P, Gaston KJ (2006) Variations on a theme: sources of heterogeneity in the form of the interspecific relationship between abundance and distribution. J Anim Ecol 75: 1426–1439.
3. Foggo A, Bilton DT, Rundle SD (2007) Do developmental mode and dispersal shape abundance-occupancy relationships in marine macroinvertebrates? J Anim Ecol 76: 695–702.
4. Webb TJ, Tyler EHM, Somerfield PJ (2009) Life history mediates large-scale population ecology in marine benthic taxa. Mar Ecol Prog Ser 396: 293–306.
5. Buckley HL, Freckleton RP (2010) Understanding the role of species dynamics in abundance-occupancy relationships. J Ecol 98: 645–658.
6. Verberk WCEP, van der Velde G, Esselink H (2010) Explaining abundance-occupancy relationships in specialists and generalists: a case study on aquatic macroinvertebrates in standing waters. J Anim Ecol 79: 589–601.
7. Holt AR, Gaston KJ, He F (2002) Occupancy-abundance relationships and spatial distribution: a review. Basic Appl Ecol 3: 1–13.

8. He F, Gaston KJ (2000) Estimating species abundance from occurrence. Am Nat 156: 553–559.
9. French FE, Hagan DV (1995) Two-tier box trap catches *Chrysops atlanticus* and *C. fuliginosus* (Diptera: Tabanidae) near a Georgia salt marsh. J Med Entomol 32: 197–200.
10. Mullen GR, Mullen G, Durden L (2009) Medical and veterinary entomology. San Diego: Academic Press. 259 p.
11. Teskey HJ (1990) The insects and arachnids of Canada. Part 16: The horse flies and deer flies of Canada and Alaska (Diptera: Tabanidae). Ottawa: Agriculture Canada. 381 p.
12. Quinn GP, Keough MJ (2002) Experimental design and data analysis for biologists. Cambridge: Cambridge University Press. 520 p.
13. Wilson PD (2011) The consequences of using different measures of mean abundance to characterize the abundance-occupancy relationship. Global Ecol Biogeogr 20: 193–202.
14. Krebs CJ (1999) Ecological methodology. Menlo Park: Benjamin Cummings. 624 p.

15. Guo Q, Brown JH, Valone TJ (2000) Abundance and distribution of desert annuals: are spatial and temporal patterns related? J Ecol 88: 551–560.

16. VanDerWal J, Shoo LP, Johnson CN, Williams SE (2009) Abundance and the environmental niche: environmental suitability estimated from niche models predicts the upper limit of local abundance. Am Nat 174: 282–291.

17. Weller DE (1987) A reevaluation of the $-3/2$ power rule of plant self-thinning. Ecol Monogr 57: 23–43.

18. Weller DE (1989) The interspecific size-density relationship among crowded plant stands and its implications for the $-3/2$ power rule of self-thinning. Am Nat 133: 20–41.

15

How Are Academic Age, Productivity and Collaboration Related to Citing Behavior of Researchers?

Staša Milojević*

School of Library and Information Science, Indiana University, Bloomington, Indiana, United States of America

Abstract

References are an essential component of research articles and therefore of scientific communication. In this study we investigate referencing (citing) behavior in five diverse fields (astronomy, mathematics, robotics, ecology and economics) based on 213,756 core journal articles. At the macro level we find: (a) a steady increase in the number of references per article over the period studied (50 years), which in some fields is due to a higher rate of usage, while in others reflects longer articles and (b) an increase in all fields in the fraction of older, foundational references since the 1980s, with no obvious change in citing patterns associated with the introduction of the Internet. At the meso level we explore current (2006–2010) referencing behavior of different categories of authors (21,562 total) within each field, based on their academic age, productivity and collaborative practices. Contrary to some previous findings and expectations we find that senior researchers use references at the same rate as their junior colleagues, with similar rates of re-citation (use of same references in multiple papers). High Modified Price Index (MPI, which measures the speed of the research front more accurately than the traditional Price Index) of senior authors indicates that their research has the similar cutting-edge aspect as that of their younger colleagues. In all fields both the productive researchers and especially those who collaborate more use a significantly lower fraction of foundational references and have much higher MPI and lower re-citation rates, i.e., they are the ones pushing the research front regardless of researcher age. This paper introduces improved bibliometric methods to measure the speed of the research front, disambiguate lead authors in co-authored papers and decouple measures of productivity and collaboration.

Editor: Santo Fortunato, Aalto University, Finland

Funding: The author has no funding or support to report.

Competing Interests: The author has declared that no competing interests exist.

* E-mail: smilojev@indiana.edu

Introduction

Communication of scientific results is an integral part of modern science, making scientific results "visible" to other scientists and to society as a whole. Through the years science has perfected acceptable genres and discourses of disseminating its results [1,2]. Journals and journal articles serve as primary vehicles not only for disseminating the findings, but also for reinforcing the common paradigms of scientific fields and disciplines. Each journal article is written for a particular audience and is adhering to certain rhetorical devices in order to establish trust and authority. Modern science is based on trust [3]. One important way of establishing trust and authority is through lineage of methods, theories and problems used. That lineage is manifested in the lists of bibliographic references, which have become a staple of every scientific article since the end of the 19th century [4,5]. Thus, referencing, in addition to providing a context for a study, is a necessary device to persuade editors, referees, and ultimately the scientific audience of the study's credibility. The references themselves have undergone a transformation from mentions of authors and their work in the text to footnotes. Further transformation has occurred from footnotes to endnotes [1].

References have been exploited extensively to evaluate and describe science and trends in its development (especially since the development of Citation Indexes in the 1960s and 1970s). Research efforts that focus on references are primarily of two types: (a) studies that try to develop theories of citation by analyzing why and how people use references [6–11] and (b) studies that explore the development of science and which use references to study moving research fronts, aging and obsolescence of scientific literature [12–18]. Less attention has been paid to authors' referencing *behavior* itself, especially using large bibliographic data sets, which forms the focus of this study. Thus we are interested in the act of citing and not its consequences, performing what Ajiferuke and colleagues call "citer-centered analysis" [19].

We distinguish three levels at which one can study referencing behavior: macro, meso and micro. At the macro, or global level, the units of analysis are entire fields, or even multiple fields, either at some fixed time or through time. Macro studies have recognized that different disciplines will *on average* have very different referencing practices that result in different characteristics of references [20,21]. These differences are the result of the traditions and practices in a given field and to a large degree reflect the fact that "science is a social process" [22]. Furthermore, the referencing practices evolve, again on a global level, usually on the time scales of decades, but sometimes faster [14]. In the first part of the Results section we explore the macro characteristics of the references in five fields that are chosen for this study, setting the stage for the analysis at the meso level. At the opposite end of the spectrum many studies approach referencing behavior at the micro level, i.e., as a deeply personal act done by individual authors. Such studies have often focused on the nature and the context of usage of references. For example, whether references are perfunctory or are essential to the paper [23], whether researchers cite to give credit [7] or to persuade [6], whether references are given in positive or negative connotations [10], or even whether references can be a product of deliberate "gaming" of a system [24]. Such micro-level behavior, although interesting, does not address how authors cite as a group. Namely, science is not only a personal, but also a communal activity. Our study fills the space between macro and micro studies because it focuses on the citing behavior of different *categories* of authors currently publishing. In order to better understand which meso characteristics are important for the scientific process in general, we explore the referencing characteristics of authors in five diverse scientific fields noting trends that transcend individual fields.

We explore the referencing behavior of authors classified according to the following characteristics: academic "age," productivity, and collaborative activity. Biological age of a researcher has been considered an important factor when it comes to creativity, productivity, and collaborative practices [25]. Age has also been considered a factor when it comes to referencing behavior [12,13]. However, we are more interested in the cohorts of scientists who are at the same stage in their academic careers, having the same "academic" age, regardless of their biological age. Thus, we explore whether researcher seniority is correlated with certain properties of his/her references and referencing behavior. The reasons senior scientists might have different practices can be manifold. These researchers were enculturated into the practices of the field at a different time. Therefore, even if they are keeping pace with new developments in science, their referencing practices may contain "remnants" of those older times. Some studies have suggested that older authors tend to read less, and thus refer to older literature, which they recall from earlier times [26]. Because they have more experience, older authors may also be more selective when it comes to what they cite, leading to a smaller number of references. In this study we explore if such trends with academic age are actually present among the authors who publish in core journals in their fields.

The second characteristic of authors we are interested in is their current productivity. Scientific productivity, especially evidenced by a large number of publications, has become a holy grail of sorts for many scientists. Productivity has been closely tied to rewards in science. It has also been related to high quality of work [27]. Do more productive authors tend to have different referencing practices and strategies? Do prolific authors tend to cut down on the number of references (as suggested by [26]), or, does their increased productivity expose them to a larger body of literature that they cite? Our study will provide an answer to this question.

One of the most notable changes in science is a trend toward team science. Both the number of coauthored papers and the number of coauthors on a single paper have been increasing [28,29]. Collaboration in research is considered essential for tackling problems of present-day science and society. The above studies have also found that teams produce higher impact research. The increased importance of collaboration raises the question of whether scientists who tend to collaborate more have different referencing behaviors. Note that certain definitions of productivity make it automatically correlated to the level of collaboration, so in this study we will define measures which are independent from each other.

To be able to compare the characteristics of scientists with the appropriate reference group, we will follow each characteristic for each discipline separately.

Guided by the previous studies, we identify the following characteristics of references and referencing behavior as relevant: the number and age of references and the instances of re-citation. As pointed out, these characteristics have been studied primarily at a macro level, with special emphasis on trends over time. Price [30] was the first to write about the exponential growth in the number of references per paper. This observation has led to other studies of growth [31]. All the studies agree that reference lists are getting longer. Age of references has been considered an important indicator for the vitality of a discipline and the pace of its advancement – the moving research front [32]. There has been a sense that the "younger" (more contemporaneous) the references are, the faster the advancement of science. Researchers have come up with a number of ways to measure the age: the simple average age, the "immediacy factor" [33], the "citation half-life" [34], and the "Price Index" [32], to name a few. We will use some of these and also introduce new or modified measures. One naively expects that the exponential growth of science will automatically lead to "younger" references. What is observed is the reverse [14] – the references are getting older. Part of the trend is a simple mathematical consequence of the exponential growth of literature with finite beginning [35]. Finally, we will study authors' *re-citation* practices, first suggested by White [36]. Ajiferuke, Lu and Wolfram [19] identified two types of re-citation: (a) at the author level (the same author cited in different articles of the same citing author) and (b) at the publication level (the same publication cited in different articles by the same citing author). For White, re-citation reveals authors' intellectual histories. Previous studies of re-citation practices focused on detailed analysis of *oeuvres* of individual researchers [36,37]. Here we present for the first time the analysis of re-citation at the meso level, for categories of different authors.

To summarize, the primary objective of the current study is to establish, for authors belonging to five disciplinary affiliations, the relationships between the characteristics of these authors (academic age, productivity, and collaboration) and their referencing behavior at the present time, as manifested through the number and age of references and re-citation practices. The secondary goal, setting the stage for the primary objective, is to describe the trends of these characteristics of references over the last half century, thereby establishing the macro-level properties of the references. Thus, this paper goes beyond the motivations and behavior of a single author; to use Small's [38] words, it moves from "the author-centric to the community-centric perspective".

Methods

We analyze referencing practices in five diverse scientific fields ranging from classical to relatively new disciplines:

- Astronomy (AST; physical science; classical)
- Mathematics (MAT; mathematical science; classical)
- Robotics (ROB; technical/applied science; interdisciplinary and new)
- Ecology (ECL; life science; semi-classical)
- Economics (ECN; social science; classical)

The hope is that by examining diverse fields we will be able to sample various referencing practices and get a sense of which factors are discipline-specific and which are more generally related to the scientific process.

Data

For each field we selected up to 10 *core* journals. These are journals that are well established and usually have high impacts in their field. In addition, core journals usually publish a large fraction of the original research in a particular discipline. Finally, and what is most important for our study, they allow for "coordination of communication and access to reputation, … knowledge interchange and creation" [39] making them good representatives of that field and its practices in general. These are also journals in which most active researchers will be publishing at some time in their careers. For our study it is not at all necessary to include the majority of journals in a given field, as the characteristics that we are interested in will be expressed in the core journals, as long as they contain a statistically large number of articles. Researchers who, for whatever reason, do not publish in core journals of their fields will not be encompassed by our study. We will discuss later on whether the absence of such authors affects the results.

We used a number of studies to identify the core journals for each discipline. For astronomy, [40] identified a list of core journals; for mathematics, our starting list came from journals identified by [41]; Sabanovic (personal communication) suggested the list for robotics; the list used for ecology came from [42] and, finally, the economics journals were identified by [43]. For these 33 journals (Table 1) we downloaded bibliographic data from Thomson Reuters' Web of Science database covering the 50-year interval between 1961 and 2010. We kept only data on research articles by selecting publications classified as "article" or "conference paper". In several cases in which the journal changed its title, we collected data corresponding to all predecessor titles. In total, our data set contained records for 213,756 articles. We checked the mean number of references per article in each journal to determine whether there were any journals with "special" referencing practices. We found that *Bulletin of the American Mathematical Society* (BAMS) contains almost twice as many references as other mathematics journals. After examining the journal we realized that BAMS publishes primarily explanatory papers, not original research articles. Because genre is one of the main determinants of referencing behavior, we removed this journal from the analysis. The total number of articles published in the 2006–2010 period is 45,043. Table 1 shows the breakdown by journal.

Characteristics of authors

Our study did not follow the behavior of individual scientists through their careers, but rather focused on their collective behavior during the *current* period, which we defined as the five years from 2006 to 2010. Five years is sufficiently long enough to establish patterns of productivity and collaboration for most researchers, yet not too long to be affected by gradual changes in practice.

Referencing behavior was studied for all authors who at some point in what we call the current period were sole authors or *lead* authors on multiple-author papers. The inclusion of multiple-author papers was motivated by the model of authorship in which the lead author is primarily responsible for the published work, and therefore carries most weight in the selection of references. The identification of the lead author is not straightforward. The simplest assumption is that the first author is also the lead author. However, this assumption will be invalid for articles that intentionally list authors in alphabetical order. If some disciplines predominantly use the alphabetical scheme, treating first authors as lead authors will result in a large level of "contamination". In each field, we test for the prevalence of the alphabetical scheme by counting "alphabetical articles" and comparing them to a number expected if the author placement is not intentionally alphabetical. For a paper with n authors we expect $1/n!$ to have alphabetical ordering by chance (for example, 50% in two-authored articles). For astronomy, robotics and ecology this percentage is between 51% and 52%, i.e., these fields generally do not use the alphabetical scheme. On the other hand, in both mathematics and economics this fraction is 97%. Correspondingly, the level of contamination (fraction of authors that would be incorrectly treated as lead authors) is only several percent for astronomy, robotics and ecology, but very high 40% for mathematics and economics. Alternatively, one may treat *corresponding* authors as the lead authors. Using corresponding authors reduces the contamination rate in mathematics and economics to around 24%, which is still relatively high (for example, ~77% of two-authored papers are still alphabetical). Apparently, for mathematics and economics, we need to use a new method that will select only articles in which the lead author can be determined unambiguously. Thus we select: single authored papers, articles in which the corresponding author is not the first author and articles that are not in the alphabetical order. This filtering results in the removal of 41% of articles in mathematics and 46% in economics, which was necessary to obtain a data set with virtually no contamination. We emphasize that this removal does not preferentially affect certain categories of authors and therefore will not lead to biases in the analysis.

We identified 21,562 different lead authors in the current period. We disambiguated author names using last names and first (or first and middle) initials. For each lead author in a given field we determined three *independent* measures:

- *Time spent in the field (i.e., academic age)* – span in years, for a given author, between the first and the most recent article in the entire dataset (1961–2010), regardless of authorship placement: $a_{acad} = t_{last} - t_{first} + 1$. Maximum is 50. The academic age distribution drops exponentially, so very few authors are near this maximum (Figure 1A). Operationalization of the academic age through publications is an approximation, but we believe it is a good one, because it represents the length of an author's active engagement within a scientific community

- *Current productivity* – number of lead-author articles published between 2006 and 2010. We excluded co-authored articles not led by an author to remove correlation with the collaboration level. Distribution of author productivity is shown in Figure 1B. It is approximately power-law.

- *Current collaboration level* – number of collaborators from co-authored articles in the current period plus one. Collaborators of a given author are defined as all the different lead authors on articles on which he/she was a coauthor. This measure does not include coauthors on articles on which the author him/herself was the lead author, thus removing correlation with the productivity. Furthermore, we are interested in direct collaborative ties, therefore two authors, neither of whom was the lead author on a multi-authored article, were not considered collaborators. To get the collaboration level we added one in order to enable representation on a logarithmic scale. The distribution of collaboration level is shown in Figure 1C. It roughly follows the power-law distribution (except astronomy). The drop is steepest for mathematics and economics – these are the least collaborative fields. Ecology and robotics are moderately collaborative, while astronomy is highly collaborative (shallowest decline).

Table 1. Core journals in five fields used in this study.

Field	Journal title	Articles (1961–2010)	Articles (with references, 2006–2010)	Average number of ref (2006–2010)
AST	ASTRONOMICAL JOURNAL	14392	2022	48.6
AST	ASTRONOMY & ASTROPHYSICS	44050	9109	43.8
AST	ASTROPHYSICAL JOURNAL	70661	12297	49.3
AST	MONTHLY NOTICES OF THE ROYAL ASTRONOMICAL SOCIETY	27768	8042	47.2
MAT	ACTA MATHEMATICA	667	56	35.5
MAT	AMERICAN JOURNAL OF MATHEMATICS	2515	253	27.2
MAT	ANNALS OF MATHEMATICS	2289	310	30.5
MAT	INVENTIONES MATHEMATICAE	3477	325	31.7
MAT	JOURNAL OF FUNCTIONAL ANALYSIS	5175	1301	25.0
MAT	JOURNAL OF THE AMERICAN MATHEMATICAL SOCIETY	513	171	30.9
MAT	MATHEMATICS OF COMPUTATION	4560	563	23.2
MAT	PROCEEDINGS OF THE LONDON MATHEMATICAL SOCIETY	2406	264	28.2
ROB	ADVANCED ROBOTICS	1009	424	22.3
ROB	AUTONOMOUS ROBOTS	454	208	33.8
ROB	IEEE ROBOTICS & AUTOMATION MAGAZINE	406	167	23.2
ROB	IEEE TRANSACTIONS ON ROBOTICS*	1720	447	30.9
ROB	INDUSTRIAL ROBOT-AN INTERNATIONAL JOURNAL	409	198	18.2
ROB	INTERNATIONAL JOURNAL OF ROBOTICS RESEARCH	1214	251	35.9
ROB	JOURNAL OF FIELD ROBOTICS**	1178	191	31.6
ROB	JOURNAL OF INTELLIGENT & ROBOTIC SYSTEMS	1093	333	26.4
ROB	ROBOTICS AND AUTONOMOUS SYSTEMS	867	344	29.7
ECL	ECOLOGY	9753	1662	50.4
ECL	JOURNAL OF ANIMAL ECOLOGY	3532	661	54.0
ECL	JOURNAL OF ECOLOGY	3627	596	56.5
ECL	OECOLOGIA	9990	1454	54.3
ECL	OIKOS	6150	1043	50.8
ECN	AMERICAN ECONOMIC REVIEW	4502	475	36.1
ECN	ECONOMETRICA	2907	270	32.0
ECN	JOURNAL OF ECONOMIC THEORY	2857	578	24.4
ECN	JOURNAL OF MONETARY ECONOMICS	1731	427	30.1
ECN	JOURNAL OF POLITICAL ECONOMY	2700	158	36.9
ECN	QUARTERLY JOURNAL OF ECONOMICS	2080	206	42.4
ECN	REVIEW OF ECONOMIC STUDIES	2030	237	37.6
	Total	213756	45043	

*Includes: IEEE JOURNAL OF ROBOTICS AND AUTOMATION and IEEE TRANSACTIONS ON ROBOTICS AND AUTOMATION.
**Includes: JOURNAL OF ROBOTIC SYSTEMS.
Mean number of references per article in each journal is given in the last column. Field abbreviations: astronomy (AST), mathematics (MAT), robotics (ROB), ecology (ECL), and economics (ECN).

For trends involving productivity and collaboration we presented the data using partial logarithmic binning [44] but averaging only non-zero elements. Error bars represent standard deviations of the mean. Logarithmic representation is used in order to show the full dynamic range of author-related quantities more evenly.

Characteristics of references

For each article we determined the number of references (N_{all}) by counting references in the WoS record. We introduced the measure of references per article *page*: $n = N_{all}/p$ to obtain a normalized measure, which we refer to as *reference rate*. Reference rate, by taking into account the expectation that longer papers will

have more references, is a more uniform measure of the usage of references than the bulk number of references.

Further, for each article we calculated the following age-related characteristics:

- *Average age of references*:

$$\bar{a} = \frac{1}{N_{all}} \sum (2010 - t_{pub})$$

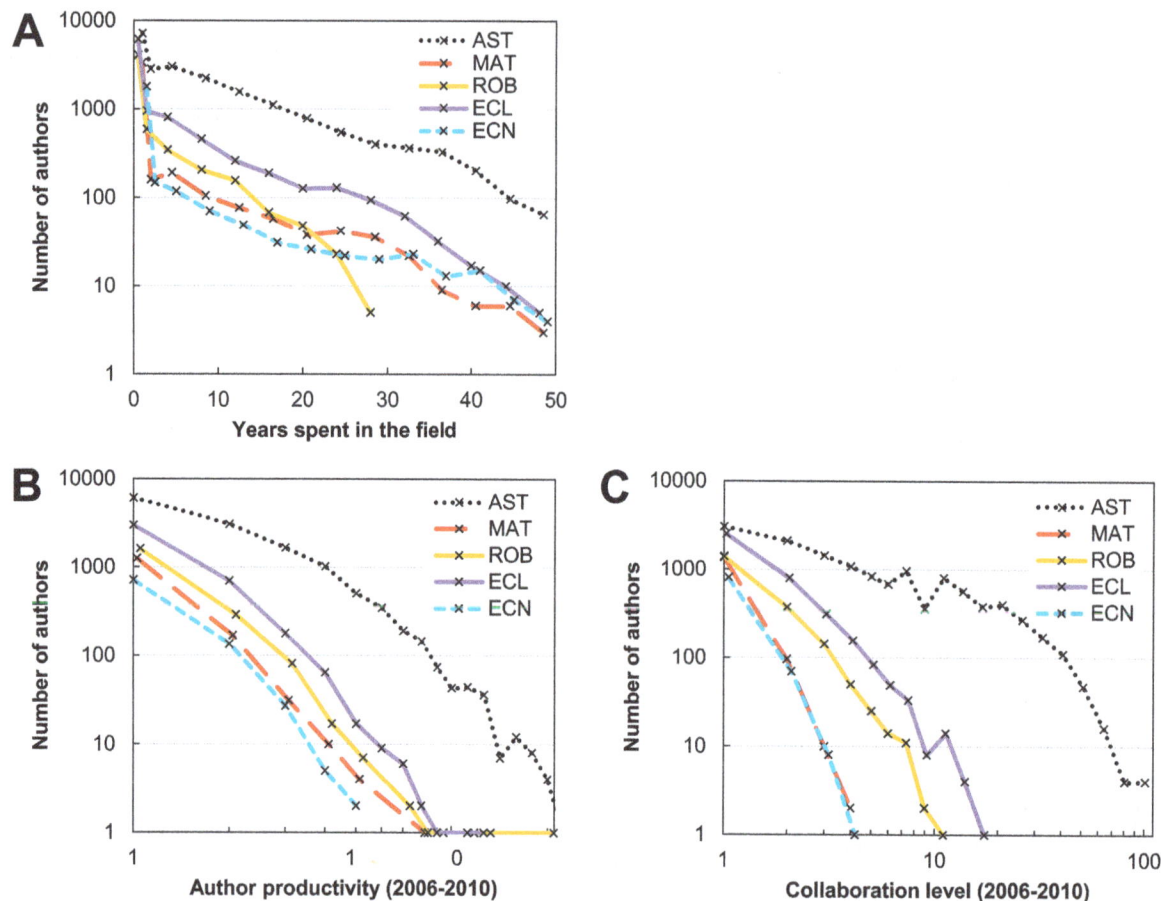

Figure 1. Distributions of authors of different academic age (A), recent productivity (B), and collaboration level (C). Note the logarithmic scale on the y axis and on the x axis for productivity and collaboration. Highest bins in every category contain very few authors. Distribution of the academic age (panel A) is per bin, and uses 4-year bins beyond year two. First bin ($x = 1$) in panel A is higher for all fields because it contains transient authors who appear with only one publication. Decline is approximately exponential in A and power-law in B and C.

- *Price Index* – the fraction of references published within five years of the article publication year:

$$PI = N_{\leq 5}/N_{all}$$

- *Modified Price Index* - the fraction of references published within five years compared to those published within 10 years:

$$MPI = N_{\leq 5}/N_{\leq 10}$$

This measure is similar to the Price Index. However, instead of comparing recent references to all references, it compares recent references only to those published within 10 years. This measure gives a better assessment of the speed of the research front than the traditional Price Index by eliminating the role of "old," foundational references.

- *Old fraction* – fraction of references older than 10 years:

$$O = N_{>10}/N_{all}$$

This measure can be thought of as the opposite of the Price Index. It indicates the contribution of *foundational* references. Another advantage of the Modified Price Index, compared to

the regular Price Index, is that it is completely independent from the old fraction.

Finally, for all current authors who had published two or more papers as lead authors in 2006–2010 we calculated the re-citation fraction at the publication level (the re-use of the same article) as:

$$r = 2\left(1 - \frac{N_{unique(1,2)}}{N_{all,1} + N_{all,2}}\right)$$

Where $N_{unique(1,2)}$ is the number of unique references in two articles and $N_{all,1}$ and $N_{all,2}$ are the total number of references in two articles respectively. If an author had published more than two articles we randomly chose two from which the re-citation fraction was calculated. To increase the accuracy of the re-citation measure we perform the two-article drawing 10 times, thus sampling many different pairs of articles of a given author.

Results

Usage and characteristics of references over the last 50 years

The purpose of this section is to provide the context for the rest of the analysis by describing the macro-level characteristics of references: how references differ from field to field, and how their characteristics evolved in each respective field over the last half century.

Trends in numbers of references. As previous studies have reported, we observe that the number of references per article had been rising in all five fields (Figure 2A). The rate of increase in all fields was similar, with a doubling time of around 30 years. Interestingly, we do not observe any change in the rate of referencing in the late 1990s, when the Internet became more widespread. Ecology and astronomy currently use around 50 references per article on average, twice as many as the other three fields.

The increase in the number of references per article over the period studied can be attributed to two distinct factors: (a) due to articles becoming longer or (b) due to a higher *rate* of referencing. The increase could also result from the combination of these two factors. To distinguish these factors we examined the trends in the number of article pages and in the number of *references per page* (Figure 2B and 2C). Article length has remained remarkably constant in astronomy, robotics and ecology. However, in mathematics and economics there has been a steady increase since at least the 1970s. On the other hand, when it comes to the rate of referencing (number of references per page), astronomy and ecology have seen a steady rise throughout the period under study, robotics has seen an overall rise, while the rate in mathematics and economics remained constant. Therefore, we conclude that

different factors contributed to the increase in the number of references per article in these fields. In astronomy, ecology, and robotics this was due to the increase in the rate at which references were being used; in mathematics and economics the articles became significantly longer over time but the number of references per page remained fairly constant. The finding regarding economics agrees with Frandsen and Nicolaisen [26], who found that in the subfield of econometrics "we expect about a 7% increase in the number of references when the number of words increases by 10%" (p. 68). However, it appears that in fields with a strong experimental/observational component there is an increase in reference rate without the increase in the amount of text.

It is interesting that astronomy and ecology articles are the shortest but have the largest number of references, hence their references rate is currently some five times higher than for mathematics, robotics, and economics. As with the distribution of the bulk number of references per paper, we observe no changes in reference rate that would correlate with the spread of the Internet.

Trends in age-related characteristics of references. Overall, in most fields (to lesser degree in ecology) the average age of references has *risen* since the 1980s (Figure 3A). In astronomy and ecology (and to some degree economics) this period was preceded by one in which the references were on average getting *"younger"* in the 1960s and 1970s. Our analysis partially confirms Larivière and colleagues' [14] finding that

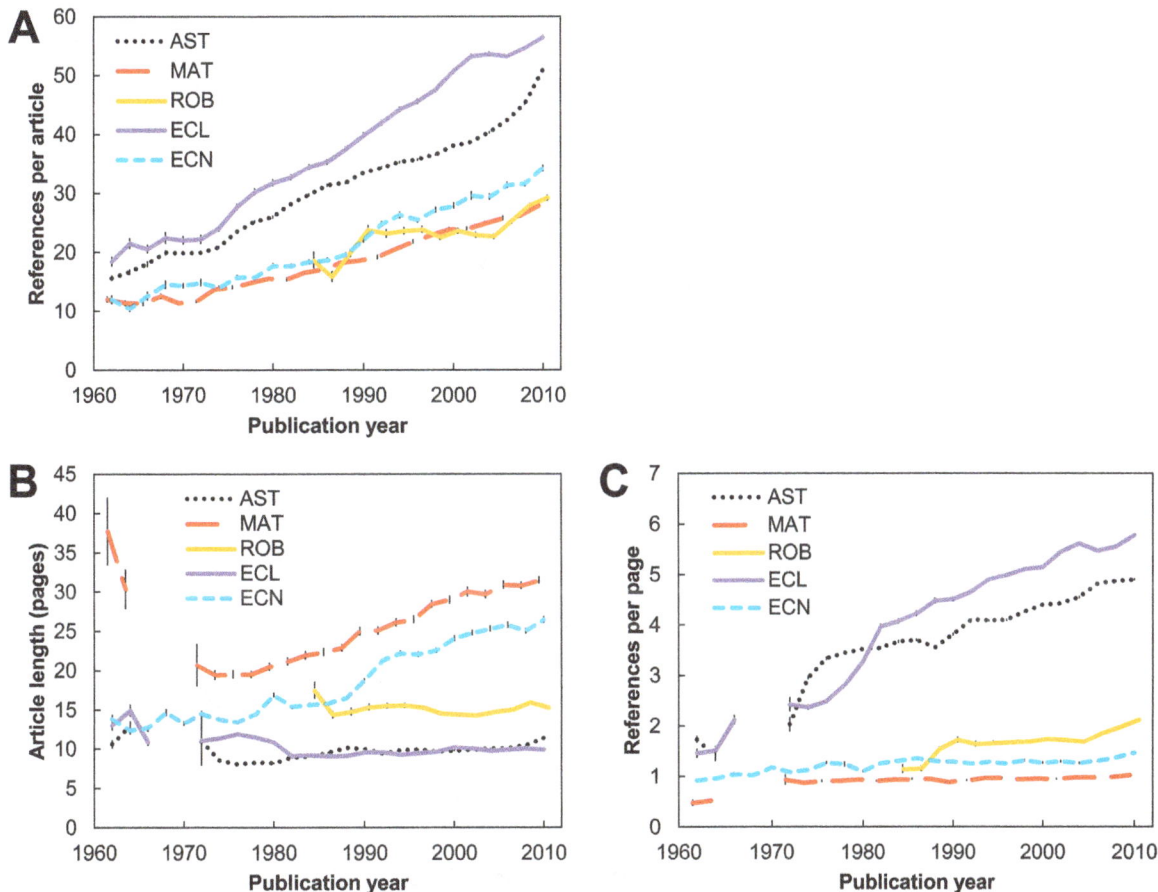

Figure 2. Trends involving numbers of references for 5 disciplines (astronomy (AST), mathematics (MAT), robotics (ROB), ecology (ECL), and economics (ECN)) from 1961–2010. A) The number of references per article. B) The article length, i.e. number of pages per article. C) Number of references per page (reference rate). Data for ROB start in 1983. Page information is missing for some articles in the late 1960s. Data points are averaged in bins of 2 years.

"contrary to a widely held belief, the age of cited material has risen continuously since the mid-1960s". Our data show that this average aging started more recently for fields in this study. Increase in the average age is to some degree a consequence of the aging of the knowledge base, which happens even with the exponential growth in literature that has started at some point in time [35]. However, the *old fraction* (fraction of references older than 10 years, Figure 3B) will be affected only moderately by this "mathematical" aging. If we assume an exponential growth in literature that started in the 1910s then, for typical growth rates, the old fraction should increase by only 0.07 over the period 1960–2010 (based on [35]). We see from Figure 3B that the trends in the old fraction are much larger than 0.07, thereby reflecting real changes in the use of references. Moreover, as with the average age the old fraction was decreasing for astronomy and ecology in the 1960s and 1970s. Overall, the trends in the old fraction generally follow those of average age, meaning that the latter measure is driven primarily by the level of use of foundational references. For mathematics and economics, (the fields that kept the referencing rate but increased article length) this fraction is higher now than it was in the 1960s.

How can we explain the trends? If we take the view of Allen and colleagues [45] that the age of the references can be viewed as the age of "persuasive communities" and can therefore be used to reveal the characteristics of scientific communities, then the fields or the time periods characterized by larger fractions of young references represent recent persuasive communities, which are characterized by very rapid change. The change is explained by paradigm shifts that lead to rapid successions of references that are considered acceptable (are able to persuade) by a larger community. On the other hand, the fields that have very old references may be "typified by highly stable, or even tradition-bound science" (p. 301). In that respect, the 1960s and 1970s (we do not have the data for the preceding periods) can be considered revolutionary for astronomy and ecology. Afterwards, these fields can be said to have moved to a more stable/normal phase of development building on the foundational work, although ecology shows some indications of entering another period of lesser use of foundational references. On the other hand, mathematics has been stable through the whole period. With the age of references constantly on the rise, the references in mathematics are now considerably older than they were prior to the 1980s. In other words, mathematics has continued to build on the foundational work that occurred prior to the 1960s. This finding is interesting in the light of what we have found regarding the number of references. While the rate of referencing in mathematics has not changed, its character (age) has. The robotics journals appeared in the early 1980s. However, the age of the references has been more or less constantly on the rise. Although it is the youngest of all fields, robotics has features of fields in their stable phase.

The most commonly used measure of the speed of the research front, the Price Index, is sensitive to old references, the presence of

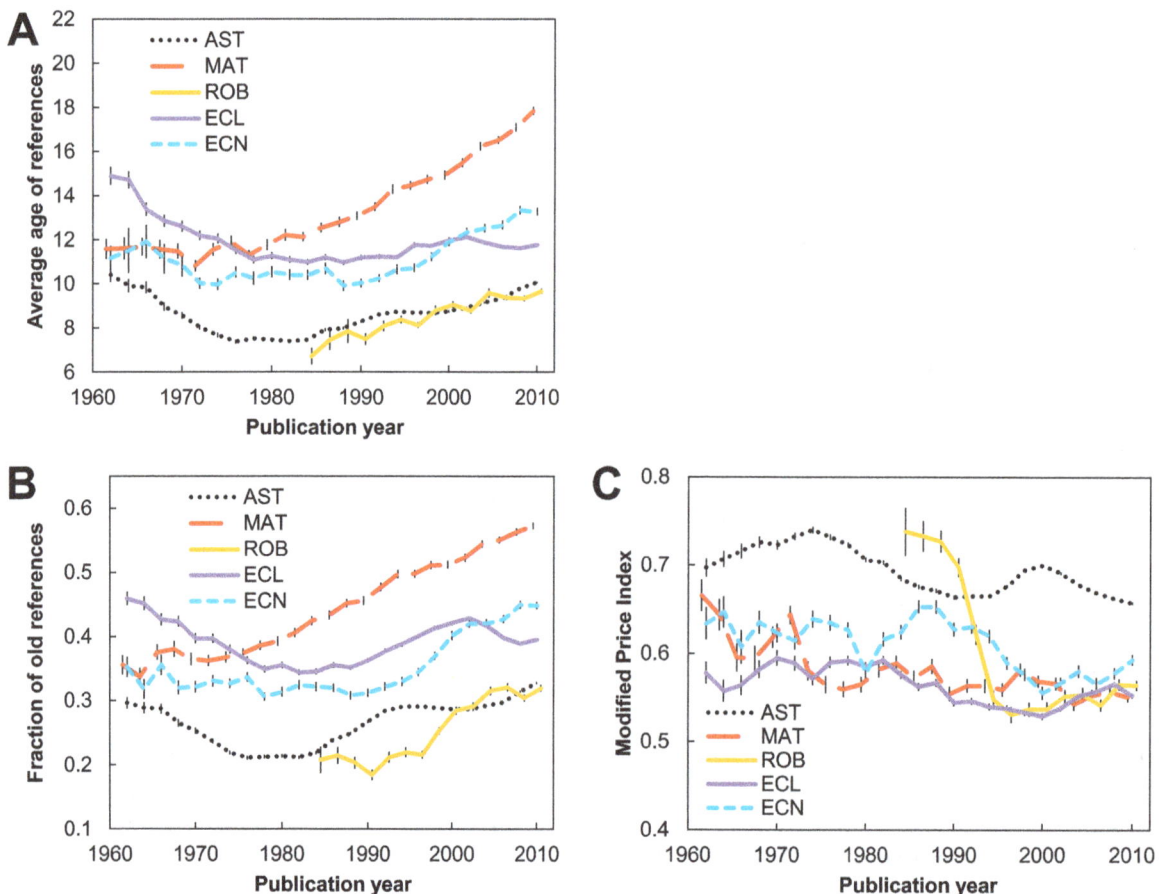

Figure 3. Trends involving ages of references for five disciplines (astronomy (AST), mathematics (MAT), robotics (ROB), ecology (ECL), and economics (ECN)) from 1961–2010. A) Average age of references. B) Fraction of old references (>10 years old). C) Modified Price Index (fraction of references published within five years compared to those published within 10 years). Data points are averaged in bins of two years.

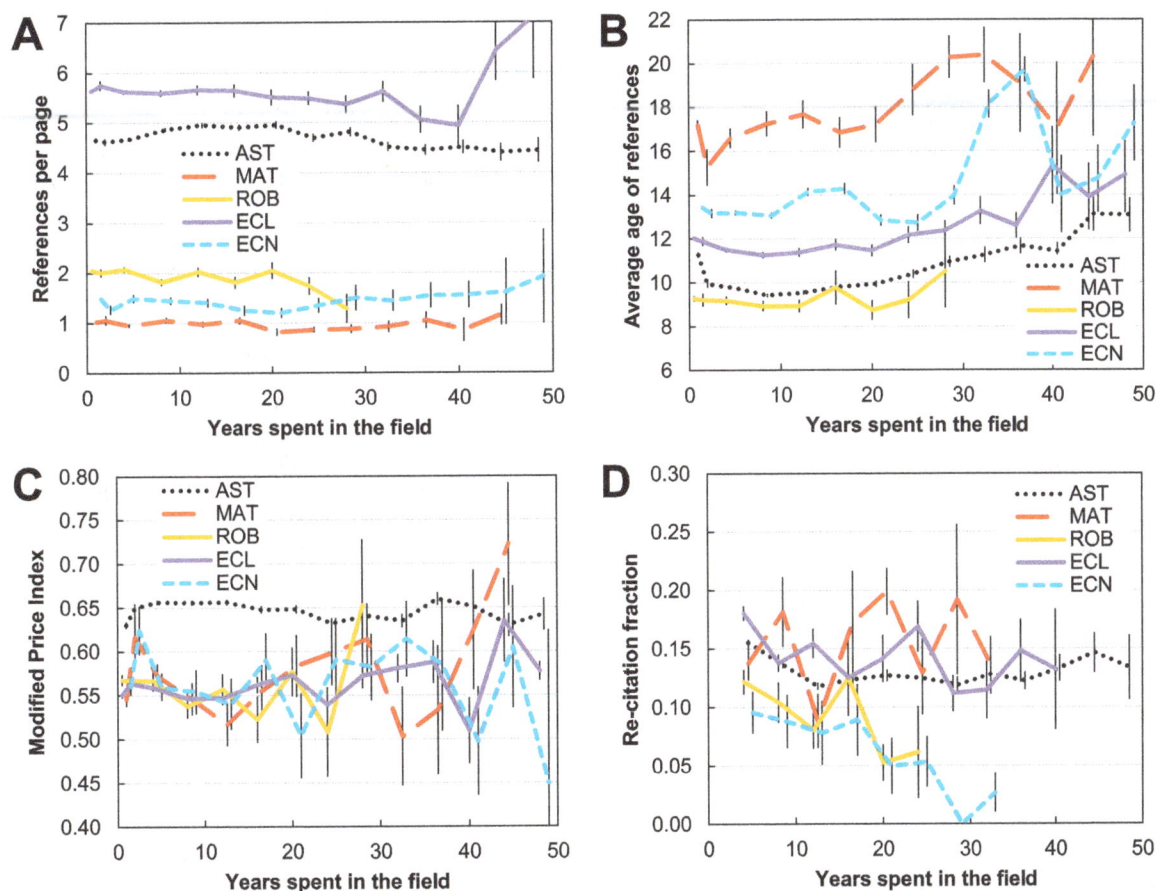

Figure 4. Referencing behavior for authors of different academic age (number of years spent in the field) in five disciplines (astronomy (AST), mathematics (MAT), robotics (ROB), ecology (ECL), and economics (ECN)). A) Number of references per page (reference rate). B) Average age of references. C) Modified Price Index (fraction of references published within five years compared to those published within 10 years). D) Re-citation fraction (repeated citations in pairs of articles). Authors who appear once have $x = 1$. First two years not binned; afterwards data are averaged in bins of four. Bins with fewer than 4 authors (rightmost, see Figure 1A) are excluded. For re-citation (D) points are not shown for the first two years because the average span between the pair of publications is less than it is for later ages.

which does not necessarily indicate a slower research front. Thus we introduce a new measure: the Modified Price Index, which by giving the share of recent (<5 years) references compared to those of intermediate age (<10 years) gives a better measure of the actual speed of changes – with higher values meaning higher speed. Figure 3C shows that there is an overall deceleration for all fields, interrupted with periods of acceleration. For robotics we see a huge drop in the 1990s, but some acceleration since then. Astronomy had a temporary boost in the mid-1990s, but is now slowing. Larivière and colleagues [14] also observed this boost in 1990s; they attribute it to the creation of the e-print repository arXiv. However, the prediction that arXiv would lead to the continual lowering in the age of citations did not come to fruition; the change was only temporary. Economics had a period of accelerated research front in the 1980s. As indicated previously, in the current period it is only ecology that exhibits some signs of accelerated development.

In the current period, astronomy and robotics references remain the youngest (∼10 years), mathematics references are the oldest (∼18 years), while both ecology and economics are in between (13–14 years). These differences in age reflect mainly the differences in the fraction of foundational references (30% for astronomy and ecology to 60% in mathematics).On the other hand, when it comes to the Modified Price Index all fields are very

similar except astronomy, whose research front moves faster than that of other fields.

The conclusion from this section is that fields, even with fast moving research fronts, will show aging in their references whenever they are in a stable phase, building on foundational work. It is only during the periods of paradigm shifts (even for classical fields) that this trend will be reversed. Currently, all the disciplines, even robotics, are in the stable, "traditional" stages of development.

Referencing behavior of authors

We now turn to the meso-level analysis of referencing behavior: the relationship between particular characteristics of an author (academic age, productivity, and collaboration) and his/her referencing behavior (rate of reference usage, age characteristics of references, and re-citation rates). To remove the dependence from global trends discussed above, from this point on we consider only the literature from the most recent five-year period (2006–2010).

Dependence on academic age. Previous studies of referencing behavior focused on the actual age of researchers and have suggested the following trends: (a) the more senior the author, the fewer references he/she uses, (b) those references are older, and (c) he/she tends to re-cite more. We focus on the academic age as a probably more relevant type of age (the two will, of course, be to large degree related).

To control for differences in article lengths, we analyzed the relationship between the academic age and number of references per page (reference rate). We find (Figure 4A) that the notion that senior authors use fewer references is generally not true in any of the five fields studied here. In astronomy, ecology and robotics there is a very slight downward trend (total decrease of ~10%) especially after 15 to 20 years spent in the field. In mathematics and economics there is no statistically significant trend at all. (In this study we discuss only the trends that are statistically significant for a given range in x, that is, when the probability of the linear trend (with error bars in y taken into account) arising by chance is smaller than 5%.) Although the reference rate in astronomy and ecology today is much higher than what it was when academically older authors were establishing their practices (Figure 2C), these authors have kept up with the times and use the references at the same rate as their junior colleagues.

On the other hand, we confirm the findings of previous studies that senior authors on average use older references (Figure 4B). However, this trend in the mean age is almost entirely due to the increase in the fraction of old, "foundational" references (older than 10 years) (graph not shown), and not due to less innovative research, as we will see shortly. The senior researchers, having spent more time in the field, have more knowledge of the older literature (having had a chance to accumulate their knowledge base longer) and tend to use it more. The exception is robotics, where the trend is not statistically significant with the current data, but note that for robotics we do not have the data beyond three decades. In most fields the difference in reference ages starts becoming pronounced for scientists active for more than a decade. The "youngest" references are used by the scientists who have been active for around a decade, which does not necessarily need to be interpreted as a sign that these researchers are at the forefront of research [13,32] (as we will confirm shortly). While they may be the ones who are the least burdened by either tradition or authority, they are also likely less familiar with the foundational work, and therefore cite it less. The age of references is only one side of the coin. In order to test whether authors who have spent long time periods in the field actually lose an "edge," as suggested in previous studies, we calculated their Modified Price Index. Gingras and colleagues [13] found that the young authors (between 28 and 40) have the highest *regular* Price Index, with the index steadily falling afterward. However, the regular Price Index, unlike our modified index, does not isolate foundational references. As can be seen in Figure 4C, all fields except astronomy show strong fluctuations in the Modified Price Index, but are generally consistent with no change. Thus, although senior researchers tend to have a higher percentage of old references, they don't fall behind their junior colleagues when it comes to citing the most recent literature.

Finally, it has also been suggested [13] that researchers actively follow the literature and accumulate references until they are about 40 and reuse their accumulated sets of references after that. This would translate into higher re-citation fractions for senior researchers. We show trends of re-citation rate for authors of different academic age in Figure 4D. In terms of the overall re-citation level, astronomy, mathematics and ecology all have similar values (13%). Robotics is somewhat lower (11%), and economics is significantly lower (7%). Again, our results do not fit the expected picture. The re-citation rates are either consistent with no overall change (astronomy and mathematics), or are even dropping (for the other three fields). In astronomy, where the data are of higher

quality (many more authors with more than a single publication), we see that the re-citation rates are slightly higher at both the beginning of their career (<10 years) and towards the end (>35 years). However, overall there is no trend that would associate older scientists with much higher re-citation rate.

While the focus of this section was mostly on older vs. younger authors, let us now focus on the authors who have just entered the field (shown at $x = 1$, the first bin in Figure 4ABC), a category that also contains a large fraction of authors who will not continue to publish (the transients, see Figure 1A). Interestingly, in all the fields except robotics the newcomers/transients tend to use somewhat older references and have a lower Modified Price Index. Such referencing behavior is to some extent because the entering researchers still depend heavily on their acceptance from the gatekeepers (who by default tend to be senior and thus have a proclivity for somewhat older references), but more likely because they need to build trust and authority by showing their knowledge of the foundations of the field. For example, many of those works are probably derived from dissertations, where the review of foundational work is more important than in regular science papers. Also, the papers based on dissertations are apparently often not at the research forefront, reflecting research conducted over a number of years and so not always including the most up-to-date references.

To summarize, analyzing the referencing behavior of authors with different academic ages led to the shattering of some of the myths regarding the practices of senior authors. Overall, we find that they use references at the same rate as their junior colleagues, with the same rate of re-citation. Although they tend to use a higher fraction of foundational references, their Modified Price Index indicates that their research does not lack very recent references. On the other hand, for authors who have just entered the field, many of whom will not continue in research careers, their MPI reflects less innovative work.

Dependence on current productivity. Do prolific authors tend to economize on references, or does their increased productivity also expose them to a larger body of literature that they cite? Previous studies have not focused on this factor. We show results in Figure 5A. For mathematics, robotics and economics neither is the case – productivity has no effect on the referencing rate. There is also no difference in the article length, so the total number of references is also constant (graphs not shown). In ecology and astronomy, on the other hand, more prolific authors have significantly *higher* referencing rates.

Figure 5B shows that more prolific authors tend to use on average "younger" references (trends are significant in astronomy, robotics and ecology). This finding may be used to support the idea that, at least in some fields, it is the more productive authors whose research is farthest from the foundational work in their field. To further test this idea, we show the trends regarding the relationship between author productivity and Modified Price Index (Figure 5C). We observe that researchers in all fields except robotics show a significant increase in the Modified Price Index as their productivity increases. This can further support our findings that the productivity (as measured here by the number of lead author publications) is on average positively related to the quality of one's research.

Figure 5D shows the relationship between re-citation and the author's productivity. If the very productive authors were to continue to work on the same topic we would expect high re-citation fractions. However, that is not the case in any of these fields. On the contrary, there is a hint of an opposite trend, which is statistically significant for astronomy (where the data are most reliable) in that we observe that the re-citation fraction actually

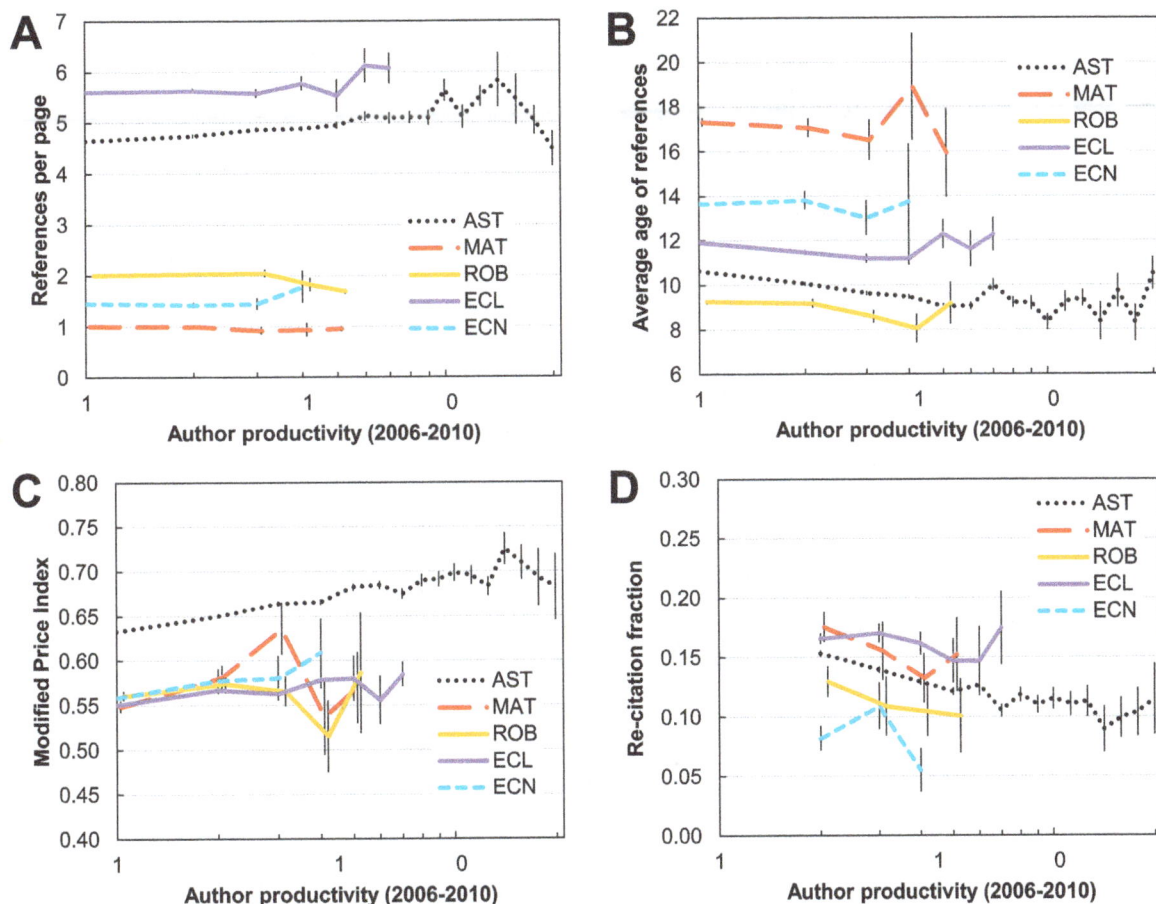

Figure 5. Referencing behavior for authors of different recent productivity in five disciplines (astronomy (AST), mathematics (MAT), robotics (ROB), ecology (ECL), and economics (ECN)). A) Number of references per page (reference rate). B) Average age of references. C) Modified Price Index (fraction of references published within five years compared to those published within 10 years). D) Re-citation fraction (repeated citations in pairs of articles). Data are binned in intervals of 0.05 decades. Bins with fewer than four elements are excluded.

decreases for more productive authors. Recall that the re-citation fraction is determined only from articles led by the author, so the lower re-citation rate is not an artifact of co-authorship.

Based on these analyses it is fairly safe to say that for most of the fields the most productive authors are the ones engaged in the most innovative and varied research, as evidenced by their usage of more recent literature and an increase in the Modified Price Index. In addition, these researchers draw from a larger pool of knowledge, indicating that they may be moving more often from topic to topic. These findings can be used to dispel the notions of most prolific authors engaging in the so-called "salami" publishing, at least for authors who publish their numerous papers in high-quality *core* journals.

Dependence on current level of collaboration. Do highly collaborative authors tend to use references differently from authors in their field who work alone or with very few people? We define collaboration through a number of recent papers to which an author contributed as a co-author, thus decoupling it entirely from the productivity (papers which the author led). Before describing the trends we want to say a few words regarding collaborative practices. In all fields studied here except astronomy most researchers have few or no collaborators (Figure 1C), which we associate with a "normal mode" of collaboration [46], a mode that is not a result of the rich-get-richer phenomenon (preferential attachment mode) or from large-team projects (hyperauthorship

mode). Only in astronomy can we follow the referencing trend for authors who have many collaborators (up to ~70) and are involved in the "big science" aspect of this field.

Generally, we find that authors of various collaborative affinities within the normal mode use references in their works at the same rate. For authors in astronomy who collaborated with more than four researchers the referencing rate increases steadily continuing well into preferential attachment mode. It is likely that the collaborative activity above some threshold exposes an author to a wider body of literature, which the author then uses in his/her publications.

We also found a universal trend (significant in all fields except economics) that the authors who collaborated more used, on average, "younger" references (Figure 6B). The trend was especially pronounced for astronomy, where there was a very marked difference between solo authors and those with ~50 collaborators (11.6 vs. 6.9 years on average). Thus, we may tentatively conclude that the authors who were more collaborative dwell the least on foundational work and push the frontiers of research. To test whether the latter is true we examine the Modified Price Index.

Similar "benefits" of collaboration are evident in the trends of the Modified Price Index (Figure 6C). In astronomy, authors who collaborated more used a higher fraction of recent references within the body of references up to 10 years old. The fact that the

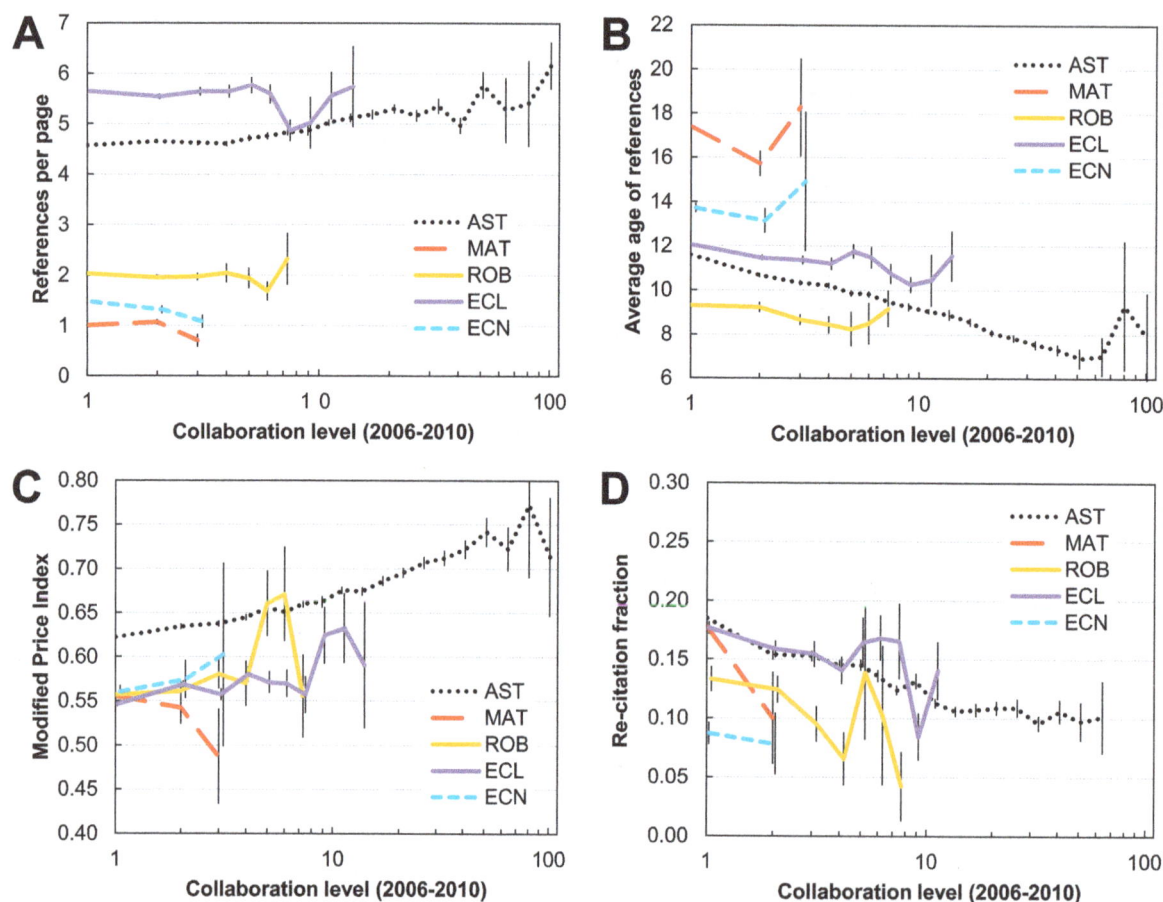

Figure 6. Referencing behavior for authors of different recent collaboration level in five disciplines (astronomy (AST), mathematics (MAT), robotics (ROB), ecology (ECL), and economics (ECN)). Authors with no collaborators are shown at collaboration level 1, those with one at x = 2, etc. A) Number of references per page (reference rate). B) Average age of references. C) Modified Price Index (fraction of references published within five years compared to those published within 10 years). D) Re-citation fraction (repeated citations in pairs of articles). Data are binned in intervals of 0.1 decades. Bins with fewer than four elements are excluded.

MPI was especially high for authors in astronomy who participated in "big science" projects supports the argument that scientists who participate in big projects, *and publish as lead authors*, are those who push the frontiers of the field. Trends are present, but weaker in less collaborative fields such as ecology and robotics. In fields where collaboration is rare, like mathematics and economics, those who do collaborate have a similar MPI to those who do not.

Lastly, we find that authors who collaborate more tend to re-cite less. In all fields, except economics, we observe a very strong decline of re-citation (Figure 6D). Thus, collaborative activity appears to go hand in hand with more varied research.

The trends in this section are qualitatively very similar to trends with respect to productivity. Like productive authors, those who were very collaborative tended to do most innovative science based on the younger references. In addition, these authors were more versatile in terms of the topics they covered, as evidenced by the decrease in their re-citation fraction. We emphasize that our measures of productivity and collaboration are completely independent in that they are based on different sets of documents (papers led by an author vs. papers not led by an author). Yet, the two may be correlated due to a common cause. Indeed, we verify that the mean productivity and collaboration are correlated in all fields. It may not be surprising that the researchers who have the

most to offer to others (and are therefore being invited to join collaborations led by others) are the ones who themselves produce the most.

Discussion

Relative impact of author-related factors on the citing behavior

The results presented in the previous section explored how a certain characteristic of a group of authors in a given discipline – the academic age, productivity or collaboration level – correlates with the references that those authors used.

For most fields, the rate at which authors currently use references (references per page) is very weakly dependent on any of the three author characteristics. The exception is astronomy, where all three factors lead to significant trends – the most important being the increase of the referencing rate for authors who collaborated more.

Author-related characteristics are much more relevant for the age of references used. The strongest trend was that authors who collaborated more used references with significantly lower average age (typically by two years, but in astronomy the effect was up to five years). References that were, on average, "younger" by several years were also characteristic of authors who recently entered the field (but not those who had just entered). Finally, more prolific authors also tend to use references that are up to two years "younger". There are also some field-specific differences. For example, in astronomy and ecology, collaboration has a somewhat larger effect than age, while the opposite is true in mathematics and economy, fields in which the authors do not collaborate extensively.

Trends involving the Modified Price Index were similar but not identical to those for the age of references. Active collaboration activity led to strong increases in the index (i.e., their research front moved faster) in astronomy, robotics and ecology. No author characteristic is strongly correlated with the MPI in economics. There was also some increase of the index for authors who were more productive. Interestingly, academic age was not a strong factor in determining the Modified Price Index – senior authors typically had similar indexes to those who had recently entered the field.

Re-citation fractions, which we interpret as being negatively correlated with the variety of topics on which scientists work, either were not dependent on the time spent in the field or actually went down for more senior researchers. For astronomy, there was some negative correlation between the re-citation fraction and productivity. However, the strongest correlation is with the number of collaborators.

Speaking generally, we conclude that the collaboration level is probably the most important factor related to citing behavior, especially in fields with extensive collaborative practices.

Possible age-related biases

Our results indicate that, in general, more senior authors do not fall behind their junior colleagues. So, is the picture of an out-of-touch older scientist entirely wrong? Obviously one can only study citing behavior of authors who publish. Furthermore, our study analyzes references in core journals, which have higher impacts. If older scientists preferentially gravitate towards lesser impact journals and if their citing behavior is rather different from those who still publish in core journals then our study will not necessarily represent their citing behavior. Whether that is actually the case is outside of the scope of this study. Rather than being a limitation of our study, the analysis only of authors who publish in core journals simply means that we explore the behavior of that section of the scientific community which is the most active in pushing the scientific frontiers.

Conclusions

It has been claimed that paradigms, research traditions and disciplines provide a framework or guidelines for problems and their possible solutions. Do they also provide frameworks on both how and what we reference, which is an essential component of reporting results in research articles and therefore of scientific communication? Do other factors, such as the traits of individual authors also play a role?

The properties of references in a given article, such as their number, average age, or fraction of recent references are the result of referencing behavior and can tell us about the character of research. In this study we investigated referencing behavior in five diverse fields (astronomy, mathematics, robotics, ecology and economics) based on 213,756 articles published in core journals of their respective fields. The study first revisited the referencing behavior at the macro level, following the trends of reference usage in the five fields over the last 50 years. We confirmed earlier results that both the discipline and the time period are strong determinants of citing behavior. We found steady increases in the number of references, but found that it arises from diverse causes: in some fields it was due to a higher rate of usage (number of references per page), while in others it reflected longer articles at more recent times. In all fields the fraction of references older than 10 years had been increasing since the 1980s beyond what is expected from the pure mathematical aging in the exponentially growing body of literature. Contrary to some expectations, the introduction of the Internet apparently resulted in no long-term changes in the referencing behavior.

Our main focus and the novel aspect of this study was to explore current referencing behavior in five fields at the meso level in order to investigate whether different categories of authors use references differently. We analyzed 21,562 authors of differing academic age, productivity and collaboration levels, who published in relatively high-impact core journals, i.e., those who define the currents of modern science. Our results may not necessarily reflect the practices of the entire population of researchers in a given field. Contrary to some previous findings and expectations, we found that senior researchers (who still published in core journals) used references at the same rate as their junior colleagues, with similar rates of re-citation. References of senior authors contained a higher fraction of older foundational works, however their fraction of new (<5 years) vs. recent (<10 years) references (the Modified Price Index) indicated that their research had a similar cutting-edge aspect. Interestingly, this was less evident for researchers who had just entered the field. We found that both the productive researchers and especially those who collaborated more extensively used a significantly lower fraction of foundational references, had a much higher Modified Price Index and used the same references less repeatedly. These author-related trends were similar in all five fields. In other words, highly collaborative and highly productive scientists (regardless of age) typically possess those characteristics as part of their overall excellence, which is also reflected in the way they use references, which suggests that they are the scientists pushing the research front.

Acknowledgments

I thank the referee for useful suggestions, Deborah Shaw for careful reading and comments on the manuscript and Colleen Martin for excellent copy editing.

Author Contributions

Conceived and designed the experiments: SM. Performed the experiments: SM. Analyzed the data: SM. Contributed reagents/materials/analysis tools: SM. Wrote the paper: SM.

References

1. Bazerman C (1988) Shaping written knowledge: The genre and activity of the experimental article in science. Madison: The University of Wisconsin Press.
2. Hyland K (2004) Disciplinary discourses: Social interactions in academic writing. Ann Arbor: The University of Michigan Press.
3. Shapin S (1994) A social history of truth: Civility and science in seventeenth-century England. Chicago: The University of Chicago Press.
4. Börner K (2010) Atlas of science: Visualizing what we know. Cambridge: MIT Press.
5. Leydesdorff L, Wouters P (1999) Between texts and contexts: Advances in theories of citation? Scientometrics 44: 169–182.
6. Gilbert GN (1977) Referencing as persuasion. Social Studies of Science 7: 113–122.

7. Kaplan N (1965) The norms of citation behavior: Prolegomena to the footnote. American Documentation 16: 179–184.

8. Small H (1978) Cited documents as concept symbols. Social Studies of Science 8: 327–340.

9. Brooks TA (1985) Private acts and public objects: An investigation of citer motivations. Journal of the American Society for Information Science 36: 223–229.

10. Chubin DE, Moitra SD (1975) Content analysis of references: Adjunct or alternative to citation counting? Social Studies of Science 5: 423–441.

11. Cronin B (1981) The need for a theory of citing. Journal of Documentation 37: 16–24.

12. Barnett GA, Fink EL (2008) Impact of the internet and scholar age distribution on academic citation age. Journal of the American Society for Information Science and Technology 59: 526–534.

13. Gingras Y, Larivière V, Macaluso B, Robitaille J-P (2008) The effects of aging on researchers' publication and citation patterns. PLoS ONE 3: e4048.

14. Larivière V, Archambault É, Gingras Y (2008) Long-term variations in the aging of scientific literature: From exponential growth to steady-state science (1900–2004). Journal of the American Society for Information Science and Technology 59: 288–296.

15. Glänzel W, Schoepflin U (1995) A bibliometric study on ageing and reception processes of scientific literature. Journal of Information Science 21: 37–54.

16. Line MB (1970) The half-life of periodical literature: Apparent and real obsolescence. Journal of Documentation 26: 46–54.

17. Glänzel W, Schoepflin U (1999) A bibliometric study of reference literature in the sciences and social sciences. Information Processing and Management 35: 31–44.

18. Moed HF, van Leeuwen TN, Reedijk J (1998) A new classification system to describe the ageing of scientific journals and their impact factors. Journal of Documentation 54: 387–419.

19. Ajiferuke I, Lu K, Wolfram D (2011) Who are the research disciples of an author? Examining publication recitation and oeuvre citation exhaustivity. Journal of Informetrics 5: 292–302.

20. Wouters P (1999) The citation culture. Unpublished doctoral dissertation, University of Amsterdam.

21. Cole JR (2000) A short history of the use of citations as a measure of the impact of scientific and scholarly work. In: Cronin B, Atkins HB, editors. The web of knowledge: A festschrift in honor of Eugene Garfield. Metford, NJ: Information Today. pp. 281–300.

22. Cronin B (1984) The citation process: The role and significance of citations in scientific communication. London: Taylor Graham.

23. Moravcsik MJ, Murugesan P (1975) Some results on the function and quality of citations. Social Studies of Science 5: 86–92.

24. Greenberg SA (2009) How citation distortions create unfounded authority: Analysis of a citation network. British Medical Journal 339: b2680.

25. Zuckerman H, Merton RK (1973) Age, aging, and age structure in science. In: Merton RK, editor. The sociology of science: Theoretical and empirical investigations. Chicago: The University of Chicago Press. pp. 497–559.

26. Frandsen TF, Nicolaisen J (2012) Effects of academic experience and prestige on researchers' citing behavior. Journal of the American Society for Information Science and Technology 63: 64–71.

27. Joy S (2006) What should I be doing, and where are they doing it? Scholarly productivity of academic psychologists. Perspectives on Psychological Science 1: 346–364.

28. Wuchty S, Jones BF, Uzzi B (2007) The increasing dominance of teams in production of knowledge. Science 316: 1036–1039.

29. Jones BF, Wuchty S, Uzzi B (2008) Multi-university research teams: Shifting impact, geography, and stratification in science. Science 322.

30. Price DJdS (1963) Little science, big science. New York: Columbia University Press.

31. van Raan AFJ (2000) On growth, ageing, and fractal differentiation of science. Scientometrics 47: 347–362.

32. Price DJdS (1970) Citation measures of hard science, soft science, technology, and nonscience. In: Nelson CE, Pollock DK, editors. Communication among scientists and engineers. Lexington: Heath Lexington Books. pp. 3–22.

33. Price DJdS (1965) Networks of scientific papers. Science 149: 510–515.

34. Burton RE, Kebler RW (1960) The half-life of some scientific and technical literatures. American Documentation 11: 18–22.

35. Egghe L (2010) A model showing the increase in time of the average and median reference age and the decrease in time of the Price Index. Scientometrics 82: 243–248.

36. White HD (2001) Authors as citers over time. Journal of the American Society for Information Science and Technology 52: 87–108.

37. Cronin B, Shaw D (2002) Identity-creators and image-makers: Using citation analysis and thick description to put authors in their place. Scientometrics 54: 31–49.

38. Small H (2010) Referencing through history: How the analysis of landmark scholarly texts can inform citation theory. Research Evaluation 19: 185–193.

39. Minguillo D (2010) Toward a new way of mapping scientific fields: Authors' competence for publishing in scholarly journals. Journal of the American Society for Information Science and Technology 61: 772–786.

40. Henneken EA, Kurtz MJ, Guenther E, Accomazzi A, Grant CS, et al. (2007) E-prints and journal articles in astronomy: A productive co-existence. Learned Publishing 20: 16–22.

41. Garfield E (1982) Journal citation studies, 36. Pure and applied mathematics journals: What they cite and vice versa. Current Contents 15: 5–13.

42. Nobis M, Wohlgemuth T (2004) Trend words in ecological core journals over the last 25 years (1978–2002). Oikos 106: 411–421.

43. Liner GH (2002) Core journals in economics. Economic Inquiry 40: 138–145.

44. Milojević S (2010b) Power-law distributions in information science - Making the case for logarithmic binning. Journal of the American Society for Information Science and Technology 61: 2417–2425.

45. Allen B, Qin J, Lancaster FW (1994) Persuasive communities: A longitudinal analysis of references in the Philosophical Transactions of the Royal Society, 1665–1990. Social Studies of Science 24: 279–310.

46. Milojević S (2010a) Modes of collaboration in modern science - Beyond power laws and preferential attachment. Journal of the American Society for Information Science and Technology 61: 1410–1423.

16

Inconsistent Results of Diagnostic Tools Hamper the Differentiation between Bee and Vespid Venom Allergy

Gunter J. Sturm[1]*, Chunsheng Jin[2], Bettina Kranzelbinder[1], Wolfgang Hemmer[3], Eva M. Sturm[4], Antonia Griesbacher[5], Akos Heinemann[4], Jutta Vollmann[6], Friedrich Altmann[2], Karl Crailsheim[6], Margarete Focke[7], Werner Aberer[1]

1 Division of Environmental Dermatology and Venerology, Department of Dermatology, Medical University of Graz, Graz, Austria, 2 Department of Chemistry, University of Natural Resources and Applied Life Sciences, Vienna, Austria, 3 Floridsdorf Allergy Center, Vienna, Austria, 4 Institute of Experimental and Clinical Pharmacology, Medical University of Graz, Graz, Austria, 5 Division of Biostatistics, Center for Medical Research, Medical University of Graz, Graz, Austria, 6 Institute of Zoology, University of Graz, Graz, Austria, 7 Institute of Pathophysiology, Medical University of Vienna, Vienna, Austria

Abstract

Background: Double sensitization (DS) to bee and vespid venom is frequently observed in the diagnosis of hymenoptera venom allergy, but clinically relevant DS is rare. Therefore it is sophisticated to choose the relevant venom for specific immunotherapy and overtreatment with both venoms may occur. We aimed to compare currently available routine diagnostic tests as well as experimental tests to identify the most accurate diagnostic tool.

Methods: 117 patients with a history of a bee or vespid allergy were included in the study. Initially, IgE determination by the ImmunoCAP, by the Immulite, and by the ADVIA Centaur, as well as the intradermal test (IDT) and the basophil activation test (BAT) were performed. In 72 CAP double positive patients, individual IgE patterns were determined by western blot inhibition and component resolved diagnosis (CRD) with rApi m 1, nVes v 1, and nVes v 5.

Results: Among 117 patients, DS was observed in 63.7% by the Immulite, in 61.5% by the CAP, in 47.9% by the IDT, in 20.5% by the ADVIA, and in 17.1% by the BAT. In CAP double positive patients, western blot inhibition revealed CCD-based DS in 50.8%, and the CRD showed 41.7% of patients with true DS. Generally, agreement between the tests was only fair and inconsistent results were common.

Conclusion: BAT, CRD, and ADVIA showed a low rate of DS. However, the rate of DS is higher than expected by personal history, indicating that the matter of clinical relevance is still not solved even by novel tests. Furthermore, the lack of agreement between these tests makes it difficult to distinguish between bee and vespid venom allergy. At present, no routinely employed test can be regarded as gold standard to find the clinically relevant sensitization.

Editor: Jan-Hendrik Niess, Ulm University, Germany

Funding: The study is supported by a grant of the Austrian Society for Dermatology and Venerology (http://www.oegdv.at). The funders had no role in study design, data collection and analysis, decision to publish, or preparation of the manuscript.

Competing Interests: The authors have declared that no competing interests exist.

* E-mail: gunter.sturm@medunigraz.at

Introduction

Personal history, skin testing, and detection of sIgE, are the mainstays of the diagnostic procedure in cases of hymenoptera venom allergy. Although sensitization to both, honeybee and vespid venom, is observed in up to 59% of patients [1], clinically relevant double sensitization (DS) is rare and patients usually react either to bee or to wasp stings. Therefore, in clinical routine it can be sophisticated to find the relevant venom for specific immunotherapy with common diagnostic tests.

There are several reasons for DS: Generally, a true DS with antibodies to different bee and vespid venom allergens should be considered. DS can also be a result of an around 50% sequence identity of the hyaluronidases in bee and vespid venom. However, a recent study revealed that the wasp hyaluronidase is only a minor allergen, and cross-reactivity between vespid and honeybee venom is not due to protein cross-reactivity, but is mainly caused by cross-reactive carbohydrate determinants (CCDs) [2]. Generally, CCDs are a frequent cause for double positivity as CCD-specific IgE (sIgE) mimics DS in vitro. Asparagine linked carbohydrate moieties of plant and insect glycoproteins are the structural basis of CCDs. In hymenoptera venom, these moieties are found in honeybee venom phospholipase A2 (Api m 1) and hyaluronidase (Api m 2), in vespid venom only in hyaluronidase (e.g. Ves v 2). CCD-sIgE is believed to be clinically irrelevant, although the underlying mechanisms are not completely understood [3,4].

In cases of double positivity, also characteristics of different methods of serum IgE determination should be regarded: Depending on the method, frequencies of double-positive test results vary and range from 10 to 59% [1,5]. In this context, affinity may play an important role. Affinity is largely determined by the stability of the allergen/IgE complex; therefore low affinity is usually correlated with a rapid dissociation of the complex. To efficiently activate mast cells or basophils, high affinity antibodies are required. Most of the current systems of IgE determination use high doses of allergen for IgE detection due to the binding competition with specific IgG. As a consequence low affinity IgE antibodies [6], which are thought to be less relevant for eliciting an allergic reaction [7], are bound as well. Nevertheless, low affinity IgE is not completely irrelevant: in the presence of high affinity IgE it may also activate basophils [8].

The intradermal test is considered not to be influenced by CCDs, as low affinity antibodies itself are not able to cause positive reactions. However, clinically irrelevant positive test results at 1,0 µg/ml are frequently observed [9] and side effects cannot be ruled out [10].

Several studies confirmed the usefulness of the CD63 based basophil activation test (BAT) as a routine diagnostic tool [11,12,13] and as a valuable test in patients with inconclusive tests and history (negative skin tests, undetectable sIgE or unknown stinging insect) [14,15]. Compared with the IgE determination in the serum, BAT has the advantage of demonstrating functional responses: Positive test results will only occur after successful cross-linking of two identical FcεRI-bound IgE antibodies and not by monovalent binding like in IgE assays.

Recently, the component resolved diagnosis (CRD) has been described as useful tool to facilitate the diagnosis of bee and vespid venom allergy [1,16]. Nevertheless, in these studies only rApi m 1 and rVes v 5 were employed to discriminate between true and CCD-based DS. But it is crucial to additionally determine Ves v 1, otherwise 10–13% of vespid venom allergic patients will not be diagnosed due to a mono-sensitization to Ves v 1 [1,2].

Treatment of double positive patients with both venoms is a pragmatic way, but frequently not justified because of asymptomatic sensitization or cross-reactions caused by CCDs. Therefore there is still need for a test which is able to discriminate between clinically relevant or irrelevant sensitization in order to reduce the burden of treatment and to keep therapy as cost-efficient as possible.

In clinical routine, we observed a high frequency of double positivity in the IgE determination by the CAP system (Im-munoCAP®, Phadia, Uppsala, Sweden) and a markedly lower frequency of double positive results obtained by the BAT. Giving this background, we initiated a prospective study to evaluate the usefulness of new diagnostic approaches for the routine diagnosis of hymenoptera venom allergy. For this purpose, we aimed to compare the outcomes of the BAT with the IDT (intradermal test) as well as with three different methods of IgE determination (CAP, ADVIA (ADVIA Centaur®, Siemens, Tarrytown, NY, USA), Immulite (Immulite 2000®, Siemens, Tarrytown, NY, USA)) regarding the frequency of double positive results. To study IgE binding patterns, western blot (WB) inhibition as well as a CRD with native and recombinant Api m1, Ves v 1 and Ves v 5 were performed in all patients with DS.

Methods

Patients

One hundred and seventeen consecutive patients, who had been admitted to our outpatient clinic because of systemic allergic reactions with at least generalized skin symptoms after a hymenoptera sting, were screened. Their personal history was taken and the current standard diagnostic procedures (intradermal tests, IgE determination by CAP) were performed. As wasp and European hornet belong to the family of Vespidae and their venoms contain the same major antigens, we did not differentiate between these genera. Additionally, sIgE was determined by

ADVIA, and the Immulite; basophil responsiveness was analyzed by a CD63 based BAT. In 72 patients showing specific IgE to honeybee and vespid venom in the CAP system, IgE patterns were determined by WB inhibition and CRD. This study was approved by the ethics committee of the Medical University of Graz.

Personal history

According to the modified classification of Ring and Messmer, generalized skin symptoms such as flush, urticaria and angioedema were classified as grade I reaction. Mild to moderate pulmonary, cardiovascular or gastrointestinal symptoms were rated as grade II reaction. Bronchoconstriction, emesis, anaphylactic shock, and loss of consciousness were classified as grade III reaction.

Reagents

All laboratory reagents were obtained from Merck (Whitehouse Station, NJ, USA) or Sigma-Aldrich (St Louis, CA, USA) unless otherwise specified. Dulbecco's modified phosphate-buffered saline (PBS; with or without Ca^{2+} and Mg^{2+}) was purchased from Gibco-Invitrogen (Carlsbad, CA, USA). CellFix and anti-CD123 (PE-conjugated) were supplied by Becton Dickinson (Franklin Lakes, NJ, USA). Antibodies to HLA-DR (PC5-conjugated), CD63 (FITC-conjugated), and monoclonal antibodies to IgE were purchased from Beckman Coulter (Fullerton, CA, USA). Honeybee and vespid venom for the skin tests and BAT were purchased from ALK-Abelló (Hørsholm, Denmark). Honeybee venom and vespid venom sac extracts (mixture of *Vespula vulgaris* and *germanica*) were kindly provided by Vespa Laboratories, PA, USA.

Skin tests

The nature of sensitization was confirmed by standardized end-point titration IDTs (0.02 mL of 0.001, 0.01, 0.1 and 1 µg/mL solution) using purified honeybee and vespid venom extracts. IDTs were considered to be positive in the presence of a wheal ≥5 mm in diameter and erythema.

Determination of sIgE and tIgE

Specific and total IgE antibody levels in the patients' serum were measured using ImmunoCAP 1000 (Phadia, Uppsala, Sweden), ADVIA Centaur, and Immulite 2000 (both: Siemens, Tarrytown, NY, USA) according to the manufacturer's instructions. The CRD with native and recombinant nApi m 1 and rApi m 1 was done on the ImmunoCAP 1000. Diagnosis with the major wasp allergens nVes v 1 and nVes v 5 as well as with nApi m 1 was done on the ADVIA Centaur platform by the Department of I+D, ALK-Abelló, Madrid, Spain.

Basophil activation test (BAT)

BAT was performed as previously described [17]. In brief, EDTA whole blood was stained with anti-CD123 PE-conjugated antibody (1:50), anti-HLA-DR PC5-conjugated antibody (1:50) and anti-CD63 FITC-conjugated antibody (1:50). Basophil reactivity was measured using serial dilutions of honeybee or vespid venom (1000, 100, 10, 1 ng/mL) or serial dilutions of anti-IgE antibody (1:10–1:1000 dilution).

Finally, cell samples were analyzed by three-color flow cytometry (FC 500, Beckman Coulter). Basophils were identified as a single population of cells that stained positive for CD123 (FL-2) and negative for HLA-DR (FL-4). Up-regulation of CD63 expression was indicated by an increase in fluorescence in the FL-1 channel. Acquisition was terminated after 500 basophil target events. An approximately 2.5-fold increase in the number of activated basophils (>25%) as compared with the negative control

(10%) at any of the test concentrations of the allergen was considered to be a positive response. This threshold was determined by ROC analysis as described earlier [12].

Western blots and western blot inhibition

Honeybee venom and vespid venom were separated by SDS-PAGE using 13.5% resolving and 5.7% stacking gels under reducing conditions using dithiothreitol and heat. Electrophoretically separated proteins were blotted onto nitrocellulose membranes and single strips (6 μg venom/strip) blocked with PBS buffer (50 mM sodium phosphate, pH 7.5, 0.5% Tween 20, and 0.05% NaN_3) containing 0.5% BSA at room temperature for 1 h. Subsequently, strips were incubated overnight with 1 mL of serum (diluted 1:5–1:10) at 4°C under continuous shaking. After washing twice with PBS buffer for 30 min, bound IgE was detected by ^{125}I-labelled rabbit anti-human IgE (Phadia, Uppsala, Sweden). After overnight incubation at room temperature, washed and dried strips were exposed to a high-performance autoradiography film (Hyperfilm MP, Amersham, England) at $-70°C$ for 5–10 days.

To discriminate between IgE specific for peptide or carbohydrate epitopes, antibody binding to CCDs was inhibited by preincubating sera with 5 μg/mL of MUXF-BSA as done in previous studies [18]. MUXF-BSA is a synthetic glycoprotein obtained by coupling purified N-glycans from pineapple stem bromelain to BSA [19], whereby MUXF (or more exactly MUXF^3) stands for the glycan structure Manα1-3(Xylβ1-2)Manβ1-4GlcNAcβ1-4(Fucα1-3)GlcNAcβ1.

Data analysis

All data are expressed as medians (25%; 75% percentiles) on the raw scale, unless otherwise indicated. Data were tested for normality using the Kolmogorov-Smirnov test. Continuous variables were analysed by the Kruskal Wallis test; categorical variables were compared by the Chi-square test or Fisher's exact test. To check agreement between the tests, Cohen's kappa coefficient was calculated. The level of significance was set at $p<0.05$. The SPSS 17.0 software (SPSS Inc, USA) was used for statistical analysis.

Results

History and demographic data

One hundred and seventeen patients with a unequivocal history of a systemic sting reaction were included in the study. Fifty-eight (49.6%) were female, and 59 (50.4%) male. Median age was 42.0 (30.5; 53.0) years; the majority of patients (45.3%) were in the age group between 30 and 50 years.

Four patients (3.4%) had a history of grade I reactions, 80 patients (68.4%) had experienced grade II reactions and 33 patients (28.2%) grade III reactions. Thirty-eight (32.5%) identified a honey bee as culprit insect, 55 (47.0%) a wasp, and 24 (20.5%) could not identify the insect. None of the patients reported systemic sting reaction after both, honeybee and wasp stings.

Double sensitization

Frequency of DS differed considerably among performed diagnostic tests and ranged from 63.7% with the Immulite to 17.1% with the BAT (Figure 1). Generally, agreement of tests was fair with 53.1% (kappa 0.318; $p<0.0001$)

Differences between mono and double sensitized as well as double negative patients

In all tests except in the BAT, tIgE levels were up to 2.3-fold higher in double sensitized patients compared to mono sensitized patients. Conversely, patients with double negative results had lower tIgE levels compared to mono or double sensitized patients (Table 1). The comparison of mean age between the three categories revealed no significant difference.

Additionally, regression analysis to check the influence of the severity of sting reaction, sex, age and tIgE on DS was performed: The frequency of DS was influenced by tIgE levels in the CAP (e^b 1.005, $p = 0.035$) and ADVIA (e^b1.003, $p = 0.048$). Additionally, higher age of the patients was associated with a lower frequency of DS in the CAP (e^b 0.966, $p = 0.038$). The rate of DS in the BAT, IDT, and Immulite was not influenced by the tested variables.

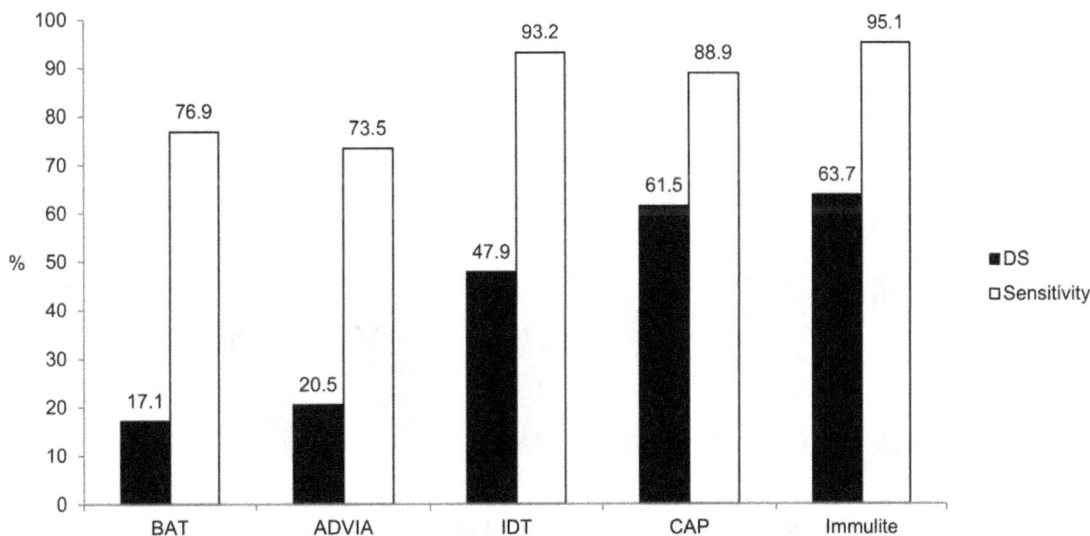

Figure 1. Frequency of double sensitization. The rate of DS in 117 consecutive patients differed significantly ($p<0.0001$) and ranged from 17.1% with the BAT to 63.7% with the Immulite.

Table 1. Correlation between total IgE (kU/L) and test results.

	Double sensitization	Mono sensitization	Double negative	p
BAT	54.3 (24.5; 217.3)	64.7 (36.9; 151.0)	58.9 (22.7; 112.0)	0.463
ADVIA	117.0 (50.9; 397.6)	51.7 (30.3; 123.4)	35.9 (10.4; 142.4)	0.008
IDT	88.7 (45.2; 252.0)	53.9 (29.4; 103.5)	35.3 (8.7; 69.0)	0.014
CAP	90.8 (48.9; 230.0)	43.3 (29.1; 64.2)	28.0 (10.6; 66.0)	0.000
Immulite	87.2 (42.6; 246.0)	51.5 (19.6; 87.5)	8.0 (2.7; 95.6)	0.006

In all tests except in the BAT, double sensitized patients showed higher levels of total IgE (tIgE) compared to mono sensitized patients. Conversely, double negative patients had lower tIgE levels compared to mono- and double sensitized patients.

Subgroup analysis of double sensitized patients in the CAP

IgE determination by CAP yielded together with the Immulite the highest frequency of double positive results. As the CAP system is widely used, and this group of 72 patients comprised virtually all patients with double positive results in supplemental tests, further analysis regarding the individual IgE pattern was done in this subgroup.

First, the rate of DS of each commercially available and experimental test was determined to identify the most specific test to reduce the high frequency of clinically not relevant DS. As expected, CRD analysis solely done with the native main allergen components nApi m 1, nVes v 1, and nVes v 5 led to a slightly reduced, but still high frequency of DS. The use of non-glycosylated rApi m 1, nVes v 1 and nVes v 5 reduced the frequency considerably by 49.0%. Similar lower rates of DS were observed with the WB inhibition, ADVIA and BAT, while the Immulite and the IDT revealed high frequencies of DS (Figure 2).

IgE patterns of CAP double sensitized patients with WB inhibition

The WB was not interpretable in 11 of 72 patients. Among the remaining patients, true DS was diagnosed in 24 of 61 patients, putative cross-reactivity due to hyaluronidase in 6 patients, and double positive results caused by CCD alone in 31 patients (typical IgE patterns see Figure 3).

CRD in CAP double sensitized patients

>As at the time when the study was performed rApi m 1 was not available for the ADVIA, and vice versa nVes v 1 and nVes v 5 not for the CAP, rApi m1 was determined with the CAP and nVes v 1 and nVes v 5 with the ADVIA. To check compatibility, nApi m 1 was determined on the CAP as well as ADVIA. In contrast to the IgE determination with bee and vespid extracts, the test results with native components were coinciding with 92.3%, assuming an almost perfect agreement.

Finally, CRD with recombinant and native allergens was performed in 64 of 72 CAP double positive sera; four patients were negative for the tested bee and vespid venom allergens (Figure 4 A+B). There was a substantial agreement between the WB and the CRD for Api m 1 with 88.5% (kappa 0.770, p<0.0001), and Ves v 5 with 87.7% (kappa 0.744, p<0.0001). The agreement for Ves v 1 was only fair with 71.9% (kappa 0.377, p = 0.005).

BAT compared to CRD and WB inhibition in CAP double sensitized patients

Beside the component-specific tests (CRD, WB) only the BAT and ADVIA showed a comparable low frequency of DS despite the use of conventional allergen extracts. As ADVIA is no longer available, further analysis was only done with the BAT in 72 patients; 11 were negative for both venoms (Figure 5). Noteworthy, in 11 patients with DS in the CRD, basophils were only activated by one venom in the BAT. Conversely, 7 BAT double positive patients showed only a mono sensitization in the CRD. There was a similar picture with the WB inhibition: 13 double

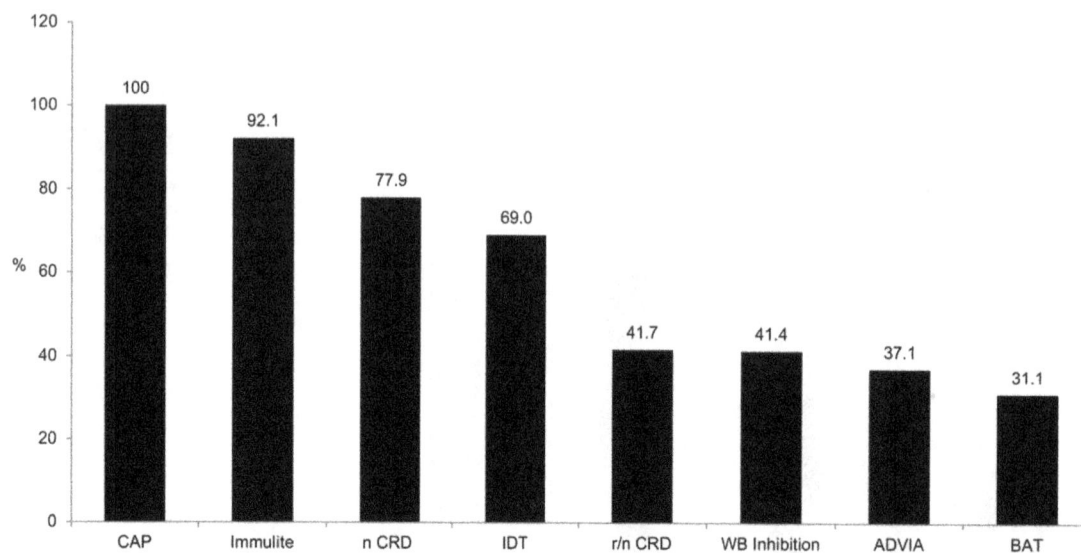

Figure 2. Frequency of double sensitization in supplemental tests in 72 CAP double positive patients. n CRD: native component resolved diagnosis with nApi m 1, nVes v 1, nVes v 5. r/n CRD: combined component resolved diagnosis with recombinant rApi m 1, and native nVes v 1, nVes v 5. BAT (p = 0.324) and ADVIA (p = 0.874) showed a similar frequency of DS compared to WB inhibition and r/n CRD, although they were performed with native venom extracts.

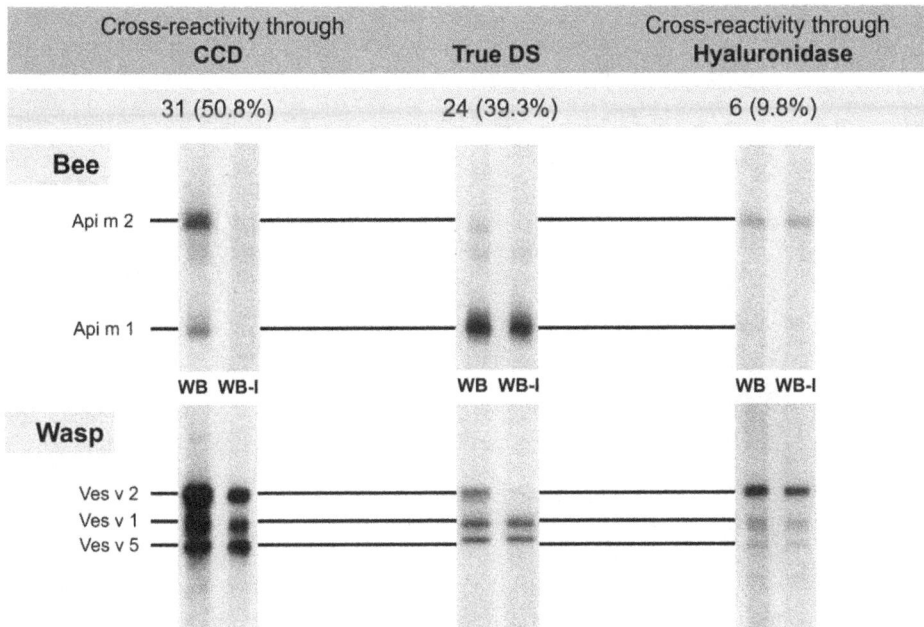

Figure 3. Frequency of typical IgE patterns obtained by western blot inhibition in CAP double sensitized patients. CCD: cross-reactive carbohydrate determinants, True DS: true double sensitization, WB: western blot, WB-I (western blot inhibition): To discriminate between IgE specific for peptide or carbohydrate epitopes, antibody binding to CCDs was inhibited by preincubating sera with MUXF-BSA. Among these patients the majority of DS was CCD-dependent. DS due to protein components of hyaluronidases played a minor role. n = 61.

positive patients in the WB inhibition were only positive to one venom in the BAT and 7 BAT double positive patients showed only a mono sensitization in the WB inhibition. Generally, results of the BAT were in fair agreement with those of the CRD (Figure 6).

sIgE to MUXF (CCD)

Determination of sIgE to MUXF in the CAP (CCD-IgE) was not appropriate to distinguish between true DS and CCD based DS. 16 of 30 patients with true DS in the WB (sensitization to

major allergens or hyaluronidase) had detectable sIgE to MUXF and conversely, only 16 of 31 patients with a verified CCD-based DS by the WB inhibition had sIgE to MUXF (Figure 7).

Additionally, also 15 of 25 (60.0%) patients with true DS verified by CRD had sIgE to MUXF.

Double positive results in CCD-dependent DS

CCD dependent DS was verified by WB inhibition in 31 patients. Depending on the test, frequency of DS ranged from 12.0% to 89.3% in these patients (Figure 8).

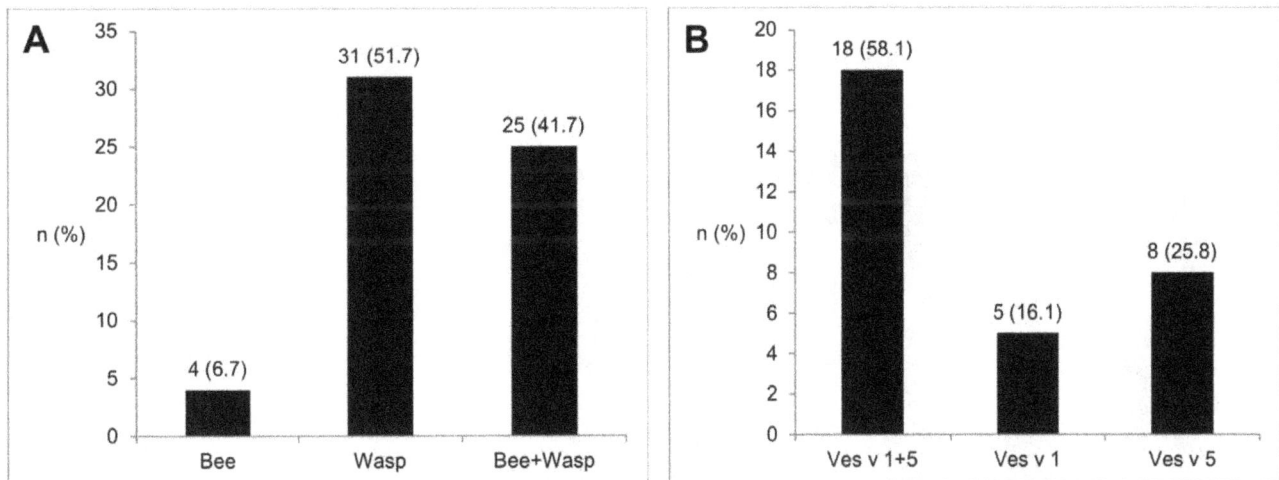

Figure 4. Component resolved diagnosis in CAP double sensitized patients. A Sensitization to bee and/or vespid venom in the component resolved diagnosis. Positive for bee venom: rApi m 1^{pos} / nVes v 1^{neg} and nVes v 5^{neg}; Positive for wasp venom: rApi m 1^{neg} / nVes v 1 and/or nVes v 5^{pos}; DS: rApi m 1^{pos} / nVes v 1 and/or nVes v 5^{pos}. n = 60. **B Sensitization pattern in vespid venom allergic patients.** The majority of patients were sensitized to both vespid major allergens (nVes v 1 and nVes v 5). Nevertheless, a considerable proportion had a mono sensitization to nVes v 1 or nVes v 5. n = 31.

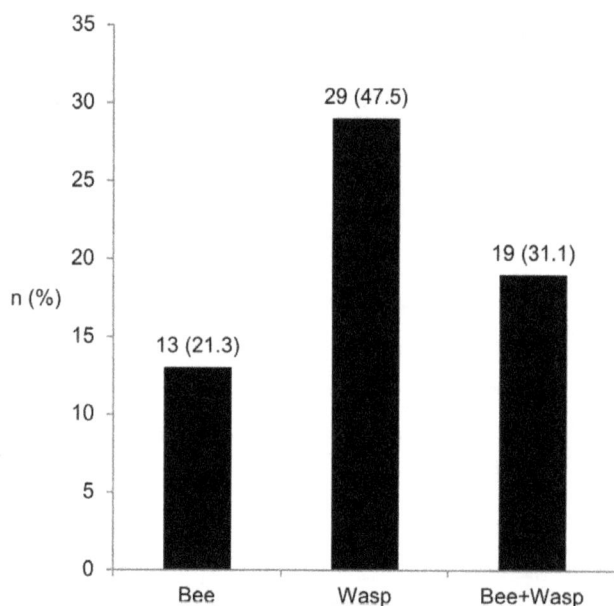

Figure 5. Frequency of sensitization to bee and/or vespid venom in the basophil activation test. n = 61.

Discussion

Positive test results to bee and vespid venom are frequently observed in the routine diagnosis of hymenoptera venom allergy and raise problems to determine the causative insect for a correct treatment. Treatment with two venoms is generally accepted in patients with severe sting reactions and inconclusive test results. Nevertheless, there is a high risk of overtreatment, and even for a novel sensitization, if positive results are unspecific and caused by weakness of diagnostic methods or by CCDs.

In the current study, we performed an extensive evaluation of various conventional, recently established, and experimental test methods. We could demonstrate that the BAT had the lowest frequency of DS and thus correlated best with the patients' history.

Nevertheless, the BAT showed double positive results in nearly one third of patients with CCD-based DS, and vice versa was sometimes only positive for one venom in patients with DS in the WB inhibition and CRD. CCDs can lead *in vitro* to a stimulation of basophils [20,21] and the question of clinical relevance of these positive results remains still unanswered. Conversely, even a true (double-) sensitization must not be clinically relevant [22,23]. In this case, the BAT as functional test may be helpful to find the culprit venom. IgE determination by the ADVIA also resulted in a low frequency of DS, even though it was slightly higher compared to the BAT. However, the ADVIA platform is no longer available for routine diagnosis as it has been taken off the market despite of its revolutionary concept of IgE determination and its excellent performance. Additionally, we could show that the intradermal test was not beneficial in the discrimination between mono- and double sensitization because it revealed DS in as much as 69% of patients. This may either reflect false-positive reactions due to histamine liberating substances or toxic effects of the venom, as well as some mast cell activation by CCDs at very high venom concentrations (1 µg/ml). As expected, the CRD with recombinant and native CCD-free allergens discriminated well between CCD based and true DS, and hence represents a clear step forward in the diagnosis of hymenoptera venom allergy. Importantly, the sensitization patterns of the CRD correlated well with those of the western blot. Nevertheless, the CRD revealed a markedly lower frequency of honey bee sensitization compared to BAT and WB which could indicate an insufficient sensitivity of rApi m 1 and the need for additional honeybee venom allergens.

Clinically relevant DS is rarely observed: in a large European (EAACI) multicenter study regarding side-effects during immunotherapy only 58 of 840 (6.9%) were treated with two venoms [24]. At the same time, asymptomatic sensitization is observed in 27.1 to 66.7% of the general population depending on the test method of IgE determination and tIgE levels [5,23].

Depending on the methods and venoms used, the specificity of serum IgE determination ranges between 60% and 94% [12]. Leading manufacturers of automated lab systems generally postulate high sensitivities and specificities for their IgE determination. However, the studies leading to these results must be viewed critically: control subjects with high tIgE levels, positive

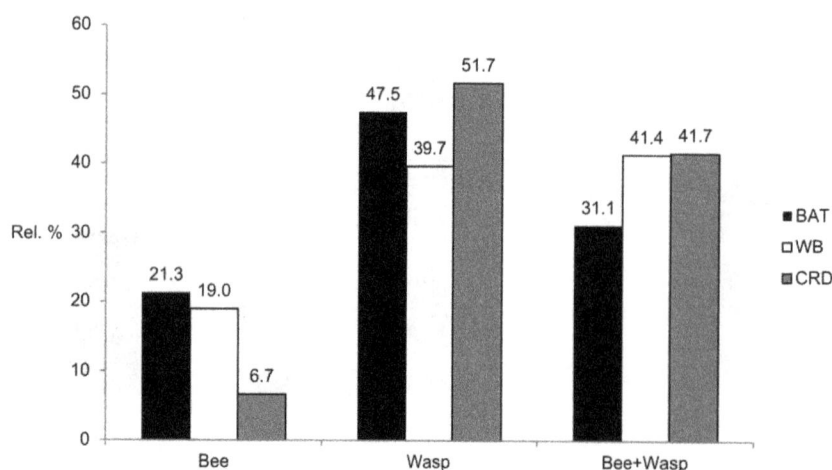

Figure 6. BAT results in relation to western blot inhibition and component resolved diagnosis. Although BAT was performed with native venom extracts, frequency of mono- and double sensitization was comparable with component based methods. Results of the BAT were in fair agreement with those of the CRD (60.0%, kappa 0.373, p<0001) and WB (59.6%, kappa 0.377, p<0001). Interestingly, the frequency of honey bee sensitization obtained with the CRD was markedly lower compared to BAT and WB, which could indicate a lower sensitivity of rApi m 1.

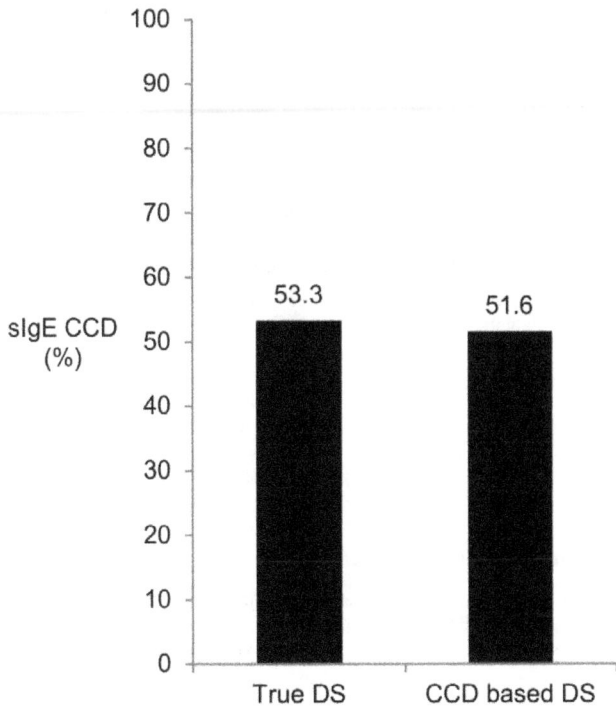

Figure 7. Determination of sIgE to MUXF (CCD) was not appropriate to distinguish between true and CCD-based double sensitization. Patients with true DS in the WB (sensitization to major allergens or hyaluronidase) had detectable sIgE to MUXF and conversely, patients with a verified CCD-based DS by the WB inhibition had no detectable sIgE to MUXF. As the coincidence of true DS and detectable sIgE to MUXF was high, results could be misinterpreted and true DS could be easily overlooked.

allergen is bound to a solid cellulose sponge matrix. After incubation with the patient's serum sIgE and also specific IgG is bound to the covalently coupled allergen. To quantify sIgE levels, sIgE is detected by enzyme-labeled anti-IgE. To minimize competition between the low quantity of IgE and the substantial quantity of IgG a very high amount of allergen is bound to the immunosorbent. Therefore also low-affinity cross-reacting sIgE like those to CCDs with questionable clinical relevance are detected. The same might be valid for the Immulite, although it depends on another principle: In brief, ligand-labeled liquid allergens first bind to anti-ligand-coated polystyrene beads; after adding the patient's serum, sIgE is bound to the allergen. Again, sIgE is detected by anti-IgE. High doses of allergen to avoid displacement of sIgE antibodies in both tests would explain the similar frequency of DS with 61.5% and 63.7%, respectively.

The concept of the ADVIA is completely different to exclude interference with non-IgE antibodies like IgG. Anti-IgE is coupled to paramagnetic particles that catch all IgE in the serum. Then biotin-labeled allergen is added and bound sIgE reacts with the allergen in suspension. Finally sIgE is detected indirectly with acridinium ester labeled streptavidin [27]. The main advantage of this approach is that much less allergen is needed and therefore the affinity of sIgE is better considered. This explains the good performance of the ADVIA despite of the native venom extracts used.

The IDT and the BAT have the advantage of demonstrating functional responses as positive results usually only occur after cross-linking of two identical cell-bound IgE antibodies. Nevertheless, we observed a considerable difference in the occurrence of DS: The IDT was positive for bee and vespid venom in 47.9% of patients compared to 17.1% double positive results obtained by the BAT. The high frequency in the IDT might be explained by the irritant effect of the venom at higher doses and, as mentioned earlier, by the activation of some mast cells by CCDs at very high venom concentrations. On the other hand, the low rate of DS in the BAT with native venom extracts supports the hypothesis, that the BAT is able to demonstrate a functional response without possible irritant reactions as seen in the IDT and without considerable influence of CCDs on test results as obtained with

skin tests and an atopic disposition are generally ruled out in order to obtain optimum specificities [25,26].

Generally, methods of serum IgE determination differ considerably and therefore results are difficult to compare. In CAP, the

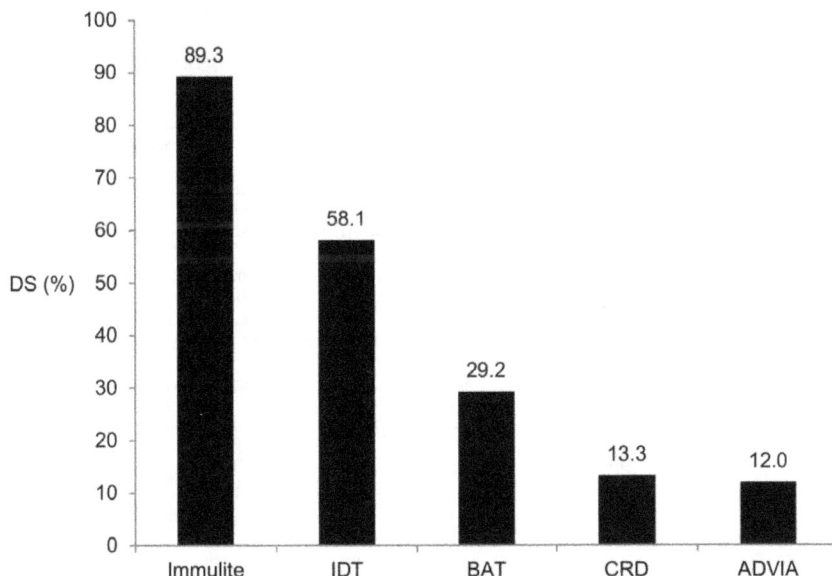

Figure 8. Double positive results in CCD-dependent double sensitization. CCD-dependent DS was verified with WB inhibition in 31 patients. The Immulite and IDT revealed the highest rates of DS in these patients (p<0.001; n = 24–31).

the CAP or Immulite. Recently, up to 67% double positive results were reported with the CD203c based BAT [28], this is contrary to our findings. This extraordinary high rate of DS might not depend on the different activation marker CD203c, but on an internationally uncommon protocol and unusual interpretation of results. Nevertheless, there still remain a few open questions: In our study, the BAT showed in 29% of patients with a verified CCD-based DS double positive results and vice versa the BAT was sometimes only positive for one venom despite that the CRD and WB inhibition revealed double positive results, respectively.

The role of CCDs for eliciting clinical symptoms is still unclear. There are several hypotheses why sIgE to CCDs are not relevant, one of them is that patients are constantly exposed to these carbohydrate structures and therefore produce blocking IgG_4 antibodies, comparable with the effect of immunotherapy [4]. This might explain that basophils can be activated in the BAT, but not in vivo.

The application of recombinant or native CCD-free allergens will be a considerable progress in the diagnosis of hymenoptera venom allergy. nApi m 1 showed clearly more positive results compared to rApi m 1, again indicating the crucial role of CCDs in DS. Thus makes it inevitable to use components which are CCD-free by nature or to produce recombinant allergens without CCDs. Importantly, the generally accepted use of sIgE to CCD as marker for CCD-based cross-reactivity has to be viewed critically and must be considered obsolete. As shown in the WB, the presence of IgE to CCDs does not exclude true DS, therefore true DS can be easily overlooked, which may result in fatal reactions.

To summarize, BAT and CRD showed the lowest rates of DS, but inconsistent results were common. Although each test alone seems to help finding the clinically relevant venom, it is still unclear which test represents the most accurate. Therefore, studies with sting challenges to check the accurate negative predictive value of the BAT and CRD in otherwise double sensitized patients would be preferable. At present, no routinely employed test can be regarded as gold standard to distinguish between clinically relevant bee and wasp venom sensitization.

Acknowledgments

We thank Karin Laipold for skilled technical assistance as well as Jesper Fromberg, Lucia Jimeno Nogales, and Agustin Galan Nieto from ALK for the opportunity to perform component resolved diagnosis on the ADVIA platform.

Author Contributions

Conceived and designed the experiments: GJS WH FA MF. Performed the experiments: CJ EMS BK. Analyzed the data: AG AH. Contributed reagents/materials/analysis tools: JV KC. Wrote the paper: GJS WA.

References

1. Muller UR, Johansen N, Petersen AB, Fromberg-Nielsen J, Haeberli G (2009) Hymenoptera venom allergy: analysis of double positivity to honey bee and Vespula venom by estimation of IgE antibodies to species-specific major allergens Api m1 and Ves v5. Allergy 64: 543–548.

2. Jin C, Focke M, Leonard R, Jarisch R, Altmann F, et al. (2010) Reassessing the role of hyaluronidase in yellow jacket venom allergy. J Allergy Clin Immunol 125: 184–190 e181.

3. Jin C, Hantusch B, Hemmer W, Stadlmann J, Altmann F (2008) Affinity of IgE and IgG against cross-reactive carbohydrate determinants on plant and insect glycoproteins. J Allergy Clin Immunol 121: 185–190 e182.

4. Altmann F (2007) The role of protein glycosylation in allergy. Int Arch Allergy Immunol 142: 99–115.

5. Egner W, Ward C, Brown DL, Ewan PW (1998) The frequency and clinical significance of specific IgE to both wasp (Vespula) and honey-bee (Apis) venoms in the same patient. Clin Exp Allergy 28: 26–34.

6. Aalberse RC, Kleine Budde I, Stapel SO, van Ree R (2001) Structural aspects of cross-reactivity and its relation to antibody affinity. Allergy 56 Suppl 67: 27–29.

7. Pierson-Mullany LK, Jackola DR, Blumenthal MN, Rosenberg A (2002) Evidence of an affinity threshold for IgE-allergen binding in the percutaneous skin test reaction. Clin Exp Allergy 32: 107–116.

8. Christensen LH, Holm J, Lund G, Riise E, Lund K (2008) Several distinct properties of the IgE repertoire determine effector cell degranulation in response to allergen challenge. J Allergy Clin Immunol 122: 298–304.

9. Georgitis JW, Reisman RE (1985) Venom skin tests in insect-allergic and insect-nonallergic populations. J Allergy Clin Immunol 76: 803–807.

10. Lockey RF, Turkeltaub PC, Olive CA, Baird-Warren IA, Olive ES, et al. (1989) The Hymenoptera venom study. II: Skin test results and safety of venom skin testing. J Allergy Clin Immunol 84: 967–974.

11. Sainte-Laudy J, Sabbah A, Drouet M, Lauret MG, Loiry M (2000) Diagnosis of venom allergy by flow cytometry. Correlation with clinical history, skin tests, specific IgE, histamine and leukotriene C4 release. Clin Exp Allergy 30: 1166–1171.

12. Sturm GJ, Bohm E, Trummer M, Weiglhofer I, Heinemann A, et al. (2004) The CD63 basophil activation test in Hymenoptera venom allergy: a prospective study. Allergy 59: 1110–1117.

13. Erdmann SM, Sachs B, Kwiecien R, Moll-Slodowy S, Sauer I, et al. (2004) The basophil activation test in wasp venom allergy: sensitivity, specificity and monitoring specific immunotherapy. Allergy 59: 1102–1109.

14. Ebo DG, Hagendorens MM, Bridts CH, De Clerck LS, Stevens WJ (2007) Hymenoptera venom allergy: taking the sting out of difficult cases. J Investig Allergol Clin Immunol 17: 357–360.

15. Korosec P, Erzen R, Silar M, Bajrovic N, Kopac P, et al. (2009) Basophil responsiveness in patients with insect sting allergies and negative venom-specific immunoglobulin E and skin prick test results. Clin Exp Allergy 39: 1730–1737.

16. Mittermann I, Zidarn M, Silar M, Markovic-Housley Z, Aberer W, et al. (2010) Recombinant allergen-based IgE testing to distinguish bee and wasp allergy. J Allergy Clin Immunol 125: 1300–1307 e1303.

17. Sturm GJ, Kranzelbinder B, Sturm EM, Heinemann A, Groselj-Strele A, et al. (2009) The basophil activation test in the diagnosis of allergy: technical issues and critical factors. Allergy 64: 1319–1326.

18. Hemmer W, Focke M, Kolarich D, Dalik I, Gotz M, et al. (2004) Identification by immunoblot of venom glycoproteins displaying immunoglobulin E-binding N-glycans as cross-reactive allergens in honeybee and yellow jacket venom. Clin Exp Allergy 34: 460–469.

19. Wilson IB, Harthill JE, Mullin NP, Ashford DA, Altmann F (1998) Core alpha1,3-fucose is a key part of the epitope recognized by antibodies reacting against plant N-linked oligosaccharides and is present in a wide variety of plant extracts. Glycobiology 8: 651–661.

20. Batanero E, Crespo JF, Monsalve RI, Martin-Esteban M, Villalba M, et al. (1999) IgE-binding and histamine-release capabilities of the main carbohydrate component isolated from the major allergen of olive tree pollen, Ole e 1. J Allergy Clin Immunol 103: 147–153.

21. Wicklein D, Lindner B, Moll H, Kolarich D, Altmann F, et al. (2004) Carbohydrate moieties can induce mediator release: a detailed characterization of two major timothy grass pollen allergens. Biol Chem 385: 397–407.

22. Schafer T, Przybilla B (1996) IgE antibodies to Hymenoptera venoms in the serum are common in the general population and are related to indications of atopy. Allergy 51: 372–377.

23. Sturm GJ, Schuster C, Kranzelbinder B, Wiednig M, Groselj-Strele A, et al. (2008) Asymptomatic Sensitization to Hymenoptera Venom Is Related to Total Immunoglobulin E Levels. Int Arch Allergy Immunol 148: 261–264.

24. Mosbech H, Muller U (2000) Side-effects of insect venom immunotherapy: results from an EAACI multicenter study. European Academy of Allergology and Clinical Immunology. Allergy 55: 1005–1010.

25. Pastorello EA, Incorvaia C, Pravettoni V, Marelli A, Farioli L, et al. (1992) Clinical evaluation of CAP System and RAST in the measurement of specific IgE. Allergy 47: 463–466.

26. Paganelli R, Ansotegui IJ, Sastre J, Lange CE, Roovers MH, et al. (1998) Specific IgE antibodies in the diagnosis of atopic disease. Clinical evaluation of a new in vitro test system, UniCAP, in six European allergy clinics. Allergy 53: 763–768.

27. Petersen AB, Gudmann P, Milvang-Gronager P, Morkeberg R, Bogestrand S, et al. (2004) Performance evaluation of a specific IgE assay developed for the ADVIA centaur immunoassay system. Clin Biochem 37: 882–892.

28. Mertens M, Amler S, Moerschbacher BM, Brehler R (2010) Cross-reactive carbohydrate determinants strongly affect the results of the basophil activation test in hymenoptera-venom allergy. Clin Exp Allergy 40: 1333–1345.

17

The Land Gini Coefficient and Its Application for Land Use Structure Analysis in China

Xinqi Zheng[1]*, Tian Xia[1], Xin Yang[2], Tao Yuan[1], Yecui Hu[1]

1 School of Land Science and Technology, China University of Geosciences (Beijing), Beijing, China, **2** Research Center for Operation and Development of Beijing, Institute of Policy and Management, Chinese Academy of Sciences, Beijing, China

Abstract

We introduce the Gini coefficient to assess the rationality of land use structure. The rapid transformation of land use in China provides a typical case for land use structure analysis. In this study, a land Gini coefficient (LGC) analysis tool was developed. The land use structure rationality was analyzed and evaluated based on statistical data for China between 1996 and 2008. The results show: (1)The LGC of three major land use types–farmland, built-up land and unused land–was smaller when the four economic districts were considered as assessment units instead of the provinces. Therefore, the LGC is spatially dependent; if the calculation unit expands, then the LGC decreases, and this relationship does not change with time. Additionally, land use activities in different provinces of a single district differed greatly. (2) At the national level, the LGC of the three main land use types indicated that during the 13 years analyzed, the farmland and unused land were evenly distributed across China. However, the built-up land distribution was relatively or absolutely unequal and highlights the rapid urbanization in China. (3) Trends in the distribution of the three major land use types are very different. At the national level, when using a district as the calculation unit, the LGC of the three main land use types increased, and their distribution became increasingly concentrated. However, when a province was used as the calculation unit, the LGC of the farmland increased, while the LGC of the built-up and unused land decreased. These findings indicate that the distribution of the farmland became increasingly concentrated, while the built-up land and unused land became increasingly uniform. (4) The LGC analysis method of land use structure based on geographic information systems (GIS) is flexible and convenient.

Editor: Rodrigo Huerta-Quintanilla, Cinvestav-Merida, Mexico

Funding: This study was supported by the following grant: the National Public Benefit (Land) Research Foundation of China (No.201111014). The funders had no role in study design, data collection and analysis, decision to publish, or preparation of the manuscript.

Competing Interests: The authors have declared that no competing interests exist.

* E-mail: zxqsd@126.com

Introduction

Land use structure is a product of human activities and natural conditions. This structure consists of area proportion, space distribution and the influence of and relationship between all land use types. Land use structure indicates the status of natural and socioeconomic development; the structure affects (e.g., restricts) the development of all aspects of society. Land use structure rationality refers to the concept that the land use structure adapts to the harmony and sustainable development of society, the economy, and ecology. Land use structure optimization involves activities to organize the land use structure more rationally, which primarily includes area proportioning and space distributing. This optimization would increase the rationality of actual land use behavior and effectively balance various land use types, which promotes balanced land ecosystems. Furthermore, optimization facilitates the coordination and sustainable development of the economy, society, and ecology [1]. Analysis of land use structure rationality is the foundation of land use structure optimization. Land use structure is a well-established research field in land planning and land resources [2].

To understand the rationality of land use structure, the present status of the land use structure must be analyzed, and the problems of land use must be made explicit. The systematic theoretical research on urban land use structure was initiated in ecology in the 1920s. Subsequently, other schools of thought, such as economic location, social behavior, and political economics, were formed, and social science theories and varied analysis methods were developed. Classical models are presented in some of these theories. For example, the concentric circles model, the fan-shaped model, and the multi-core model are presented in the theory of economic location [3]. The research perspective has shifted from ecology to humanities (i.e., political and economic aspects) and has gradually turned to the functional space of urban areas [4,5]. The research methods have shifted from traditional statistical analysis to information technology and multi-agent system modeling [6,7]. Many scholars have researched land use structure in various regions [8] and at various spatial scales [13,14]. These researchers have studied the status of one particular land use type [10], considered the eco-environment as the emphasis point of land use structure [15], analyzed land use structure on the basis of low carbon [13], used indices to understand the status of land use structure [12], or studied the order degree of land use structure [14]. As the methods, study regions, and emphases of the studies have not been consistent, the reported results have differed. Additionally, there is still no generally accepted method for assessing land use structure.

The scientific foundation, practicality and operability of analyses of land use structure rationality are critical to the viability of land use structure optimization plans and outcomes. Therefore, a suitable method to perform land use structure research should be chosen carefully. After generalizing and comparing the existing methods, it is apparent that some researchers have combined computer technology and mathematical theories to perform their studies. This technique allows for very rapid calculations and a variety of presentation options. Additionally, traditional geographical methods are still widely used and accepted because they are easy to understand and are well established. However, both methods have disadvantages. The former method can be overly complicated, and accurate programming is difficult. The latter method has analysis issues due to the lack of basic data and the unintuitive and limited methods for presenting results. Furthermore, neither of these two methods can quantify differences. Thus, the method for judging land use structure rationality needs to be improved. To combine the advantages of the methods described and to expand the research into land use structure, scholars have gradually introduced economic methods into studies of land use structure; the Gini coefficient model [16] is one of the most important of these methods [17].

Originally, the Gini coefficient was used in economics to assess the distribution of income. Recently, the coefficient was developed substantially for use in economics and applied more widely in other research fields [18,19]. For example, this method has been applied to the distribution of household incomes under various conditions and impact assessments [20,21], the distribution of medical resources [22,23], the conditions and impact of plant growth [24–25,26], and the consumption and use of material resources [27,28]. The Gini coefficient method is also applied in environmental studies to understand water pollution issues and assist policy makers [29–30,31]. These applications highlighted the universality of the Gini coefficient and popularized its use.

The Lorenz curve and Gini coefficient have been applied in land use analysis with encouraging results. The previous studies calculated numerical data, generated results and described the distribution of land use structure without considering spatial data. To improve the method for calculating the Gini coefficient, Yang (2008) explored dynamic calculation methods based on GIS spatial data. As a result, the calculation of the land Gini coefficient (LGC) was more operable and efficient [32].

Referencing previous research, this work applies the Gini coefficient to analyze land use structure and rationality. After defining the LGC, the paper combines the LGC with GIS, improves the LGC calculation tool, and builds a land use dataset from 1996 to 2008 for China. We calculate the LGC using various scales to analyze the spatiotemporal characteristics of the land use structure, evaluate the structure rationality, and gain new insights into land use in China.

Materials and Methods

Materials

We obtained land use data from 1996 to 2008 from the land use change survey managed by the land management department in China. This period was selected mainly because the statistical caliber and classification system of the data were consistent and the data were complete; the classification of land use types changed in 2009.

The population data are from the *Population Statistics of Counties in the People's Republic of China from the Year 1997 to 2009* published by Masses Press [33]. Because the publication date was a year after the data collection date, the population data correspond to the same years as the land use data.

Based on the 1:4,000,000 fundamental geographic information database, we built a dataset within the ArcGIS platform using the 13-year data by connecting the land use and population data with fundamental geographic information. This dataset contains the boundaries of cities, provinces and districts. The classification system used in this paper is that of the land use change survey during 1996 to 2008 [34]. This study analyzed the status of land use structure of the first level in the classification, which includes farmland, built-up land and unused land.

Methods

The land gini coefficient. The major advantage of the Gini coefficient is that it can quantify differences, and it is very intuitive because it is based on the Lorenz curve. The Gini coefficient, originally used for quantifying differences in income, was reintroduced as the LGC to analyze the rationality of land use structure. The LGC has a different connotation for each target and has many sub-LGCs, called basic land use Gini coefficients, such as LGCAmount, LGCSpatial, and LGCMass.

The basic LGC involved in this study primarily includes LGCSpatial and LGCAmount. LGCSpatial indicates the evenness of a single land use type distribution in the study area, whereas LGCAmount indicates the rationality of the area proportions of all land use types in that area. The LGC ranges from 0 to 1, with smaller values indicating a more balanced land use structure. The standard for assessing different levels of land use structure based on the LGC is shown in Table 1.

Calculation of the land gini coefficient. The main processes for calculating the LGC are constructing the land Lorenz curve, computing the basic LGC, and calculating the final LGC based on the basic LGC.

Because this study was mainly concerned with the distribution of land use in China, we illustrate the process for calculating LGCSpatial.

The steps of constructing the Traditional Lorenz Curve for LGCSpatial were following:

Step 1: Calculation of the location entropy of every land use type. The following formula is used:

$$Q = \frac{A_1/A_2}{A_3/A_4} = \frac{P_{K1}}{P_{T1}} \tag{1}$$

where Q is location entropy, A_1 is the area of a certain land use type in a subordinate region of the calculation unit, A_2 is area of the same land use type of the calculation unit, A_3 is area of the subordinate region of the calculation unit, A_4 is area of the calculation unit, P_{K1} is the percentage that A_1 occupies in A_2, and P_{T1} is the percentage that A_3 occupies in A_4.

Step 2: Sorting the location entropy from small to large and calculating the cumulative percentages of the areas of each land use type and total area.

Step 3: Drawing the Lorenz curve of LGCSpatial based on the calculation and sorting results using the cumulative percentage of the total area as the X-coordinate and the cumulative percentage of each type of land use area as the Y-coordinate.

Demand and productivity of various land use types differ greatly and cannot be compared without transformation. The rational area of each land use type should be comprehensively determined with consideration of the ecological environment, the population that it serves, and the aims and directions of land use in the study area. Among those related factors, the population that a land use type serves could reflect the state of other factors; quantification of the population is the easiest and most viable of the factors. Thus, this study used the population served by a land use type to improve the index (i.e., location entropy). Location entropy is used for

Table 1. The standard to assess different levels of the LGC.

LGC	Less than 0.2	0.2–0.3	0.3–0.4	0.4–0.5	Greater than 0.5
Level	Absolutely equal	Relatively equal	Reasonable	Relatively unequal	Absolutely unequal

sorting in the construction of the traditional Lorenz Curve of LGCSpatial.

The steps of constructing the Improved Lorenz Curve of LGCSpatial were following:

Step 1: Calculation of the related index using the following formula:

$$P_{ji} = \frac{\dfrac{C_{ji}/S_{ji}}{C_i/S_i}}{\dfrac{C_j/S_j}{C/S}} \quad (2)$$

where P_{ji} is the related index of land use i in region j, C_{ji} is the population that land use i serves in region j, S_{ji} is the area of land use i in region j, C_i is the total population that land use i serves in the entire study area, S_i is the area of land use i in that area, C_j is the population in region j, S_j is the area of region j, C is the population of the entire study region, and S is its area. Region j is the subordinate study district of the entire study region.

Step 2: Sorting the index P_{ji} from small to large and calculating the cumulative percentages of the areas of each land use type and the total area.

Step 3: Drawing the Lorenz curve of LGCSpatial based on calculations and sorting results using the cumulative percentage of the total area as the X-coordinate and the cumulative percentage of each type of land use area as the Y-coordinate.

The Gini coefficient calculation in GIS is based on a discrete distribution. Here, we used the method that Yang X et al [32]. proposed to improve and optimize the calculation of the basic LGC to make the calculation tool more suitable for this study.

The calculation process is as follows:

Step 1: The data are divided into different groups, the sum of each group is calculated, and the Lorenz curve is drawn as above.

Step 2: Because the connotations of the LGC and the basic LGCs are extended from the Gini coefficient in economics, and the formulas for calculating the basic LGCs and the interpretations of formulas are introduced from economics, the formula for basic LGC is as follows:

$$G_{LSB} = \frac{A}{A+B} \quad (3)$$

where G_{LSB} is LGCSpatial, A is the area between the data line and the diagonal (i.e., the perfect equality line), and B is the area between the data line and X-coordinate.

After translating formula (3), the formula to calculate G_{LSB} is as follows:

$$G_{LSB} = \sum_{i=1}^{i=1n} x_i y_i + 2 \sum_{i=1}^{i=1n} x_i (1 - s_i) - 1 \quad (4)$$

where n is the code of group, x_i is the percentage of subordinate study districts that the group shares with the entire study region

sum, y_i is the percentage of P-values that the sum of the group shares with the entire study region sum, and s_i is the cumulative percentage of P-values quantified with the formula $s_i = y_1 + y_2 + y_3 + \ldots\ldots + y_i$.

This calculation was developed into a tool using VC++ and MapObjects software. The tool can calculate the basic LGC dynamically, including the dynamic selection of assessment units, dynamic adjustment of the assessment scope and dynamic selection of the assessment index. This tool alters the traditional operation of the database and makes it possible to change the calculation objects by performing operations on the graphical data. The visual graphical data operation renders the calculation of the basic LGC more intuitive and user-friendly. The data for calculating the basic LGC in GIS are the graphical data and all of the indices of each assessment unit at each level. Figure 1 illustrates the computing flow and the connection between the GIS data operations and the basic LGC calculation. Figure 2 shows the operating interface of the LGC calculation tool.

Other basic LGCs can be calculated according to the method for calculating LGCSpatial. After all of the related basic LGCs are attained, the LGC can be calculated using a weighted sum method. To obtain the scientific LGC results, the weight of each basic LGC is determined from the impact of the basic LGCs on the LGC. The formula for this calculation is as follows:

$$G_L = G_{LB1} \times W_1 + G_{LB2} \times W_2 + \cdots + G_{LBn} \times W_n \quad (5)$$

where G_L is the LGC, G_{LB1} to G_{LBn} are the basic LGCs, and W_1 to W_n are the weights of each basic LGC. The weight of each basic LGC can be comprehensively determined by considering the suggestions from experts and the main issues regarding land use structure in China.

Results and Discussions

After calculating and processing the data in ArcGIS 9.3 and using the aforementioned calculation tool, we obtained the results presented in the following sections.

National Level

Results. In this paper, the study area was the Chinese mainland and Hainan Island. This area was divided into four regions–East, Northeast, Middle, and West–according to the four economic development districts. The Eastern district included 10 provinces: Beijing, Tianjin, Hebei, Shandong, Jiangsu, Shanghai, Zhejiang, Fujian, Guangdong, and Hainan. The Northeast district included three provinces: Heilongjiang, Jilin, and Liaoning. The West included twelve provinces based on the Western Development Strategy: Chongqing, Yunnan, Sichuan, Guizhou, Tibet, Guangxi, Xinjiang, Qinghai, Ningxia, Gansu, Shaanxi, and Inner Mongolia. The remaining provinces were in the Middle district [35].

LGCSpatial was calculated using districts and provinces as calculation units; the results are shown in Table 2.

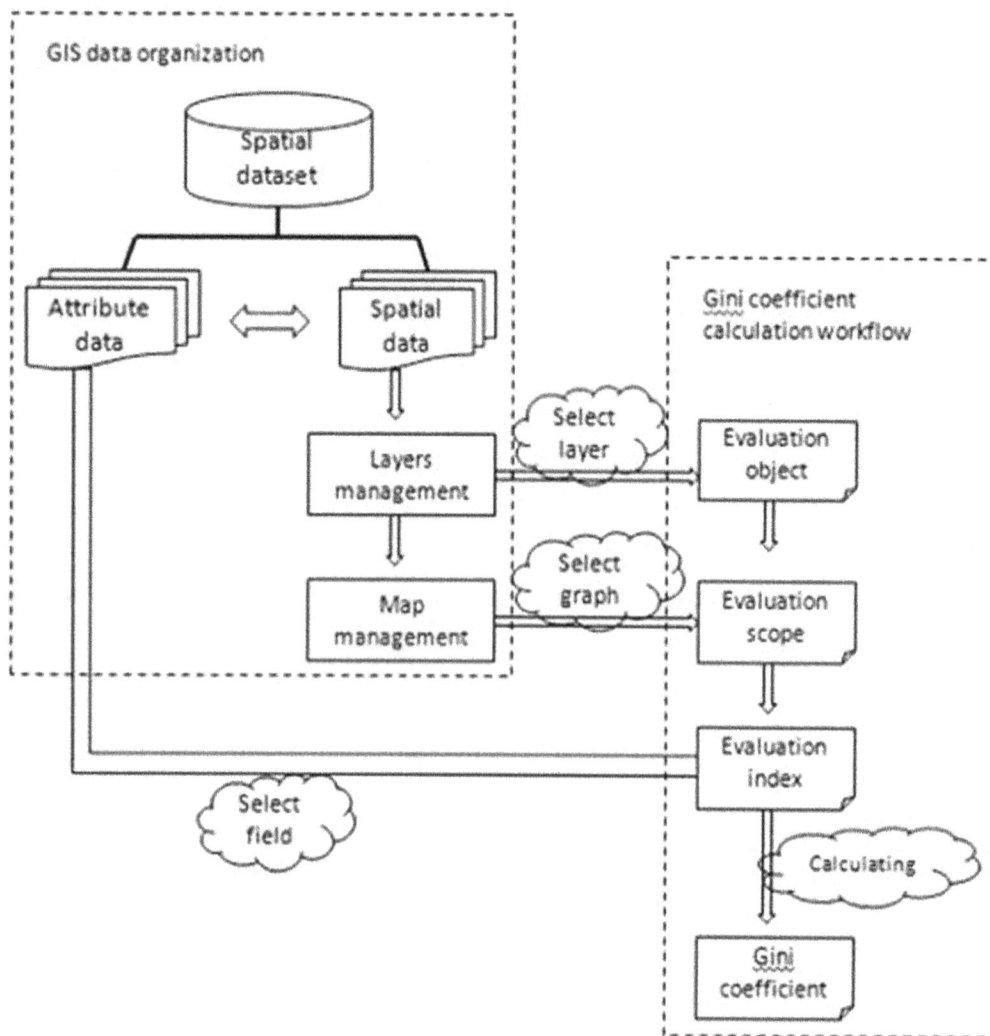

Figure 1. The dynamic calculation flow of the basic LGC based on GIS. Figure 1 illustrates the computing flow and the connection between the GIS data operations and the basic LGC calculation. It shows the dynamic selection of assessment units, dynamic adjustment of the assessment scope and dynamic selection of the assessment index.

As there are some mistakes in the data for 1997 and it is not viable to recalculate statistics, the results for 1997 are treated as noise and are excluded from the analyses.

Discussions. (1) Using a District as the Calculation Unit. The national LGCSpatial calculations using district units are presented in Table 2.

When four districts were used as the calculation units, the farmland and unused land were evenly distributed with LGCSpatial less than 0.2, but the distribution of the built-up land was relatively concentrated with LGCSpatial between 0.4 and 0.5. The trends in the distribution of the three land use types were very similar in that all became increasingly concentrated as LGCSpatial increased. The changes are reasonable, and the change in the farmland was less than the changes in the built-up and unused land. This finding indicates that the farmland decreased at the expense of the occupied built-up land and from destruction caused by disasters, among other activities. Although many activities have been put into practice to supplement farmland, the trend towards the increasingly concentrated distribution of farmland has not ceased. Nevertheless, stringent government policies protecting farmland, especially cultivated land, have had significant results.

For example, the government aims to protect farmland as "18 million hectares of cultivated land" and protect capital farmland. This policy has ensured that the quantity of farmland has not decreased below 18 million hectares and that the quality of farmland (particularly cultivated land) improved over the past 13 years. LGCSpatial reflects the effect of these policies on slow growth. In the 13 years analyzed, considerable manpower and materials have been invested in reclamation projects and development of unused land, such as the Gorges Reservoir Area Fertilizing Project in Chongqing. Such projects have put extensive unused land into rational use and have led to the reutilization of abandoned land. These activities have reduced the amount of unused land and increased its concentration, as indicated by the increasing LGCSpatial statistic. The remaining unused land is difficult to develop into usable land. The trend toward the concentration of built-up land reflects economic development in China. Chinese development strategies include using cities to help rural areas develop and improving the agricultural modernization process (i.e., industrialization). These factors have led to rapid economic improvement and urbanization in the East District. In this region, the area of built-up land has increased greatly, and this

Figure 2. The operating interface of the LGC calculation tool. Figure 2 illustrates the operating interface of the LGC calculation tool. The visual graphical data operation renders the calculation of the basic LGC more intuitive and user-friendly. The data for calculating the basic LGC in GIS are the graphical data and all of the indices of each assessment unit at each level.

land has had an obvious agglomeration effect in rapidly developing districts.

(2) Using a Province as the Calculation Unit.

The national LGCSpatial values using provinces as the calculation unit are shown in Table 2. During the 13 years analyzed, the farmland was very evenly distributed across the country with LGCSpatial less than 0.2, and the distribution of unused land was relatively equal with LGCSpatial between 0.2 and 0.3. However, the distribution of the built-up land was absolutely unequal with LGCSpatial greater than 0.8. The trends in distribution vary with land use types. The farmland became increasingly concentrated as LGCSpatial increased, while the

built-up land and unused land became more equal as LGCSpatial decreased. The built-up land changed the quickest, whereas the unused land changed slowly. These results indicate that at the beginning of the study period, the percentages of farmland area were approximately the same in most provinces, whereas the percentages of built-up land area varied greatly (e.g., in the Zhejiang Province). Meanwhile, the farmland was much more abundant than the built-up land, but the percentage of built-up land increased while the farmland decreased. These changes caused the evenly distributed farmland to become increasingly concentrated, whereas the built-up land spread out. However, the distribution of built-up land remained absolutely unequal.

Table 2. National LGCSpatial results using district and province calculation units.

Year	Calculation Unit	1996	1997	1998	1999	2000	2001	2002	2003	2004	2005	2006	2007	2008
Farmland	District	0.043	0.041	0.044	0.045	0.045	0.045	0.045	0.045	0.047	0.047	0.048	0.048	0.048
	Province	0.125	0.149	0.125	0.126	0.132	0.134	0.137	0.136	0.136	0.137	0.133	0.135	0.136
Built-up land	District	0.462	0.161	0.464	0.461	0.459	0.461	0.469	0.468	0.472	0.472	0.472	0.473	0.473
	Province	0.825	0.287	0.825	0.825	0.825	0.826	0.824	0.826	0.816	0.813	0.813	0.812	0.812
Unused land	District	0.127	0.173	0.136	0.136	0.13	0.132	0.133	0.134	0.135	0.137	0.144	0.145	0.145
	Province	0.249	0.341	0.28	0.281	0.283	0.283	0.281	0.281	0.275	0.276	0.277	0.278	0.278

Adjustment of the structure and distribution of the built-up land will be a major problem requiring government attention. Some provinces, such as Gansu Province, had greater proportions of unused land that could be developed, excluding snow mountains; the development is associated with greater comprehensive benefits compared to other land use types. Furthermore, these provinces performed more research on developing the unused land and used more advanced technology compared with other provinces. This difference led to the uneven distribution of development. The provinces with greater areas of unused land developed more than the other provinces, which evened out the distribution of unused land.

(3) Comparison of Results Using Different Calculation Units.

Comparing the results using the different calculation units listed in Table 2, we observe that LGCSpatial values calculated using province units are larger than those using district units. Therefore, LGCSpatial is higher when the area of the calculation unit is smaller. However, the principal situation and the resultant trend are essentially the same regardless of the calculation unit. This finding shows that LGCSpatial is inversely affected by scale (i.e., it decreases with a larger calculation unit). Comparing these results also reveals that land use activities differ greatly in individual provinces within a district and are more apparent with built-up land and unused land. Therefore, individual provinces within a district are at varying developmental stages and conditions.

Additionally, the results using district units show that the three main land use types are becoming increasingly concentrated. However, when province units were used, the farmland became more concentrated, whereas the built-up land and unused land were reduced. This finding indicates that the development speed is not balanced between the districts. A district at an advanced developmental stage is developing faster. The development speed in a district is balanced between the individual provinces. The underdeveloped provinces develop slightly faster than the provinces at advanced development stages within a district. This observation indicates that the regional economic development strategy to develop the East and Northeast Districts first, followed by the West District, and the Middle District last is well-implemented.

District Level

Results. After performing calculations using province units at the district level, we obtained LGCSpatial of the three main land use types for each study year in the four districts (Figure 3). Figure 4 depicts a sample of the results from 2008, which demonstrates that the LGC can be visualized and portrayed using the ArcGIS platform.

As there are some mistakes in the data for 1997 and it is not viable to recalculate statistics, the results for 1997 are treated as noise and are excluded from analyses.

Analysis and Discussions. The district LGCSpatial values (in province units) reveal that the distributions and trends of the three land use types are very different at the district level.

The results indicate that the LGCSpatial values of built-up land in the West District are larger than 0.70 (Figure 5); therefore, the distribution of built-up land in this region is completely unequal. In the East, these values are between 0.29 and 0.35, indicating that the distribution of built-up land is reasonable. In the Northeast, the LGCSpatial values are between 0.20 and 0.22, which indicate that the distribution of this type of land is relatively equal. In the Middle District, the values are less than 0.2, demonstrating that the distribution is absolutely equal. The trends varied across districts. The built-up land is increasingly concentrated in the East with an approximate 0.05 increase in LGCSpatial, more evenly

distributed in the Middle and West Districts with approximate decreases in LGCSpatial of 0.01 and 0.02, respectively, and stable in the Northeast.

As observed in Figure 6, the LGCSpatial values of unused land in the West District are between 0.3 and 0.4, indicating that the distribution of unused land is reasonable. In the Middle District, the values are between 0.18 and 0.22, suggesting that the distribution of unused land is relatively equal. In the East and Northeast Districts, the values are less than 0.2, indicating an absolutely equal distribution. The trend observed in the East District is unique, as it repeatedly changed: first tending to disperse, then returning to the original level with an approximate LGCSpatial change of 0.03. The Middle District is becoming increasingly concentrated with a LGCSpatial increase of approximately 0.03. The Northeast District is becoming less concentrated with a LGCSpatial decrease of approximately 0.02. The concentration of the distribution of unused land in the West District has slightly increased.

As observed in Figure 7, the distributions and trends of farmland are different than those of built-up and unused land. The LGCSpatial values of farmland are less than 0.2 in every district, indicating that farmland is absolutely equally distributed in all districts. In the Northeast, Middle and West Districts, farmland remained virtually stable, while in the East, it remained stable with slight fluctuations.

The most important reasons for these distributions are the environmental conditions and population distributions, but the economy and policies also play roles.

Concerning built-up land, the West District is associated with a larger area, smaller population and worse eco-environmental conditions than other districts; the population and suitable areas for built-up land are very concentrated, and the need for built-up land is lower than that in other districts (thus, any built-up land in the West District would be concentrated). However, the associated LGCSpatial value is larger than 0.70 (i.e., absolutely unequal); the excessive concentration does not support sustainable development and requires improvement. The most developed district is the East. Because the built-up land has an agglomeration effect, the distribution of this type of land is reasonable. The policy is to develop the East and West first and the Middle District last. The development of the Northeast Dis faster and more stable than that of the Middle District. The mountains and hills of the Middle District, unlike the other districts, make it difficult to concentrate built-up land. The development of the West District is not sufficient to cause an agglomeration effect of the built-up land. Overall, these factors have caused the observed trends in built-up land.

Within farmland, crop types depend on local geography, climate and other eco-environmental conditions; methods for crop production meet local needs. The production of all crops in each district is essential; therefore, the distributions and trends of farmland are inherently rational. The East is the most developed district, the West has the most severe eco-environment conditions, and the Middle and Northeast Districts have moderate development; it is reasonable that the LGCSpatial values of the East and West Districts are higher than those of the Middle and Northeast Districts.

Many types of unused land, such as snow mountains, are mainly centralized and distributed throughout the West District, whereas extreme terrain is sparse in other districts. Therefore, the distribution of unused land in the West District is more concentrated, along with larger LGCSpatial values. Given the limitations imposed by the eco-environmental conditions, technology, and the shortage of economic development, the develop-

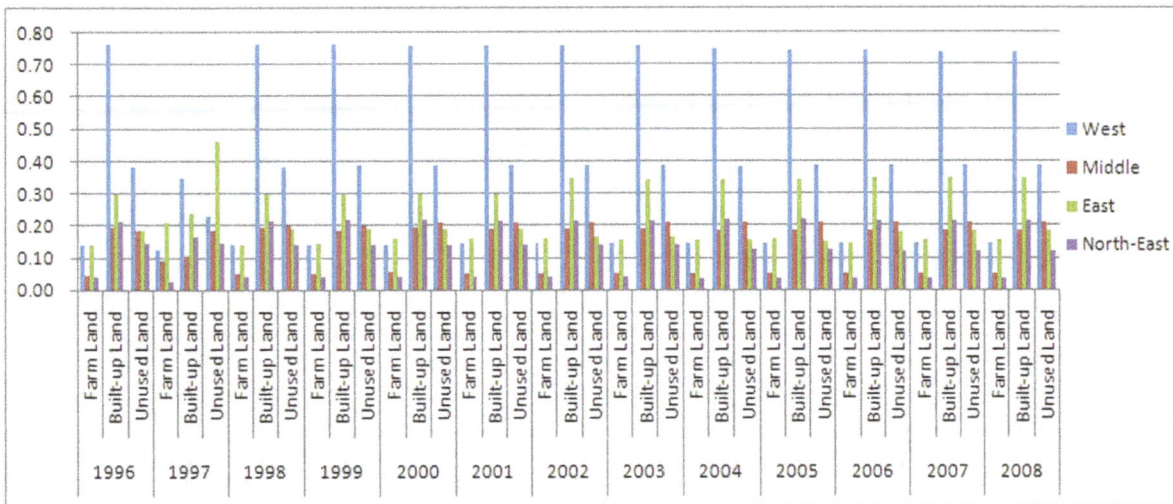

Figure 3. District LGCSpatial in province units. Figure 3 is the LGCSpatial of the three main land use types for each study year in the four districts using province units at the district level.

ment of unused land is difficult to implement in the West. Because only a small area of unused land is developed, its distribution in the West is slightly centralized. The distribution and trend of unused land in the East are mainly influenced by economic development. As the East District is the most developed and the eco-

environmental conditions are conducive for developing unused land, developing large areas of this land has an agglomeration effect. Therefore, development tends to disperse, and the LGCSpatial value decreases with the development of large areas of unused land in the East District, which is eventually utilized in

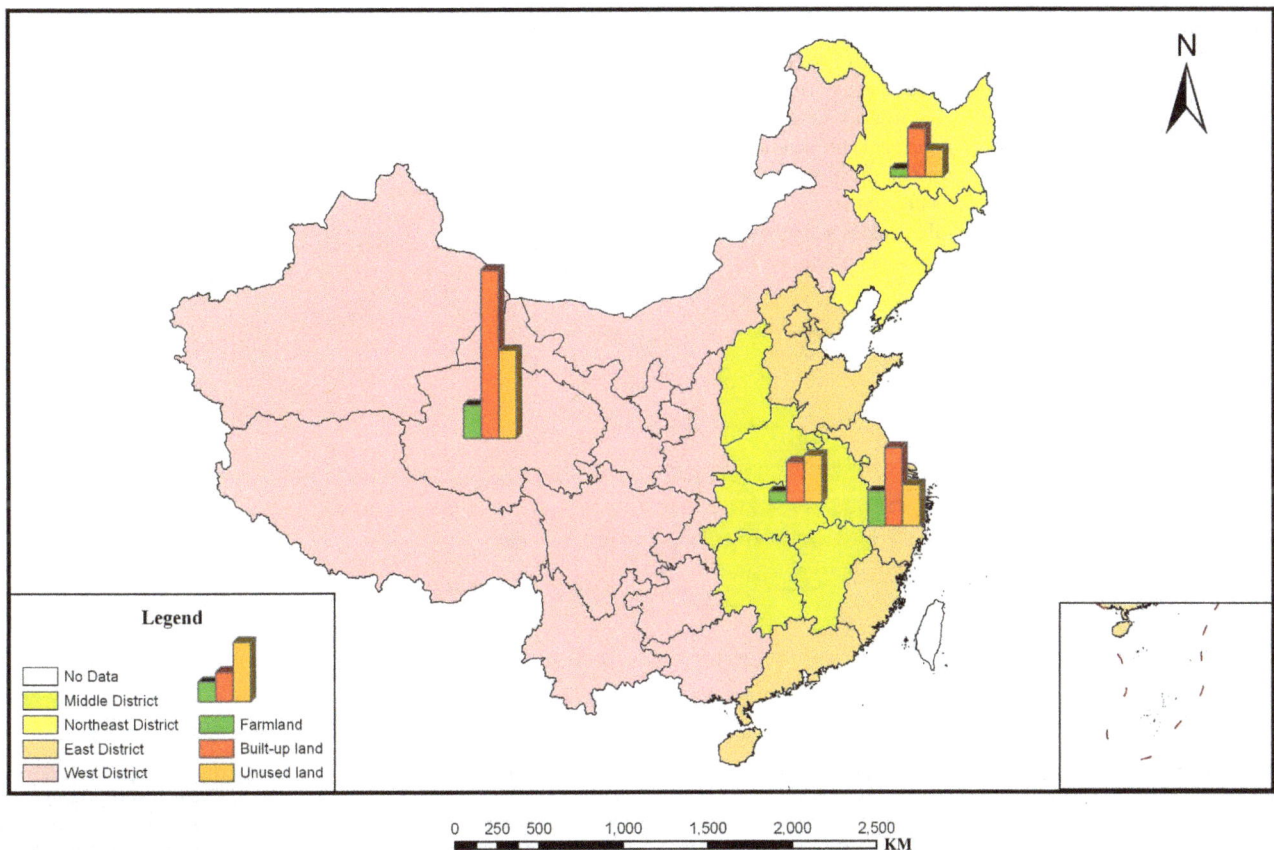

Figure 4. District LGCSpatial of the three main land use types in 2008. Figure 4 depicts a sample of the results from 2008, which demonstrates that the LGC can be visualized and portrayed using the ArcGIS platform.

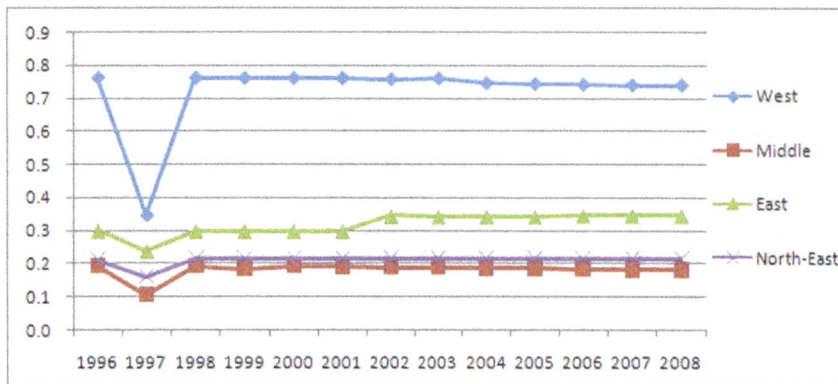

Figure 5. The district LGCSpatial values and associated trends of built-up land. The results indicate that the LGCSpatial values of built-up land in the West District are larger than 0.70; therefore, the distribution of built-up land in this region is completely unequal. In the East, these values are between 0.29 and 0.35, indicating that the distribution of built-up land is reasonable. In the Northeast, the LGCSpatial values are between 0.20 and 0.22, which indicate that the distribution of this type of land is relatively equal. In the Middle District, the values are less than 0.2, demonstrating that the distribution is absolutely equal. The trends varied across districts. The built-up land is increasingly concentrated in the East with an approximate 0.05 increase in LGCSpatial, more evenly distributed in the Middle and West Districts with approximate decreases in LGCSpatial of 0.01 and 0.02, respectively, and stable in the Northeast.

the first stage. The remaining unused land is smaller in scale, and its development causes an increasingly centralized distribution with an increasing LGCSpatial value. In the Middle and Northeast Districts, the distributions and trends of unused land are determined by eco-environmental and economic factors.

Inspiration of LGC for Improving the Land Use Spatial Structure

The LGCSpatial results indicate that the built-up land is sufficiently concentrated. We suggest that the government take the necessary steps to stop or slow down the transformation of built-up land distribution. Built-up land, at a certain centralization level, is beneficial for social development, but if the distribution is too concentrated, then it has a negative effect on society.

The distribution of farmland should remain stable considering its present concentration. Farmland production varies by region

and local conditions; thus, the LGCSpatial values of farmland should be at an absolutely equal level (i.e., LGCSpatial less than 0.2).

The distribution of unused land should remain unchanged or become slightly concentrated to maintain reasonable proportions and ensure its protection.

Conclusions

This paper presents the concept of the LGC and basic LGCs. The GIS-based LGC calculation tool can efficiently calculate the LGC and visualize the results for simplified comprehension. We calculated and analyzed land use data from 1996 to 2008 using two different calculation units for various land use types in China. We discerned the temporal and spatial changes of LGCSpatial values and analyzed the associated policies and economic backgrounds. Based on this research, suggestions on land use

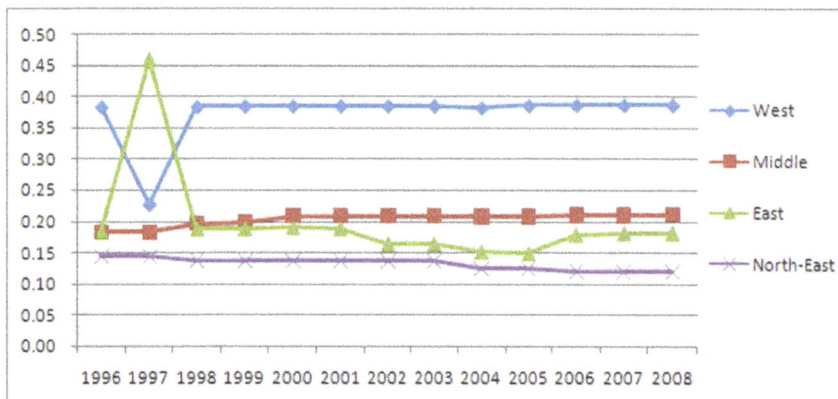

Figure 6. The district LGCSpatial values and trends of the unused land. The LGCSpatial values of unused land in the West District are between 0.3 and 0.4, indicating that the distribution of unused land is reasonable. In the Middle District, the values are between 0.18 and 0.22, suggesting that the distribution of unused land is relatively equal. In the East and Northeast Districts, the values are less than 0.2, indicating an absolutely equal distribution. The trend observed in the East District is unique, as it repeatedly changed: first tending to disperse, then returning to the original level with an approximate LGCSpatial change of 0.03. The Middle District is becoming increasingly concentrated with a LGCSpatial increase of approximately 0.03. The Northeast District is becoming less concentrated with a LGCSpatial decrease of approximately 0.02. The concentration of the distribution of unused land in the West District has slightly increased.

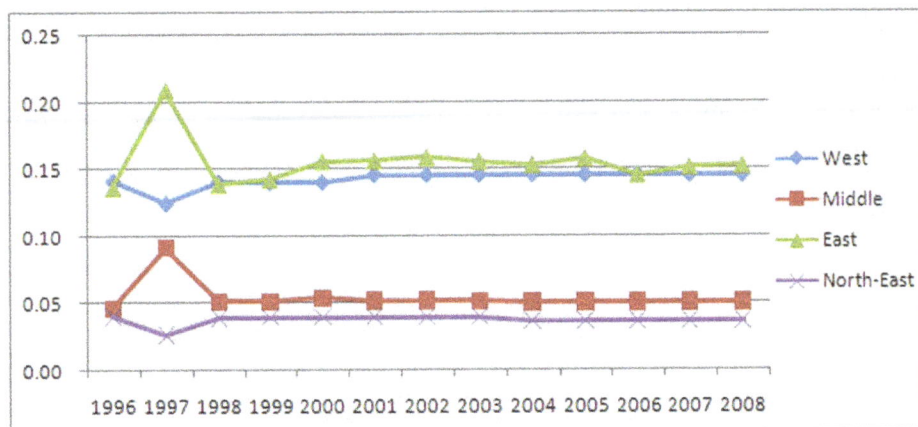

Figure 7. The district LGCSpatial values and trends of the farmland. The distributions and trends of farmland are different than those of built-up and unused land. The LGCSpatial values of farmland are less than 0.2 in every district, indicating that farmland is absolutely equally distributed in all districts. In the Northeast, Middle and West Districts, farmland remained virtually stable, while in the East, it remained stable with slight fluctuations.

structure optimization using the LGC were presented. The technology and methodology reported can help to analyze and optimize land use structure, including spatial distribution and temporal arrangement. This work indicates that it is necessary to incorporate the Gini coefficient into land use research and that this coefficient will be helpful in related research.

We have also shown that LGCSpatial values are affected by scale. Use of differing calculation units yields varying LGCSpatial results. The LGCSpatial values increase when the scale of the calculation unit decreases.

Based on our method and calculation tool using the LGC to analyze land use structure, cities and counties can be used as calculation units in research. This method can be used to analyze the distribution of each land use type at a smaller scale and elicit more specific and effective suggestions for optimizing land use structure. We could also calculate and examine the LGC of the first, second and third levels of land use types, rather than the first level only, to optimize the land use structure (i.e., more detailed data). Boundaries are very important for analyzing the results and should be varied based on the calculation unit; future research may consider this aspect. Additionally, research on other basic LGCs will gradually be put into practice to improve this method and calculation tool.

Author Contributions

Conceived and designed the experiments: XQZ TX XY TY YCH. Performed the experiments: XQZ TX XY. Analyzed the data: XQZ TX. Contributed reagents/materials/analysis tools: XQZ TX XY TY YCH. Wrote the paper: XQZ TX XY.

References

1. Herold M, Couclelis H, Clarke CK (2005). The role of spatial metrics in the analysis and modeling of urban land use change. Coumputers, Environment and Urban Systems, 29: 369–399.
2. Lu CY, Yang QY, Qi DX (2010) Research progress and prospects of the researches on urban land use structure in China. Progress in Geography 29: 861–868.
3. Xu XQ, Zhou YX, Ning YM (1997) Urban geography. Beijing: High Education Press.
4. Thinh NX, Arlt G, Heber B, Hennersdorf J, Lehmann I (2002) Evaluation of urban land-use structures with a view to sustainable development. Environmental Impact Assessment Review 22: 475–492.
5. Chuai XW, Huang XJ, Lai L, Wang WJ, Peng JW, et al. (2013) Land use structure optimization based on carbon storage in several regional terrestrial ecosystems across China. Environmental Science & Policy 25: 50–61.
6. Pan JH; Shi PJ; Zhao RF (2010) Research on optimal allocation model of land use structure based on LP-MCDM-CA Model:the case of Tianshui. Journal of Mountain Science 28: 407–414.
7. Miao ZH, Chen Y, Zeng XY (2011) CA Model of optimization allocation for land use spatial structure based on Genetic Algorithm. Taiyuan, China. Second International Conference of Artificial Intelligence and Computational Intelligence 7002: 671–678.
8. McDonald RI, Forman RTT, Kareiva P (2010) Open space loss and land inequality in united states' cities, 1990–2000. Plos One 5: e9509.
9. Wang T, Lv CH (2010) Quantitative structural analysis on land use change in Beijing-Tianjin-Hebei Region. Journal of Shanxi University(Nat. Sci. Ed.) 33: 473–478.
10. Hu CR, Jiang D, Tang X, Zhang LY, Liu YL (2009) Sampling analysis of national urban land use status based on Lorentz Curves. China Land Science 23: 44–50.
11. Barr LM, Pressey RL, Fuller RA, Segan DB, McDonald-Madden E, et al. (2011) A new way to measure the world's protected area coverage. Plos One 6: e24707.
12. Fang LN, Song JP, Yue XY (2009) Analysis of Land Use Structure in Urban Fringe Area: A Case Study in Daxing District of Beijing. Ecological Economy 2: 329–334.
13. Sun YJ, Zhang YC, Li P (2011) Study on the reasonable land use evaluation under the background of low carbon. Areal Research and Development 30: 93–96, 117.
14. Tan YZ, Wu CF (2003) The laws of the information entropy values of land use composition. Journal of Natural Resources 18: 112–117.
15. Wang R (2006) Study on optimization of land use structure based on ecological footprint in Huanghe mouth delta area: A case of Dongying. Jinan, Shandong, People's Republic of China: Shandong Normal University.
16. Yitzhaki S (1979) Relative deprivation and the Gini Coefficient. The Quarterly Journal of Economics 93: 321–324.
17. Zheng XQ, Sun YJ, Fu MC, Hu X (2008) Study on the analytical method of rationality of urban construction land structure. China Land Science 22: 4–10.
18. Norheim OF (2010) Gini impact analysis: measuring pure health inequity before and after interventions. Public Health Ethics 3: 282–292.
19. Liu Y, Xie M, Ding Y (2004) The comparison and thinking on the method of Gini coefficient. Statistics and Decision 9: 15–16.
20. Kamanga P, Vedeld P, Sjaastad E (2009) Forest incomes and rural livelihoods in Chiradzulu District,Malawi. Ecological Economics 68: 613–624.
21. Li J, Feldman MW, Li SZ, Daily GC (2011) Rural household income and inequality under the sloping land conversion program in western China. Proceedings of the National Academy of Sciences of the United States of America 108: 7721–7726.
22. Lee W (1996) Analysis of seasonal data using the Lorenz Curve and the associated Gini index. International Journal of Epidemiology 25: 426–434.
23. Matsumoto M, Inoue K, Noguchi S, Toyokawa S, Kajii E (2009) Community characteristics that attract physicians in japan: a cross-sectional analysis of community demographic and economic factors. Human Resources for Health 7. DOI: 10.1186/1478-4491-7-12.

24. Martinez E, Santelices B (1992) Size hierarchy and the -3/2 "power law" relationship in a coalescent seaweed. Journal of Phycology 28: 259–264.

25. Jurik TW (1991) Population distributions of plant size and light environment of giant ragweed (ambrosia trifida l.) at three density. Oecologia 87: 539–550.

26. He ZX, Ma Z, Brown KM, Lynch JP (2005) Assessment of inequality of root hair density in arabidopsis thaliana using the gini coefficient: a close look at the effect of phosphorus and its interaction with ethylene. Annals of Botany-London 95: 287–293.

27. Steinberger JK, Krausmann F, Eisenmenger N (2010) Global patterns of materials use: a socioeconomic and geophysical analysis. Ecological Economics 69: 1148–1158.

28. White TJ (2007) Sharing resources: The global distribution of the Ecological Footprint. Ecological Economics 64: 402–410.

29. Wang J, Guo W, Chen J, Sheng Y (2011) Study on the allocation of the water pollutants in the basin by the Gini coefficient method. Xi'an, China. International Symposium on Water Resource and Environmental Protection: 798–801.

30. Sun T, Zhang HW, Wang Y, Meng XM, Wang CW (2010) The application of environmental Gini coefficient (EGC) in allocating wastewater discharge permit: the case study of watershed total mass control in Tianjin, China. Resources, Conservation and Recycling 54: 601–608.

31. Wang ML, Luo B, Zhou WB, Huang Y, Liu L (2011) The application of Gini Coefficient in the total load allocation in Jinjiang river, China. Xi'an, China. International Symposium on Water Resource and Environmental Protection: 992–995.

32. Yang X, Li ZJ, Zheng XQ (2008) Dynamic calculation method of Gini coefficient based on GIS. Available: http://www.paper.edu.cn/index.php/default/releasepaper/content/200808-113. Accessed 2013 Aug 25.

33. Ministry of Public Security Administration(1997–2009)Demographic Data for Counties in China, Public Press.(In Chinese)

34. Nanning Municipal Bureau of Land and Resources (2007) "Land use classification" and "National Land Classification" correspondence table. Available: http://www.nnland.gov.cn/show.aspx?id = 1050&cid = 47. Accessed 2013 Aug 20.

35. The Central People's Government of the People's Republic of China (2011) "Two Strategies" constitute the strategy of territory development. Available: http://www.gov.cn/wszb/zhibo453/content_1879391.htm. Accessed 2013 Aug 15.

18

Effects of Infection on Honey Bee Population Dynamics: A Model

Matt I. Betti[1], Lindi M. Wahl[1], Mair Zamir[1,2]*

1 Department of Applied Mathematics, Western University, London, Ontario, Canada, **2** Department of Medical Biophysics, Western University, London, Ontario, Canada

Abstract

We propose a model that combines the dynamics of the spread of disease within a bee colony with the underlying demographic dynamics of the colony to determine the ultimate fate of the colony under different scenarios. The model suggests that key factors in the survival or collapse of a honey bee colony in the face of an infection are the rate of transmission of the infection and the disease-induced death rate. An increase in the disease-induced death rate, which can be thought of as an increase in the severity of the disease, may actually help the colony overcome the disease and survive through winter. By contrast, an increase in the transmission rate, which means that bees are being infected at an earlier age, has a drastic deleterious effect. Another important finding relates to the timing of infection in relation to the onset of winter, indicating that in a time interval of approximately 20 days before the onset of winter the colony is most affected by the onset of infection. The results suggest further that the age of recruitment of hive bees to foraging duties is a good early marker for the survival or collapse of a honey bee colony in the face of infection, which is consistent with experimental evidence but the model provides insight into the underlying mechanisms. The most important result of the study is a clear distinction between an exposure of the honey bee colony to an environmental hazard such as pesticides or insecticides, or an exposure to an infectious disease. The results indicate unequivocally that in the scenarios that we have examined, and perhaps more generally, an infectious disease is far more hazardous to the survival of a bee colony than an environmental hazard that causes an equal death rate in foraging bees.

Editor: Olav Rueppell, University of North Carolina, Greensboro, United States of America

Funding: The work was funded by the Natural Sciences and Engineering Research Council of Canada. The funder had no role in study design, data collection and analysis, decision to publish, or preparation of the manuscript.

Competing Interests: The authors have declared that no competing interests exist.

* Email: zamir@uwo.ca

Introduction

The widespread collapse of honey bee colonies has been the subject of much discussion and research in recent years [1–3]. Aside from their ecological importance [4], honey bee populations have a large economical impact on agriculture in North America, Europe, the Middle East, and Japan [5–7].

The focus of research has been largely on environmental factors outside the hive, such as pesticides or insecticides, which may cause death or injury to foraging bees and jeopardize their return to the hive. The reduced number of foraging bees then leads to younger hive bees being recruited prematurely to perform foraging duties and this chain reaction ultimately leads to a disruption in the dynamics of the colony as a whole. Examples of this scenario would be produced by the effects of various pesticides to which foraging bees are exposed in the course of their duties [2, 8]. Other factors in the same category include possible disruptions to the bees' navigation system by mobile phones or other electronic devices, again to the effect of jeopardizing their return to the hive and thereby reducing their numbers [9].

A key element in this category of disruption to honey bee population dynamics is the untimely *death* of a certain proportion of foraging bees outside the hive and the consequences of this on the colony as a whole. An important question here concerns the threshold in the death rate of foraging bees that would determine the survival or collapse of the bee colony. This was examined recently in two papers by Khoury et al. [10, 11].

In the present paper we consider a different category of disruption to the healthy dynamics of a bee colony, namely one in which the key hazard is an *infection* by a communicable disease acquired by foraging bees outside the hive. The key difference here is that foraging bees that have been infected would then transport the disease into the hive and go on to infect other members of the colony *within the hive*. Here too the affected bees will ultimately suffer an untimely death, but the effects on the dynamics of the colony are clearly more complex because the infection in this case may now involve all members of the colony. We sought a model that would allow a comparison between the effects of these two categories of hazards (pesticide versus infection) on the ultimate fate of the bee colony.

Disease in honey bee colonies has been studied previously by Sumpter et al. [12] who modeled the effects of Varroa mites on the brood and on the adult worker bees. The focus of the model was on the relationship between the mite population within a hive and its role in virus transmission within the hive. A study by Ratti et al. [13] examined the transmission of viruses via Varroa mites, using an SIR-framework with the mites as vectors for transmission.

In the present paper we propose a more general model which combines the normal dynamics of a honey bee colony with the dynamics of an infectious disease which is acquired outside the hive but ultimately spreads to the rest of the colony. As a working example, we use a disease known as "Nosema" which is a common disease affecting both hive bees and foraging bees [14]. Nosema is caused by a microsporidian parasite with two common strains: *Nosema ceranae* and *Nosema apis*. The former was first discovered in Asian honey bees (*Apis ceranae*) and the latter is common among European honey bees (*Apis mellifera*). A key factor in the collapse of honey bee colonies in recent years is thought to be the introduction of *Nosema ceranae* to *Apis mellifera* [15].

The main aim of the model is to provide a general tool for determining the ultimate fate of a honey bee colony under this fairly common hazard. In particular, we identify key variables that determine the collapse or survival of the bee colony, namely the severity of the disease and the rate of transmission, and examine different scenarios using different combinations of these variables. Winter is an important phase in the normal demographic dynamics of a bee colony; the queen lays fewer eggs and foraging bees return to and remain within the hive [16, 17]. Therefore, the time interval between the onset of disease and the onset of winter may play a critical role in the ultimate survival or collapse of the colony in the face of an infection. We show that the model can be used to explore potential markers of the presence of the disease within the bee colony and of the ultimate fate of the colony under different scenarios.

Background

2.1 Normal Demographics of a Honey Bee Colony

Honey bee colonies are complex societies in which different members of the colony have specialized functions that serve the entire colony, thus making members of the colony highly dependent on each other.

The queen can live up to three years, is responsible for laying eggs, and during peak season may lay up to 2000 eggs per day [18]. In this function the queen is dependent on worker bees [19]. The worker bees emerge from fertilized eggs of the queen and consist of females who maintain the hive and gather resources, and males who mate with the queen to produce more eggs [20]. Drones are born from unfertilized eggs of the queen [20] and typically making up less than 5% of the hive population [20, 21]. Because they do not contribute to the colony work force, and because of their small numbers, they are generally neglected when considering the dynamics of the colony as a whole.

Female hive bees, following a transition period, leave the hive to start foraging duties and usually forage until their death. The age at which they start foraging duties is variable, depending on the state of the colony and its needs. If the number of forager bees is lower than is required for meeting the colony needs, hive bees will begin foraging duties at a younger age [22]. If the number of forager bees is higher than required, behavioural maturation of hive bees will be regulated by a pheromone, ethyl oleate, produced by the foragers. This process is usually referred to as "social inhibition" [23]. Similarly, if the number of hive bees is too low, it is possible for foragers to revert back to hive bee duties [22].

As the temperature drops outside the hive, foraging becomes less frequent, the queen begins to lay fewer eggs [19], and drones are expelled from the hive to save hive resources [20]. When the temperature drops below a certain threshold, the colony enters a winter phase in which the queen will cease to lay eggs [16] and any remaining foraging bees will return to the hive. During winter the entire hive population surrounds the queen in order to maintain a temperature of 34–36°C within the hive [20].

2.2 Nosema Infection

Nosema, also known as "Nosemosis", is an infection affecting honey bees that is spread by the microsporidian parasites in the Nosema family. *Nosema ceranae* is of particular interest, as it is thought to be linked to colony collapse incidents [15, 24]. We use this disease only as an example to illustrate the utility of the model. The choice was motivated by the availability of parameter values which allowed us to examine some realistic scenarios of the dynamics of the bee colony in the presence of infection.

Within the bee colony, Nosema is typically spread via fecal-oral transmission. Adult bees will contract Nosema either from eating food contaminated by infected bees, or while ridding the hive of infected fecal matter [25]. There is also evidence that Nosema can be spread via oral-oral transmission, through feeding [26].

While it is typically asymptomatic at the level of individual bees, Nosema has some symptoms that can be observed at the colony level [14, 27]. Stevanovic et al. [27] observed in 2013 that colonies infected by the parasite *Nosema ceranae* exhibited many of the classic signs that precede colony collapse.

Much of the experimental research linking Nosema infection to colony collapse is based on correlated observations, but direct cause and effect evidence is lacking [14]. Our model aims to provide a possible mechanism for this linkage in terms of the interplay between the dynamics of the infection and the normal dynamics of the honey bee colony.

Mathematical Model

In what follows we present a mathematical model that combines the normal demographic dynamics of a honey bee colony with the dynamics of an infection affecting foraging bees outside the hive at first and then spreading to the rest of the colony. We follow a model for the basic dynamics of a bee colony in the absence of disease presented recently by Khoury et al. [10, 11], in which the adult bee population is divided into a number of hive bees H, and a number of foraging bees F. In the model to be described below we extend this division into four categories, namely susceptible hive bees H_S, infected hive bees H_I, susceptible foraging bees F_S, and infected foraging bees F_I. Equations governing each of these four populations during the active and winter seasons are presented in the following section.

3.1 Governing Equations: Active Season

The rate of change in time t (days) of the susceptible hive bee population H_S during the active season is assumed to be governed by

$$\frac{dH_S}{dt} = LS - H_S R - (\beta_{HH} H_I + \beta_{HF} F_I) H_S. \quad (1)$$

In the first term on the right L is the queen's egg laying rate per day and S is the proportion of those eggs that survive both larval and pupal stages to yield mature bees. This proportion is a function of the total number of hive bees and of the amount of food f available within the hive because the brood requires food as well as a sufficient number of supporting hive bees in order to survive [28]. Following [11] we take

$$S = \left(\frac{H_S + H_I}{w + H_S + H_I}\right)\left(\frac{f}{b + f}\right). \quad (2)$$

This function is constructed such that the value of S saturates at 1.0 in the limiting case when the amount of food f and the total number of hive bees $H_S + H_I$ are sufficiently large to ensure the survival of 100% of the eggs laid by the queen. The parameters b and w determine at what values of f and $H_S + H_I$ this saturation occurs and they will be discussed later.

In the second term on the right of Eq. 1, R is the proportion of maturing hive bees H_S that are being recruited to foraging duties. As discussed earlier, and following [11], we assume that recruitment is increased when either food stores or forager populations are low and recruitment is reduced when food stores and forager populations are in excess. Note that in an overabundance of foragers, R may become negative, which implies that foragers are reverting to hive duties.

$$R = R_b + \alpha_f \left(\frac{b}{b+f} \right) - \alpha_F \left(\frac{F_I + F_S}{N} \right) \qquad (3)$$

where R_b is the baseline recruitment rate in the absence of foragers but sufficient food stores, α_f is a weighting of the effect of low food, α_F is a weighting of the effect of excess foragers on recruitment, and $N = F_I + F_S + H_S + H_I$ is the colony adult population size. The Average Age of Recruitment to Foraging (AARF) at any point in time is equal to $1/R$.

The last term in Eq. 1 determines the rate at which susceptible hive bees become infected. The transmission rate per day per susceptible hive bee is given by $(\beta_{HH} H_I + \beta_{HF} F_I)$, where β_{HH} is the contact rate between hive bees and β_{HF} is that between hive bees and foraging bees.

Hive bees are safe within the hive environment under normal circumstances, surviving up to 6 months over winter [10, 20]. It is therefore assumed that the natural death rate of hive bees is negligible compared to their recruitment rate to foraging duties.

For the rate of change of the infected hive bee population, we take

$$\frac{dH_I}{dt} = (\beta_{HH} H_I + \beta_{HF} F_I)H_S - H_I R - d_H H_I. \qquad (4)$$

Infected hive bees continue to be recruited to foraging duties but, unlike their healthy counterparts, they are at risk of dying from the disease before they do so; d_H is the rate at which this occurs.

Susceptible foragers are recruited from susceptible hive bees and may subsequently suffer natural death, at a rate m, or become infected. Their rate of change is therefore governed by

$$\frac{dF_S}{dt} = H_S R - mF_S - (\beta_{HF} H_I + \beta_{FF} F_I)F_S. \qquad (5)$$

Infected foragers are recruited from infected hive bees or are susceptible foragers that have become infected. If the death rate from the infection is assumed to be d_F then their rate of change is governed by

$$\frac{dF_I}{dt} = H_I R + (\beta_{HF} H_I + \beta_{FF} F_I)F_S - (m + d_F)F_I. \qquad (6)$$

Food is brought into the hive by foragers, either healthy or infected. Although infected foragers may forage less efficiently, for

simplicity we assume the same foraging rate, c (gm/day) per forager. The collected food is then consumed by both foragers and hive bees and for simplicity again we assume the same consumption rate, γ_A (gm/day). The amount of food consumed by the larvae is substantial. We assume that the number of larvae is proportional to the number of surviving eggs and that the larvae consume food at a rate of γ_L (gm/day). The amount of food available at time t is thus given by

$$\frac{df}{dt} = c(F_S + F_I) - \gamma_A N - \gamma_L LS. \qquad (7)$$

The full dynamics of the bee colony are thus governed by Eqs. 1, 4, 5, 6 and 7 to be solved simultaneously. A compartmental diagram of these dynamics is shown in Figure 1.

3.2 Governing Equations: Winter

During winter the rate of egg laying by the queen is considerably diminished, and in harsh climates the queen may cease laying eggs completely [16]. For simplicity, in our model simulations therefore we take $L = 0$ for the winter season.

Foraging resources become scarce in winter and foraging bees return to the hive to join hive bees in their effort to keep the hive warm [16]. The two groups thus perform the same duties in winter and there is no longer any recruitment from hive to foraging duties. We therefore set $R = 0$, although we maintain the separate identities of the two groups in the model in order to track the behaviors of bees that were foraging before winter against those that were hive bees.

Since there is no foraging in winter, food production halts and we set $c = 0$. Also, bees are able to survive longer in winter than they do outside the hive during the active season [29]. Thus the new natural death rate for both hive bees and foraging bees during the winter season is set to be m_W.

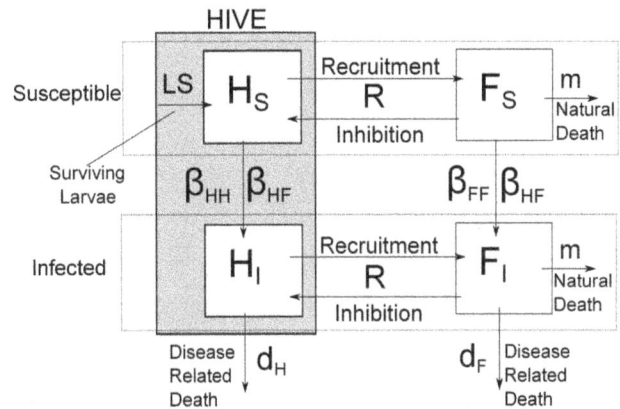

Figure 1. A compartmental diagram of the dynamics of the honey bee colony combined with the dynamics of an infectious disease. The susceptible and infected hive bees, H_S and H_I live within the hive. New susceptible hive bees are generated by surviving brood through the survival function, S. New infected hive bees are generated through interactions of susceptible hive bees with infected hive bees and infected foragers at rates β_{HH} and β_{HF}. Hive bees are recruited to foraging duties through the recruitment function R, which also allows for the reversal of duties, from foraging to hive duties. Foragers move into the infected compartment via interactions with infected hive bees and infected foragers at rates β_{HF} and β_{FF}. All infected bees die at rates d_H or d_F, and foragers die naturally at rate m.

Introducing these changes into the equations governing the dynamics of the colony (Eqs. 1, 4, 5, 6, 7) we obtain the corresponding equations for the winter season:

$$\frac{dH_S}{dt} = -m_W H_S - (\beta_{HH} H_I + \beta_{HF} F_I) H_S \tag{8}$$

$$\frac{dF_S}{dt} = -m_W F_S - (\beta_{HF} H_I + \beta_{FF} F_I) F_S \tag{9}$$

$$\frac{dH_I}{dt} = (\beta_{HH} H_I + \beta_{HF} F_I) H_S - (m_W + d_H) H_I \tag{10}$$

$$\frac{dF_I}{dt} = (\beta_{HF} H_I + \beta_{FF} F_I) F_S - (m_W + d_F) F_I \tag{11}$$

$$\frac{df}{dt} = -\gamma_A N - \gamma_L L S \tag{12}$$

3.3 Parameter Values

The model presented in Section 3 contains a total of 13 parameters. Of these, 10 parameters relate to the baseline demographic dynamics of a honey bee colony, in the absence of disease, for which empirical estimates are available in the literature. In particular, we consider a bee colony in which the maximum rate (L in Eq. 2) of egg laying by the queen is 2000 eggs/day and take $w = 5000$ [10]. Hive bees spend, on average, a minimum of 4 days in the hive before being recruited to foraging duties [30], and foragers will not revert to hive duties unless one-third of the bee population is foraging [10]. Based on these values, and following [10], we take $R_b = 0.25$ and $\alpha_f = 0.75$. In the complete absence of food, recruitment of foragers will double [31], thus we take $\alpha_F = 0.25$. Foraging bees are estimated to live approximately 6.76 days outside the hive [32], thus we set $m = 0.14$ deaths per bee per day.

The parameter b in Eq. 2 is the amount of food required to ensure the survival of half of the eggs to maturation. Based on the observation that the effects of low food stores become evident when there is less than 1 kg of stored food [10], we take $b = 500$. It is estimated that as long as the hive is in an environment that provides sufficient food resources, a forager will return with $c = 0.1$ g of food per day [33, 34]. It is also estimated that the daily food requirement of each member of the brood is $\gamma_L = 0.018$ g and that of an adult hive or foraging bee is $\gamma_A = 0.007$ g [10, 11, 34].

We assume that both the rate of food consumption and the transmission rate of the disease remain the same during the active and winter seasons. However, empirical evidence indicates that bees live longer in winter, surviving up to six months [29], and on that basis we take the natural death rate in winter, $m_W = 1/180$ deaths per bee per day.

The remaining parameters relate to the dynamics of the disease and, as stated earlier, we have chosen *Nosema ceranae* particularly because of the availability of parameter values. The effect of *Nosema ceranae* infection is estimated to double the mortality rate of adult bees [35]. On that basis we take $d_H = d_F = m = 0.14$ deaths per bee per day. For the rates of transmission at first we considered different values of β_{HH}, β_{HF}, β_{FF}. Following some preliminary

simulations, however, we found these different values have only a marginal qualitative effect on the overall dynamics of the disease. Accordingly, and in the absence of any field values on which to base a meaningful examination of this issue, the simulations which we present in this paper are based on taking $\beta_{HH} = \beta_{HF} = \beta_{FF} = \beta$. Generally, transmission of the disease is mediated via the food stores [26], which makes it difficult in practice to measure the rate of transmission from an infected bee to a susceptible bee.

A summary of all the parameter values we used is provided in Table 1.

Results

In what follows we present the results of numerical simulations of key scenarios that illustrate the main dynamics of the bee colony in the presence of disease.

To simulate the dynamics of the bee colony, we integrate the governing equations (Eqs. 1, 4, 5, 6, 7) numerically, with initial conditions $H_I(0) = F_I(0) = 0$ and $H_S(0)$, $F_S(0)$ based on steady state values for the disease free equilibrium which can be determined analytically. The food stores, f, continue to grow throughout the active season, and we have found that the results are not sensitive to the initial value of food in the hive. We present scenarios in which the dynamics of the disease begin at day 100. The initial onset of infection is simulated by turning 10% of the susceptible foragers into infected foragers.

Scenario 0

In this scenario we illustrate the baseline demographic dynamics of the colony in the absence of disease, particularly to highlight the natural seasonal variations. Thus, for this purpose, in this case we introduce winter after the initial 100 days of integration. The results are shown in Figure 2. The figure shows that both the hive and the foraging bee populations decrease (from natural death) over winter, but sufficient numbers remain (because of a lower death rate within the safety of the hive) after a fairly long winter of 100 days. At day 200, the active season resumes and the colony rebounds to the pre-winter equilibrium.

Scenario 1

In this scenario, after the initial 100 days we introduce infected foragers into the system, followed by winter 100 days later. The results are shown in Figure 3 based on $\beta = 5 \times 10^{-5}$ and $d_H = d_F = m = 0.14$. The figure shows that within about 5 days the susceptible bee population suffers a drastic drop and the majority of the hive bees have become infected. The infection greatly reduces the overall size of the colony but a new equilibrium is reached, with about 65% of the total population sustaining the infection. At the onset of winter, the size of the colony is not sustainable and within 50 days of winter the colony has collapsed.

Scenario 2

In this scenario we examine the effect of a more severe infection in which the transmission rate is unchanged but the mortality rates from the disease are increased to $d_H = d_F = 4m = 0.56$. The results, in Figure 4, show that after an initial drastic drop, the population of susceptible bees begins to recover approximately 10 days after the onset of the infection. The small numbers of infected hive and forager bees lead to their quick demise soon after the onset of winter, and the disease is eradicated from the hive within 25 days of the onset of winter. Thus, in this case while the colony has sustained heavy losses from the infection, it survives winter with a viable number of bees and no disease. A more severe infection, in

Table 1. Parameter values and references.

L	maximum rate of egg laying	2000 eggs/day	[10]
W	number of hive bees for 50% egg survival	5000 bees	[10]
R_b	baseline recruitment rate	25%/day	[30]
α_f	maximum additional recruitment in absence of food	25%/day	[31]
α_F	effect of excess foragers on recruitment	75%/day	[10]
m	natural death rate of foragers (active season)	14%/day	[32]
m_w	natural death rate of foragers and hive bees (winter)	0.56%/day	[29]
b	mass of food stored for 50% egg survival	500 g	[11]
c	food gathered per day per forager	0.1 g/day	[33]
γ	daily food requirement per adult bee	0.007 g	[11]
d_H	death rate of hive bees due to infection	14%/day	[35]
d_F	death rate of foragers due to infection	14%/day	[35]
β_{HH}	disease transmission rate: hive bee to hive bee	variable	
β_{HF}	disease transmission rate: hive bee to forager	variable	
β_{FH}	disease transmission rate: forager to hive bee	variable	
β_{FF}	disease transmission rate: forager to forager	variable	

the sense that it kills faster, can therefore lead to the survival of the colony as a whole.

Scenario 3

In this scenario we examine the effect of an increased rate of transmission, setting $\beta = 5 \times 10^{-3}$ and $d_H = d_F = 2m = 0.28$. The results are shown in Figure 5. The infection spreads quickly through the colony, the susceptible population is almost immediately eradicated, and within 30 days the colony drops drastically to

<10% of its size before infection. Thereafter, the colony population continues to dwindle slowly, and at the onset of winter it collapses within 10 days. For comparison, with the same natural death rate but in the absence of infection, the colony survives through winter and rebounds to its pre-winter level at the onset of the next active season as seen in Figure 2.

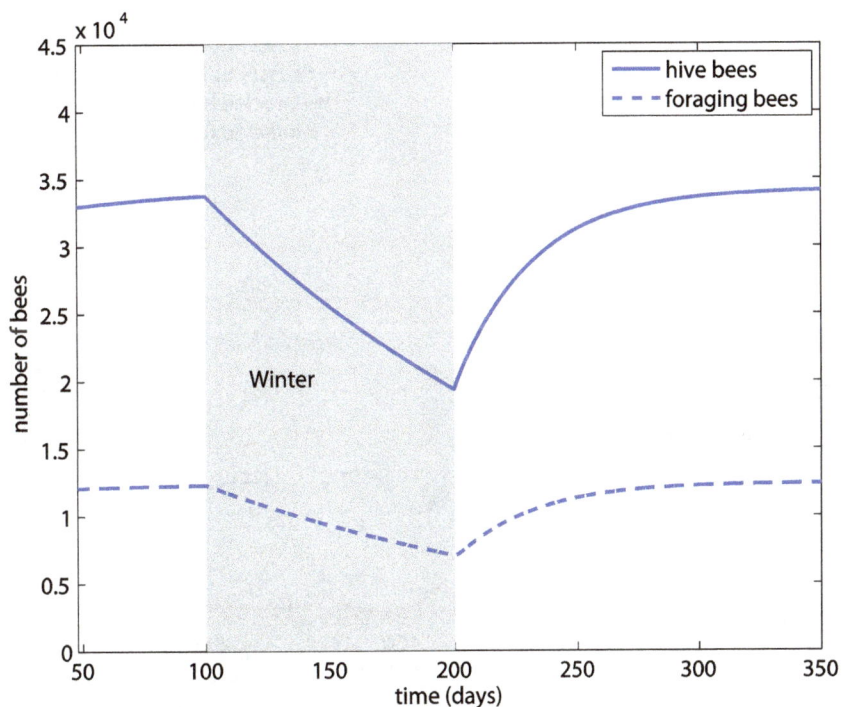

Figure 2. Baseline demographic dynamics of the honey bee colony in the absence of disease.

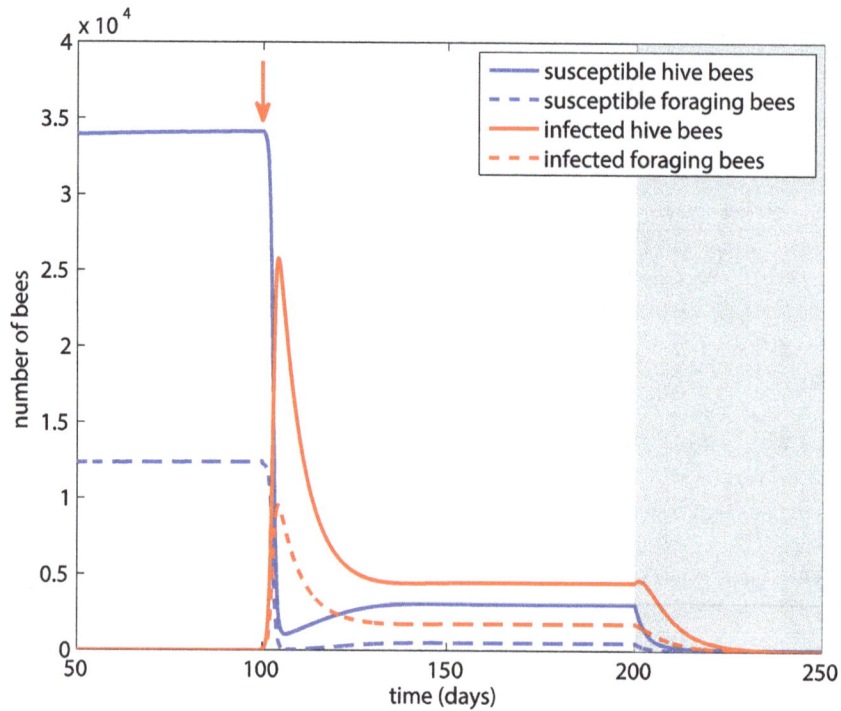

Figure 3. Scenario 1: Colony dynamics in the presence of disease with $\beta = 5 \times 10^{-5}$, $d_H = d_F = 0.14$. Red arrow = onset of infection, grey shading = winter.

Age of Recruitment to Foraging Duties

The average age at which hive bees are recruited to foraging duties (AARF) under the three scenarios is shown in Figure 6. The figure shows that AARF is an important marker of the health of the colony in the sense that a colony with a younger workforce can be taken as a sign of disease within the colony. In Scenario 1,

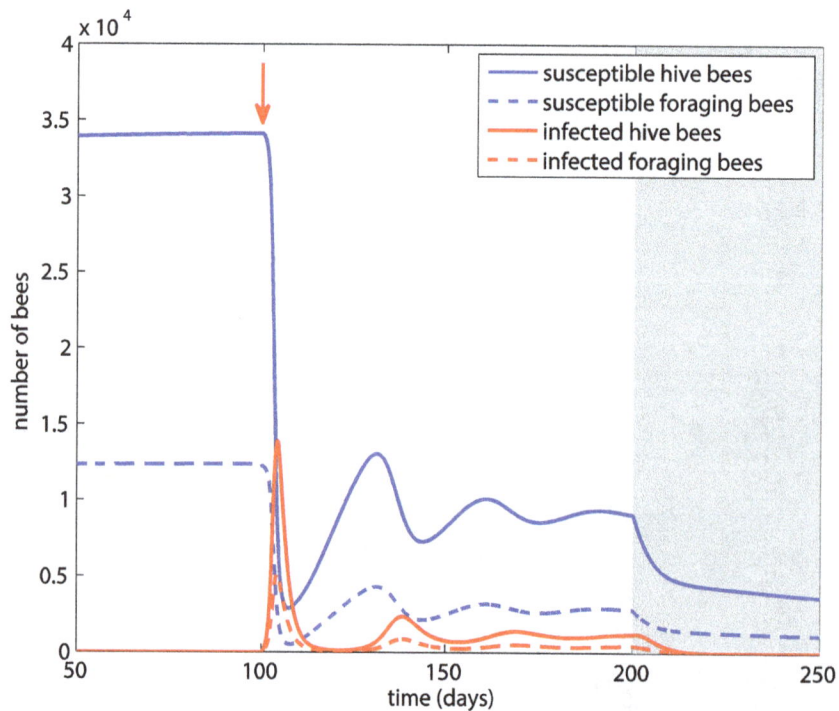

Figure 4. Scenario 2: Colony dynamics under a more severe infection represented by a higher death rates from the disease, with $\beta = 5 \times 10^{-5}$, $d_H = d_F = 0.56$. Red arrow = onset of infection, grey shading = winter.

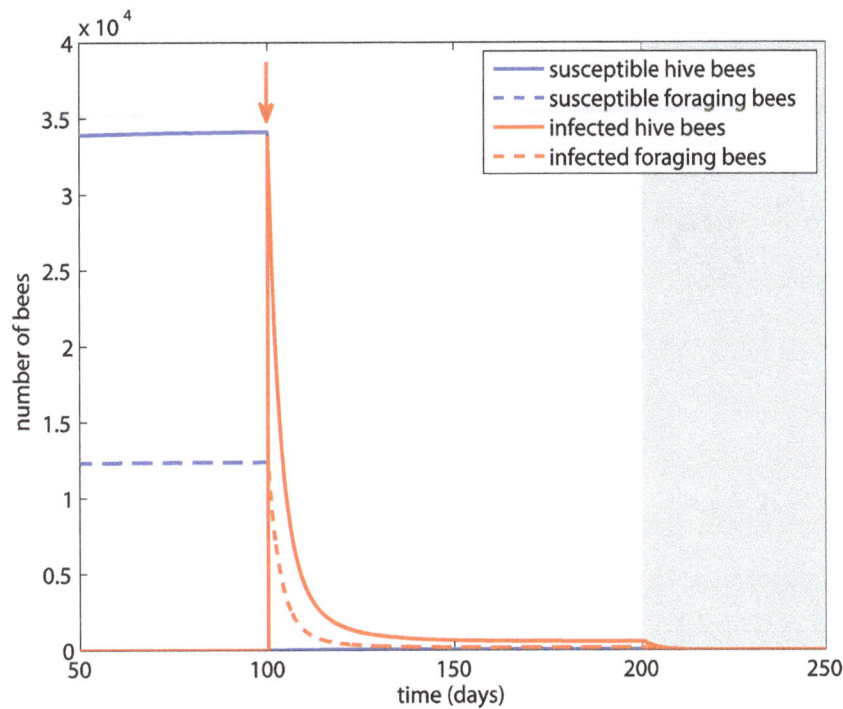

Figure 5. Scenario 3: Colony dynamics under a higher rate of transmission of the disease, with $\beta = 5 \times 10^{-3}$, $d_H = d_F = 0.56$. Red arrow = onset of infection, grey shading = winter.

AARF is reduced from 19.6 days before the onset of infection to 13.16 after the infection. In Scenario 2, with a higher disease-induced death rate, AARF is reduced to about 14.6 days, though fluctuating between 10 days and 16 days at first. In Scenario 3, with a higher rate of transmission of the infection, AARF is reduced drastically to 9.7 days.

Figure 7 shows the complex relationship between the rate of transmission β and disease-induced the death rates d_H, d_F in their effects on the AARF. The figure shows that a combination of small β and large d_H is favorable in that it leads to a higher value of the AARF. At higher values of β, however, the AARF becomes less sensitive to the value of β (as indicated by the clumping of the curves in that region). The position of the three scenarios in this relationship as shown in the figure, and their ultimate fate as described earlier, shows again that the AARF is an early marker of colony collapse, which has been supported by experimental evidence [36].

Scenario 4

In this scenario we examine the effect of the timing of the infection in relation to the onset of winter. Figure 8 shows the effect of infection occurring only 10 days before the onset of winter, compared with 100 days in earlier scenarios. The results, compared with those in Scenario 2, show that the disease is eradicated sooner by early winter. This is clearly because healthy bees live longer in the safety of the hive in winter, while the death rate from infection is unchanged.

Another important indicator of the ultimate fate of the bee colony is the size of the bee population at the end of winter. While under all scenarios winter is taken to last 100 days, the size of the bee population at the end of winter is influenced by the severity of the disease (d_H, d_F), the transmission rate (β), and the time interval between the onset of infection and the beginning of winter which

we shall denote Δt. This complex relationship is shown first in Figure 9 for Scenarios 1, 2, 3 where $\Delta t = 100$ days in all three cases. Again, we see a decrease in sensitivity to β at higher values of β. Furthermore, an increase in the value of d_H initially has an unfavorable effect on the colony size at end of winter, but at high values of d_H this effect is reversed. The region of fractional values is included in Figure 9 only for (mathematical) completeness of the figure. Biologically, the region represents colonies that do not survive. By comparison, in Scenario 4 where $\Delta t = 10$ the size of the bee population at end of winter is reduced by 38% from that in Scenario 2 where the values of other parameters are the same. A more general indication of the dependence of the size of the bee population at end of winter on Δt is shown in Figure 10. The figure shows that for $\Delta t < 20$ days or so, there is very high sensitivity to the value of Δt, but for $\Delta t > 20$ days or so this sensitivity is considerably diminished. This indicates that in the three weeks or so before winter the bee colony is most vulnerable to the risk of infection.

Finally, in Figure 11 we compare the effects of two major types of hazards faced by a honey bee colony, one in which there is a simple increase in the death rate of foragers because of exposure to an environmental hazard and another in which the bees are exposed to an infectious disease. Specifically, in this figure we contrast the dynamics of Scenario 3 with the dynamics of an environmental hazard scenario in which the hive is disease-free but the death rate from the environmental hazard is *the same as the total death rate in Scenario 3*. Specifically, in Scenario 3 we had $d_F = 0.28$, $d_H = 0.28$, $m = 0.14$ for a total death rate of 0.7, thus, for a comparable environmental hazard scenario we take $m = 0.7$ and $d_F = d_H = 0$. The figure shows clearly that the survival of the colony is almost guaranteed in the environmental hazard scenario, while the collapse of the colony is almost guaranteed in the disease scenario.

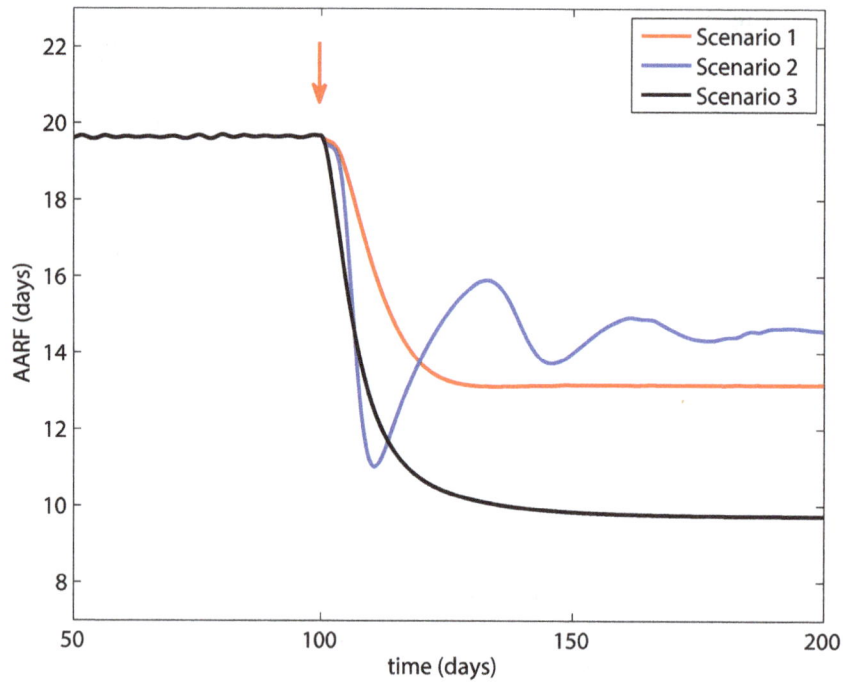

Figure 6. Average age of recruitment to foraging duties (AARF) under the three scenarios in Figures 3, 4, 5. Red arrow = onset of infection.

This comparison is clearly approximate because the three components of death rate in the infectious disease case (d_F, d_H, m) are independent of each other and therefore their sum is not accurately comparable to the total death rate in the environmental hazard case. For this reason, in Figure 12 we consider another comparison in which the dynamics of the two hazards are such

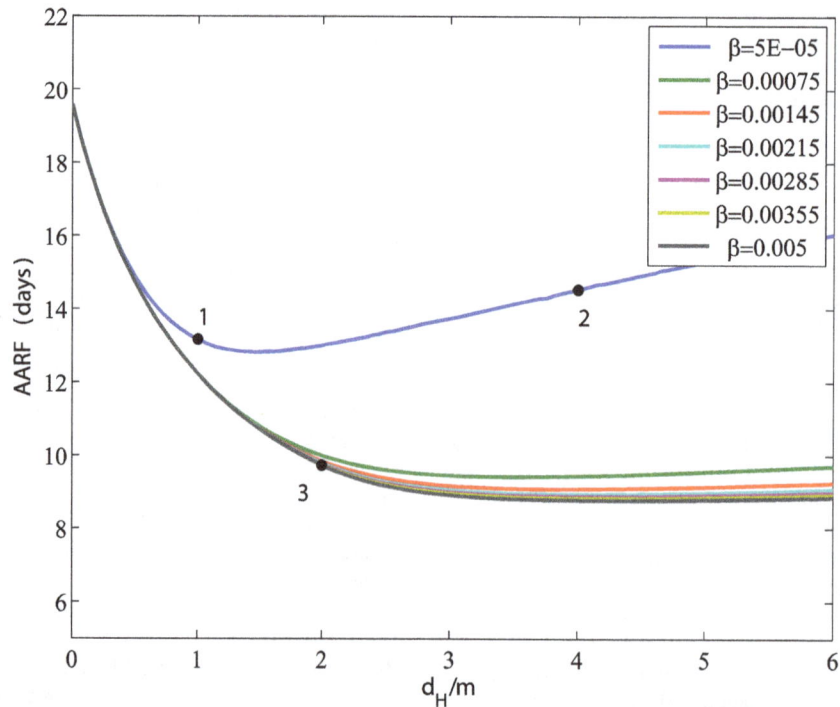

Figure 7. Relationship between the rate of transmission β and disease-induced death rates $d_H = d_F$ in their effects on the Average Age of Recruitment to Foraging. The figure shows the effects of an increase of β and d_H on the AARF. Note that the AARF becomes less sensitive to changes in β as β is increased. Meanwhile, for small β, an increase in d_H can have a favourable effect on the AARF.

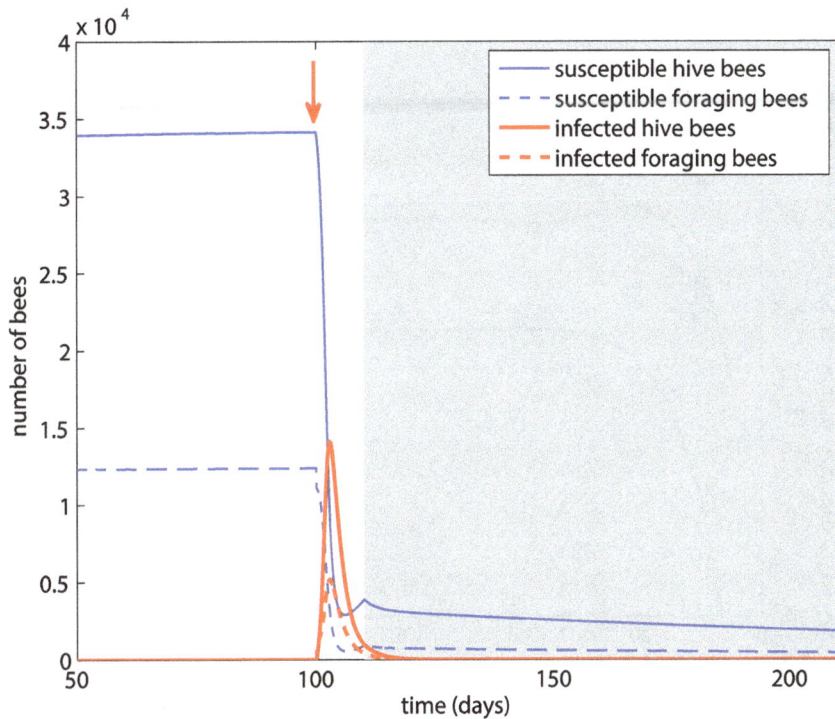

Figure 8. Scenario 4: $\beta = 5 \times 10^{-5}$, $d_H = d_F = 0.56$. Effect of the proximity of the onset of infection to the onset of winter. Red arrow = onset of infection, grey shading = winter.

that the *average lifespan of bees is the same in both cases*. The results again show that the colony survives under the environmental hazard.

Discussions and Conclusions

The main aim of this study was to construct a model for examining the way in which the dynamics of a honey bee colony

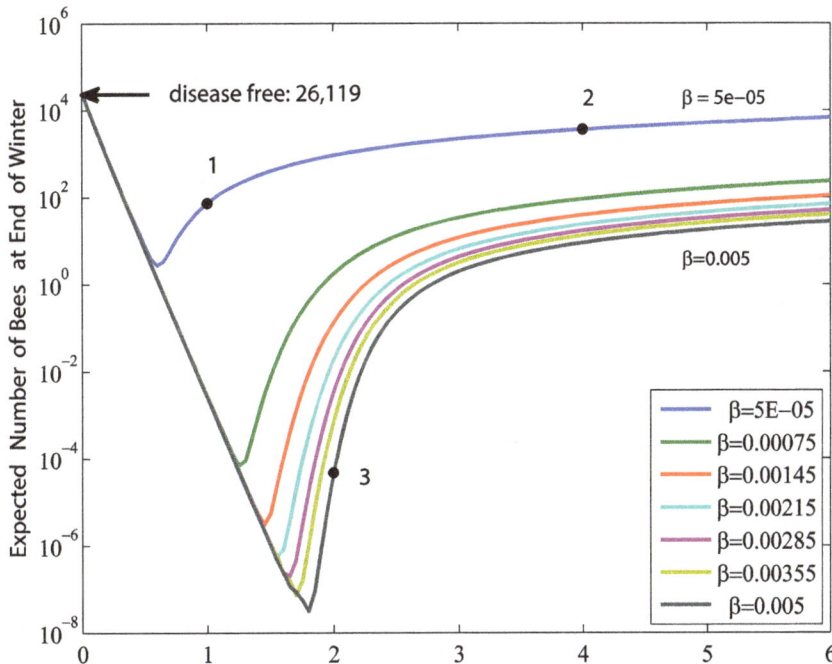

Figure 9. The expected size of the bee population at the end of winter as influenced by the severity of the disease ($d_H = d_F$) and the transmission rate of the disease (β). For comparison, the black arrow indicates the population size at end of winter in the absence of disease (Figure 2). The figure illustrates the different sensitivity to β and d_H. Note that d_H has a favourable effect for small β and d_H large enough.

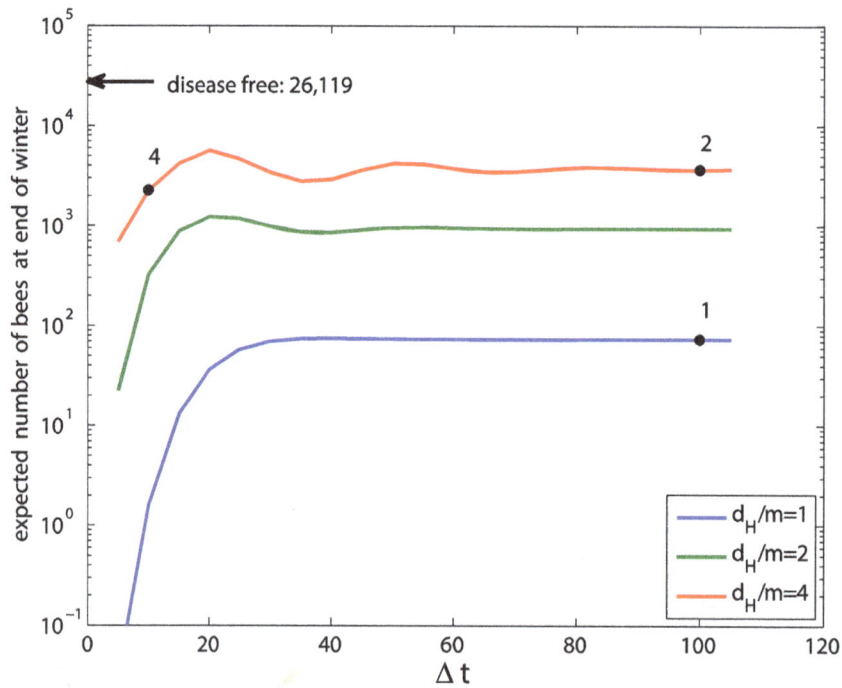

Figure 10. The expected size of the bee population at the end of winter as influenced by the time interval between the onset of infection and the beginning of winter (Δt), **with** $\beta = 5 \times 10^{-5}$**.** For comparison, the black arrow indicates the population size at end of winter in the absence of disease (Figure 2).

are affected by an infection. We present this model in terms of a set of governing equations representing the interplay between the dynamics of the spread of the disease and the demographic dynamics of the bee colony. Up to this point the model is fairly general in regard to the specific characteristics of the colony or the disease and can thus be adapted to a variety of specific cases by an

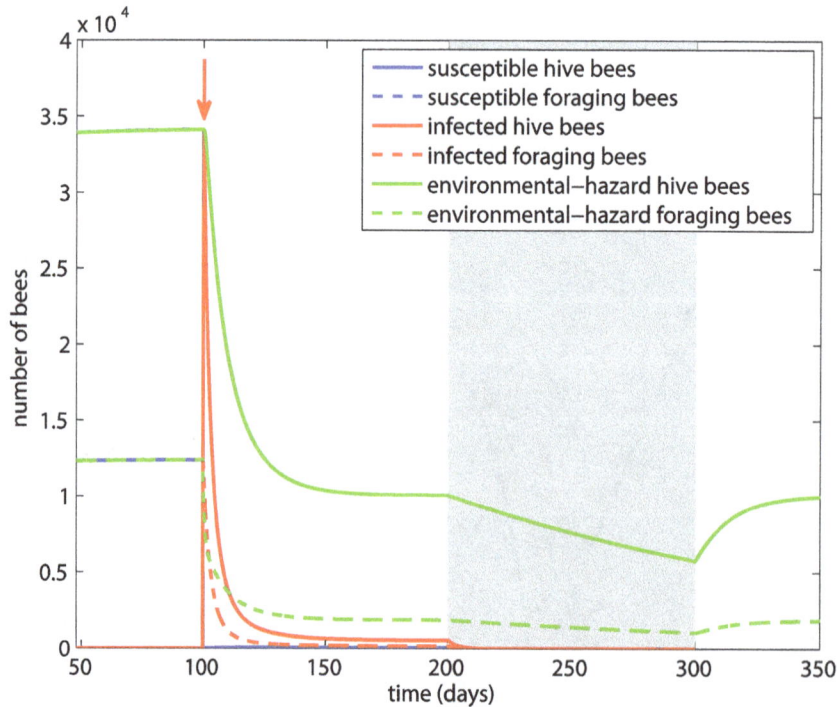

Figure 11. The stark difference between the dynamics of Scenario 3 with an environmental hazard scenario in which the death rate is increased (by the effects of pesticides, for example) *to equal the total death rate in Scenario 3.* The survival of the colony is almost guaranteed in the environmental hazard scenario while the collapse of the colony is almost guaranteed in the disease scenario.

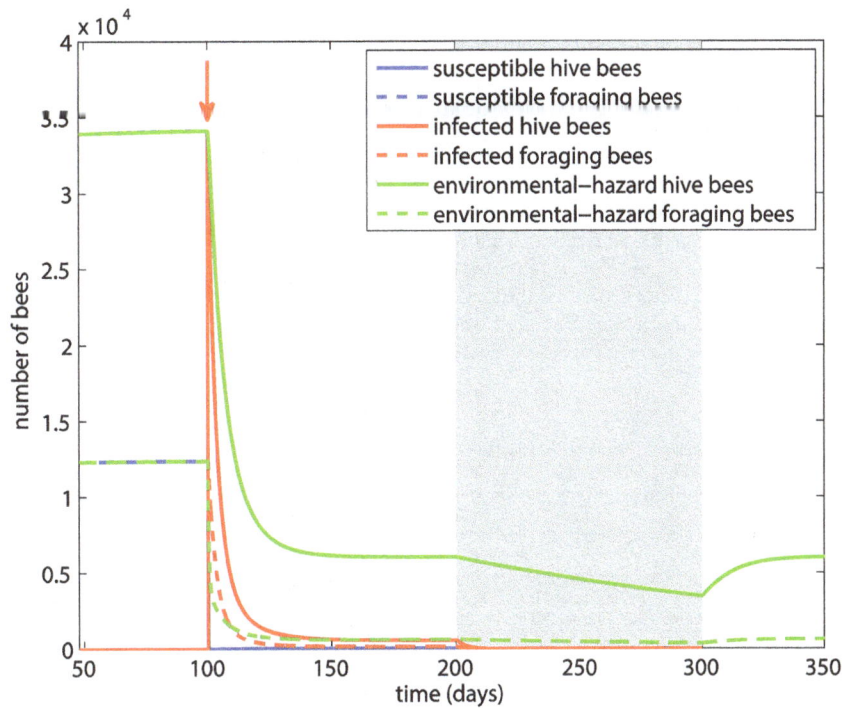

Figure 12. An alternative comparison of the dynamics of Scenario 3 with an environmental hazard scenario in which the comparison between the two hazards is based not on the total death rate as in Figure 11 but on the average lifespan of bees being the same in both cases.

appropriate choice of parameter values. To illustrate the utility of the model, we chose parameter values associated with *Nosema ceranae* which has been well studied experimentally. Our findings, compared with those found experimentally are summarized in Table 2.

The model suggests that key factors in the survival or collapse of a honey bee colony in the face of an infection are the rate of transmission of the infection β and the disease-induced death rates, d_H and d_F. An increase in the disease-induced death rates, which can be thought of as an increase in the severity of the disease, may actually help the colony overcome the disease and survive through winter (Scenario 2), which is consistent with SIR models of epidemics. By contrast, an increase in the transmission rate, which means that bees are being infected at an earlier age, has a drastic deleterious effect (Scenario 3).

Another important finding relates to the timing of infection in relation to the onset of winter. The results (Figure 10) suggest that in a time interval of approximately 20 days before the onset of winter the colony is most affected by the onset of infection. An infection during this "dangerous" time period is more likely to lead

to colony collapse because the number of bees surviving through winter becomes unviable for a rebound of the colony in the new active season. Outside this dangerous time period, i.e. for $\Delta t > 20$ days, the survival of the colony is no longer critically affected by the timing of infection. It must be emphasized that the numerical value of 20 days for this dangerous time period is likely not a "universal" value but one that is specific to the choice of parameter values we used both for the colony and the disease. With other combinations of colony and disease parameters, the model can be used to find the corresponding critical time period.

Our results (Figures 6 and 7) suggest that the AARF is a good early marker for the survival or collapse of a honey bee colony in the face of infection. This is consistent with experimental evidence in [36] but the model and the results in Figures 6 and 7 provide an insight into the underlying mechanisms for this.

Finally, an important result of this study is the clear distinction between two major types of hazards faced by a honey bee colony, namely, one in which there is a simple increase in the death rate of bees because of exposure to an environmental hazard such as pesticides or insecticides, and another in which the bees are

Table 2. Tabulated results from the model scenarios 1, 2, and 3 and experimental data from [36] and [35].

	Exp.	1	2	3
AARF- Healthy	14–21	19.6	19.6	19.6
AARF- Infected	7–16	13.1	14.6	9.7
$(H_I + F_I)/N$	45%	64%	12.2%	92%

The last row shows the percentage of the population infected at the endemic equilibrium, and the experimental value is the threshold value which leads to over-winter colony collapse.

exposed to an infectious disease. The results in Figure 11 show that an exposure to an infectious disease is almost guaranteed to lead to colony collapse while under an environmental hazard the colony has a good chance of survival. This conclusion is confirmed by the results of Figure 12 in which the comparison between the two hazards is based not on the total death rate but on the average lifespan of bees being the same in both cases. Since an environmental hazard in the first place affects only forager bees, the comparison in this case is equivalent to considering a more severe environmental hazard than that in Figure 11, or to considering the long term consequences of an environmental hazard as it affects the demographics of the colony. Together, the two comparisons lead us to suspect that, under comparable death

rates and the range of disease transmission rates which we have considered, an infectious disease may typically be more hazardous to the survival of a bee colony than an exposure to pesticide or insecticide.

Acknowledgments

We thank Mary Myerscough and Gloria DeGrandi-Hoffman for insightful comments.

Author Contributions

Analyzed the data: MIB LMW MZ. Wrote the paper: MIB LMW MZ.

References

1. vanEngelsdorp D, Evans JD, Saegerman C, Mullin C, Haubruge E, et al. (2009) Colony collapse disorder: A descriptive study. PLoS ONE 4: e6481.
2. Watanabe ME (2008) Colony collapse disorder: Many suspects, no smoking gun. BioScience 58: 384–388.
3. Ho MW, Cummins J (2007) Mystery of disappearing honeybees. Science in Society 34: 35–36.
4. Devillers J (2002) The ecological importance of honey bees and their relevance to ecotoxicology. In: Devillers J, Pham-Delègue M, editors, Honey Bees: Estimating the Environmental Impact of Chemicals, London: Taylor and Francis London. pp. 1–11.
5. Calderone NW (2012) Insect pollinated crops, insect pollinators and US agriculture: Trend analysis of aggregate data for the period 1992–2009. PLoS ONE 7: e37235.
6. Southwick EE, Southwick Jr L (1992) Estimating the economic value of honey bees (Hymenoptera: Apidae) as agricultural pollinators in the United States. Journal of Economic Entomology 85: 621–633.
7. Neumann P, Carreck NL (2010) Honey bee colony losses. Journal of Apicultural Research 49: 1–6.
8. Henry M, Beguin M, Requier F, Rollin O, Odoux JF, et al. (2012) A common pesticide decreases foraging success and survival in honey bees. Science 336: 348–350.
9. Favre D (2011) Mobile phone-induced honeybee worker piping. Apidologie 42: 270–279.
10. Khoury DS, Myerscough MR, Barron AB (2011) A quantitative model of honey bee colony population dynamics. PLoS ONE 6: e18491.
11. Khoury DS, Barron AB, Myerscough MR (2013) Modelling food and population dynamics in honey bee colonies. PLoS ONE 8: e59084.
12. Sumpter DJT, Martin SJ (2004) The dynamics of virus epidemics in Varroa-infested honey bee colonies. Journal of Animal Ecology 73: 51–63.
13. Ratti V, Kevan PG, Eberl HJ A mathematical model for population dynamics in honeybee colonies infested with Varroa destructor and the Acute Bee Paralysis Virus. Canadian Applied Mathematics Quarterly: accepted.
14. Fries I (2010) Nosema ceranae in European honey bees (Apis mellifera). Journal of invertebrate pathology 103: S73–S79.
15. Higes M, Martin-Hernandez R, Garrido-Bailon E, Gonzalez-Porto AV, Garcia-Palencia P, et al. (2009) Honeybee colony collapse due to Nosema ceranae in professional apiaries. Environmental Microbiology Reports 1: 110–113.
16. Kauffeld NM (1980) Beekeeping in the United States Agriculture Handbook Number 335. U.S. Department of Agriculture.
17. Seeley TD, Visscher PK (1985) Survival of honeybees in cold climates: the critical timing of colony growth and reproduction. Ecological Entomology 10: 81–88.
18. Cramp D (2008) A Practical Manual of Beekeeping. London: How To Books.
19. Winston M (1987) The biology of the honey bee. Harvard University Press.
20. Seeley TD (2010) Honeybee Democracy. Princeton University Press.
21. Jay S (1974) Seasonal development of honeybee colonies started from package bees. Journal of Apicultural Research 13: 149–152.
22. Huang ZY, Robinson GE (1996) Regulation of honey bee division of labor by colony age demography. Behavioral Ecology and Sociobiology 39: 147–158.
23. Leoncini I, Le Conte Y, Costagliola G, Plettner E, Toth AL, et al. (2004) Regulation of behavioral maturation by a primer pheromone produced by adult worker honey bees. Proceedings of the National Academy of Sciences of the United States of America 101: 17559–17564.
24. Higes M, Martin-Hernandez R, Botias C, Bailon EG, Gonzalez-Porto AV, et al. (2008) How natural infection by Nosema ceranae causes honeybee colony collapse. Environmental microbiology 10: 2659–2669.
25. Chen Y, Evans JD, Smith IB, Pettis JS (2008) Nosema ceranae is a long-present and wide-spread microsporidian infection of the European honey bee (Apis mellifera) in the United States. Journal of Invertebrate Pathology 97: 186–188.
26. Smith ML (2012) The honey bee parasite Nosema ceranae: Transmissible via food exchange? PLoS ONE 7: e43319.
27. Stevanovic J, Simeunovic P, Gajic B, Lakic N, Radovic D, et al. (2013) Characteristics of Nosema ceranae infection in Serbian honey bee colonies. Apidologie 44: 522–536.
28. Jones JC, Helliwell P, Beekman M, Maleszka R, Oldroyd B (2005) The effects of rearing temperature on developmental stability and learning and memory in the honey bee, Apis mellifera. Journal of Comparative Physiology A 191: 1121–1129.
29. Sakagami S, Fukuda H (1968) Life tables for worker honeybees. Researches on Population Ecology 10: 127–139.
30. Fahrbach S, Robinson G (1996) Juvenile hormone, behavioral maturation and brain structure in the honey bee. Developmental Neuroscience 18: 102–114.
31. Schulz DJ, Huang ZY, Robinson GE (1998) Effects of colony food shortage on behavioral development in honey bees. Behavioral Ecology and Sociobiology 42: 295–303.
32. Dukas R (2008) Mortality rates of honey bees in the wild. Insectes Sociaux 55: 252–255.
33. Russell S, Barron AB, Harris D (2013) Dynamic modelling of honey bee (Apis mellifera) colony growth and failure. Ecological Modelling 265: 158–169.
34. Harbo JR (1993) Effect of brood rearing on honey consumption and the survival of the worker honey bees. Journal of Apicultural Research 32: 11–17.
35. Goblirsch M, Huang ZY, Spivak M (2013) Physiological and behavioral changes in honey bees (Apis mellifera) induced by Nosema ceranae infection. PLoS ONE 8: e58165.
36. Botias C, Martin-Hernandez R, Barrios L, Meana A, Higes M (2013) Nosema spp. infection and its negative effects on honey bees (Apis mellifera iberiensis) at the colony level. Veterinary research 44: 1–15.

19

Production and Robustness of a Cacao Agroecosystem: Effects of Two Contrasting Types of Management Strategies

Rodolphe Sabatier*¤, Kerstin Wiegand, Katrin Meyer

Department of Ecosystem Modelling, Büsgen-Institute, Georg-August-University of Göttingen, Göttingen, Germany

Abstract

Ecological intensification, i.e. relying on ecological processes to replace chemical inputs, is often presented as the ideal alternative to conventional farming based on an intensive use of chemicals. It is said to both maintain high yield and provide more robustness to the agroecosystem. However few studies compared the two types of management with respect to their consequences for production and robustness toward perturbation. In this study our aim is to assess productive performance and robustness toward diverse perturbations of a Cacao agroecosystem managed with two contrasting groups of strategies: one group of strategies relying on a high level of pesticides and a second relying on low levels of pesticides. We conducted this study using a dynamical model of a Cacao agroecosystem that includes Cacao production dynamics, and dynamics of three insects: a pest (the Cacao Pod Borer, *Conopomorpha cramerella*) and two characteristic but unspecified beneficial insects (a pollinator of Cacao and a parasitoid of the Cacao Pod Borer). Our results showed two opposite behaviors of the Cacao agroecosystem depending on its management, i.e. an agroecosystem relying on a high input of pesticides and showing low ecosystem functioning and an agroecosystem with low inputs, relying on a high functioning of the ecosystem. From the production point of view, no type of management clearly outclassed the other and their ranking depended on the type of pesticide used. From the robustness point of view, the two types of managements performed differently when subjected to different types of perturbations. Ecologically intensive systems were more robust to pest outbreaks and perturbations related to pesticide characteristics while chemically intensive systems were more robust to Cacao production and management-related perturbation.

Editor: Nicholas J. Mills, University of California, Berkeley, United States of America

Funding: This study was part of the DFG-funded research project "Environmental and land-use change in Sulawesi, Indonesia: Socioeconomic and ecological perspectives" (PAK 569, WI1816/12). KMM was partly funded by the State of Lower Saxony (Ministry of Science and Culture; Cluster of Excellence "Functional Biodiversity Research"). The funders had no role in study design, data collection and analysis, decision to publish, or preparation of the manuscript.

Competing Interests: The authors have declared that no competing interests exist.

* E-mail: rodolphe.sabatier@agroparistech.fr

¤ Current address: INRA, UMR 1048 SADAPT, Paris, France

Introduction

New paradigms in agriculture based on ecological intensification such as natural farming [1], agroecology [2,3], the evergreen revolution [4], or the doubly green revolution [5] are presented as challenging alternatives to more conventional farming relying on a high level of chemical input. They are presented as being more respectful of the environment while ensuring a high level of production. A central idea behind these paradigms is that associated biodiversity is strongly impacted by the use of chemicals in conventional farming [6,7] while the associated biodiversity could provide a large range of ecosystem services that often have the same effect as the chemical used (e.g. pest regulation, [8]). According to these paradigms, reducing the amount of chemicals used would maintain biodiversity and ecosystem services at a high level and ensure high yields with lower economic costs. Moreover, based on the ecological concept of stability of ecosystems [9], agroecosystems with high levels of biodiversity are considered more robust, stable and resilient toward perturbations. In other words, such systems have the advantages of a high productivity

due to the maintenance of high levels of ecosystem services [10] and of a strong capacity to resist to perturbations [11].

The three concepts of resilience, stability and robustness, although related, slightly differ: resilience is "the persistence of relationships within a system and is a measure of the ability of these systems to absorb changes of state variables, driving variables, and parameters, and still persist" [12], stability is "the ability of a system to return to an equilibrium state after a temporary disturbance" [12] and robustness is: "the ability to maintain performances in the face of perturbation and uncertainty" [13]. In this study we consider the issue of robustness that addresses the performance of the system when the perturbation occurs.

The overall production of an agroecosystem is quite easy to quantify but its behavior in face of a perturbation is much more difficult to assess as it strongly depends on the perturbation considered [14].

In this study our aim is to assess productive performance and robustness toward diverse perturbations of an agroecosystem managed with two contrasting strategies. More precisely, we address the following two questions:

1: What is the shape of the relationship between robustness to perturbations and yield of an agroecosystem under a broad range of different management schemes?

2: Where are the management strategies based on ecological processes (hereafter called ecological strategies, ES) and the ones based on pesticide use (hereafter called chemical-based strategies, CBS) located within the range of outcomes found in 1) and which of these two types of strategies performs better with respect to yield and robustness?

To answer these questions, we developed a model based on a case study of the Cacao agroecosystem in Central Sulawesi (Indonesia). The aim of this model was to capture some general patterns of the interactions between an agroecosystem, its environment and its management by the farmer. Hence, the product of this study is not intended for direct application in the field, but rather for informing management decisions on a general basis. To calibrate and parameterize this model, the Cacao case study was chosen as it captures the main above-ground ecosystem services and disservices and therefore represents a wide variety of agroecosystems well. Cacao crops were introduced in South East Asia 200 years ago. First records of Cacao production in Indonesia date back to 1848 (15) but production remained low (<5000 t.year^{-1} until the late 1970's; [15,16]). Production strongly increased in the last decades to reach more than 800 000 t.year^{-1} in 2010 [16]. In Indonesia, Cacao production is continuous but yield is not constant through the year and shows a main peak in January and in some cases a minor peak six months later. Two main pests impact Cacao production: *Helopeltis theobromae* and the Cacao Pod Borer *Conopomorpha cramerella* [17]. Cacao pollination is done by midges (Ceratopogonidae) and is a limiting factor for production [18].

After briefly describing the model that we used, we determine the relationship between ecosystem functioning and the number of spraying events and define two groups of strategies, ES and CBS. Then we determine the relationship between yield and robustness for the set of all possible management strategies and six types of perturbations. We then look more precisely at the relative positions of the two subsets of management strategies ES and CBS on the production-robustness relationship. Our results show that ecosystem functioning is strongly negatively correlated to the number of spraying events and show how both production and robustness of the two extreme types of strategies depend on both the type of pesticide used and the type of perturbation.

Materials and Methods

Model overview

Robustness of agroecosystems is difficult to address in the field due to the high complexity of agroecosystems and their low reproducibility for experimental purposes. In this context, modeling approaches are powerful tools. *In silico* experiments can cope with the complexity and their high level of reproducibility makes them useful frameworks to represent complex agroecosystems and to study their robustness to perturbations.

The study that we present here is based on a model developed and presented in [19]. This model gives an agroecological representation of a Cacao plantation (Figure 1). It is a discrete time model with a time step of one month and a time horizon of 20 years (T = 240 months). It links the Cacao pod dynamics to the population dynamics of a pest species (the Cacao Pod Borer,

Conopomorpha cramerella) and two characteristic but unspecified beneficial insect populations (a pollinator of Cacao and a parasitoid of the Cacao Pod Borer). The Cacao Pod Borer and parasitoid parts of the model were inspired by both the Cacao model of [20] and the more general Nicholson and Bailey host-parasitoid model [21].

The insects impact the Cacao yield in several ways. The pollinators N_{Pol} positively affect the number of pods of age 0 $Pods_0$ by pollinating the obligately outcrossing Cacao plants. The Cacao Pod Borer population reduces the amount of Cacao beans eventually harvested. The Cacao Pod Borer population N_{CPB} is regulated by a parasitoid N_{Par}. All three insect populations are affected by the use of pesticides by the farmer. We distinguish two effects of spraying: the efficiency η (the effect on the Cacao Pod Borer) and its selectivity θ (the ratio of effects on beneficial and on pest populations). Efficiency and selectivity range from 0 to 1, $\eta = 1$ means that 100% of the Cacao Pod Borer are killed by the pesticides, $\theta = 1$ means that the effect of pesticide application on beneficial insects is as strong as its effect on the Cacao Pod Borer. Since the aim of pesticide application in the field is to control the Cacao Pod Borer, we limited the study to pesticides having a stronger effect on the target species (Cacao Pod Borer) than on the other species (beneficial insects). We therefore implicitly assume that farmers would not use pesticides that have a net negative impact. In this sense, we avoided the trivial situation where pesticides should be banned due to net negative effects. The model computes the Cacao yield dynamics Y through time as well as the dynamics of the three insect populations for different types of pesticides (characterized by their selectivity and efficiency), and for different timings of spraying.

The model can be described by the following system:

$$\begin{cases} Pods_0(t+1) = f_1(t, N_{Pol}(t)) \\ Y(t) = f_2(Pods_0(t-5), N_{CPB}(t-2)) \\ N_{CPB}(t+1) = (1 - \mathbb{1}_{\{Spray\}}(t)\eta)f_3(Pods_0(t-3), N_{Par}(t), N_{CPB}(t)) \ (1) \\ N_{Par}(t+1) = (1 - \mathbb{1}_{\{Spray\}}(t)\eta(1-\theta))f_4(N_{Par}(t), N_{CPB}(t)) \\ N_{Pol}(t+1) = (1 - \mathbb{1}_{\{Spray\}}(t)\eta(1-\theta))f_5(N_{Pol}(t)) \end{cases}$$

Cacao Pod Borers preferentially attack pods of age 3, pods are harvested at age 5. $\mathbb{1}_{\{Spray\}}$, the characteristic function related to

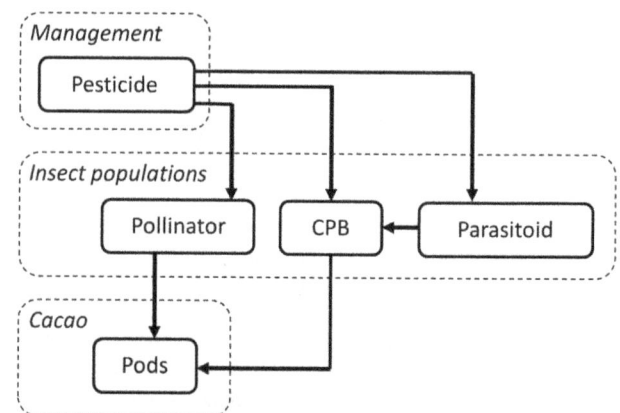

Figure 1. Overview of the model structure. CPB stands for Cacao Pod Borer. Bold black arrows stand for the processes considered in the ecosystem functioning index (EFI; eqn 4).
doi:10.1371/journal.pone.0080352.g001

the spraying event is defined as follows:

$$\begin{cases} 1_{\{Spray\}}(t) = 1 \ \textit{if spraying is true at time } t \\ 1_{\{Spray\}}(t) = 0 \ \textit{if spraying is false at time } t \end{cases} \quad (2)$$

f_1–f_5 are the functions related to the different natural dynamics (see Appendix S1, eqn S.1.1, S.1.11, S.1.8, S.1.9 and S.1.6)

Model calibration and sensitivity analysis

Most parameter values were taken from the literature, when possible from studies conducted in central Sulawesi (Table 1); [19].

Data on pollinators of the Cacao (ceratopogonid midges) are very scarce and we could not calibrate this dynamics on data from any of the pollinating species. Therefore we relied on available data from another tropical species of the same family: *Leptoconops albiventris*. Although it concerned a different species than the pollinators of our system, we considered these data suitable enough regarding the ambitions of the model. These data [22] provide a 6-months survey of a population of midges subjected to regular pesticide applications and were used to calibrate the equation corresponding to pollinator dynamics. Pesticide applications are followed by an instantaneous drop of the midge population followed by a fast recovery of the population. Given

Table 1. List of parameters.

Name	Description	Value	Unit	Reference
Intra-annual dynamic of Cacao pod production (eqn S.1.2)				
α_1	Mean sinusoid 1	7.176	pods.ha^{-1}	[23]*,[24]*
β_1	Amplitude sinusoid 1	1.425	pods.ha^{-1}	[23]*,[24]*
Ω	Time period sinusoids	12	months	[23]*,[24]*
Inter-annual trend of Cacao pod production (eqn S.1.3)				
α_A	Age dependence parameter 1	3.82	-	[42]
β_A	Age dependence parameter 2	0.086	-	[42]
γ_A	Age dependence parameter 3	1.33	-	[42]
A_0	Initial age	120	months	[42]
φ_A	Standardization coefficient	$2.5\ 10^{-3}$	-	[42]*
Pollination effect (eqn S.1.4, eqn S.1.5)				
α_P	Pollination factor 1	−0.83	kg. ha^{-1}.	[18]
β_P	Pollination factor 2	0.34	kg. ha.$^{-1}$ pollinated flower^{-1}	[18]
γ_P	Standardization coefficient	2.36	ha.kg^{-1}	[18]*
α_ψ	Pollinated flowers per pollinator unit	40	pollinated flower.pollinator^{-1}	[18]
Effects of management (eqn S.1.6)				
η	Spraying effect on pests (efficiency)	$\in[0,1]$	-	-
θ	Spraying effect on beneficial insects (selectivity)	$\in[0,1]$	-	-
Pollinator dynamics (eqn S.1.6)				
λ_P	Growth rate	100	-	[22]*
c_P	Competition rate	99	pollinator^{-1}	[22]*
Cacao Pod Borer (CPB) dynamics (eqn S.1.8)				
α_{CPB}	Number of eggs per female	130	egg.female CPB^{-1}	[20]
β_{CPB}	Egg predation rate	0.15	-	[20]
S_0	Larvae survival	0.1	CPB.egg^{-1}	[20]*
σ_{CPB}	Adult sex-ratio	0.5	female CPB. CPB^{-1}	Wielgoss & Clough (unpub)
δ_{CPB}	Adult predation rate	0.41	-	[20]
μ	Density dependence coefficient 1	0.69	pod.CPB$-^1$	[20]
ρ	Density dependence coefficient 2	1.92	-	[20]
Parasitoid dynamics (eqn S.1.9, eqn S.1.10)				
a_{Par}	Parasitism probability	0.003	-	[20]*
Yield function (eqn S.1.12)				
ω	Weight of dry beans per pod	0.03	kg.pod^{-1}	Clough (unpub)*
α_Y	Yield loss parameter 1	0.01	CPB.pod^{-1}	[20]
β_Y	Yield loss parameter 2	6.33	-	[20]

*refers to parameters calibrated from the cited reference.

that we do not aim at quantitative prediction but at qualitative understanding, we only modeled relative abundances and standardized the midge population in the absence of spraying to 1000 at equilibrium.

A sensitivity analysis was then conducted to test the effect of a variation of +/− 10% of each parameter of the model on the average Cacao yield and on the average population sizes of each of the three insects (for details, see [19]). The sensitivity analysis showed a high sensitivity of the model to the two parameters of the intra-annual dynamics of Cacao pod production (parameters α_1, and β_1 in appendix S1). Calibration of these key parameters was then adjusted using data from a survey of the Cacao yield of two plots in central Sulawesi [23,24]. Initially conducted to compare rainfall treatments, this database provided us with six control subplots that we used for calibration. The survey was conducted from January to December 2007 with a two-week time step. Calibration was obtained by minimizing the Root Mean Square Error (RMSE) of the full model compared to the Cacao yield data. To ensure a periodic pattern of the dynamics, calibration was made on two successive years. This sensitivity analysis made it possible to isolate the most sensitive parameters. Once the sensitivity to input data had been tested, all simulations were run in a deterministic manner, to keep the number of simulation within feasible limits.

Simulations

Indices. The model that we used made it possible to compute the agroecosystem dynamics for the whole set of possible management strategies. For each simulated spraying strategy, we recorded the average yearly yield and the number of spraying events as well as two specific indices to record information on robustness and ecological functioning of the system. The Robustness index records the average deviation (in absolute values) of the productive output for a given set of perturbations. It reads as follows:

$$Robustness = \frac{Y_M - M(|Y - y_i|)}{Y_M} \quad (3)$$

with Y_M the maximum reachable yield (750 kg.ha^{-1}), Y the yield without perturbation and y_i the yield with perturbation i (i belongs to the set of studied perturbations). Function M stands for the mean and $| |$ for the absolute value.

The Ecosystem Functioning Index (EFI) relates to the ecological functioning of the agroecosystem. It synthesizes the different ecological processes at stake and reads as follows:

$$EFI = P_F + P_{CPB} + P_{Pods} \quad (4)$$

with P_F the pollination rate, P_{CPB} rate of parasitism of the Cacao Pod Borers and P_{Pods} the infestation rate of the Cacao pods. This index synthesizes all ecological functions at stake in our modeled agroecosystem and does not only focus on ecosystem services. We decided to consider all functions regardless of their effects on production to reflect information available in real systems. Indeed, in real systems it is difficult to put a number on services specifically, especially when the distinction between services and disservices is not clear [25] or when the services are not known well.

Typology of the management strategies. To identify spraying strategies that combine extreme intensities of management and ecosystem functioning we built two contrasting groups of strategies, depending on their position in the ecosystem functioning (EFI) - number of spraying events (N_S) - plane. The set of

Ecological Strategies (ES) encompass all strategies that are within both the 10% EFI upper quantile and the 10% N_S lower quantile. The set of Chemical Based Strategies, i.e. the CBS-management strategies encompass all strategies that are within both the 10% EFI lower quantile and the 10% N_S upper quantile.

Production and robustness of the management strategies. We analyzed the relationship between yield and robustness of the system under six types of perturbations related to management or environmental conditions. More precisely, we subjected the system to the following perturbations.

Perturbations related to modifications of the environmental conditions.

- Pest outbreak: For each management strategy, we simulate two alternative perturbations via a sudden increase in the Cacao Pod Borer population by 500 and 1000 individuals.ha^{-1} at month 3 (time of the year where the number of pods sensitive to this pest is the highest).

- Variation in Cacao production: For each management strategy, we simulate four alternative situations with each of the two parameters of the Cacao pod dynamics (mean and amplitude of the pod production sinusoid) increased or decreased by 10%. This perturbation could reflect diverse environmental variations, including climatic ones.

Perturbations related to modifications of the management strategies.

- Event shift: For each management strategy, we simulate 12 alternative, modified strategies involving the shift of one of the 12 spraying/non-spraying events (i.e. 1 spraying event was replaced by a non-spraying event and *vice versa*).

- Temporal shift: For each management strategy, we simulate two alternative strategies by shifting the entire spraying sequence by one month (either shifted one month earlier or one month later).

Perturbations related to the pesticide characteristics.

- Pesticide efficiency: This scenario corresponds to a modification of the pesticide efficiency. For each management strategy, we simulate two alternative situations with a pesticide whose efficiency is either increased or decreased by 20%.

- Pesticide selectivity: This scenario corresponds to a modification of the pesticide selectivity. For each management strategy, we simulate two alternative situations with a pesticide whose selectivity is either increased or decreased by 20%.

These six perturbations were tested on the 2^{12} possible management strategies. Then the relative performances of the ES and CBS defined were assessed. To compare the performances of ES and CBS, we compared all possible pairs of strategies (always one from the ES set and one from the CBS set) and computed the proportion of ES that performed better than CBS (hereafter called Comparison Index, CInd).

We first give a detailed analysis of the response of the Cacao agroecosystems to these perturbations for a single type of pesticide (Efficiency = 0.5, Selectivity = 0.5). Then, we conduct the same analysis for 81 different pesticide types (Efficiency and Selectivity ranging from 0.1 to 0.9 with a step of 0.1) so as to test the range of validity of these first findings.

Numeric computations and statistical analyses were performed with Python 2.7.2 (http://www.python.org/).

Results

Calibration

After calibration, our model gave a reasonable visual fit (Figure 2). Comparison of the model outputs with the Cacao yield data showed that we managed to capture the general behavior of the system.

Typology of the management strategies

We logically observed a strong relationship between the number of spraying events N_S and the functioning of the agroecosystem EFI (Figure 3; $EFI = 1.63 - 0.05 \; N_S$, $p < 10^{-3}$, $R^2 = 0.68$). The management strategies built a continuum. At the two extremes of this continuum, we distinguished two subsets of trajectories: the Chemical Based Strategies (CBS) that corresponded to a high number of spraying events and a low functioning of the ecosystem and the Ecological Strategies (ES) that corresponded to a low number of spraying events and a high ecosystem functioning (Figure 3).

Production and robustness of the management strategies

The relationship between yield and robustness differed between the different types of perturbations (Table 2). Negative relationships were found for environment- and management-related perturbations while a positive relationship was found for pesticide-related perturbations.

On average, the yield obtained with ES was higher than the yield obtained with CBS (Figure 4, $CInd = 0.71$). The robustness of CBS was lower than the one of the ES under perturbations due to a pest outbreak (Figure 4.a.; $CInd = 1.00$) as well as under pesticide-related perturbations (Figure 4e, $CInd = 0.99$ and Figure 4f, $CInd = 0.80$). However, robustness of CBS was higher than the robustness of ES under perturbations related to management

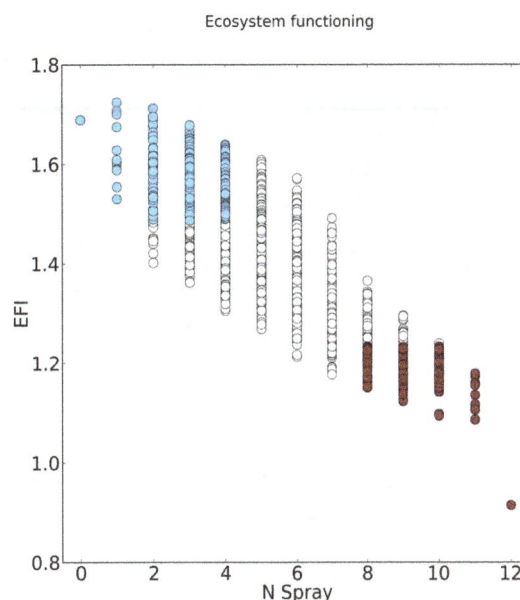

Ecosystem functioning

Figure 3. Intensity of spraying and ecosystem functioning of the different spraying strategies. The x-axis is the Ecosystem Functioning Index (EFI) and the y-axis is the number of spraying events per year. Each symbol represents one of the possible spraying strategies. Black symbols stand for the subset of Chemical Based Strategies (CBS; high spraying intensity and a low ecosystem functioning), white symbols stand for the subset of Ecological Strategies (ES; low spraying intensity and a high ecosystem functioning). Gray symbols stand for all other strategies.

(Figure 4c, $CInd = 0.12$ and Figure 4d, $CInd = 0.32$) or to production (Figure 4b, $CInd = 0.02$).

These results can be qualitatively explained in the following way. Systems managed with ES have a high level of ecosystem functioning. This gives a higher capacity of self-regulation to the system, which explains its high robustness to pest outbreaks and to the pesticide characteristics. However ES are very specific strategies aimed at "driving" the system instead of "controlling" it. This explains their low robustness to both production- and management-related perturbations.

Effect of the type of pesticide

Comparison of the robustness of CBS and ES showed similar results for the different types of pesticides as for the first pesticide detailed in the former section (Figure 5). Due to the non-linearity of model dynamics, the patterns observed are not necessarily

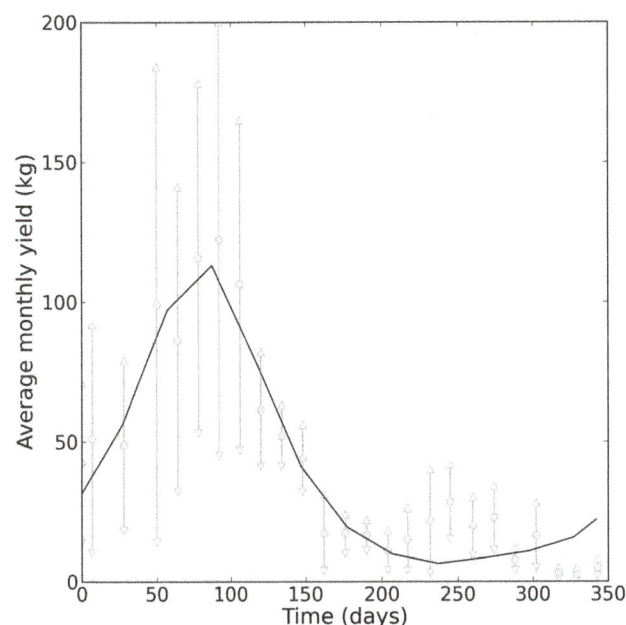

Figure 2. Model output and data from the study area (zoom on year 10). The solid black curve corresponds to the model output, the gray dots to the average monthly yields observed in the six plots of the study area. The grey bars around these dots correspond to the average values +/− standard deviation.

Table 2. Relationship between yield and robustness for four types of perturbation (linear models).

Perturbation	Trend	Intercept	Slope	R^2	P-value
Environment (pest outbreak)	Negative	600	−160	0.03	<10-3
Environment (production)	Negative	1158	−1018	0.65	<10-3
Management (event shift)	Negative	2478	−2053	0.14	<10-3
Management (temporal shift)	Negative	859	−413	0.03	<10-3
Pesticide (selectivity)	Positive	−270	747	0.37	<10-3
Pesticide (efficiency)	Positive	−478	944	0.19	<10-3

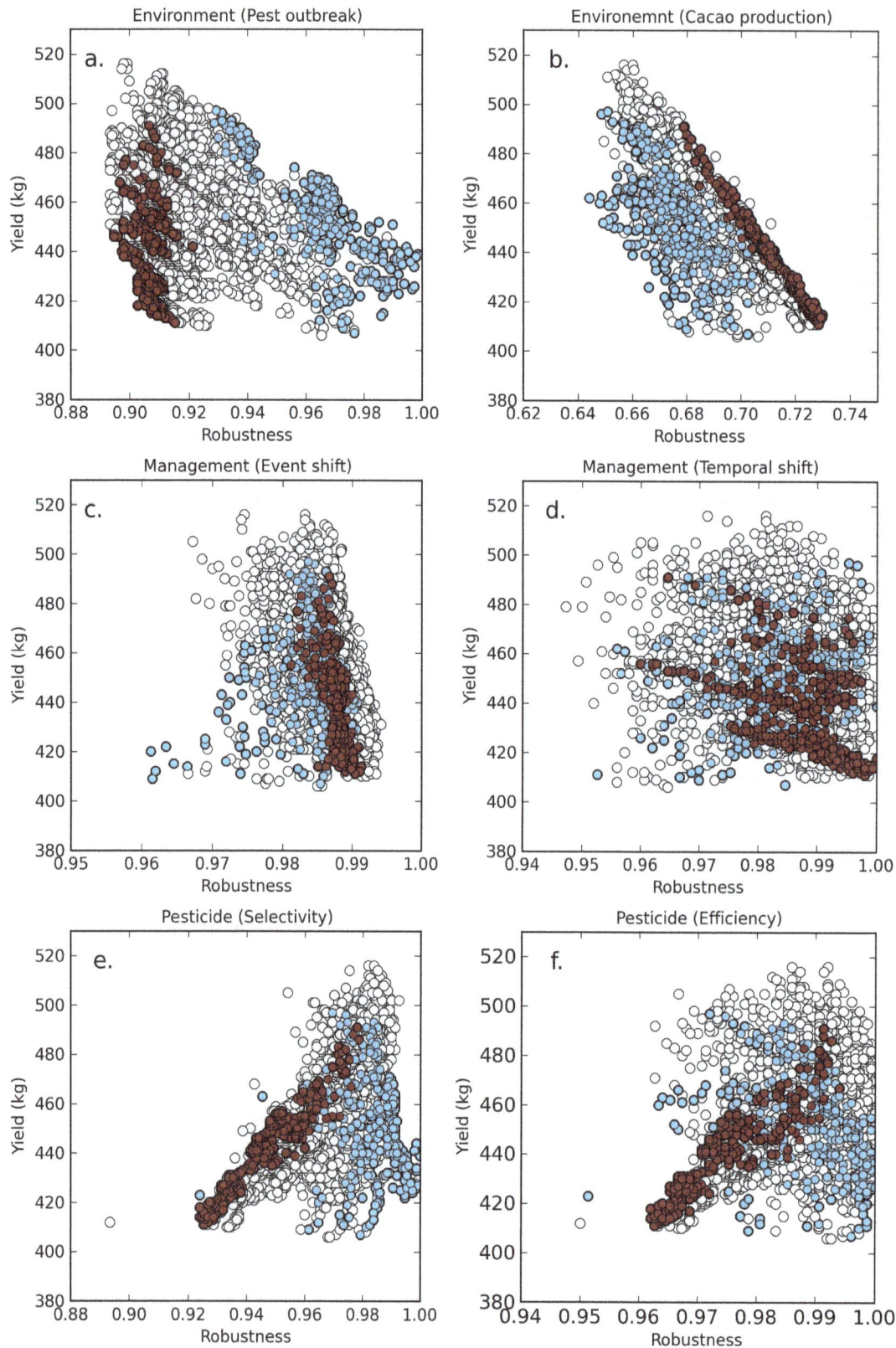

Figure 4. Relationship between yield and robustness under different perturbations. Under environmental perturbation (a: pest outbreak; b: production variation), under a modification of the management strategy (c: shift in one of the spraying/non spraying events; d: temporal shift) or under a perturbation linked to the type of pesticide (e: selectivity; f: efficiency). Each white symbol represents one of the 4096 possible spraying strategies. Dark red symbols stand for the subset of Chemical Based Strategies (CBS, high spraying intensity and a low ecosystem functioning), light blue symbols stand for the subset of Ecological Strategies (ES, low spraying intensity and a high ecosystem functioning).

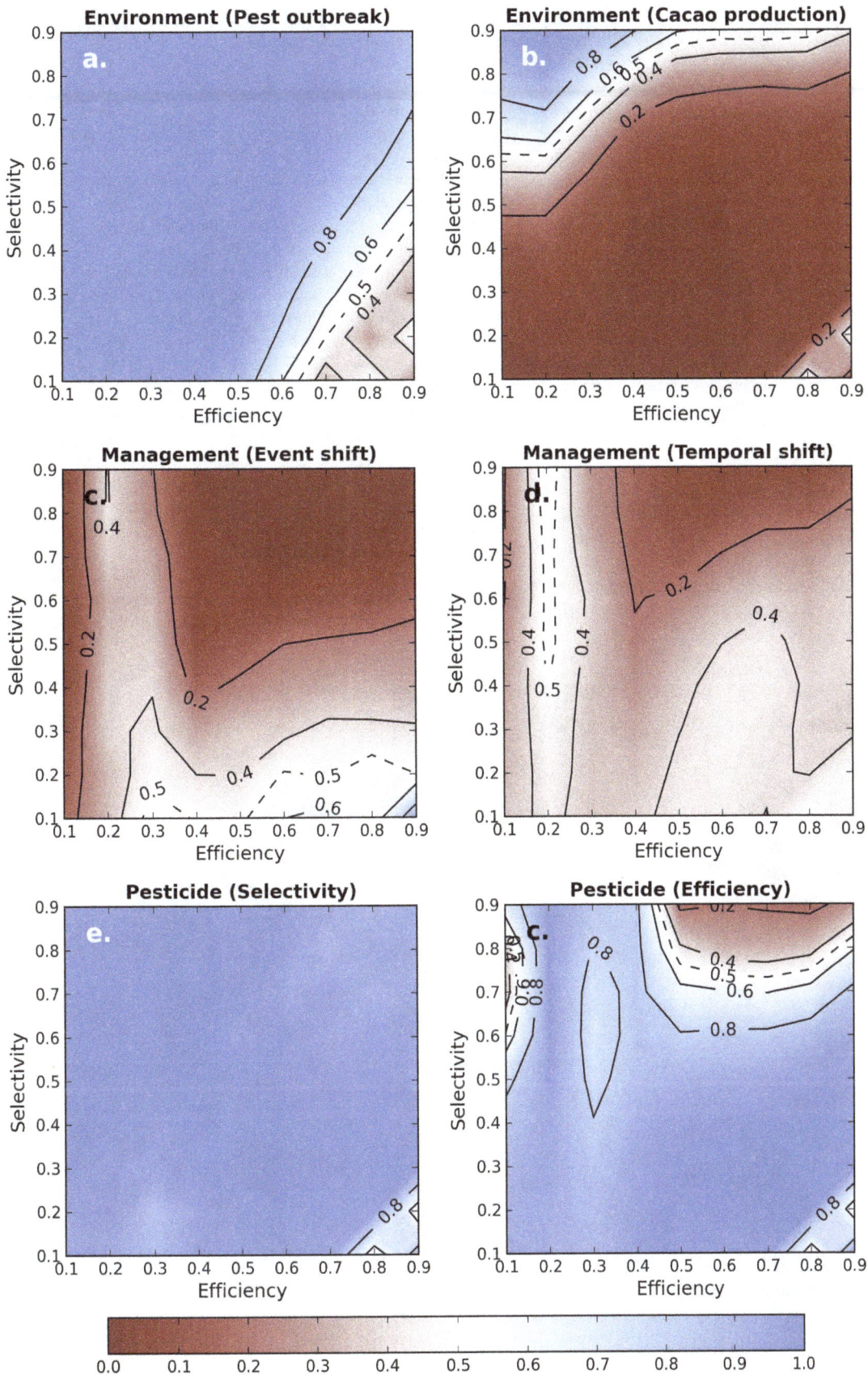

Figure 5. Comparison of the robustness of the agroecosystem managed by Chemical Based Strategies (CBS) or Ecological Strategies (ES) under 6 different perturbations. Figures a and b correspond to an environmental perturbation (a: pest outbreak, b: production perturbation). Figures c and d correspond to perturbation of the management (c: shift of one spraying/non spraying event, d: temporal shift of the

management strategy). Figures e and f correspond to perturbations of the pesticide characteristics (e: selectivity; f: efficiency). Color indicates the comparison index calculated, i.e. the percentage of pairs of strategies in which the ES performs better than the CBS.

smooth but general conclusions can still be drawn. For most pesticides, CBS were more robust than ES to Cacao production-related perturbations (average $CInd = 0.22$) as well as to management-related perturbations (event shift, average $CInd = 0.24$, temporal shift, average $CInd = 0.32$) and ES were more robust than CBS to pest outbreaks (average $CInd = 0.88$) as well as to pesticide-related perturbations (efficiency, average $CInd = 0.80$, selectivity, average $CInd = 0.96$).

However, yield showed a more balanced pattern (Figure 6, average $CInd = 0.51$). ES showed higher yields with pesticides of high efficiency and low selectivity while CBS showed higher yields with pesticides of high selectivity and low efficiency. This illustrates the non-trivial role of pesticides in a complex agroecosystem.

Discussion

Our results showed two opposite behaviors of the Cacao agroecosystem that we modeled depending on their management: First, Cacao agroecosystems relying on a high input of pesticides and showing low ecosystem functioning that fit to the conventional model of chemical intensification inspired by the Green Revolution; second, Cacao agroecosystems relying on ecosystem functioning with a low level of chemical input that fit to the model of

ecological intensification inspired by agroecology-like paradigms. These two types of management led to different levels of production as well as different robustness. From the production point of view, no system clearly outclassed the other and their ranking depended on the type of pesticide used. From the robustness point of view, the two types of systems performed differently when subjected to different types of perturbations. Ecologically intensive systems were more robust to pest outbreaks and pesticide-related perturbation while chemically intensive systems were more robust to management perturbation and production-related perturbations.

Generality of the results

Our results were obtained with a simplified model of a Cacao agroecosystem and the generality of our results should be discussed. The agroecosystem that we modeled only includes three insects, which is far less than what can be found in the Cacao agroecosystem of the study area [26] and in agroecosystems in general (e.g. [27]). However, ecosystem properties depend much more on functional diversity than on species richness *per se* [28] and agroecosystems are no exception [29]. Choosing these three species, we focused on the three main above-ground ecosystem services and disservices commonly found in agroecosystems (pest

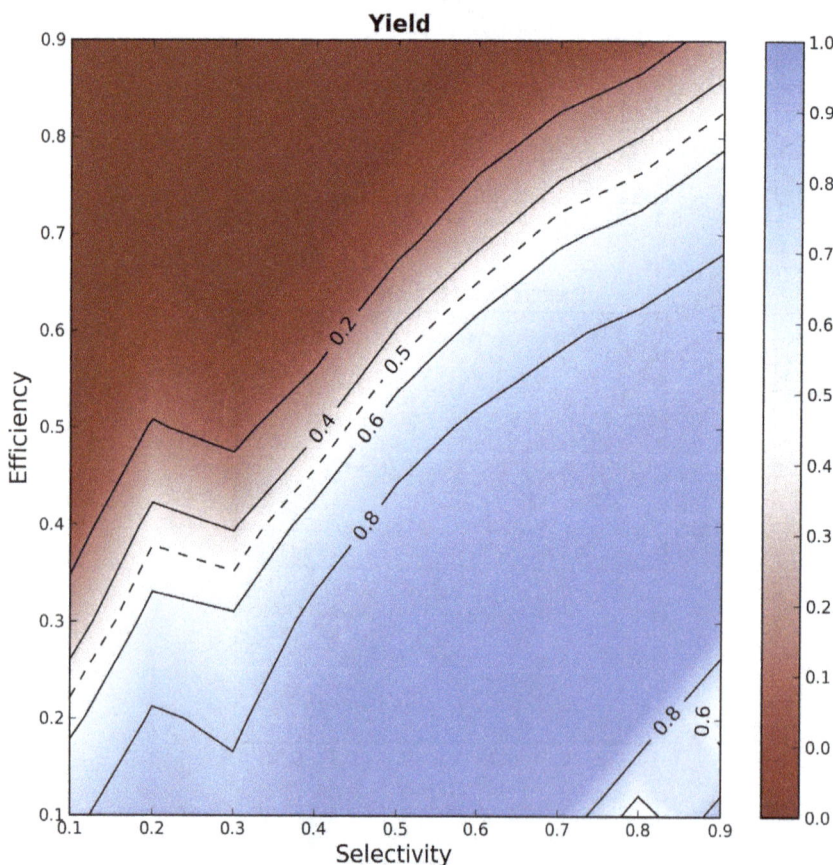

Figure 6. Comparison of the yield obtained in agroecosystems managed by Chemical Based Strategies (CBS) or Ecological Strategies (ES). Color indicates the comparison index calculated, i.e. the percentage of pairs of strategies in which the ES performs better than the CBS.

damage, pest regulation and pollination; [10]). In this sense, even though the quantitative outputs of our models may differ when applied to other agroecosystems, the following qualitative results should remain valid to a broad range of agroecosystems:

- There is a strong negative correlation between the frequency of spraying events and the functioning of the agroecosystem.
- Higher yields can be reached with farming practices based on ecosystem functioning when broad-spectrum pesticides (high efficiency and low selectivity) are used.
- Farming practices based on ecosystem functioning are more robust to pest outbreak perturbations than farming practices based on chemical inputs.
- Farming practices based on ecosystem functioning are more sensitive to management perturbation than farming practices based on chemical inputs.

We see two ways of validating these general findings through modelling. These general results could be tested by applying our modeling framework to other types of agroecosystems (e.g. orchards, oil palm plantations,...). The application of other models of agroecosystems (e.g. [30–33]) to the question raised in this article would also be a way of testing the generality of these results. However, most of these models would first have to be extended to explicitly include crop dynamics and the effects of insects on production. The use of such models developed in different scientific contexts would contribute to the cross-validation of our results.

Limits and perspectives

In this article, we only focused on Cacao agroecosystem management at the field scale through the use of pesticides. Therefore, our model could be developed in two main directions to improve our understanding of the mechanisms of ecological intensification: adding new dimensions to management and increasing spatial scale.

With respect to new dimensions, the dynamics of shade trees could be added to the model. In the specific case of agroforestry systems such as Cacao, the management of shade trees is a strong driver of ecosystem services and disservices [34,35]. Including this aspect in our model would improve its predictive power and allow us to address the paradox raised by [36]: on the one hand, several studies highlight the strong importance of shade trees for the sustainability of the production system, but on the other hand,

farmers tend to remove shade trees to improve yields and do not notice any major drawback.

With respect to scaling-up, several studies have emphasized the importance of the landscape scale when considering ecological dynamics of agroecosystems [37]. Especially, the spatial distribution of insects involved in ecosystem services such as pest control has a strong effect on pest populations [38,39]. In our study area, parasitism rate, for instance, has been shown to depend strongly on the distance to forest [40] and pollination also depends strongly on landscape structure [41]. Transferring our model to the landscape scale would make it possible to consider both ecological and economic interactions between different fields. Refining the management component of our model and transferring it to greater spatial scales would increase the number of tools available to balance between ecosystem services and disservices and give a better understanding of how ecological intensification could be put into practice.

Conclusion

We modeled a Cacao agroecosystem under two management scenarios. The Cacao agroecosystem managed in an ecologically intensive way strongly differed from the one managed in a more conventional way using high quantities of chemical inputs. The ecologically intensive Cacao agroecosystem was more robust to pest outbreaks. It also showed higher yields when broad spectrum pesticides were used. However, the ecologically intensive Cacao agroecosystem was more sensitive to management-related perturbations and Cacao production perturbations, which confirms the high level of expertise needed to conduct such a management.

Acknowledgments

We thank Andreas Spangenberg for farm surveys and Gerald Moser for providing us the Cacao yield databases.

Author Contributions

Conceived and designed the experiments: RS. Performed the experiments: RS. Analyzed the data: RS. Wrote the paper: RS KW KM.

References

1. Fukuoka M (2010) The one-straw revolution: An introduction to natural farming. New York Review Books. 229 p.
2. Altieri MA, Nicholis CI (2005) Agroecology and the search for a truly sustainable agriculture. Mexico D.F., Mexico: United Nations Environmental Programme, Environmental Training Network for Latin America and the Caribbean.
3. Gliessman SR (2012) Agroecology: Researching the ecological basis for sustainable agriculture. Springer London, Limited. 400 p.
4. Swaminathan MS (1996) Sustainable Agriculture: Towards an evergreen revolution. Konark Publishers. 219 p.
5. Conway G (1998) The doubly green revolution: food for all in the twenty-first century. Comstock Pub Associates. 356 p.
6. Kremen C, Williams NM, Aizen MA, Gemmill-Herren B, LeBuhn G, et al. (2007) Pollination and other ecosystem services produced by mobile organisms: a conceptual framework for the effects of land-use change. Ecol Lett 10: 299–314. doi:10.1111/j.1461-0248.2007.01018.x.
7. Bommarco R, Miranda F, Bylund H, Björkman C (2011) Insecticides suppress natural enemies and increase pest damage in cabbage. J Econ Entomol 104: 782–791.
8. Altieri MA, Ponti L, Nicholls CI (2005) Manipulating vineyard biodiversity for improved insect pest management: case studies from northern California. Int J Biodivers Sci Manag 1: 191–203.
9. Ives AR, Carpenter SR (2007) Stability and diversity of ecosystems. Science 317: 58–62. doi:10.1126/science.1133258.
10. Zhang W, Ricketts TH, Kremen C, Carney K, Swinton SM (2007) Ecosystem services and dis-services to agriculture. Ecol Econ 64: 253–260. doi:10.1016/j.ecolecon.2007.02.024.
11. Tilman D, Downing JA (1994) Biodiversity and stability in grasslands. Nature 367: 363–365. doi:10.1038/367363a0.
12. Holling CS (1973) Resilience and stability of ecological systems. Annu Rev Ecol Syst 4: 1–23. doi:10.1146/annurev.es.04.110173.000245.
13. Stelling J, Sauer U, Szallasi Z, Doyle FJ 3rd, Doyle J (2004) Robustness of cellular functions. Cell 118: 675–685. doi:10.1016/j.cell.2004.09.008.
14. Carpenter S, Walker B, Anderies JM, Abel N (2001) From metaphor to measurement: resilience of what to what? Ecosystems 4: 765–781. doi:10.1007/s10021-001-0045-9.
15. Ruf F (1995) Booms et crises du cacao: les vertiges de l'or brun. KARTHALA Editions. 504 p.
16. FAO website (2013) Available: www.faostat.fao.org, Accessed 2013 July 17.
17. Entwistle PF, Philip F (1972) Pests of cocoa. [London]: Longman.
18. Groeneveld JH, Tscharntke T, Moser G, Clough Y (2010) Experimental evidence for stronger Cacao yield limitation by pollination than by plant resources. Perspect Plant Ecol Evol Syst 12: 183–191. doi:10.1016/j.ppees.2010.02.005.
19. Sabatier R, Meyer K, Wiegand K, Clough Y (2013) Non-linear effects of pesticide application on biodiversity-driven ecosystem services and disservices in

a Cacao agroecosystem: A modeling study. Basic Appl Ecol 14: 115–125. doi:10.1016/j.baae.2012.12.006.

20. Day R (1985) Control of the cocoa pod borer (Conopomorpha cramerella) London: University of London.

21. Hassell MP (1978) The spatial and temporal dynamics of host-parasitoid interactions. Oxford; New York: Oxford University Press.

22. Séchan Y, Faaruia M, Tetuanui A (1996) Lutte contre le nono blanc des plages: rapport d'exécution et de fin des travaux. Papeete: ITRMLM. Available: http://www.documentation.ird.fr/hor/fdi:010013638. Accessed 8 August 2013.

23. Moser G, Leuschner C, Hertel D, Hölscher D, Köhler M, et al. (2010) Response of cocoa trees (Theobroma cacao) to a 13-month desiccation period in Sulawesi, Indonesia. Agrofor Syst 79: 171–187. doi:10.1007/s10457-010-9303-1.

24. Schwendenmann L, Veldkamp E, Moser G, Hölscher D, Köhler M, et al. (2010) Effects of an experimental drought on the functioning of a Cacao agroforestry system, Sulawesi, Indonesia. Glob Change Biol 16: 1515–1530. doi:10.1111/j.1365-2486.2009.02034.x.

25. Wielgoss A, Clough Y, Fiala B, Rumede A, Tscharntke T (2012) A minor pest reduces yield losses by a major pest: plant-mediated herbivore interactions in Indonesian Cacao: Indirect interaction between Cacao pests. J Appl Ecol 49: 465–473. doi:10.1111/j.1365-2664.2012.02122.x.

26. Kessler M, Abrahamczyk S, Bos M, Buchori D, Putra DD, et al. (2011) Cost-effectiveness of plant and animal biodiversity indicators in tropical forest and agroforest habitats. J Appl Ecol 48: 330–339. doi:10.1111/j.1365-2664.2010.01932.x.

27. Duelli P, Obrist MK, Schmatz DR (1999) Biodiversity evaluation in agricultural landscapes: above-ground insects. Agric Ecosyst Environ 74: 33–64. doi:10.1016/S0167-8809(99)00029-8.

28. Hooper DU, Chapin Iii FS, Ewel JJ, Hector A, Inchausti P, et al. (2005) Effects of biodiversity on ecosystem functioning: a consensus of current knowledge. Ecol Monogr 75: 3–35.

29. Moonen A-C, Bàrberi P (2008) Functional biodiversity: An agroecosystem approach. Agric Ecosyst Environ 127: 7–21. doi:10.1016/j.agee.2008.02.013.

30. Chatterjee S, Isaia M, Venturino E (2009) Spiders as biological controllers in the agroecosystem. J Theor Biol 258: 352–362. doi:10.1016/j.jtbi.2008.11.029.

31. Ives AR, Settle WH (1997) Metapopulation dynamics and pest control in agricultural systems. Am Nat 149: 220–246.

32. Drechsler M, Settele J (2001) Predator–prey interactions in rice ecosystems: effects of guild composition, trophic relationships, and land use changes—a model study exemplified for Philippine rice terraces. Ecol Model 137: 135–159.

33. Bambaradeniya CNB, Edirisinghe JP (2008) Composition, structure and dynamics of arthropod communities in a rice agro-ecosystem. Ceylon J Sci Biol Sci 37: 23–48.

34. Steffan-Dewenter I, Kessler M, Barkmann J, Bos MM, Buchori D, et al. (2007) Tradeoffs between income, biodiversity, and ecosystem functioning during tropical rainforest conversion and agroforestry intensification. Proc Natl Acad Sci 104: 4973–4978. doi:10.1073/pnas.0608409104.

35. Tscharntke T, Clough Y, Bhagwat SA, Buchori D, Faust H, et al. (2011) Multifunctional shade-tree management in tropical agroforestry landscapes – a review. J Appl Ecol 48: 619–629. doi:10.1111/j.1365-2664.2010.01939.x.

36. Ruf FO (2011) The myth of complex cocoa agroforests: the case of Ghana. Hum Ecol 39: 373–388. doi:10.1007/s10745-011-9392-0.

37. Tscharntke T, Klein AM (2005) Landscape perspectives on agricultural intensification and biodiversity – ecosystem service management. Ecol Lett 8: 857–874.

38. Bianchi F, Schellhorn NA, Buckley YM, Possingham HP (2010) Spatial variability in ecosystem services: simple rules for predator-mediated pest suppression. Ecol Appl 20: 2322–2333.

39. Ricci B, Franck P, Toubon J-F, Bouvier J-C, Sauphanor B, et al. (2009) The influence of landscape on insect pest dynamics: a case study in southeastern France. Landsc Ecol 24: 337–349. doi:10.1007/s10980-008-9308-6.

40. Klein A-M, Steffan-Dewenter I, Tscharntke T (2006) Rain forest promotes trophic interactions and diversity of trap-nesting Hymenoptera in adjacent agroforestry. J Anim Ecol 75: 315–323. doi:10.1111/j.1365-2656.2006.01042.x.

41. Priess JA, Mimler M, Klein AM, Schwarze S, Tscharntke T, et al. (2007) Linking deforestation scenarios to pollination services and economic returns in coffee agroforestry systems. Ecol Appl Publ Ecol Soc Am 17: 407–417.

42. Juhrbandt J (2011) Economic valuation of of land use change-A case study on rainforest conversion and agroforestry intensification in Central Sulawesi, Indonesia. Available: https://ediss.uni-goettingen.de/handle/11858/00-1735-0000-0006-AB32-C. Accessed 7 August 2013.

20

Indigenous Burning as Conservation Practice: Neotropical Savanna Recovery amid Agribusiness Deforestation in Central Brazil

James R. Welch[1]*, Eduardo S. Brondízio[2,3], Scott S. Hetrick[3], Carlos E. A. Coimbra Jr[1]

1 Escola Nacional de Saúde Pública, Fundação Oswaldo Cruz, Rio de Janeiro, Rio de Janeiro, Brazil, **2** Department of Anthropology, Indiana University, Bloomington, Indiana, United States of America, **3** Anthropological Center for Training and Research on Global Environmental Change, Indiana University, Bloomington, Indiana, United States of America

Abstract

International efforts to address climate change by reducing tropical deforestation increasingly rely on indigenous reserves as conservation units and indigenous peoples as strategic partners. Considered win-win situations where global conservation measures also contribute to cultural preservation, such alliances also frame indigenous peoples in diverse ecological settings with the responsibility to offset global carbon budgets through fire suppression based on the presumed positive value of non-alteration of tropical landscapes. Anthropogenic fire associated with indigenous ceremonial and collective hunting practices in the Neotropical savannas (cerrado) of Central Brazil is routinely represented in public and scientific conservation discourse as a cause of deforestation and increased CO_2 emissions despite a lack of supporting evidence. We evaluate this claim for the Xavante people of Pimentel Barbosa Indigenous Reserve, Brazil. Building upon 23 years of longitudinal interdisciplinary research in the area, we used multi-temporal spatial analyses to compare land cover change under indigenous and agribusiness management over the last four decades (1973–2010) and quantify the contemporary Xavante burning regime contributing to observed patterns based on a four year sample at the end of this sequence (2007–2010). The overall proportion of deforested land remained stable inside the reserve (0.6%) but increased sharply outside (1.5% to 26.0%). Vegetation recovery occurred where reserve boundary adjustments transferred lands previously deforested by agribusiness to indigenous management. Periodic traditional burning by the Xavante had a large spatial distribution but repeated burning in consecutive years was restricted. Our results suggest a need to reassess overreaching conservation narratives about the purported destructiveness of indigenous anthropogenic fire in the cerrado. The real challenge to conservation in the fire-adapted cerrado biome is the long-term sustainability of indigenous lands and other tropical conservation islands increasingly subsumed by agribusiness expansion rather than the localized subsistence practices of indigenous and other traditional peoples.

Editor: Brock Fenton, University of Western Ontario, Canada

Funding: Financing was provided by the Brazilian National School of Public Health (INOVA-ENSP program), the Fulbright Commission (Fulbright-Hays DDRAF no. P022A040016), and the Brazilian Research Council (CNPq grants 475674/2008-1, 403569/2008-7, and 500072/2010-8). The funders had no role in study design, data collection and analysis, decision to publish, or preparation of the manuscript.

Competing Interests: The authors have declared that no competing interests exist.

* E-mail: welch@ensp.fiocruz.br

Introduction

Efforts to reduce the negative environmental and biodiversity impacts of commercial agriculture and pasture activities are increasingly recognized to benefit from locally based knowledge and practices [1]. The new United Nations Intergovernmental Science-Policy Platform on Biodiversity and Ecosystem Services (IPBES) program, for instance, recognizes explicitly the need to include local and indigenous knowledge as part of its assessment and policy reach mandates [2]. Nevertheless, international conservation discourse generates important misconceptions by presuming the destructiveness of human alteration of tropical landscapes and, specifically, overgeneralizing about the effects of anthropogenic fire in diverse cultural and ecological settings.

Indigenous reserves and other types of protected areas have become the most important policy mechanism for controlling deforestation [3,4] and fire [5] associated with the booming expansion of large-scale agriculture in Northern and Central Brazil. Today these areas represent a mosaic of conservation islands amid expanding monoculture landscapes and are often considered win-win situations where global conservation goals and reduction in carbon emissions also contribute to cultural preservation [3,6,7]. Nevertheless, the generalized association between fire impacts on conservation, CO_2 emissions, and climate change is contributing to important misrepresentations when extrapolated without attention to regional and local contexts.

Increasingly part of the widespread implementation of Reducing Emissions from Deforestation and Forest Degradation (REDD) and REDD+ programs [8], these ideas regarding anthropogenic fire are also widely incorporated by interest groups as political narratives contesting indigenous rights to land in Brazil. They promote the notion that indigenous burning activities represent a destructive mentality ("culturally endorsed pyromania," to borrow an expression applied to indigenous practices in northern Australia

[9]) out of line with the conservation agenda set forward at global, national, and local levels to revert biodiversity losses and climate change. Thus, localized traditional landscape management practices are subjected to the scrutiny of unexpected alliances (e.g., large commercial ranchers using conservation arguments). Increasingly, indigenous and other traditional peoples in Brazil are framed with the responsibility to offset a substantial portion of the national and global carbon budget through conservation stewardship and REDD+ projects financed by high-emission industrialized countries [10]. Nevertheless, industrial agriculture and ranching expansion since the late 1960s has transformed the Central Brazilian cerrado into one of the most threatened biomes in the country [11,12].

Biodiversity and Fire in the Cerrado

The cerrado, included among the 35 most important "hotspots" in the world [13,14], is a diverse but highly threatened tropical savanna-like biome covering over 2 million hectares, approximately 24% of the total area of Brazil, mainly in the Central Brazilian Plateau. According to one estimate, of the approximately 10,000 plant species identified in the cerrado, about 4,400 are endemic [15]. The agricultural potential of the cerrado and its positive economic benefits for Brazil have been widely publicized for decades [16]. At least 35% of the cerrado region has been completely converted to intensive human use since 1960, predominantly for pasture, intensive monoculture and, more recently, eucalyptus cultivation [11]. Recent land conversion to soybean and sugarcane biofuel production in the cerrado results in large carbon debts estimated to require 17 to 37 years to repay [17]. In lieu of buffering the Amazon from further expansion of agricultural commodities and biofuels, in coming years the cerrado region is prone to increasing pressures with associated climatic

impacts [18,19] and escalation of conflicts between agribusiness and indigenous interests [20].

Differently from the Brazilian rainforests – Amazonian and Atlantic forests, adaptation to periodic surface fires is an evolutionary characteristic of cerrado ecology independently from human action [21,22]. Lightning, which is common during the rainy season, is considered the principal environmental ignition factor in the cerrado [23] and explains evidence of cerrado wildfires as early as 32,000 BP, long before human presence in the region [24,25]. The recurrence of fire in a determined cerrado landscape is thought to have resulted in over 70% of the total cerrado biomass being subterranean, as well as other anatomical adaptations providing resistance to fire damage and favoring vigorous vegetation regrowth immediately after the fires, even before first rains [21,26,27]. The diversity of structural forms in the cerrado biome is considerable. Depending on the environmental factors that predominate in a given region, such as soil depth and drainage, level of acidity, and aluminum saturation, vegetation can range from dense forests to open grasslands, with many intermediate configurations in close proximity [27].

The presence and potential positive ecological effects of anthropogenic fire in the cerrado has been noted by botanists travelling through the region since the nineteenth century. In 1892, botanist Eugene Warming observed that plant form and vegetation structure in the cerrados of Minas Gerais, Brazil, benefitted from fire, with most plants undergoing vigorous sprouting and flowering shortly after being burned [28]. As he wrote, "the most beautiful, also the most blooming and freshest green field I have ever seen was precisely one that was burned in October" [28]. More recently, based on observations of cerrados in Xavante territory, botanist George Eiten wrote that frequencies of indigenous fires greater than every one or two years had no evident long-term effects on plant physiognomy and recently

Figure 1. Xavante elder during collective hunt with fire, Pimentel Barbosa Indigenous Reserve, Brazil, 2005. Photograph by James R. Welch. The legal representative of the depicted individual provided informed written consent for the publication of his image.

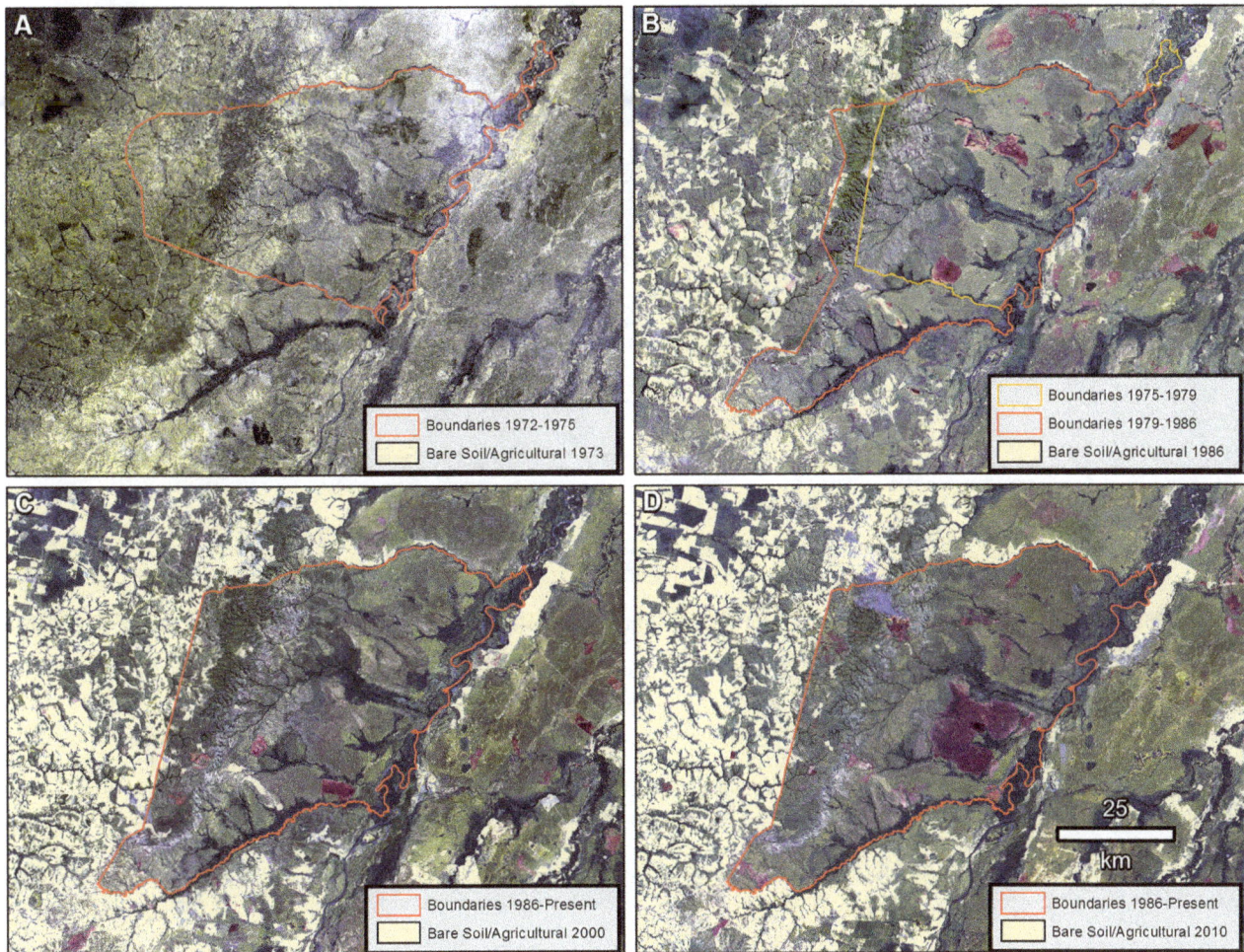

Figure 2. Mosaic of calibrated satellite imagery overlaid with reserve boundaries and deforestation mapped from 1973 to 2010, Pimentel Barbosa Indigenous Reserve, Brazil, 1973–2010. Landsat MSS and TM imagery used in our analyses was obtained courtesy of the United States Geological Survey (USGS) and from the Brazilian National Institute for Space Research (INPE) under Creative Commons Attribution-ShareAlike 3.0 Unported License.

burned areas attracted game animals through the creation of mineral licks and the abundant production of tender plant shoots [27].

Recent ecological studies show that cerrado landscapes may be even more highly resistant to the impacts of fire, including anthropogenic burning, than previously thought [21,22]. Also, fire regimes that vary in intensity, frequency, and seasonality may have significantly different effects on biomass and vegetation structure [26]. Periodic burning has been found to result in less intense and more fragmented, and therefore less destructive, fires than occur in protected cerrado areas where fire suppression causes the accumulation of large amounts of dead aboveground biomass [22,23,29]. Conversely, sporadic fires produce positive effects on cerrado biodiversity [25,30]. According to botanists, fires benefit cerrado plants due to adaptations such as fire-dependent sexual reproduction and dispersal, as well as rapid regeneration and increased flowering and germination [29,30]. These adaptations in turn benefit wildlife through increased availability of such foods as pollens, nectars, shoots, and fruits [29].

Although fire can effect major short term landscape change through its impact on cerrado vegetation, the resilience of fauna populations is indicated by scant evidence of local animal extinctions [22,31]. Additionally, the low temperatures and patchy distribution of cerrado surface fires allow animals ample escape routes and areas of cover. Ecological evidence also shows that moderate intensity fires increase the availability of foods for some taxa, including small vertebrates and arthropods, by stimulating vegetative growth, flowering, and fruiting [32–34]. Ecological studies specifically evaluating the impacts of indigenous burning activities in the cerrado also found no evidence that animal populations, including those of economically important game species, decreased following fires [35–39].

Cultural Aspects of Hunting with Fire

Hunting with fire is a characteristic subsistence activity of many indigenous inhabitants of the Central Brazilian cerrado, including such Gê-speaking indigenous groups as the Kayapó, Krahô, Canela, and Xavante [25]. Contrary to journalistic representations of Xavante hunting with fire as a cause of cerrado degradation and deforestation [40,41] or an impediment to vegetation recovery [42], ethnographic accounts indicate periodic burning by these indigenous peoples has minimal or desirable effects on cerrado vegetation structure and productivity [43,44]. Our previous ethnographic research [45] revealed that the Xavante people,

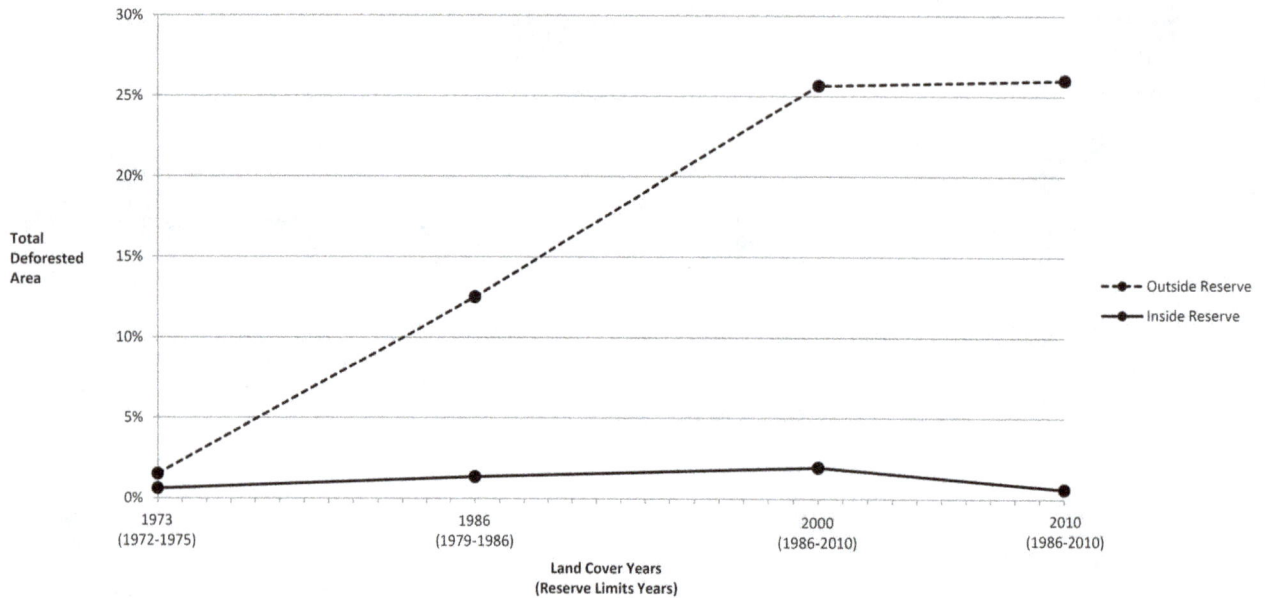

Figure 3. Total deforested area (%) relative to total area inside and outside (20 km buffer) corresponding limits of Pimentel Barbosa Indigenous Reserve, Brazil, 1973–2010.

eastern Mato Grosso State, use fire for hunting annually during the dry season (primarily July to September and less frequently the moister months of May and June) in widely dispersed locations throughout the reserve.

Xavante collective hunting with fire often occurs in conjunction with such ceremonial occasions as weddings and adolescent rites of passage [45]. Hunting excursions follow a ritualized format,

whereby representatives from opposite exogamous moieties set fire to the vegetation as they run footraces in two wide arcs towards a distant finishing place. As the fire expands in a shifting mosaic pattern, numerous hunters act individually or in collaboration with other hunters to dispatch fleeing game animals (Figure 1). At the conclusion of hunting excursions, the acquired game meat is often pooled and ceremoniously presented to designated recipients, such

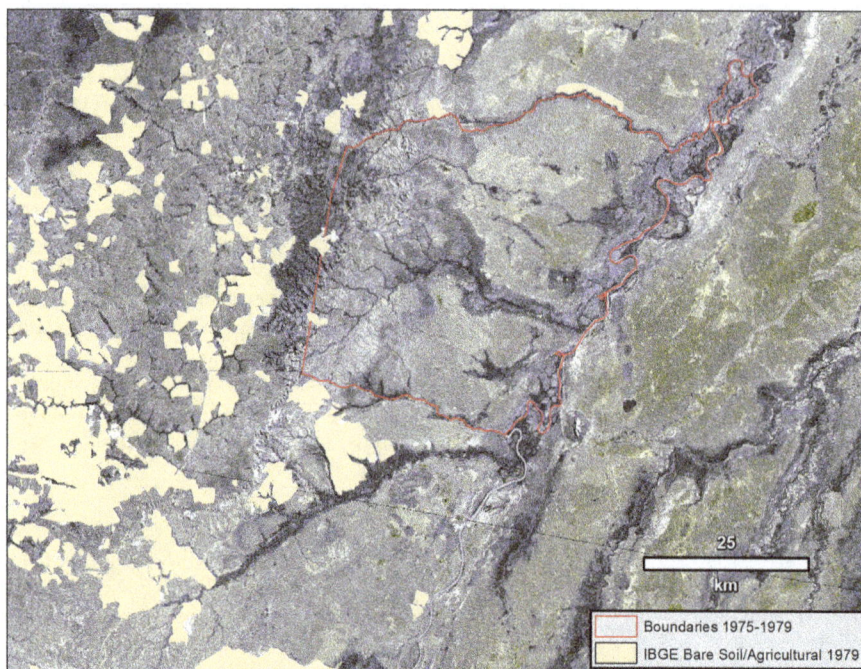

Figure 4. Mosaic of calibrated satellite imagery (1976) overlaid with reserve boundaries (1975–1979) and areas of bare soil used for livestock pasture and agricultural fields (1979), Pimentel Barbosa Indigenous Reserve, Brazil. Landsat MSS imagery used in our analyses was obtained from the Brazilian National Institute for Space Research (INPE) under Creative Commons Attribution-ShareAlike 3.0 Unported License. Cartographic data were digitized with permission from IBGE topographic maps [54].

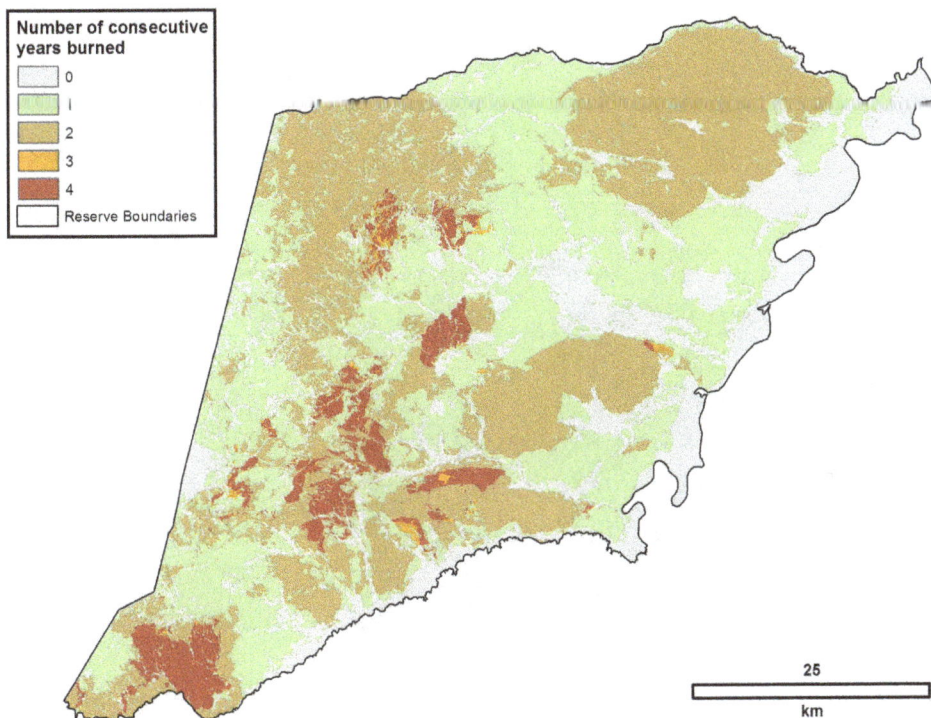

Figure 5. Burn scar transition matrix analysis showing land area by number of sequential years burned, Pimentel Barbosa Indigenous Reserve, Brazil, 2007–2010. Landsat TM and ETM+ imagery used in our analyses were obtained courtesy of the United States Geological Survey (USGS).

as in-laws and initiation leaders. From the Xavante perspective, collective hunting with fire is an essential component of the ceremonial activities that distinguish them ethnically and causes no undesirable ecological changes when practiced according to traditional protocols. To the contrary, Xavante cultural knowledge recognizes that traditional periodic burning accelerates the vegetative growth of some plants, increases the availability of certain animal foods, and reduces the intensity of future fires.

Although the Xavante practice agriculture and in the 1970s sold a small portion of rice yields as part of a government development project, today this activity is largely restricted to domestic production in small rotating patches, primarily in narrow gallery

Figure 6. Agribusiness deforestation (right) along western border of Pimentel Barbosa Indigenous Reserve (left), Brazil, 2009. Photograph by James R. Welch.

Figure 7. Remains of mechanized tree-chaining on private property adjacent to western boundary of Pimentel Barbosa Indigenous Reserve, Brazil, 2012. Photograph by Carlos E. A. Coimbra Jr.

forests along river margins [46]. In contrast to agriculture, periodic hunting with fire has a large spatial distribution and therefore greater potential for causing landscape-scale ecological effects within reserve boundaries. However, Xavante strategies for hunting with fire are highly regulated by specific traditional protocols and knowledge aimed to maximize both short-term hunting productivity and long-term sustainability. Our ethnographic research [45] shows that when planning fires, Xavante hunters discuss such diverse factors as season, weather, winds, soil and foliage moisture, type of vegetation, and natural barriers in order to control their intensity, spatial distribution, and frequency through time (see also [47–49]). Among these considerations ample attention is given to the potential negative ecological effects of burning the same patches of vegetation in consecutive years, especially during extended droughts.

Aims of this study

In this paper, we evaluate conservation narratives regarding the relative long-term ecological impacts of indigenous and agribusiness management of Brazilian neotropical savanna landscapes based on estimates of diachronic environmental changes and indigenous burning patterns in a cerrado landscape in Mato Grosso State. Specifically, we compared land cover change over four decades inside and within a 20 km buffer area around Pimentel Barbosa Indigenous Reserve (Xavante ethnic group) in order to ascertain their respective deforestation patterns. Multitemporal spatial analyses of changes in vegetation cover (1973–2010) were performed utilizing satellite and georeferenced field data, as well as archival research for reconstructing changes in reserve boundaries since 1972. Additionally, in order to evaluate the contemporary Xavante burning regime contributing to observed results at the end of this sequence, we analyzed the spatial and temporal distribution of anthropogenic fire within the reserve utilizing a four year sample (2007–2010). This integrated study expands upon our long-term ethnographic and interdisci-

plinary research carried out in collaboration with the Xavante people since the 1990s [45,50–53].

Methods

Ethics statement

Permission for the project was provided by the National Indian Foundation (FUNAI) and the National Ethics Committee (CONEP, permission #652-2011).

Location and population

Research was conducted in the Pimentel Barbosa Indigenous Reserve, Mato Grosso State, Brazil. The majority of this territory pertains to the Araguaia River Basin, although a small portion at the western boundary pertains to the Xingu River Basin. The ecological biome is cerrado, with a division occurring between western portions of the reserve, in which cerrado *sensu stricto* and dry forests predominate, and eastern portions, in which inundated grasslands are more prevalent. The boundaries of this reserve underwent a series of alterations during the study period (through decrees in 1972, 1975, and 1979, and ratification in 1986).

The 2010 population of Pimentel Barbosa Indigenous Reserve was approximately 1,400 individuals residing in nine villages [45]. The entire population pertains to the Xavante ethnic group and all individuals speak the Xavante language (Gê language family).

Georeferenced field data

Field data was collected for 117 georeferenced training samples inside and outside the reserve. These included multiple locations for 16 land cover classes (water, bare soil, agricultural field/pasture, fire scar, open grassland, arboreous (*murundu*) grassland, scrub grassland, cerrado *sensu stricto*, dense cerrado, rupicolous cerrado, *cerradão* woodland, dry upland forest, gallery forest, bamboo forest, *buriti* palm forest, and *ipuca* flooded forest). Data collected included vegetation type, land cover, land use history, regrowth status, notable species, and photographs. Elder members of the indigenous community accompanied all fieldwork activities and assisted with the identification of land use histories at all training sample locations.

Multi-temporal remote sensing: classification and deforestation analysis

Using Landsat MSS (WRS I Path/Row −240/69; 240/70; 80 meter spatial resolution) and TM (WRS II Path Row −223/69; 223/70; 224/69; 224/70; 30 meter spatial resolution) imagery from the United States Geological Survey (USGS) and Brazilian National Institute for Space Research (INPE), a multi-temporal data set was established for the image dates 1973, 1986, 2000, and 2010. Each date was created using a mosaic of two (1973) and four (1986, 2000, 2010) contiguous images. All images were subjected to radiometric and atmospheric calibrations, georeferenced, and image subsets were layer stacked to form multitemporal images. Image classification was carried out using ERDAS IMAGINE 11.0.4. Deforestation analysis utilized a land cover classification system consisting of water, bare soil/agricultural, grassland/cerrado, and forest.

Thematic data extraction for reserve boundaries and a 20 km buffer zone around the reserve was carried out using the tabulate area tool in ArcGIS 10.1. Classification procedure for the 1973 image was based on unsupervised procedures only; for 1986, 2000, and 2010 classification was based on a hybrid approach (i.e., involving integration of unsupervised and supervised classifiers). Images were independently classified. Image classification benefited from field collected georeferenced training samples. Field sites representative of natural vegetation were used as references for all dates, while recently altered sites were used as references for 2010. A layer of field sites was created and referenced into the multi-temporal image dataset.

Unsupervised analysis using ISODATA started with high-dimensional unsupervised clustering (50 classes) of the whole image area. Classes were analyzed according to spectral-structural differences (relationship between spectral data and indicators of vegetation structure), spatial distribution, and statistical values (mean, standard deviation, and covariance) for all spectral bands except thermal. Using supervised procedures, training samples with known vegetation and history of land use were created and used to develop spectral signatures for the classes of interest in each data set for 1986, 2000, and 2010. Training samples and unsupervised classes were combined into a single spectral signature file. Analysis of training samples and unsupervised signatures include a combination of Transformed Divergence separability analysis, spectral signature comparison, and correlation analysis of spectral-vegetation data. Based on these procedures, signatures were merged (or eliminated) to produce a final signature set representing a gradient of vegetation structure, from bare soil/agricultural to high forest. Signature sets were submitted to a probability-based Gaussian Maximum-Likelihood classifier. Classification accuracy of the 2010 classified image was assessed for aggregated classes (i.e., bare soil/agricultural, grassland/cerrado, and forests) used in transition matrix analysis. A set of 30 test fields of known features (i.e., training samples not used for supervised classification) was used to assess the accuracy of each class. Classification accuracy ranged from 91% for bare soil/agricultural, 86% for grassland/cerrado, and 84% for forests. Confusion between classes occurred mainly in areas of transitional vegetation structure with mixed elements of soil, herbaceous, arbustive, and arboreal coverage.

Legal descriptions of historical reserve boundaries (dates 1972–1975, 1975–1979, 1979–1986, and 1986-present) were digitally reconstructed from archival maps and descriptive documents of delimiting landmarks and overlaid on the multi-temporal image dataset in ArcGIS. Analyses of deforestation trajectories were developed by calculating transition matrices in ERDAS IMAGINE for each subsequent pair of images starting with the earliest date (1973–1986, 1986–2000, and 2000–2010). For the transition matrix analysis grassland and cerrado classes were aggregated. Deforestation was considered when a shrub or arboreous class converted into bare soil or agricultural fields/pasture) at a subsequent date. Areas of bare soil/agricultural that returned into grassland/cerrado or forest classes were recoded as such to indicate a reforestation process. The final thematic layer produced eight classes, including those that remained unaltered across dates (i.e., water, bare soil/agricultural, grassland/cerrado, and forest) and those involving change, including deforestation before 1973, deforestation between 1973 and 1986, deforestation between 1986 and 2000, and deforestation between 2000 and 2010. The final layer was recoded to show the area deforested for each period. We used cross-tabulation in ArcGIS to allow estimation of deforestation during each period within (respective reserve boundaries) and outside (respective buffer zones) the reserve.

We used ancillary data to verify the expansion of agriculture and pasture within the reserve during the late 1970s, years not represented in the analysis sequence described above but described by Xavante observers and historical sources as the apex of agribusiness encroachment and deforestation in and around Pimentel Barbosa Indigenous Reserve [45,50]. A thematic map of land cover for 1979 was produced based on cartographic maps by the Brazilian Institute for Geography and Statistics (IBGE) printed at a scale of 1:100,000 [54]. These IBGE maps were developed from aerial photographs of the area from 1967 and fieldwork from 1979. The map was digitized to produce a thematic layer, which was overlaid to a Landsat MSS image (1976) for correction. This map discriminates cerrado and areas opened for agriculture and pasture. The resulting thematic layer provides important supporting evidence for the deforestation of cerrado areas within reserve boundaries and the buffer zone towards the end of the 1970s.

Fire matrix analysis

Burned areas within reserve boundaries in the years 2007–2010 were mapped using Landsat TM and ETM+ imagery from the United States Geological Survey (USGS). Images free of clouds were acquired for each month from July-September for each of the four years to capture the primary annual burning season. As with the land-cover classifications, the imagery used for burned area mapping were coregistered to ensure spatial accuracy. Each image was classified using the ISODATA clustering algorithm in ERDAS IMAGINE. Fifty clusters were initially created using a convergence threshold of 0.995. Burned areas were isolated from other land covers through spectral signature analysis and visual interpretation. Mixed pixels indicating burning scars were manually digitized if necessary. Classifications from all months in the same year were then mosaicked and manually cleaned before running a 3×3 majority filter to complete the burned area classification. Each pair of years was subjected to a transition matrix analysis to estimate areas of consecutive burnings: first the transition from 2007 to 2008, second the transition from 2007/2008 to 2009, and third the transition from 2007/2008/2009 to 2010. A final transition matrix image 2007–2010 comprised of 5 classes was created to illustrate land area by number of consecutive years burned (i.e., burning in no more than 0, 1, 2, 3, and 4 consecutive years).

Results

The evolution of reserve boundaries overlaid on deforestation maps for 1973, 1986, 2000, and 2010 show the progression of deforestation and vegetational recovery inside and outside the reserve (Figure 2). Deforestation outside the reserve indicates a pattern of encroachment from the west with its eastern limit shifting through time as reserve boundaries were adjusted. Comparison of intermediate reserve boundaries (Figure 2B:1975–1979 and 1979–1986) indicate that most bare soil and agricultural fields or pasture inside the reserve in the years 1986 and 2000 was in locations previously outside or bordering the reserve limits. Additionally, large deforested patches evident inside the reserve in 1986 and 2000, especially near its western and southern boundaries, were greatly reduced in size by 2010.

Considering the historical reserve boundaries associated with each image year (Figure 3), the deforested area inside the reserve remained stable at 0.6% between 1973 (1,645.9 ha) and 2010 (1,989.0 ha). However, intermediate years showed higher proportions (1.3% in 1986 and 1.9% in 2000) associated with land previously impacted by agricultural operations being incorporated into the reserve and, to a lesser extent, with agricultural projects promoted by the federal government inside the reserve. The deforested area inside the reserve decreased by 68.9% (4,412.5 ha) from its peak in 2000 to the end of the sequence in 2010. The proportion of deforested land in a 20 km buffer area outside historical reserve limits increased continually from 1.5% (9,589.6 ha) in 1973 to 26.0% (175,412.4 ha) in 2010, with the largest increase observed in the period 1973–1986 (718.4%; 76,470.8 ha) and the smallest in 2000–2010 (1.4%; 2,383.4 ha).

Data from Brazilian topographic maps based on 1979 aerial photography (Figure 4) identify the distribution of commercial livestock pasture and agricultural fields when reserve boundaries were most restricted (1975–1979). These boundaries are also shown without temporally corresponding satellite imagery in Figure 2B. Figure 4 shows that many of the locations first appearing as bare soil/agricultural within the reserve in 1986 (Figure 2B) were in fact previously deforested by agribusiness before they were incorporated into the reserve by legal changes to the reserve limits.

Vegetation burning matrix analyses show the distribution of lands inside the reserve by number of consecutive years burned from 2007 to 2010 (Figure 5). The majority of the reserve burned at least once during the four-year period (83.2%; 273,979.26 ha) but only 40.7% (134,043.84) burned in two or more consecutive years. The largest areas were burned one (42.5%; 139,935.42 ha) or two (35.2%; 115,829.01 ha) years sequentially. Only 0.6% (2,019.24 ha) and 4.9% (16,195.59 ha) of the reserve was burned in three and four consecutive years, respectively. Additionally, 16.8% (55,496.52 ha) did not burn in any year.

Implications for cerrado conservation

The proportion of deforested land inside reserve boundaries increased in 1973–1986 and 1986–2000, but decreased in 2000–2010. Considering the entire period 1973–2010, the proportion of deforestation remained constant at an extremely low level (0.6%) inside the reserve. Most deforestation inside the reserve occurred along the western and southern boundaries, which suffered the greatest impacts of regional eastward economic expansion since the mid-1970s (Figure 6).

A notable factor in this progression is the evolution of reserve boundaries through time. In particular, Xavante political action [45] resulted in substantial increases in the reserve size (62.5% between 1975 and 1986), mainly through reclaiming traditional territory from adjacent agribusiness lands to the west and south. By returning these areas to indigenous management, previously deforested lands subsequently underwent vegetational recovery, thereby restoring the proportion of deforested land inside the reserve to the value observed four decades earlier (0.6%). This interpretation is corroborated by evidence from topographic maps generated in 1979, when the reserve was reduced to its smallest size, showing extensive agribusiness deforestation in areas which were subsequently incorporated or reincorporated into the reserve. A substantial portion of these areas were still classified as bare soil/agricultural in 1986 and 2000 but returned to grassland/cerrado or forest by 2010.

In contrast, outside reserve boundaries the proportion of deforested land increased in all periods, reaching 26% in 2010. The comparatively smaller increase observed in 2000–2010 contrasts with a regional pattern of rapid agribusiness expansion in the same period [55]. This difference may be attributable to decreased economic advantage of agricultural expansion along the western boundary of Pimentel Barbosa Indigenous Reserve given the extensive availability of lands nearby that were less intensively deforested before 2000.

The temporal burn scar pattern observed during the annual burning seasons of the last four years of this four decade sequence (2007 to 2010) indicates that almost 60% of the reserve was never burned or not burned in consecutive years. Only a very small portion of the reserve, about 5%, was burned in three or more consecutive years. Comparison of Figures 2D and 5 reveals that vegetation cover was maintained or recovered even in areas of high fire periodicity within the reserve. Thus, the observed Xavante burning pattern involved predominantly low fire periodicity and did not cause deforestation even where fires occurred in multiple consecutive years. Although these data do not allow us to draw direct comparisons with non-anthropogenic cerrado fire regimes, differences are likely to occur in terms of seasonality and frequency. In particular, whereas we observed the Xavante hunt with fire most often during the dry season, most cerrado fires not ignited through human agency are caused by lightning during rainy or transitional seasons [23,29,56]. However, in general terms, both scenarios involve highly varied surface fire periodicity and patchy distributions.

Ecological studies demonstrate that periodic cerrado fires can have positive effects on species diversity, as well as other environmental measures, and ought to be considered a strategic technique for environmental management in this biome [21,22,25,26,29,30]. This practice has yet to be incorporated in environmental management policies by the Brazilian government as it is in some other parts of the world with fire-adapted landscapes [57–59]. This oversight may result from a tendency for the Brazilian rainforests, which are not fire adapted, to receive more attention from conservationists and environmental policies than the cerrado biome and to frame narratives of conservation and climate change mitigation in Brazil.

Additionally, some scientific research on climate change [60] conflates rainforests with cerrados, despite their enormous ecological distinctions, because they are both included in Brazil's geopolitical division "Legal Amazon" (Amazônia Legal). However, the cerrado landscape and the indigenous reserves in Central Brazil face different threats than the Brazilian rainforests due to their distinct economic potentials for commercial agricultural development and ecological responses to anthropogenic fire [3,15,22,61]. For the past 50 years, the cerrado has been massively exploited through large scale clearing techniques largely absent in the Brazilian rainforests, involving mechanized tree-chaining that uproots trees in such a manner that regrowth is completely averted

(Figure 7). Conversely, commercial exploitation of Amazonian rainforests (in some cases involving indigenous peoples) involves some deforestation activities that are less frequent in the cerrado of Central Brazil, including illegal timber logging, large-scale industrial burning, and gold mining. As the Brazilian government and the agribusiness sector respond to international pressures to curb carbon emission from deforestation in the Amazonian rainforests, economic pressure on the cerrado and associated forest clearing activities tend to increase.

Much of policy and public discourse on tropical conservation presumes the positive value of non-alteration of tropical land-scapes. For example, although international initiatives such as REDD and REDD+ promote the inclusion of indigenous peoples and their interests in adopting policies for reducing carbon emissions, these programs nevertheless often aim at halting traditional subsistence activities presumed to be responsible for CO_2 emissions, irrespective of their particular social and ecological contexts [62]. Similarly, scientific conservation discourse in Brazil often assumes that anthropogenic burning in the cerrado, whether practiced by indigenous peoples or others, is uniformly destructive, contributes to global warming, and threatens the value of indigenous reserves and other conservation areas [63–68]. However, as demonstrated by the results of the present study, indigenous Xavante landscape management practices over four decades, including periodic collective hunting with fire and political advocacy for federal recognition of traditional lands, maintained the integrity of the cerrado landscape and sustained vegetational recovery as compared to adjacent lands under non-indigenous management.

Our findings call into question the widespread assumption, evident in international conservation narratives such as those associated with the implementation of REDD and REDD+ programs, that the non-alteration of tropical landscapes through the suppression of indigenous anthropogenic burning is a beneficial strategy for reducing deforestation in indigenous reserves in the cerrado biome. This point is especially pertinent because rapid plant regrowth in the cerrado quickly re-assimilates carbon emissions from appropriate fire regimes [22,26]. Furthermore, our results suggest a need to reassess the implications of overreaching conservation and reduction in carbon emission narratives, increasingly shared by contrasting interest groups in Brazil and elsewhere, about the purportedly destructive nature of indigenous land use, particularly landscape management with fire [40,41]. In this context, the real challenge of cerrado conservation is to address the long-term sustainability of 'conservation islands,' such as indigenous reserves, and the carbon offset approach to deal with the inter-connected nature of land use systems that now characterizes Brazil and elsewhere, whereby conservation units are increasingly subsumed by the impacts of the larger landscapes to which they pertain.

Acknowledgments

Welch and Coimbra thank the members of the Xavante villages Pimentel Barbosa and Etênhiritipá for their participation and support during research activities. In particular, our field activities benefitted from assistance by Tsuptó Bruprewem Wairi Xavante, José Paulo Seriuwarão Xavante, Marco Aurelio Serenho Ihi Xavante, Tsidowi Wai'adzatse', Wahipó Xavante, and Pari'õwá Xavante. Brondízio and Hetrick express thanks for the support of Indiana University and the Anthropological Center for Training and Research on Global Environmental Change (ACT), the Institut d'Études Avancées (IEA), Paris, and the Institut des Hautes Études de l'Amérique Latine at Université Sorbonne Nouvelle, Paris.

Author Contributions

Conceived and designed the experiments: JRW SSH ESB CEAC. Analyzed the data: SSH ESB. Wrote the paper: JRW SSH ESB CEAC. Conducted the fieldwork: JRW CEAC.

References

1. Turnhout E, Bloomfield B, Hulme M, Vogel J, Wynne B (2012) Conservation policy: Listen to the voices of experience. Nature 488: 454–455.

2. United Nations (2012) Report of the second session of the plenary meeting to determine modalities and institutional arrangements for an intergovernmental science-policy platform on biodiversity and ecosystem services. Nairobi: United Nations Environment Programme.

3. Brondízio ES, Ostrom E, Young OR (2009) Connectivity and the governance of multilevel socio-ecological systems: The role of social capital. Annu Rev Environ Resour 34: 253–278.

4. Soares-Filho B, Moutinho P, Nepstad D, Anderson A, Rodrigues H, et al. (2010) Role of Brazilian Amazon protected areas in climate change mitigation. Proc Natl Acad Sci USA 107: 10821–10826.

5. Nepstad D, Schwartzman S, Bamberger B, Santilli M, Ray D, et al. (2006) Inhibition of Amazon deforestation and fire by parks and indigenous lands. Conserv Biol 20: 65–73.

6. Ricketts TH, Soares-Filho B, Fonseca GAB, Nepstad D, Pfaff A, et al. (2010) Indigenous lands, protected areas, and slowing climate change. PLoS Biol 8: e1000331.

7. Adeney JM, Christensen Jr NL, Pimm SL (2009) Reserves protect against deforestation fires in the Amazon. PLoS ONE 4: e5014.

8. Fairhead J, Leach M, Scoones I (2012) Green grabbing: A new appropriation of nature? J Peasant Stud 39: 237–261.

9. Whitehead PJ, Bowman DMJS, Preece N, Fraser F, Cooke P (2003) Customary use of fire by indigenous peoples in northern Australia: Its contemporary role in savanna management. Int J Wildland Fire 12: 415–425.

10. Birrell K, Godden L, Tehan M (2012) Climate change and REDD+: Property as a prism for conceiving Indigenous peoples' engagement. Journal of Human Rights and the Environment 3: 196–216.

11. Klink CA, Moreira AG (2002) Past and current human occupation, and land use. In: Oliveira PS, Marquis RJ, editors. The Cerrados of Brazil: Ecology and Natural History of a Neotropical Savanna. New York: Columbia University Press. 69–88.

12. Phalan B, Bertzky M, Butchart SHM, Donald PF, Scharlemann JPW, et al. (2013) Crop expansion and conservation priorities in tropical countries. PLoS ONE 8: e51759.

13. Myers N, Mittermeier RA, Mittermeier CG, Fonseca GAB, Kent J (2000) Biodiversity hotspots for conservation priorities. Nature 403: 853–858.

14. Mittermeier RA, Robles Gil P, Hoffman M, Pilgrim J, Brooks T, et al, editors (2004) Hotspots Revisited: Earth's Biologically Richest and Most Endangered Terrestrial Ecoregions. Chicago: Conservation International.

15. Klink CA, Machado RB (2005) Conservation of the Brazilian cerrado. Conserv Biol 19: 707–713.

16. Abelson PH, Rowe JW (1987) A new agricultural frontier. Science 235: 1450–1451.

17. Fargione J, Hill J, Tilman D, Polasky S, Hawthorne P (2008) Land clearing and the biofuel carbon debt. Science 319: 1235–1238.

18. Georgescu M, Lobell DB, Field CB, Mahalov A (2013) Simulated hydroclimatic impacts of projected Brazilian sugarcane expansion. Geophys Res Lett 40: 972–977.

19. Davidson EA, de Araujo AC, Artaxo P, Balch JK, Brown IF, et al. (2012) The Amazon basin in transition. Nature 481: 321–328.

20. Garfield S (2001) Indigenous Struggle at the Heart of Brazil: State Policy, Frontier Expansion, and the Xavante Indians, 1937–1988. Durham, NC: Duke University Press.

21. Simon MF, Grether R, Queiroz LP, Skema C, Pennington RT, et al. (2009) Recent assembly of the cerrado, a neotropical plant diversity hotspot, by in situ evolution of adaptations to fire. Proc Natl Acad Sci USA 106: 20359–20364.

22. Pivello VR (2011) Use of fire in the cerrado and Amazonian rainforests of Brazil: Past and present. Fire Ecol 7: 24–39.

23. Ramos-Neto MB, Pivello VR (2000) Lightning fires in a Brazilian savanna national park: Rethinking management strategies. Environ Manage 26: 675–684.

24. Ledru M–P (2002) Late quaternary history and evolution of the cerrados as revealed by palynological records. In: Oliveira PS, Marquis RJ, editors. The Cerrados of Brazil: Ecology and Natural History of a Neotropical Savanna. New York: Columbia University Press. 33–50.

25. Pivello VR (2006) Fire management for biological conservation in the Brazilian cerrado. In: Mistry J, Berardi A, editors. Savannas and Dry Forests: Linking people with Nature. Aldershot, England: Ashgate Publishing. 129–154.

26. Castro EA, Kauffman JB (1998) Ecosystem structure in the Brazilian Cerrado: A vegetation gradient of aboveground biomass, root mass and consumption by fire. J Trop Ecol 14: 263–283.

27. Eiten G (1972) The cerrado vegetation of Brazil. Bot Rev 38: 201–341.

28. Warming E (1892) Lagoa Santa: Et Bidrag til den biologiske Plantegeografi. Copenhagen: Bianco Luno.

29. Coutinho LM (1990) Fire in the ecology of the Brazilian cerrado. In: Goldammer JG, editor. Fire in the Tropical Biota: Ecosystem Processes and Global Challenges. New York: Columbia University Press. 82–105.

30. Miranda HS, Bustamante MMC, Miranda AC (2002) The fire factor. In: Oliveira PS, Marquis RJ, editors. The Cerrados of Brazil: Ecology and Natural History of a Neotropical Savannah. New York: Columbia University Press. 51–68.

31. Frizzo TLM, Bonizário C, Vasconcelos HL (2011) Revisão dos efeitos do fogo sobre a fauna de formações savânicas do Brasil. Oecologia Australis 15: 365–379.

32. Cavalcanti RB, Alves MAS (1997) Effects of fire on savanna birds in Central Brazil. Ornitologia Neotropical 8: 85–87.

33. Prada M, Marinho-Filho JS, Price PW (1995) Insects in flower heads of *Aspilia foliacea* (Asteraceae) after a fire in a central Brazilian savanna: Evidence for the plant vigor hypothesis. Biotropica 27: 513–518.

34. Vitt LJ, Caldwell JP (1993) Ecological observations on cerrado lizards in Rondônia, Brazil. J Herpetol 27: 46–52.

35. Prada M (2001) Effects of fire on the abundance of large mammalian herbivores in Mato Grosso, Brazil. Mammalia 65: 55–61.

36. Prada M, Marinho-Filho JS (2004) Effects of fire on the abundance of Xenarthrans in Mato Grosso, Brazil. Austral Ecol 29: 568–573.

37. Villalobos MP (2002) Efeito de Fogo e da Caça na Abundância de Mamíferos na Reserva Xavante do Rio das Mortes, MT, Brasil [Ph.D. dissertation]. Brasília: Universidade de Brasília. 81 p.

38. Briani DC, Palma ART (2004) Post-fire succession of small mammals in the Cerrado of Central Brazil. Biodivers Conserv 13: 1023–1037.

39. Leite DLP (2007) Efeitos do Fogo sobre a Taxocenose de Lagartos em Áreas de Cerrado *Sensu Stricto* no Brasil Central [Master thesis]. Brasília: Universidade de Brasília.

40. Oliveira M (2001) Fazendeiros controlam fogo perto de reserva: Área Pimentel Barbosa teve parte de seu território destruída pelo fogo. Diário de Cuiabá. 14 August 2001. Cuiabá. Available: http://www.diariodecuiaba.com.br/detalhe.php?cod=64004. Accessed 2013 Jan 12.

41. Kassu, Milanez F (2010) Terra Indígena de Areões: Queimadas continuam. Água Boa News. 12 December 2010. Água, MT. Available: http://www.aguaboanews.com.br/portal/index.php?option=com_content&view=article&id=10599. Accessed 2013 Jan 12.

42. Nascimento IJ (2013) Incêndios atingem 31 mil hectares de terra indígena xavante em MT. G1 Mato Grosso. 20 August 2013. Cuiabá, MT. Available: http://g1.globo.com/mato-grosso/noticia/2013/08/incendios-atingem-31-mil-hectares-de-terra-indigena-xavante-em-mt.html. Accessed 2013 Aug 22.

43. Hecht SB (2009) Kayapó savanna management: Fire, soils, and forest islands in a threatened biome. In: Woods WI, Teixeira WG, Lehmann J, Steiner C, WinklerPrins A, et al., editors. Amazonian Dark Earths: Wim Sombroek's Vision. Berlin: Springer. 143–162.

44. Mistry J, Berardi A, Andrade V, Krahô T, Krahô P, et al. (2005) Indigenous fire management in the cerrado of Brazil: The case of the Krahô of Tocantíns. Hum Ecol 33: 365–386.

45. Welch JR, Santos RV, Flowers NM, Coimbra CEA Jr. (2013) Na Primeira Margem do Rio: Território e Ecologia do Povo Xavante de Wedezé. Rio de Janeiro: Museu do Índio/FUNAI.

46. Gross DR, Eiten G, Flowers NM, Leoi FM, Ritter ML, et al. (1979) Ecology and acculturation among native peoples of Central Brazil. Science 206: 1043–1050.

47. Leeuwenberg FJ, Robinson JG (2000) Traditional management of hunting by a Xavante community in Central Brazil: The search for sustainability. In: Robinson JG, Bennett EL, editors. Hunting for Sustainability in Tropical Forests. New York: Columbia University Press. 375–394.

48. Melo MM, Saito CH (2011) Regime de queima das caçadas com uso do fogo realizadas pelos Xavante no cerrado. Biodiversidade Brasileira 1: 97–109.

49. Melo MM (2013) The practice of burning savannas for hunting by the Xavante Indians based on the stars and constellations. Soc Natur Resour 26: 478–487.

50. Coimbra CEA Jr., Flowers NM, Salzano FM, Santos RV (2002) The Xavánte in Transition: Health, Ecology, and Bioanthropology in Central Brazil. Ann Arbor: University of Michigan Press.

51. Santos RV, Coimbra CEA Jr., Welch JR (2013) A half-century portrait: Health transition in the Xavante Indians from Central Brazil. In: Brondízio ES, Moran EF, editors. Human-Environment Interactions: Current and Future Directions. Dordrecht: Springer. 29–52.

52. Flowers NM (1983) Seasonal factors in subsistence, nutrition, and child growth in a Central Brazilian community. In: Hames RB, Vickers WT, editors. Adaptive Responses of Native Amazonians. New York: Academic Press. 357–390.

53. Welch JR, Ferreira AA, Santos RV, Gugelmin SA, Werneck G, et al. (2009) Nutrition transition, socioeconomic differentiation, and gender among adult Xavante Indians, Brazilian Amazon. Hum Ecol 37: 13–26.

54. Instituto Brasileiro de Geografia e Estatística (IBGE) (1981) Topographic Maps (1:100,000) Corixão da Mata Azul (SD-22-X-A-IV), Bandeirantes (SD-22-X-C-IV), Canarana (SD-22-V-D-IV), Cascalheira (SD-22-V-B-V), Luiz Alves (SD-22-X-C-I), Matinha (SD-22-V-D-V), Riberão Água Preta (SD-22-V-D-VI), Rio Darro ou Feio (SD-22-V-B-IV), Rio das Mortes (SD-22-V-D-III), São João Grande (SD-22-V-B-VI), Rio Tanguro (SD-22-V-D-I), Rio Turvo (SD-22-V-D-II). Brasília: IBGE.

55. VanWey LK, Spera S, de Sa R, Mahr D, Mustard JF (2013) Socioeconomic development and agricultural intensification in Mato Grosso. Philos Trans R Soc Lond B Biol Sci 368: 20120168.

56. Mistry J (1998) Fire in the cerrado (savannas) of Brazil: An ecological review. Prog Phys Geog 22: 425–448.

57. McGregor S, Lawson V, Christophersen P, Kennett R, Boyden J, et al. (2010) Indigenous wetland burning: Conserving natural and cultural resources in Australia's World Heritage-listed Kakadu National Park. Hum Ecol 38: 721–729.

58. Boer MM, Sadler RJ, Wittkuhn RS, McCawa L, Grierson PF (2009) Long-term impacts of prescribed burning on regional extent and incidence of wildfires: Evidence from 50 years of active fire management in SW Australian forests. For Ecol Manage 259: 132–142.

59. van Wilgen BW, Forsyth GG, Klerk Hd, Das S, Khuluse S, et al. (2010) Fire management in Mediterranean-climate shrublands: A case study from the Cape Fynbos, South Africa. J Appl Ecol 47: 631–638.

60. Aragão LEOC, Shimabukuro YE (2010) The incidence of fire in Amazonian forests with implications for REDD. Science 328: 1275–1278.

61. Ratter JA, Ribeiro JF, Bridgewater S (1997) The Brazilian cerrado vegetation and threats to its biodiversity. Ann Bot 80: 223–230.

62. van Dam C (2011) Indigenous territories and REDD in Latin America: Opportunity or threat? Forests 2: 394–414.

63. Leonel MM (2000) O uso do fogo: O manejo indígena e a piromania da monocultura. Estudos Avançados 14: 231–250.

64. Diaz MCV, Nepstad D, Mendonça MJC, Motta RS, Alencar A, et al. (2002) O Preço Oculto do Fogo na Amazônia: Os Custos Econômicos Associados às Queimadas e Incêndios Florestais. Belém: Instituto de Pesquisa Ambiental da Amazônia.

65. Pereira CA, Fiedler NC, Medeiros MB (2004) Análise de ações de prevenção e combate aos incêndios florestais em unidades de conservação do cerrado. Revista Floresta 34: 95–100.

66. Soares RV, Santos JF (2002) Perfil dos incêndios florestais no Brasil de 1994 a 1997. Revista Floresta 32: 219–232.

67. Falleiro RM (2011) Resgate do manejo tradicional do cerrado com fogo para proteção das terras indígenas do oeste do Mato Grosso: um estudo de caso. Biodiversidade Brasileira 1: 86–96.

68. Nascimento DTF, Araújo FM, Ferreira Jr LG (2011) Análise dos padrões de distribuição espacial e temporal dos focos de calor no bioma cerrado. Revista Brasileira de Cartografia 63: 577–589.

21

Assembly of Recombinant Israeli Acute Paralysis Virus Capsids

Junyuan Ren, Abigail Cone, Rebecca Willmot, Ian M. Jones*

School of Biological Sciences, University of Reading, Reading, United Kingdom

Abstract

The dicistrovirus Israeli Acute Paralysis Virus (IAPV) has been implicated in the worldwide decline of honey bees. Studies of IAPV and many other bee viruses in pure culture are restricted by available isolates and permissive cell culture. Here we show that coupling the IAPV major structural precursor protein ORF2 to its cognate 3C-like processing enzyme results in processing of the precursor to the individual structural proteins in a number of insect cell lines following expression by a recombinant baculovirus. The efficiency of expression is influenced by the level of IAPV 3C protein and moderation of its activity is required for optimal expression. The mature IAPV structural proteins assembled into empty capsids that migrated as particles on sucrose velocity gradients and showed typical dicistrovirus like morphology when examined by electron microscopy. Monoclonal antibodies raised to recombinant capsids were configured into a diagnostic test specific for the presence of IAPV. Recombinant capsids for each of the many bee viruses within the picornavirus family may provide virus specific reagents for the on-going investigation of the causes of honeybee loss.

Editor: Eric Jan, University of British Columbia, Canada

Funding: The work was funded by the UK Biotechnology and Biological Sciences Research Council (BBSRC), www.bbsrc.ac.uk. The funders had no role in study design, data collection and analysis, decision to publish, or preparation of the manuscript.

Competing Interests: The authors have declared that no competing interests exist.

* Email: i.m.jones@rdg.ac.uk

Introduction

An annual loss of honey bee colonies has been reported across the United States and more recently Europe, a proportion of which has been termed Colony Collapse Disorder (CCD) [1]. The consequential loss to agriculture as a result of reduced bee enabled pollination has been widely reported [2]. Honey bee loss and CCD may have multifactorial origins including loss of rural habitat, pesticide use, mite infestation and pathogen load [3–5]. Pathogenic bee viruses may play a role but as a number of common viruses are found, Kakugo virus; Varroa Destructor Virus; Sacbrood Virus; Deformed Wing Virus, Kashmir Bee Virus and Israeli Acute Paralysis Virus [6–8], a definitive link between any one infection and CCD has been difficult to establish. An early metagenomic analysis of hives that suffered CCD identified Israel Acute Paralysis Virus (IAPV) as a highly correlated risk factor [9] although other viruses, notably Deformed Wing Virus, have since been implicated. Infection is linked to Varroa mite infestation [10,11]. While the precise causes of CCD remain unknown the horizontal transfer of pathogenic viruses from mites to bees is a possible factor in CCD with parallels in other cases of viruses crossing the species barrier [12,13]. Direct evidence from areas of emerging CCD occurrence [14] or from model studies of Varroa stimulated virus transmission [15,16] lends support to such a possibility. IAPV can be vectored by Varroa [17] and has recently been shown to concentrate in the heads of bees and alter foraging behaviour, plausibly leading to bee disorientation and loss [18].

Many of the known bee viruses belong to the *Picornaviridae* and *Dicistroviridae* families, both of which are characterised by icosahedral particles of a pseudo T = 3 symmetry and have similar protein coding and virus assembly pathways. In the environment multiply infected hives are common [19–21] which, with virus strain variation and limited permissive honey bee cell culture [22,23], make studies on the pathogenic contribution attributable to any single infectious virus difficult.

In contrast, the assembly of non-infectious picornavirus empty capsids has been reported for several picornaviruses [24–27] and we showed recently that the level of such capsids could be improved by regulating the level of the 3C protease required for capsid protein maturation [28,29]. Here we apply the same expression strategy to IAPV to show that empty IAPV capsids can be assembled in insect cells following expression using recombinant baculoviruses. Capsids allowed the generation of monoclonal antibodies (MAbs) whose epitopes were subsequently mapped. Two non-competing MAbs were configured as a capture ELISA assay that detected IAPV antigen in expressing cells. We suggest that recombinant empty capsid synthesis may be a generally useful tool for the generation of specific reagents for fundamental and applied studies of many honeybee viruses.

Results

First reported in 2007 [30], IAPV has since been found worldwide with at least 2 strains co-circulating in the United States and Canada [31]. IAPV has been independently isolated in France in a screen of apiary overwintering losses [32] and more recently in Argentina [33], Japan [34] China [20] and Spain [21,35]. Despite this, there are relatively few specific regents for the non-nucleic acid based detection of IAPV particles. In order to allow detection of IAPV expression in a heterologous expression system we first generated a specific antibody for IAPV following a high throughput screen for stable expression of the predicted individual mature structural proteins of IAPV as His-tagged proteins in *E.coli* [36]. A synthetic IAPV ORF2 (NC_009025, codon optimised for *Spodoptera* cells) was used as template for the amplification of fragments encoding VP2+VP4, VP2, VP3 and VP1 with junctions based on an alignment of IAPV ORF2 with the known endpoints in *Solenopsis invicta* virus 1 where they have been formally mapped [37] (Figure 1A and 1B). Of the fragments screened, only that encoding VP2 gave rise to detectable amounts of protein (Figure 1C). Recombinant VP2 was purified by immobilised metal chromatography (Figure 1C) and the purified protein used to generate a high titre polyvalent serum in rabbits which did not cross react with insect cells (not shown).

To assess expression of IAPV ORF2 as an intact polyprotein and following cleavage into mature virus structural proteins we assembled baculovirus transfer vectors with the sequence encoding ORF2 as a sole open reading frame and, in other variants appended at the ORF2 carboxyl–terminus to the sequence encoding the putative IAPV 3C like protease responsible for ORF2 maturation, normally part of the ORF1 polyprotein. The 3C protease activity in many picornaviruses is associated with host cell shutdown through the cleavage of cellular targets e.g. [38] so we attenuated the level of 3C expression by the introduction of a ribosomal frameshift sequence between the sequence encoding ORF2 and that encoding IAPV 3C as described recently for Foot and Mouth Disease Virus [28]. In an additional modification we reduced the frameshift rate further through shortening of the "slippery sequence" where ribosomal slippage occurs from six to five uracils which reduces the frequency of the -1 frameshifting event [39,40]. These constructions (Figure 2A) are predicted to reduce the level of 3C enzyme while maintaining the same level of ORF2 translation. Following sequence verification recombinant baculoviruses were produced as described [41] and high titre stocks were used to infect *Spodoptera frugiperda* (Sf9) cells. Infected cells were harvested at two days post infection and total cell extracts assessed by SDS-PAGE and western blot for the presence of VP2 related antigen, by probing with the generated rabbit anti-VP2 serum, and for level of infection by probing for the baculovirus major surface glycoprotein gp64. Expression of ORF2

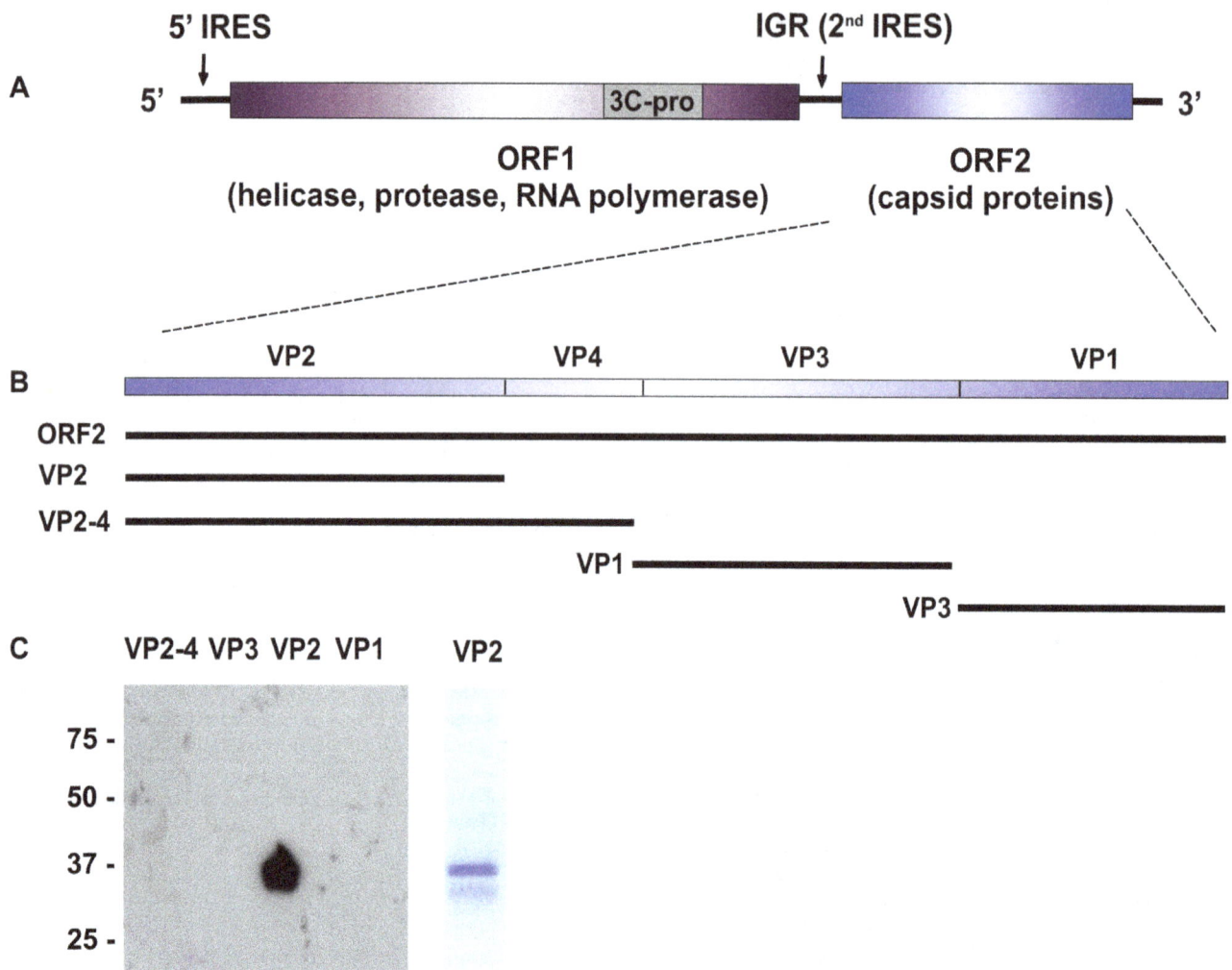

Figure 1. The IAPV genome and origin of sequences used for IAPV fragment expression. A. Genome structure of IAPV dicistrovirus with the two open reading frames and their constituent mature proteins shown. IRES – internal ribosome entry site, IGR – intergenic region. **B.** The fragments targeted for expression in *E.coli* and their encoded products. The precise endpoints used were: VP2, VTH→MQC; VP2-4, VTH→FGW; VP3, SKP→ELQ; VP1, INI→ISR. Expression screening for the proteins predicted from the fragments shown in B. **C-** left. Western blot with anti-His antibody. **C-**right. Purified VP2 used for the generation of a rabbit serum. Numbers to the left of the blot are protein molecular mass markers and are in kilodaltons.

alone resulted in very intense staining of a band at ~100 kDa with a considerable number of smaller products, the result of non-specific proteolysis within the expressing Sf9 cells (Figure 2B track 1). When ORF2 was coupled in frame to the 3C protease the band at ~100 kDa was wholly processed to a single species of ~39 kDa consistent with the expected molecular weight of the mature VP2 product (Figure 2B track 3). However the level of signal was lower than ORF2 alone. When the 3C protease was coupled to the C terminus of ORF2 in the −1 frame via a ribosomal frameshift site precursor processing to VP2 was improved when compared to that observed following direct fusion (Figure 2B track 2). When the ribosomal frameshift frequency was reduced by use of a 5 U slippage sequence the level of the VP2 signal approached that of the unprocessed precursor (Figure 2B track 4). The level of baculovirus infection by these viruses was broadly similar as defined by the level of the baculovirus encoded gp64 protein (Figure 2B middle panel). Thus IAPV 3C processes the ORF2 precursor following expression in Sf9 cells and the level of cleavage observed is affected by the ratio of enzyme to substrate (Figure 2B lower panel).

To confirm IAPV protein expression as a result of Sf9 cell infection by recombinant baculoviruses, infected cells at two days post infection were fixed and permeablised and incubated with anti VP2 serum followed by a fluorescent anti-rabbit conjugate and visualised by fluorescence microscopy. Expression of ORF2 alone gave rise to bright fluorescence distributed throughout the cytoplasm (Figure 3A). The signal from the optimised (frameshift with 5Ts) recombinant, ORF2-5T-3C, was somewhat reduced but also distributed within the expressing cell although occasional punctate staining was apparent (Figure 3B). These data suggest that processing of the IAPV structural precursor protein may be associated with some redistribution of recombinant antigen within the expressing cell, plausibly as a result of assembly of the cleaved structural proteins into discistrovirus like capsids. To confirm that capsid assembly was occurring in this heterologous expression system, cells infected with the optimised recombinant baculovirus expressing ORF2-5T-3C were lysed at two days post infection and the cytoplasmic contents fractionated on sucrose velocity gradients. VP2 reactive antigen was found in the middle fractions, typically around 40% sucrose, reflecting a higher molecular weight than would be expected for VP2 alone (Figure 4A). When peak fractions from the gradient were absorbed onto carbon coated formvar grids and analysed by transmission electron microscopy, particles with the shape and dimensions typical of assembled dicistrovirus capsids were observed in addition to baculovirus nucleocapsids which co-sedimented in the same sucrose fractions (Figure 4B). Many recombinant dicistrovirus capsids had taken up stain showing them to be hollow as expected of assembly in the absence of a packageable genome (Figure 4B arrowed). Control extracts processed from baculovirus only infected cells showed no such particles, only baculovirus nucleocapsids (not shown). Some particles appeared damaged and other material of regular size

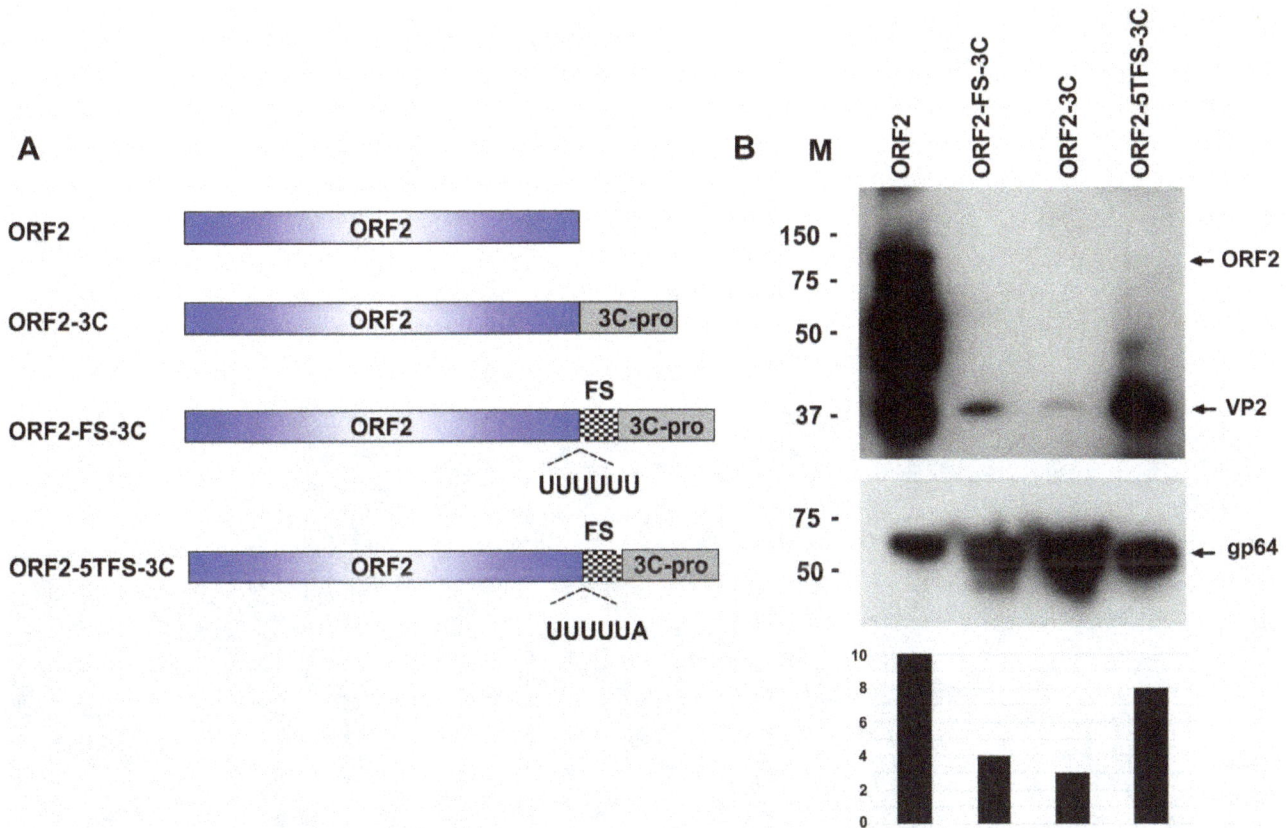

Figure 2. Design and test of baculovirus expression cassettes. A. Cartoon representations of the various genetic constructs used to assess IAPV antigen expression in infected Sf9 cells. The sequences used are identified. FS- the HIV-1 frameshift site, the polypyrimidine tract of which is indicated. 3C-pro – the IAPV 3C like protease. **B**. Western blots of Sf9 cells infected for 2 days with recombinant baculoviruses constructed with the cassettes shown in A. In B the upper panel was blotted with rabbit anti-VP2 serum, the middle panel with anti-gp64 and the lower panel is the relative VP2:gp64 level. Numbers to the left of the blots are protein molecular mass markers (M) and are in kilodaltons. The expected position of the blotted antigens is indicated.

would be consistent with disassembled capsids. A high level of disassembled capsid material has also been observed in EM grids of Triatoma virus [42]. Well defined particles did not show the obvious angular profile typical of picornavirus icosahedra reflecting the fact that dicistrovirus structures, while possessing icosahedral symmetry, are generally more spherical, shown classically for cricket paralysis virus [43] and more recently for Infectious flacherie virus [44] although projections around the 5 fold axes were observed in the recently solved Triatoma dicistrovirus structure [45].

The predominant diagnostic test for IAPV presence is nucleic acid amplification using specific primers e.g. [46], a very sensitive but technologically demanding assay. To enable the development of an antibody based diagnostic assay as an alternative tool that could be used in field conditions, we generated monoclonal antibodies to part purified recombinant IAPV empty capsids, screening hybridomas in the first instance on purified VP2 protein. Four strongly reactive antibodies, IAPVMAb8, IAPVMAb12, IAPVMAb17 and IAPVMAb27 were selected for formal epitope mapping using a library of 20 residue peptides overlapping by 10 made to the same IAPV VP2 sequence. Interestingly all four antibodies reacted with the N-terminal extended tail of VP2 visualised in the available dicistrovirus structures [43,45]. Three antibodies co-mapped to the extreme N-terminal sequence TMPGDSQQES and one to an adjacent sequence ASST-SENSVE (Figure 5A). As predicted by the mapping, IAPVMAbs 8, 12 and 17 competed when used as pairs in a twin site ELISA assay with ORF2-5T-3C infected Sf9 cell extracts as the test antigen but a combination of IAPVMAbs 12 and 27 were found to allow a dose dependent detection of IAPV antigen in the test lysates (Figure 5B). The antibody pairs had no reaction to Sf9 cells that were not infected with an IAPV antigen expressing baculovirus or bee lysate (not shown).

Discussion

We have assembled empty IAPV capsids in insect cells following expression of IAPV structural proteins using a recombinant baculovirus. The mature structural proteins were produced by cleavage of a precursor protein, ORF2, by the authentic 3C-like processing enzyme which was relocated from ORF1 and expressed as a C-terminal fusion protein. Direct fusion of 3C to ORF2 showed only low levels of cleaved VP2 whereas lowering the expression level of 3C through a translational control strategy resulted in higher levels of mature capsid expression. This is consistent with IAPV 3C activity exhibiting some cytotoxicity as has been described for picornavirus 3C where moderation of 3C activity through a mixture of translational control and 3C mutation resulted in higher empty capsid yields [28]. In the case of IAPV, a structure of the 3C-like enzyme is not available so translational control through the use of a frameshift sequence known to function in insect cells [47] was the sole option. It is likely a similar strategy will work for other members of the *Picornaviridae* or *Dicistroviridae* where reduced 3C activity is required to maximise capsid cleavage. Recombinant empty IAPV particles with typical dicistrovirus appearance were observed by EM of negatively stained sucrose gradient fractions and were immunogenic upon immunization. Monoclonal antibodies selected to VP2 were found to map exclusively to the extended N-terminal tail of the molecule. Mapping the polyvalent response also showed this region to be immunodominant, consistent with a poorly immunogenic well folded jelly-roll capsid structure (not shown). Notwithstanding a restricted repertoire in the immune response, VP2 specific monoclonal antibody pairs were validated for ELISA detection of IAPV antigen expressed in *Spodoptera* cells. The two epitopes recognised are IAPV specific which suggests that a rapid test for IAPV presence, such as a lateral flow device, would be highly specific and suitable for use in the field. Such devices have been described for diagnostic tests for other bee pathogens and viruses [48,49]. The number of viruses infecting agriculturally important insects such as the honeybee continues to grow and mixed infections are common [7,10,11,15,50]. Methods to provide the equivalent of a pure virus culture in the absence of the requirements for infectious plaque assay could be useful to generate highly specific protein based assays. Similarly, differences in individual variants of one virus, such as the recently described Varroa transmitted virulent form of DWV [16] could be assessed for capsid function through such technology. Through such studies, baculovirus expression of empty capsids following the co-expression of the structural precursor and a suitably attenuated

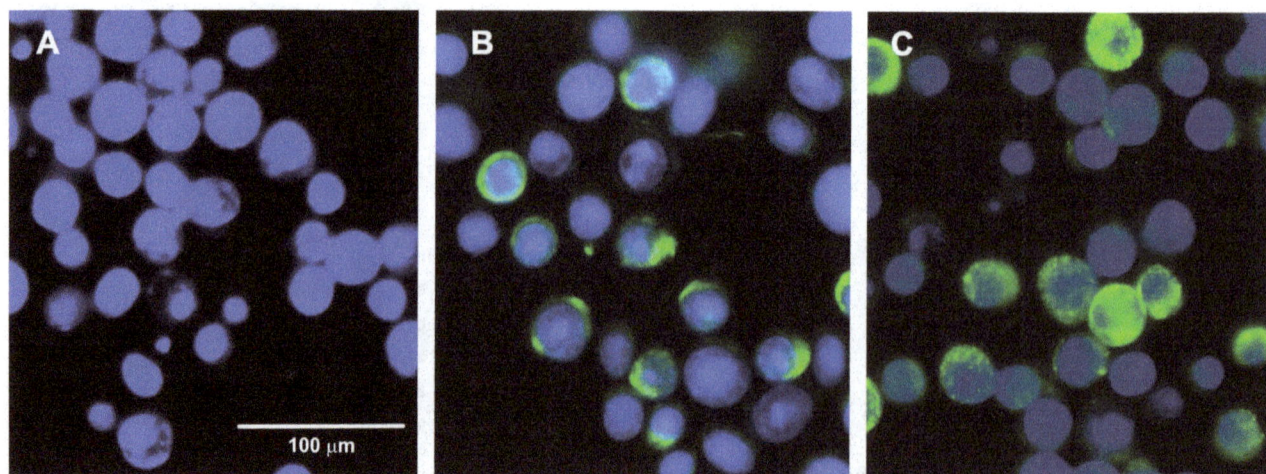

Figure 3. Immunofluorescence of Sf9 cells infected for 2 days with constructs ORF2 and ORF2-5TFS-3C stained with anti IAPV VP2 and an anti-rabbit Alexa Fluor 488 conjugate. A – Sf9 mock infected control. **B** – Sf9 cells infected with construct ORF2 expressing ORF2 only. **C** – Sf9 cells infected with construct ORF2-5TFS-3C expressing ORF2 fused with an attenuated 3C like protease. Occasional punctate staining is apparent in C.

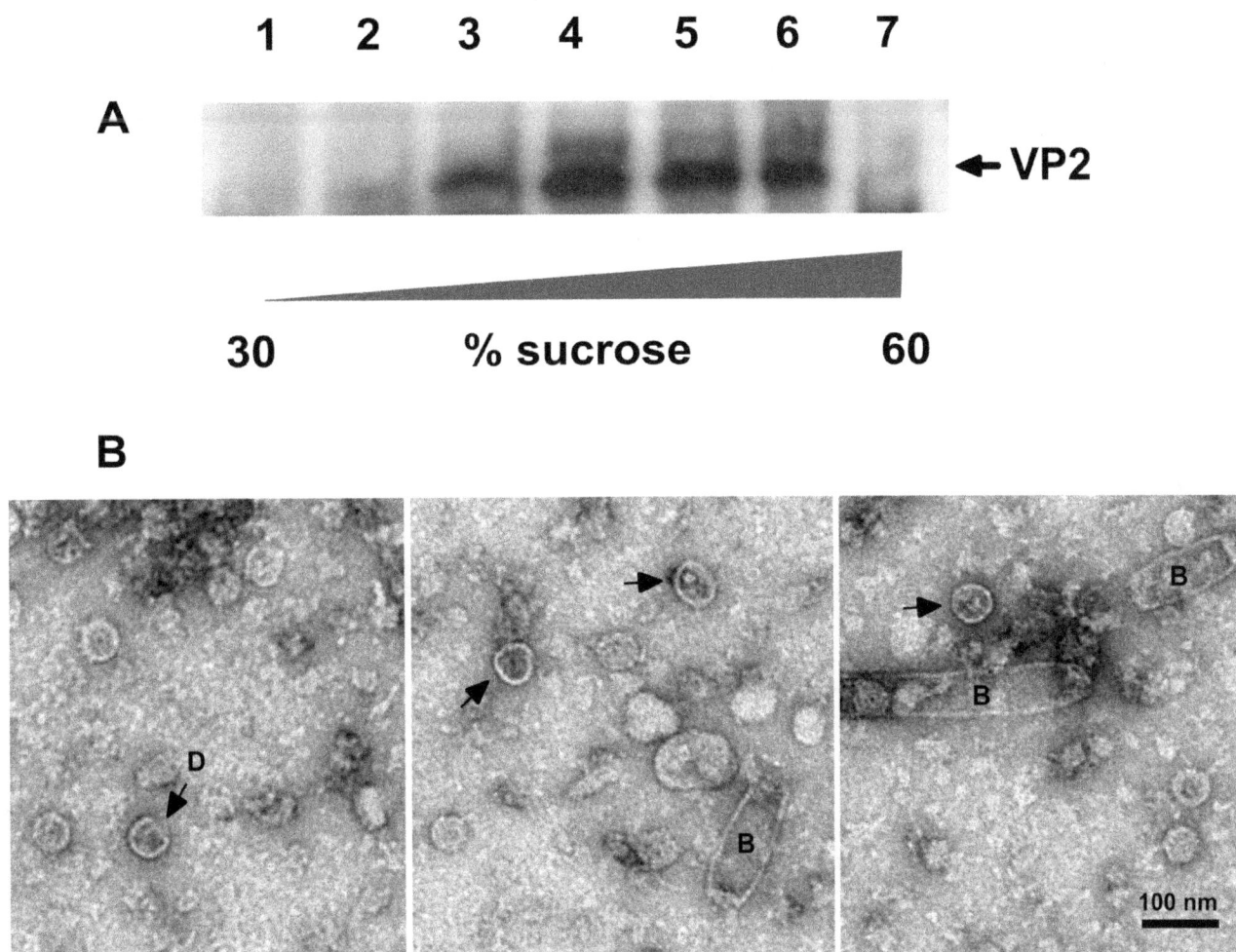

Figure 4. Analysis of empty IAPV capsids. A. Sucrose gradient analysis of Sf9 cell extracts infected with recombinant baculovirus ORF2-5TFS-3C. Each fraction represents a 5% step from 30% to 60% sucrose w/v and is blotted with the IAPV anti VP2 serum. B. EM analysis of peak fractions from the sucrose gradient shown in A. Empty capsids are indicated. A damaged particle is labelled D. Baculovirus nucleocapsids are labelled B. The bar is 100 nm.

3C-like enzyme could contribute to better understanding the honeybee-virus landscape at a molecular level.

Materials and Methods

Cell culture and virus growth

Sf9 cells (ATCC) were cultured in BioWhittaker Insect-Xpress supplemented with 2% FCS, 100units/ml penicillin, 100 µg/ml streptomycin and 2.5 µg/ml amphotericin B. Cells were grown at 28°C as monolayers or in suspension with agitation at 100 rpm. Baculoviruses were generally amplified in monolayer cultures but large scale infections for capsid isolation were done in suspension. Virus stocks were titred using plaque assay on Sf9 monolayers.

Sequences and cloning

Sequences for IAPV were taken from the databases (NC_009025) and synthesised *de novo* to include codon optimisation for *Spodoptera* cells (Geneart). The transfer vector used for all expression was based on pTriEx1.1 (EMD Bioscience) and fragments encoding ORF2 and 3C were cloned downstream of the P10 promoter via unique restriction sites introduced during gene synthesis. Recombinant baculoviruses were constructed using

recombination with AcMNPV bacmid KO1629 in insect cells as described [41]. Routine DNA procedures used standard protocols or, when kits were used, those recommended by the vendor. All vectors were confirmed by DNA sequencing prior to use for expression.

Electrophoresis and Western blotting and Quantitation

Protein samples were separated on pre-cast 10% Tris.HCl SDS-polyacrylamide gels (BioRad) and transferred to Immobilon-P membranes (Millipore) using a semi-dry blotter. For 10 well gels each sample loaded represented 5×10^4 cells. Following transfer, filters were blocked for one hour at room temperature using PBS containing 0.1% v/v Tween-20 (PBS-T), 5% w/v milk powder. Primary antibodies were used at a dilution of 1:5000 in PBS-T, 5% w/v milk powder. Following several washes with PBS-T the membranes were incubated for 1 hour with the appropriate HRP-conjugates and the bound antibodies detected by BM chemiluminescence (Roche). Bands were imaged using a Syngene G:Box chemiluminescence imager and the image file processed for pixel density by Genetools software (Syngene).

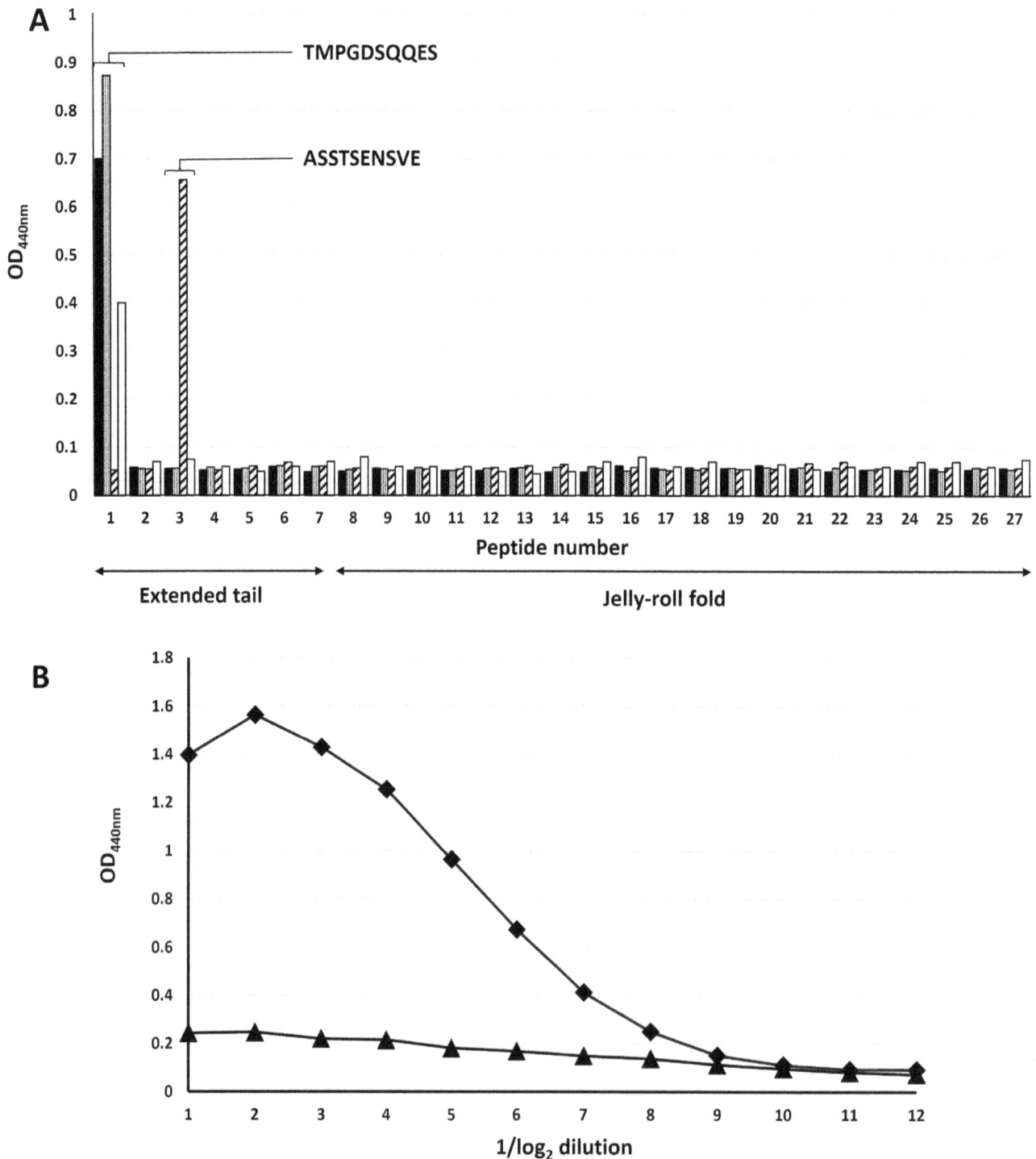

Figure 5. Development of an IAPV capture ELISA. A. Pepscan of 4 monoclonal antibodies to IAPV VP2 on an overlapping peptide library. Filled bars IAPVMAb8; stipple bars IAPVMAb12; striped bars IAPVMAb17; open bars IAPVMAb 27. The relationship between peptide identity and VP2 structure is shown. **B**. Twin site capture ELISA using capture MAbs 8 (triangles) and 27 (diamonds) as capture layer and probing with HRP-labelled IAPV MAb12. The test sample was a lysate of Sf9 cell infected with the baculovirus expressing ORF-2-5TFS-3C.

Purification of empty capsids

Infected insect cell cultures (typically 1 L at ~10^6/ml) were harvested at 3 days pi and lysed by resuspension in 1/20th of the original volume of 1% Triton X-100 in PBS and held at 4°C for 30 minutes with occasional agitation. Unbroken cells and nuclei were removed by centrifugation (4500 rpm, 30 min) and the clarified lysate layered onto a 30% sucrose (w/v in PBS) cushion and particles recovered by centrifugation at 100,000 g for 100 minutes. The gradient was discarded and the pellet resuspended in 1/10th tube volume of PBS containing 3500 units of benzonase. After 30 min at room temperature the solution was clarified and applied to a preformed sucrose gradient made up of seven steps of

5% from 30% to 60% w/v. The gradient was developed by centrifugation at 100,000 g for 16 hrs and fractionated from the top.

Electron microscopy

Samples were allowed to adhere to carbon coated formvar grids for 5 min at room temperature followed by a brief water wash (1 min) before staining with 1% uranyl acetate for 1 min. Excess stain was removed by blotting and the grids examined on a Philips CM20 operating at 80 kV.

Monoclonal antibody production

Monoclonal antibodies were produced under subcontract at FERA, Sand Hutton, UK. Briefly, three mice were immunised with 0.1 ml of recombinant IAPV empty capsids at 300 μg/ml four times over an 8 week period. Following the last immunisation, animals were sacrificed and splenocytes fused with myeloma cells using polyethylene glycol 1500 as the fusion agent. Fusion cell products were cultured in multi-well plates and cell culture supernatants (CCS) harvested for screening on IAPV VP2 protein at 12 days post fusion. Cell lines producing strongly reactive antibody were cloned by limiting dilution and re-screened by ELISA to produce pure hybridomas. Following scale –up, the antibody in the CCS from selected cell lines was purified by Protein G chromatography, concentrated to ~10 mgs/ml and stored at −80 degrees centigrade. Immunisation and work with vertebrate animals (mice) was performed under a UK Home Office project license in accordance with the UK Animals (Scientific Procedures) Act 1986 (UK Government 1986) using the principles outlined in the Home Office guidance 'Antibody Production: Principles for Protocols of Minimal Severity' (Home Office 2000). The work adhered to the principles of the National Centre for the Replacement, Refinement and Reduction of Animals in Research see http://www.nc3rs.org.uk/page.asp?id=871.

Antibody labelling

To allow competition assays selected monoclonal antibodies were labelled with horse radish peroxidase using the Lightning-Link HRP Antibody Labeling Kit (Novus Biologicals).

ELISA assay

Immulon II ELISA plates (Nunc) were coated with antigen or capture antibody at 10 μgs/ml in 200 mM sodium bicarbonate overnight at 4°C. Plates were blocked with 1% bovine serum albumin (BSA) in PBS containing 0.1% v/v Tween-20 (PBS-T) and then washed with PBS-T. Subsequent layers were incubated for 1 hour at 25°C and washed between layers with PBS-T. The plate was developed with 50 μl per well of tetramethylbenzidine (Europa Bioproducts UK) and the reaction stopped by the addition of 50 μl of 0.1 M H_2SO_4. Absorbance was measured at 440 nm using a GENios plate reader (Tecan UK). Experiments were performed in duplicate and the average plotted.

Epitope mapping

Overlapping peptides of 20 residues overlapping by 10 residues to the VP2 sequence of IAPV and biotinylated at the N-terminus were supplied by Think Peptides (UK). Peptides were dissolved in DMSO at ~10 mgs/ml and added to streptavidin coated plates at 10 μg/ml. After 1 hr incubation to allow peptide capture the plate was blocked and treated as described for ELISA.

Acknowledgments

We thank Giles Budge and Gaynor Johnson of The Food and Environment Research Agency for the gift of bee reagents and for monoclonal antibody generation respectively.

Author Contributions

Conceived and designed the experiments: IMJ. Performed the experiments: JR AC RW IMJ. Analyzed the data: JR AC RW IMJ. Wrote the paper: IMJ.

References

1. Anderson D, East IJ (2008) The latest buzz about colony collapse disorder. Science 319: 724–725; author reply 724–725.
2. Levy S (2011) The pollinator crisis: What's best for bees. Nature 479: 164–165.
3. Genersch E, Aubert M (2010) Emerging and re-emerging viruses of the honey bee (Apis mellifera L.). Vet Res 41: 54.
4. Evans JD, Schwarz RS (2011) Bees brought to their knees: microbes affecting honey bee health. Trends Microbiol 19: 614–620.
5. Nazzi F, Brown SP, Annoscia D, Del Piccolo F, Di Prisco G, et al. (2012) Synergistic parasite-pathogen interactions mediated by host immunity can drive the collapse of honeybee colonies. PLoS Pathog 8: e1002735.
6. Chen YP, Siede R (2007) Honey bee viruses. Adv Virus Res 70: 33–80.
7. de Miranda JR, Cordoni G, Budge G (2010) The Acute bee paralysis virus-Kashmir bee virus-Israeli acute paralysis virus complex. J Invertebr Pathol 103 Suppl 1: S30–47.
8. Cornman RS, Boncristiani H, Dainat B, Chen Y, vanEngelsdorp D, et al. (2013) Population-genomic variation within RNA viruses of the Western honey bee, Apis mellifera, inferred from deep sequencing. BMC Genomics 14: 154.
9. Cox-Foster DL, Conlan S, Holmes EC, Palacios G, Evans JD, et al. (2007) A metagenomic survey of microbes in honey bee colony collapse disorder. Science 318: 283–287.
10. Highfield AC, El Nagar A, Mackinder LC, Noel LM, Hall MJ, et al. (2009) Deformed wing virus implicated in overwintering honeybee colony losses. Appl Environ Microbiol 75: 7212–7220.
11. Dainat B, Evans JD, Chen YP, Gauthier L, Neumann P (2012) Predictive markers of honey bee colony collapse. PLoS One 7: e32151.
12. Mockel N, Gisder S, Genersch E (2011) Horizontal transmission of deformed wing virus: pathological consequences in adult bees (Apis mellifera) depend on the transmission route. J Gen Virol 92: 370–377.
13. Gisder S, Aumeier P, Genersch E (2009) Deformed wing virus: replication and viral load in mites (Varroa destructor). J Gen Virol 90: 463–467.
14. Martin SJ, Highfield AC, Brettell L, Villalobos EM, Budge GE, et al. (2012) Global honeybee viral landscape altered by a parasitic mite. Science 336: 1304–1306.
15. Moore J, Jironkin A, Chandler D, Burroughs N, Evans DJ, et al. (2011) Recombinants between Deformed wing virus and Varroa destructor virus-1 may prevail in Varroa destructor-infested honeybee colonies. J Gen Virol 92: 156–161.
16. Ryabov EV, Wood GR, Fannon JM, Moore JD, Bull JC, et al. (2014) A Virulent Strain of Deformed Wing Virus (DWV) of Honeybees (Apis mellifera) Prevails after Varroa destructor-Mediated, or In Vitro, Transmission. PLoS Pathog 10: e1004230.
17. Di Prisco G, Pennacchio F, Caprio E, Boncristiani HF, Jr., Evans JD, et al. (2011) Varroa destructor is an effective vector of Israeli acute paralysis virus in the honeybee, Apis mellifera. J Gen Virol 92: 151–155.
18. Li Z, Chen Y, Zhang S, Chen S, Li W, et al. (2013) Viral infection affects sucrose responsiveness and homing ability of forager honey bees, Apis mellifera L. PLoS One 8: e77354.
19. Baker A, Schroeder D (2008) Occurrence and genetic analysis of picorna-like viruses infecting worker bees of Apis mellifera L. populations in Devon, South West England. J Invertebr Pathol 98: 239–242.
20. Ai H, Yan X, Han R (2012) Occurrence and prevalence of seven bee viruses in Apis mellifera and Apis cerana apiaries in China. J Invertebr Pathol 109: 160–164.
21. Granberg F, Vicente-Rubiano M, Rubio-Guerri C, Karlsson OE, Kukielka D, et al. (2013) Metagenomic detection of viral pathogens in Spanish honeybees: co-infection by Aphid Lethal Paralysis, Israel Acute Paralysis and Lake Sinai Viruses. PLoS One 8: e57459.
22. Kitagishi Y, Okumura N, Yoshida H, Nishimura Y, Takahashi J, et al. (2011) Long-term cultivation of in vitro Apis mellifera cells by gene transfer of human c-myc proto-oncogene. In Vitro Cell Dev Biol Anim 47: 451–453.
23. Hunter WB (2010) Medium for development of bee cell cultures (Apis mellifera: Hymenoptera: Apidae). In Vitro Cell Dev Biol Anim 46: 83–86.
24. Brautigam S, Snezhkov E, Bishop DH (1993) Formation of poliovirus-like particles by recombinant baculoviruses expressing the individual VP0, VP3, and VP1 proteins by comparison to particles derived from the expressed poliovirus polyprotein. Virology 192: 512–524.

25. Ko YJ, Choi KS, Nah JJ, Paton DJ, Oem JK, et al. (2005) Noninfectious virus-like particle antigen for detection of swine vesicular disease virus antibodies in pigs by enzyme-linked immunosorbent assay. Clin Diagn Lab Immunol 12: 922–929.

26. Li Z, Yi Y, Yin X, Zhang Z, Liu J (2008) Expression of foot-and-mouth disease virus capsid proteins in silkworm-baculovirus expression system and its utilization as a subunit vaccine. PLoS ONE 3: e2273.

27. Oem JK, Park JH, Lee KN, Kim YJ, Kye SJ, et al. (2007) Characterization of recombinant foot-and-mouth disease virus pentamer-like structures expressed by baculovirus and their use as diagnostic antigens in a blocking ELISA. Vaccine 25: 4112–4121.

28. Porta C, Xu X, Loureiro S, Paramasivam S, Ren J, et al. (2013) Efficient production of foot-and-mouth disease virus empty capsids in insect cells following down regulation of 3C protease activity. J Virol Methods.

29. Porta C, Kotecha A, Burman A, Jackson T, Ren J, et al. (2013) Rational engineering of recombinant picornavirus capsids to produce safe, protective vaccine antigen. PLoS Pathog 9: e1003255.

30. Maori E, Lavi S, Mozes-Koch R, Gantman Y, Peretz Y, et al. (2007) Isolation and characterization of Israeli acute paralysis virus, a dicistrovirus affecting honeybees in Israel: evidence for diversity due to intra- and inter-species recombination. J Gen Virol 88: 3428–3438.

31. Palacios G, Hui J, Quan PL, Kalkstein A, Honkavuori KS, et al. (2008) Genetic analysis of Israel acute paralysis virus: distinct clusters are circulating in the United States. J Virol 82: 6209–6217.

32. Blanchard P, Schurr F, Celle O, Cougoule N, Drajnudel P, et al. (2008) First detection of Israeli acute paralysis virus (IAPV) in France, a dicistrovirus affecting honeybees (Apis mellifera). J Invertebr Pathol.

33. Reynaldi FJ, Sguazza GH, Tizzano MA, Fuentealba N, Galosi CM, et al. (2011) First report of Israeli acute paralysis virus in asymptomatic hives of Argentina. Rev Argent Microbiol 43: 84–86.

34. Kojima Y, Toki T, Morimoto T, Yoshiyama M, Kimura K, et al. (2011) Infestation of Japanese native honey bees by tracheal mite and virus from non-native European honey bees in Japan. Microb Ecol 62: 895–906.

35. Antunez K, Anido M, Garrido-Bailon E, Botias C, Zunino P, et al. (2012) Low prevalence of honeybee viruses in Spain during 2006 and 2007. Res Vet Sci 93: 1441–1445.

36. Pengelley SC, Chapman DC, Mark Abbott W, Lin HH, Huang W, et al. (2006) A suite of parallel vectors for baculovirus expression. Protein Expr Purif 48: 173–181.

37. Valles SM, Hashimoto Y (2008) Characterization of structural proteins of Solenopsis invicta virus 1. Virus Res 136: 189–191.

38. Ghildyal R, Jordan B, Li D, Dagher H, Bardin PG, et al. (2009) Rhinovirus 3C protease can localize in the nucleus and alter active and passive nucleocyto-plasmic transport. J Virol 83: 7349–7352.

39. Wilson W, Braddock M, Adams SE, Rathjen PD, Kingsman SM, et al. (1988) HIV expression strategies: ribosomal frameshifting is directed by a short sequence in both mammalian and yeast systems. Cell 55: 1159–1169.

40. Dinman JD (2012) Control of gene expression by translational recoding. Adv Protein Chem Struct Biol 86: 129–149.

41. Zhao Y, Chapman DA, Jones IM (2003) Improving baculovirus recombination. Nucleic Acids Res 31: E6–6.

42. Agirre J, Aloria K, Arizmendi JM, Iloro I, Elortza F, et al. (2011) Capsid protein identification and analysis of mature Triatoma virus (TrV) virions and naturally occurring empty particles. Virology 409: 91–101.

43. Tate J, Liljas L, Scotti P, Christian P, Lin T, et al. (1999) The crystal structure of cricket paralysis virus: the first view of a new virus family. Nat Struct Biol 6: 765–774.

44. Xie L, Zhang Q, Lu X, Dai X, Li K, et al. (2009) The three-dimensional structure of Infectious flacherie virus capsid determined by cryo-electron microscopy. Sci China C Life Sci 52: 1186–1191.

45. Squires G, Pous J, Agirre J, Rozas-Dennis GS, Costabel MD, et al. (2013) Structure of the Triatoma virus capsid. Acta Crystallogr D Biol Crystallogr 69: 1026–1037.

46. Yang B, Peng G, Li T, Kadowaki T (2013) Molecular and phylogenetic characterization of honey bee viruses, Nosema microsporidia, protozoan parasites, and parasitic mites in China. Ecol Evol 3: 298–311.

47. Adamson CS, Nermut M, Jones IM (2003) Control of human immunodeficiency virus type-1 protease activity in insect cells expressing Gag-Pol rescues assembly of immature but not mature virus-like particles. Virology 308: 157–165.

48. Ferris NP, Nordengrahn A, Hutchings GH, Paton DJ, Kristersson T, et al. (2010) Development and laboratory evaluation of a lateral flow device for the detection of swine vesicular disease virus in clinical samples. J Virol Methods 163: 477–480.

49. Tomkies V, Flint J, Johnson G, Waite R, Wilkins S, et al. (2009) Development and validation of a novel field test kit for European foulbrood. Apidologie 40: 63–72.

50. Cornman RS, Tarpy DR, Chen Y, Jeffreys L, Lopez D, et al. (2012) Pathogen webs in collapsing honey bee colonies. PLoS One 7: e43562.

22

Repression and Recuperation of Brood Production in *Bombus terrestris* Bumble Bees Exposed to a Pulse of the Neonicotinoid Pesticide Imidacloprid

Ian Laycock[1]*, James E. Cresswell[1,2]

1 College of Life & Environmental Sciences, Biosciences, University of Exeter, Exeter, United Kingdom, **2** Centre for Pollination Studies, University of Calcutta, Kolkata, India

Abstract

Currently, there is concern about declining bee populations and some blame the residues of neonicotinoid pesticides in the nectar and pollen of treated crops. Bumble bees are important wild pollinators that are widely exposed to dietary neonicotinoids by foraging in agricultural environments. In the laboratory, we tested the effect of a pulsed exposure (14 days 'on dose' followed by 14 days 'off dose') to a common neonicotinoid, imidacloprid, on the amount of brood (number of eggs and larvae) produced by *Bombus terrestris* L. bumble bees in small, standardised experimental colonies (a queen and four adult workers). During the initial 'on dose' period we observed a dose-dependent repression of brood production in colonies, with productivity decreasing as dosage increased up to 98 µg kg^{-1} dietary imidacloprid. During the following 'off dose' period, colonies showed a dose-dependent recuperation such that total brood production during the 28-day pulsed exposure was not correlated with imidacloprid up to 98 µg kg^{-1}. Our findings raise further concern about the threat to wild bumble bees from neonicotinoids, but they also indicate some resilience to a pulsed exposure, such as that arising from the transient bloom of a treated mass-flowering crop.

Editor: Nicolas Desneux, French National Institute for Agricultural Research (INRA), France

Funding: IL was supported by a studentship from the Natural Environment Research Council (http://www.nerc.ac.uk/). The funders had no role in study design, data collection and analysis, decision to publish, or preparation of the manuscript.

Competing Interests: The authors have declared that no competing interests exist.

* E-mail: il219@exeter.ac.uk

Introduction

Currently, there is concern about declines in bee populations [1,2] and some implicate neonicotinoid pesticides as culprits [3,4]. Neonicotinoids disrupt the insect nervous system [5] and their dietary intake can reduce the expected performance of bees [6,7]. For example, neonicotinoids may increase worker losses while reducing reproductive output and foraging performance in bumble bees, *Bombus* spp. [8,9], and induce homing failure and suppress colony growth in honey bees, *Apis mellifera* L. [10] (and see [11,12] for further discussion). Whether neonicotinoids are a principal cause of bee declines is unclear [13,14], but in regions where they are not banned [4] bees are certainly exposed to them on a massive spatial scale by foraging from treated agricultural crops. For example, oilseed rape (or canola), *Brassica napus* L., is the principal mass-flowering crop in many areas of North America (>8 million hectares [15,16]) and Northern Europe (e.g. ~0.7 million hectares in the UK [17]) and many of its fields are protected from pests by neonicotinoids [18,19]. Neonicotinoids are systemic pesticides, so they are distributed throughout the plant following application [18] and bees are exposed to dietary residues by consuming nectar and pollen [20]. For oilseed rape in the USA, residues of a widely used neonicotinoid, imidacloprid, have been detected in nectar at 0.8 parts per billion (ppb) and in pollen at 7.6 ppb [21]. Other bee-attractive crops such as sunflower and alfalfa are often protected with neonicotinoids [18,21], and so the exposure of bees to these pesticides is widespread. To understand

whether a widespread exposure to neonicotinoids is capable of causing bee populations to decline, we must understand their demographic toxicity, which occurs when a toxic agent detrimentally affects the birth and death rates of the exposed species [22].

The lethality of imidacloprid to bees appears to be dependent on the time of exposure [23,24]. However, in some laboratory trials the trace levels of imidacloprid typically found in nectar and pollen (≤10 ppb [21], but see [25]) have negligible effects on mortality in honey bees [7] and bumble bees [26,27], but they can substantively affect birth rates in bumble bees [28]. Specifically, dietary imidacloprid at levels as low as one ppb may reduce the number of eggs and larvae produced by adult bumble bee workers by one third [28], but the demographic implications of this are unclear because queens are principally responsible for a colony's reproductive output [29]. Because the number of new queens and males that a bumble bee colony produces depends on its size [30,31], the number of workers produced by a queen during a colony's development can determine colony fitness. We therefore examined the effects of dietary imidacloprid on brood production (specifically, the numbers of eggs and larvae destined to become workers) by queen bumble bees at dosages that spanned the environmentally realistic range.

We investigated the effects of a 14-day exposure to dietary imidacloprid on the performance of small, standardised experimental colonies of the buff-tailed bumble bee, *Bombus terrestris* L., in the laboratory. We found a dose-dependent decrease in brood production up to 98 ppb imidacloprid (see Results) and so we extended our experiment to create a pulsed exposure, feeding bees for an additional 14 days on an imidacloprid-free diet, because a scenario such as this may be relevant to wild bumble bee colonies. For example, a pulsed exposure may be caused by the synchronized bloom of imidacloprid-treated oilseed rape fields that normally flower for approximately four weeks in April or May [32] (where the crop is winter-sown) and the exposure subsides when the bees subsequently switch to foraging on pesticide-free wildflowers [33]. Recuperation from some imidacloprid-induced effects has been reported following an exposure in honey bees [34], coccinellids [35], aphids [36], whitefly [37], and the aquatic larvae of midge [38], but our study is the first to explore the potential for such a recovery in bumble bees.

Materials and Methods

Ethics statement

The protocol reported here conforms to the regulatory requirements for animal experimentation in the UK and was approved by the Biosciences Ethics Committee at the University of Exeter.

Bees, experimental colonies and imidacloprid diets

We obtained colonies of *B. terrestris* (subspecies *audax*) at an early stage of development (Biobest, Westerlo, Belgium). In order to create small, standardised experimental colonies for testing, we removed each queen and randomly chose four of her adult workers from their pre-experimental source colony and placed them together in a softwood box ($120 \times 120 \times 45$ mm) fitted with two 2 mL microcentrifuge tubes (Simport, Beloeil, Canada) that were punctured so as to function as syrup (artificial nectar) feeders [28]. Experimental colony size (a queen and four adult workers) was chosen to simulate early-stage bumble bee colonies, consistent with those used in similar studies [8]. We maintained these experimental colonies for 28 days in a semi-controlled environment (23–$27°C$, 21–47% relative humidity).

We obtained imidacloprid as a solution in acetonitrile (Dr. Ehrenstorfer GmbH, Ausberg, Germany). Acetonitrile was removed by evaporation and the imidacloprid was dissolved in purified water before being mixed into feeder syrup (Attracker: 1.27 kg L^{-1} fructose/glucose/saccharose solution; Koppert B.V., Berkel en Rodenrijs, Netherlands) to produce our most concentrated dosage of 125 µg imidacloprid L^{-1} (or 98.43 µg kg^{-1} = ppb). By serial dilution from 125 µg L^{-1} (dilution factor = 0.4) we produced the following nine experimental dosages: 125.00, 50.00, 20.00, 8.00, 3.20, 1.28, 0.51, 0.20, and 0.08 µg imidacloprid L^{-1} (= 98.43, 39.37, 15.75, 6.30, 2.52, 1.01, 0.40, 0.16, and 0.06 µg imidacloprid kg^{-1}). A fresh dilution series containing all nine concentrations was produced at the beginning of each pulsed exposure trial (see below) and kept inside a dark fridge at $5°C$. Dosed syrup from the second pulsed exposure trial was used in the continuous exposure experiment (below).

Exposure to dietary imidacloprid

To create a pulsed exposure, the 28-day experimental period was split into two successive periods of 14 days. During the 'on dose' period (days 1–14), 60 experimental colonies were provided *ad libitum* with either undosed control syrup (6 control colonies) or dosed syrup (6 colonies per dosage treatment, listed above). Fresh syrup at the appropriate dosage was provided to colonies daily. For the 'off dose' period (days 15–28), the bees were transferred to new softwood boxes and fed *ad libitum* with only undosed control syrup. At the beginning of each 14-day period, each experimental colony was provided with a fresh ball of undosed pollen (Biobest,

Westerlo, Belgium) to which bees had *ad libitum* access. Pollen balls (mean mass = 6.1 g, SE = 0.02) were prepared from ground pollen pellets mixed with water to form dough and were weighed before and after placement in colonies to quantify pollen consumption. We corrected for evaporation of water from syrup and pollen based on the mass change of several feeders and pollen balls kept in empty colony boxes under experimental conditions. Experimental colonies were kept in darkness except when monitored daily for the appearance of wax covered egg cells (indicating that oviposition had occurred), syrup consumption and individual mortality. To minimise disturbance to bees, we assayed brood production by collecting all laid eggs and larvae from experimental colony boxes only at the end of each 14-day period, (i.e. on days 14 and 28). The experiment was conducted in two replicate trials, one between October–November 2011 and the other between January–February 2012. Each trial comprised 30 experimental colonies and treatment groups were equally represented in both (3 colonies per treatment).

To establish that the observed recuperation from imidacloprid-induced effects under pulsed exposure (see Results) was caused by the removal of dietary imidacloprid rather than from acclimation to exposure over elapsed time, we conducted a separate continuous exposure experiment. Using the same husbandry techniques described above, we randomly assigned 12 experimental colonies to either 28 days feeding on control syrup (7 colonies) or 28 days feeding on syrup dosed at 98.43 µg imidacloprid kg^{-1} (5 colonies) and we used the same interruption to collect brood on days 14 and 28. This continuous exposure trial was conducted between March–April 2012. This protocol is an adequate test because the highest level of recuperation was observed at 98.43 µg kg^{-1} in the previous experiments (see Results).

To verify the concentration of imidacloprid in our doses, we first dissolved the dosed syrup in liquid chromatography-mass spectrometry (LCMS)-grade water (Fisher Scientific UK Ltd, Loughborough, UK) spiked with a reference standard of imidacloprid-d$_4$ (Dr. Ehrenstorfer GmbH, Augsburg, Germany) at 100 µg L^{-1} (ratio of syrup to water = 5:7). We used solid phase extraction (SPE) to extract imidacloprid and imidacloprid-d$_4$ from the syrup as follows. Diluted dosed syrup samples were processed through 1 mL Discovery® DSC-18 SPE tubes (Sigma-Aldrich, Gillingham, UK) under positive pressure. We first conditioned the SPE tube with 1 mL pure LCMS-grade methanol (Fisher Scientific UK Ltd, Loughborough, UK) followed by 1 mL pure LCMS-grade water. A 1 mL sample was passed through the tube, before the tube was washed with 1 mL pure LCMS-grade water and the imidacloprid was eluted from the column with three separate, but equivalent, aliquots of pure LCMS-grade methanol totalling 450 µL. We removed the methanol by evaporation and the remaining imidacloprid was dissolved in 500 µL of pure LCMS-grade water. Imidacloprid samples were analysed in an Agilent 1200 series liquid chromatograph interfaced via an electrospray ionisation source to an Agilent 6410 triple quadrupole mass spectrometer (Agilent Technologies, Santa Clara, CA, USA) using methods described in Laycock et al. [28]. The instrument response was linear over the range 0.06–125 µg L^{-1} for imidacloprid and imidacloprid-d$_4$ and we found that dosages in all trials contained appropriate levels of imidacloprid (pulsed exposure trial 1, *measured imidacloprid* = 0.989 × *nominal dosage* + 0.204, $R^2 > 0.99$; pulsed exposure trial 2 and continuous exposure trial, *measured imidacloprid* = 1.035 × *nominal dosage* − 0.205, $R^2 > 0.99$).

Statistical analyses

In our analyses, 'brood' represents the total number of eggs and larvae produced in an experimental colony in a given period. We tested whether the 'brood' dose-response relationships differed between our two pulsed exposure trials by analysis of covariance (ANCOVA), with 'dosage' (dosage of imidacloprid in $\mu g\ kg^{-1}$) log-transformed to log('dosage' + 1) as the covariate and 'trial' as the fixed factor, and detected no significant difference between the two trials and so the data were pooled for further analysis (ANCOVA: 'on dose' brood, dosage × trial, $F_{1,\ 56} = 0.99$, $P = 0.32$; 'off dose' brood, dosage × trial, $F_{1,\ 56} = 0.03$, $P = 0.86$; total brood, dosage × trial, $F_{1,\ 56} = 0.34$, $P = 0.56$). The size of the pre-experimental source colony (mean number of workers = 16.4, SE = 1.1; mean number of brood = 101.8, SE = 7.5) from which the members of an experimental colony (queen and four workers) originated did not explain variation in brood production among the experimental colonies and it was disregarded in the analyses below (Spearman's correlation: 'on dose' brood vs. source colony size, $\rho = -0.10$, $N = 60$, $P = 0.44$; 'off dose' brood vs. source colony size, $\rho = 0.07$, $N = 60$, $P = 0.59$; total brood vs. source colony size, $\rho = -0.01$, $N = 60$, $P = 0.91$).

We tested for dose-dependent brood production, timing of oviposition and food consumption during each period of the pulsed exposure using Spearman's correlation analyses. We tested for dose-dependent recuperation by analysing the differences in performance in experimental colonies between the 'on dose' and 'off dose' periods as follows. For a given variable X, denote the 'on dose' performance of a colony by X_{on} and its 'off dose' performance by X_{off}. For each colony we calculated ($X_{off} - X_{on}$), so that a positive value indicates that a colony produced more brood during the 'off dose' period, i.e. it showed recuperation. We investigated recuperation by testing whether ($X_{off} - X_{on}$) increased with imidacloprid dosage using Spearman's correlation analysis.

For brood production, once the statistical significance of the dose-response relationship was established by correlation we used Bayesian Hierarchical Models (BHM) to fit a relationship between 'brood' and 'dosage'. In each BHM, we fitted: brood ~ Poisson(μ); log(μ) ~ $\alpha + \beta \times$ log(dosage+1)+λ. Here, α and β are fitted coefficients analogous to the conventional regression coefficients of slope and intercept, and λ is a 'random effects' term to accommodate overdispersion (λ has a normal distribution with a mean of zero). Each model was fitted with 40,000 iterations of Bayesian inference using a Markov Chain-Monte Carlo method with Gibbs sampling after a burn-in period that discarded the first of 7000 iterations on each chain. We obtained confidence intervals on this relationship as follows. The pairs of α and β values from the final 40,000 iterations of the Bayesian inference estimate the posterior joint probability distribution of the two coefficients; we therefore plotted the 40,000 relationships corresponding to these pairs and extracted the upper and lower percentiles (2.5%, 97.5%) of the fitted brood values that corresponded to each imidacloprid intake across the range of interest. For brood production, we estimated the EC_{50} (half maximal effective concentration) and EC_{10} using the BHM best-fit relationships. We estimated EC values for the imidacloprid-induced reduction in food consumption by using GraphPad Prism v6.0c and evaluated the goodness of fit based on R^2. BHM procedures were implemented in WinBUGS v1.4.3 [39], while all other statistical analyses were conducted in R v3.0.0 [40].

Results

In both pulsed and continuous exposure experiments, *B. terrestris* queens in experimental colonies began producing eggs after approximately two days and some brood progressed to a larval stage within the 14-day periods. No queens died during the experiments and there was negligible worker mortality (one dead

worker at 98 ppb, two dead at 39 ppb in the same colony, two dead at 16 ppb in separate colonies).

During the 14-day 'on dose' period of pulsed exposure, colonies exhibited dose-dependent repression of brood production such that fewer brood were produced as dosage increased up to 98 ppb imidacloprid (Spearman's correlation: 'on dose' brood vs. dosage, $\rho = -0.45$, $N = 60$, $P < 0.001$; Figure 1). The dose-response relationship for brood and imidacloprid dosage during the 'on dose' period was given by brood = exp[2.002 − 1.788×log(dosage+1)] and the standard deviation of the overdispersion parameter was SD(λ) = 1.89 (Figure 2). Based on this relationship, the EC_{50} and EC_{10} values for imidacloprid's affect on brood production were 1.44 ppb and 0.15 ppb, respectively.

During the 14-day 'off dose' period, brood production showed dose-dependent recuperation (Spearman's correlation: ($Brood_{off} - Brood_{on}$) vs. dosage, $\rho = 0.32$, $N = 60$, $P = 0.01$; Figure 3). Dosage did not significantly affect brood production during the 'off dose' period (Spearman's correlation: 'off dose' brood vs. dosage, $\rho = 0.10$, $N = 60$, $P = 0.47$; Figure 1) and, taken over the entire 28-day pulsed exposure, total brood production was not significantly correlated with imidacloprid dosage (Spearman's correlation: total brood vs. dosage, $\rho = -0.13$, $N = 60$, $P = 0.32$; Figure 1). However, we note that based on the 28-day dose-response relationship for brood and imidacloprid, given by brood = exp[2.770 − 0.198×log(dosage+1)] with SD(λ) = 1.25 (Figure 2), recuperation of brood production was incomplete at higher dosages. For example, a 32% reduction remained apparent in colonies dosed with imidacloprid at 98 ppb (Figure 2). The EC_{50} value for reduced brood production over the entire 28-day pulsed exposure was beyond our tested dosage range (>98 ppb), while the EC_{10} was estimated at 2.5 ppb.

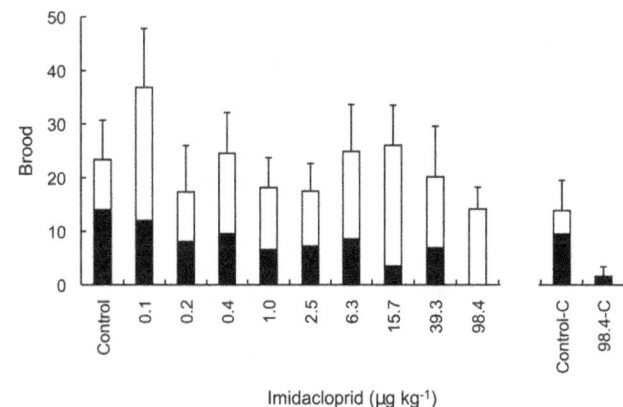

Figure 1. Brood production in *Bombus terrestris* colonies during a pulsed or continuous exposure to imidacloprid. Mean number of brood produced in standardised *Bombus terrestris* colonies ($N = 60$) during 28-day pulsed or continuous exposure to dietary imidacloprid. For pulsed exposure (from left to right, 'Control' to '98.4'): brood produced during the 14-day 'on dose' period (black bars), during which colonies were exposed to imidacloprid in syrup at the specified dosage (in $\mu g\ kg^{-1}$ = parts per billion); and brood produced during the subsequent 14-day 'off dose' period (white bars), during which all colonies fed exclusively on control syrup. For continuous exposure ('Control-C' and '98.4-C'): brood produced during first 14 days of exposure (black bars) and brood produced during second 14 days of exposure (white bars). Where a column does not contain a black bar or a white bar, zero brood were produced during days 1–14 or days 15–28, respectively. Error bars indicate ± SE of mean brood production over 28 days.

A

B

Imidacloprid (μg kg⁻¹)

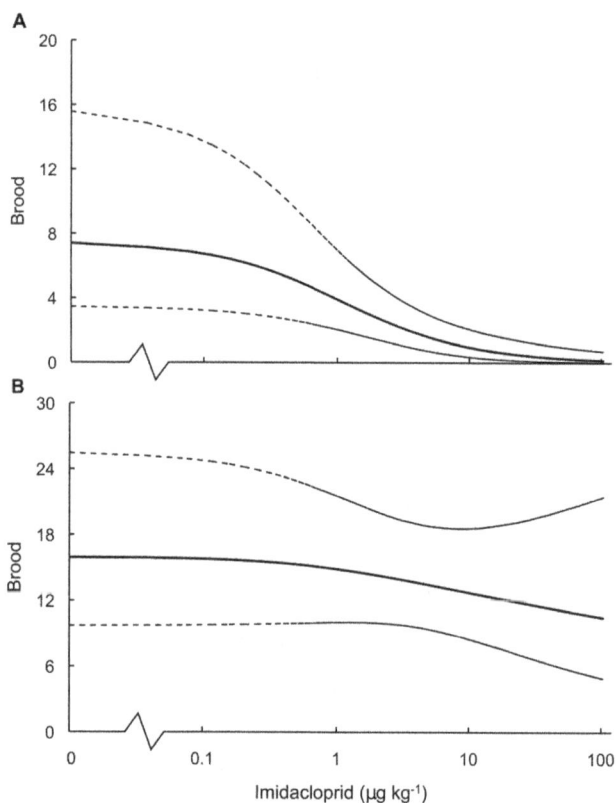

Figure 2. Best-fit dose-response relationships of brood production in *Bombus terrestris* colonies under pulsed exposure to imidacloprid. Dose-response relationships of brood production in standardised *Bombus terrestris* colonies ($N=60$) following a 28-day pulsed exposure to dietary imidacloprid in syrup. Specifically, (A) brood production during the 14-day 'on dose' period of pulsed exposure in which bees fed on syrup dosed with imidacloprid and (B) total brood production taken over the entire 28-day pulsed exposure (including brood produced during the 14-day 'on dose' period and during the subsequent 14-day 'off dose' period in which imidacloprid was removed from the bees' diet). Solid lines indicate the best-fit dose response relationship (obtained using Bayesian Hierarchical Modelling of the data summarized in Figure 1, see Methods) and dashed lines indicate the relationship's 95% confidence intervals.

Based on the fitted dose-response relationships (Figure 2), we estimate that 14-day exposures to dietary imidacloprid at environmentally realistic levels of between 0.3 ppb and 10 ppb may reduce brood production in *B. terrestris* colonies by between 18–84% (Table 1). However, the effects of recuperation in this residue range are such that given a further 14 days without exposure the drop in brood is ameliorated to between 2–19% (Table 1).

Recuperation is unlikely to be attributable to acclimation over time because brood production remained repressed under continuous exposure at 98.4 ppb over 28 days (Figure 1). Specifically, colonies dosed at 98.4 ppb imidacloprid exhibited significantly reduced brood production over 28-days compared to control colonies (ANOVA: dosage, $F_{1, 21}=6.33$, $P<0.05$), but brood production did not differ between successive 14-day periods (days 1–14 and 15–28) of continuous exposure (ANOVA: period, $F_{1, 21}=2.22$, $P=0.15$).

Where brood were produced, imidacloprid did not affect the timing of first oviposition during the 'on dose' period (Spearman's correlation: days until oviposition *vs.* dosage, $\rho=0.11$, $N=35$,

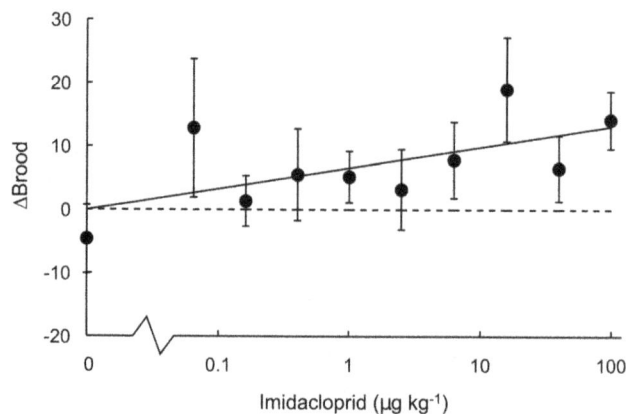

Figure 3. Recuperation of brood production in *Bombus terrestris* colonies during a pulsed exposure to imidacloprid. Recuperation of brood production in standardised *Bombus terrestris* colonies ($N=60$) during the 14-day 'off dose' period of pulsed exposure, wherein bees fed exclusively on undosed control syrup. The 'off dose' period followed a 14-day 'on dose' period during which bees' fed on syrup dosed with imidacloprid at the given concentrations (in μg kg⁻¹ = parts per billion). Recuperation ($\Delta Brood$) is determined by analyzing the difference in brood production between the 'on dose' (days 1–14) and 'off dose' (15–28) periods, specifically: $\Delta Brood = Brood_{off} - Brood_{on}$, with a positive value indicating increased production of brood when 'off dose'. Data represent the means and error bars indicate ± SE. The solid line indicates the following logarithmic trend: $\Delta Brood = 1.428 \times \ln(dosage) + 6.533$, $R^2=0.38$. Dashed line indicates $\Delta Brood=0$.

$P=0.5$; Table 2), but it delayed oviposition in the subsequent 'off dose' period (Spearman's correlation: days until oviposition *vs.* dosage, $\rho=0.53$, $N=45$, $P<0.001$; Table 2).

During pulsed exposure, we observed dose-dependent reductions in the daily consumption of syrup and pollen by experimental colonies whilst they were 'on dose' (Spearman's correlation: 'on dose' syrup consumption *vs.* dosage, $\rho=-0.59$, $N=60$, $P<0.001$; 'on dose' pollen consumption *vs.* dosage, $\rho=-0.77$, $N=60$, $P<0.001$; Figure 4). Based on these results, the EC$_{50}$ and EC$_{10}$ values for reduced pollen consumption were 4.4 ppb ($R^2=0.95$) and 0.2 ppb ($R^2=0.96$), respectively, while the equivalent values for reduced syrup consumption were >98 ppb ($R^2=0.90$) and 23.6 ppb ($R^2=0.97$).

During the 'off dose' period, colonies demonstrated dose-dependent recuperation of both syrup consumption (Spearman's correlation: ($Syrup_{off} - Syrup_{on}$) *vs.* dosage, $\rho=0.60$, $N=60$, $P<0.001$) and pollen consumption (Spearman's correlation: ($Pollen_{off} - Pollen_{on}$) *vs.* dosage, $\rho=0.81$, $N=60$, $P<0.001$). Dosage did not significantly affect syrup consumption during the 'off dose' period (Spearman's correlation: 'off dose' syrup consumption *vs.* dosage, $\rho=0.21$, $N=60$, $P=0.11$; Figure 4), but pollen consumption significantly increased among colonies previously exposed to higher dosages (Spearman's correlation: 'off dose' pollen consumption *vs.* dosage, $\rho=0.40$, $N=60$, $P=0.001$; Fig. 4).

Taken over the entire 28-day pulsed exposure period, the amount of syrup and pollen consumed in experimental colonies declined as imidacloprid dosage increased (Spearman's correlation: syrup consumption *vs.* dosage, $\rho=-0.47$, $N=60$, $P<0.001$; pollen consumption *vs.* dosage, $\rho=-0.25$, $N=60$, $P=0.05$; Figure 4), demonstrating that recuperation of food consumption was incomplete. From these results, EC$_{50}$ values were calculated to be 43.7 ppb ($R^2=0.50$) for reduced pollen consumption and >98 ppb ($R^2=0.68$) for reduced consumption of syrup, while

Table 1. Estimated decrease in brood production exhibited by *Bombus terrestris* colonies during pulsed exposure to realistic imidacloprid residues, equivalent to those previously detected in nectar of treated crops.

Realistic exposure scenario	Imidacloprid residue (ppb)	14-day 'on dose' brood reduction (%)[a]	28-day pulsed exposure brood reduction (%)[b]
OSR–Europe[c]	0.3	18 (14–24)	2 (0–6)
OSR–USA[c]	0.8	37 (30–45)	5 (0–12)
Mean max. level[d]	1.9	56 (51–64)	9 (0–19)
Gill et al.[e]	10.0	84 (84–86)	18 (9–27)

Reductions are relative to the number of brood produced in undosed control colonies and were obtained using the appropriate BHM best-fit dose-response relationship from Figure 2. The reduction's 95% confidence intervals, given in parentheses, were also obtained from BHMs in Figure 2.
[a]Refers to the estimated decrease in brood production expected after a 14-day exposure to imidacloprid at the given dosage.
[b]Refers to the estimated total decrease in brood after a 28-day pulsed exposure at the given dosage (14 days 'on dose', 14 days 'off dose').
[c]Maximum imidacloprid residues detected in the nectar of oilseed rape [21]. Data originates from studies conducted only in Member States of the European Union (OSR–Europe) and from studies including North America (OSR–USA).
[d]Mean maximum level of neonicotinoid residues in nectar calculated from 20 studies [56].
[e]Residues in dosed syrup used in a semi-field trial conducted by Gill et al. [8].

EC_{10} values were 16.2 ppb ($R^2 = 0.60$) and 32.4 ppb ($R^2 = 0.78$) for pollen for syrup, respectively.

After using partial correlation analysis to control for the effects of dosage, brood production in experimental colonies increased with higher daily consumption of both syrup and pollen (Pearson's partial correlation: brood *vs.* syrup consumption, $r = 0.32$, $df = 58$, $P = 0.01$; brood *vs.* pollen consumption, $r = 0.59$, $df = 58$, $P < 0.001$).

Discussion

Under pulsed exposure to dietary imidacloprid, standardized colonies of *B. terrestris* bumble bees 'on dose' for 14 days exhibited dose-dependent repression of brood production, such that their productivity decreased as dosage increased up to 98 ppb. The removal of imidacloprid from colonies during the subsequent 14-day 'off dose' period produced dose-dependent recuperation of brood production to the extent that total productivity under pulsed exposure was not correlated with dosage up to 98 ppb. Pulsed exposure of colonies to dietary imidacloprid at 98 ppb produced

the largest observed recuperation, but continuous exposure to the same concentration repressed brood production without recuperation during a separate experiment of equal duration. We therefore argue that recuperation is primarily achieved by the reversibility of imidacloprid-induced effects rather than acclimation to imidacloprid over time.

The dose-dependent decrease in brood production we observed in queenright colonies mirrors the effect on brood production in queenless microcolonies of *B. terrestris* workers over the same period of time [28]. Similarly, our EC_{50} value for a 14-day exposure (1.44 ppb) is comparable to the EC_{50} for imidacloprid's effect on drone production in *B. terrestris* microcolonies exposed over eleven weeks (3.7 ppb) [26]. However, the recuperation of brood production in bumble bee colonies we observed under pulsed exposure is a new finding. Other insects show recuperation from some imidacloprid-induced effects during pulsed exposure [35–38], but we are the first to demonstrate the resilience of an important demographic endpoint in bees. In our study, when imidacloprid exposure ceased, the ameliorating effect of recuper-

Table 2. Mean number of days taken by *Bombus terrestris* queens to undertake oviposition during pulsed exposure to dietary imidacloprid.

Imidacloprid dosage (μg kg^{-1} = ppb)	On dose: day of first oviposition (± SE)	Off dose: day of first oviposition (± SE)
Control	4.2 (1.1)	1.3 (0.3)
0.1	2.6 (1.1)	2.8 (0.9)
0.2	5.0 (1.5)	6.0 (1.9)
0.4	2.8 (1.2)	1.5 (0.4)
1.0	3.0 (1.3)	4.2 (1.4)
2.5	10.3 (0.3)	6.0 (2.1)
6.3	3.8 (1.9)	6.0 (2.1)
15.7	11.0 (0.0)	5.7 (1.6)
39.4	2.3 (1.0)	7.8 (1.7)
98.4	–[a]	7.2 (1.2)

Oviposition occurred in standardised experimental colonies (queen and four workers) during either the 14-day 'on dose' period of pulsed exposure (during which bees fed on syrup dosed with dietary imidacloprid at the given concentration) or the subsequent 14-day 'off dose' period (when all imidacloprid dosages were removed from the bees' diet).
[a]Oviposition did not occur during the 'on dose' period in colonies exposed at 98.4 ppb.

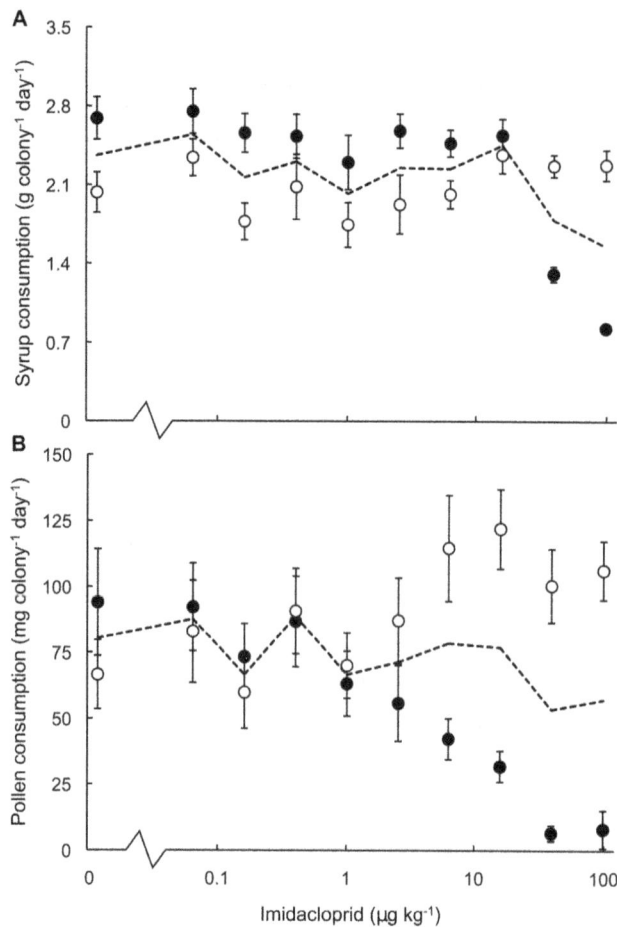

Figure 4. Food consumption in *Bombus terrestris* colonies during a pulsed exposure to imidacloprid. Feeding responses of standardised *Bombus terrestris* colonies ($N = 60$) during a 28-day pulsed exposure to dietary imidacloprid. Specifically, (A) mean daily syrup and (B) mean daily pollen consumption during the initial 14-day 'on dose' period feeding on imidacloprid dosed syrup (filled circles) and during the subsequent 14-day 'off dose' period feeding on undosed control syrup (unfilled circles). Dashed lines connect the mean consumption rates of colonies over the entire 28-day pulsed exposure. Error bars indicate $\pm$ SE. Control data (zero $\mu g \ kg^{-1}$) are displayed slightly displaced on the x-axis for ease of inspection.

ation on bumble bee brood production was such that the EC_{50} for a 28-day pulsed exposure was raised beyond 98 ppb. However, we note that recuperation remained incomplete at higher doses, with overall brood productivity still reduced by between 19–32% at dosages between 10–98 ppb. According to a recent guidance document for the risk assessment of plant protection products on bees [41], a reduction in this range would constitute a 'medium' colony-level-impact and could translate into a similar effect on colony size. Additionally, we found that oviposition was delayed during the 'off dose' period of pulsed exposure in colonies that were first presented with imidacloprid at higher dosages. Our results suggest that where bumble bees experience a pulsed exposure to residues of imidacloprid above 10 ppb [25], incomplete recuperation of brood production and delayed oviposition could detrimentally impact colony size and thereby influence colony fitness [30,31].

Consumption of syrup and pollen in our experimental colonies also underwent dose-dependent repression and recuperation

during the 'on dose' and 'off dose' periods of pulsed exposure, respectively. Repression was most severe in pollen consumption, with an EC_{50} of just 4.4 ppb, and both feeding endpoints showed incomplete recuperation at the two highest dosages (39 and 98 ppb). This result is somewhat consistent with a previous study of recovery in honey bees, in which recuperation of foraging activity was incomplete in colonies exposed to imidacloprid at 48 ppb [34]. Since the pollen in our experiment was not dosed, the imidacloprid in the syrup reduced the bees' overall ability or desire to feed during the 'on dose' period. In a previous study, *B. terrestris* workers exposed to dietary imidacloprid in microcolonies exhibited dose-dependent feeding reductions that were also linked to reductions in brood productivity [28]. Consequently, it was hypothesized that imidacloprid-induced nutrient limitation might play some part in repressing bumble bee egg production during exposure [28]. Our data lends support to this hypothesis because it demonstrates that: a) queenright colonies that consumed more syrup and pollen produced more brood; b) bees showed dose-dependent reductions in feeding whilst 'on dose'; c) repression of brood production coincided with repressed feeding. Additionally, recuperation of food consumption and brood production in colonies occurred simultaneously when exposure ceased and we therefore suggest that removal of imidacloprid from the bees' diet caused feeding rates to recover, which re-established sufficient nutrient intake to facilitate reproduction in bumble bee queens. Although the mechanism for recuperation of food consumption was not studied here, we speculate that it has its basis in the metabolic elimination of the toxicant [42], which in a previous study appeared to take place within 48 hours in bumble bees fed imidacloprid at 98 ppb [43].

Comparison with results of semi-field trials

In our study, a two-week exposure to dietary imidacloprid at 10 ppb in syrup substantively reduced brood production in *B. terrestris* colonies. In a semi-field trial, Gill et al. [8] found that *B. terrestris* colonies also dosed with 10 ppb imidacloprid solely in artificial nectar produced significantly fewer workers at the end of a four-week exposure, without suffering elevated levels of in-colony worker mortality. Although they did not measure egg production, Gill et al. found that imidacloprid-dosed colonies accumulated fewer larvae and pupae over 4 weeks and speculated that this was due to imidacloprid's effect on brood survival. Based on our findings, we hypothesize that repressed brood production may have been an important cause of Gill et al.'s observations.

In a second semi-field study, Whitehorn et al. [9] exposed *B. terrestris* colonies to field-realistic dosages of dietary imidacloprid for two weeks in the laboratory and monitored colony development for a further six weeks in the field. We exercise caution when comparing our observations to Whitehorn et al.'s because pollen was their principle delivery vehicle for imidacloprid. However, following a similar exposure duration and an extended imidacloprid-free period, Whitehorn et al. found no significant effect of imidacloprid on the number of pupae and workers in colonies, but a strong negative effect on the number of queens. Potentially, recuperation of brood and worker production occurred in Whitehorn et al.'s colonies when exposure ceased, but for some unknown reason any recovery was insufficient to sustain normal levels of queen production. Their observations may originate in either increased intoxication of the existing queen caused by consumption of contaminated pollen during lab exposure or the impact of a longer exposure to imidacloprid in the stored nectar and pollen within in the nest, which is important for successful development of new queens [44]. Additionally, if imidacloprid reduces the foraging efficiency of workers [8] then exposed colonies may lack sufficient resources to produce the normal quota of queens, each of which comprises almost twice the biomass of a male bumble bee [30]. Furthermore, brood and worker production in bumble bee colonies may recover better following imidacloprid exposure than other important endpoints. We therefore suggest that the potential for recuperation of performance in demographically important endpoints other than brood production is an area requiring further research in bumble bees.

Environmental relevance

Whilst our study raises further concerns about the threat to wild bumble bees from imidacloprid it also indicates some resilience to a pulsed exposure that could arise during the synchronized bloom of a treated mass-flowering crop. However, when interpreting the environmental relevance of our findings we recognize the limitations of our study, which are as follows. First, the pollen consumed in our colonies was not dosed. There is no reason to suspect different levels of toxicity arising due to ingestion of imidacloprid in nectar *vs.* pollen, but a bumble bee queen is likely to eat a substantial pollen load whilst producing eggs [45] and consequently her exposure in the wild may be more severe than tested here.

Second, the duration of exposure in the environment may differ from our experiment. Exposure for 14 days is a reasonable first approximation because, for example, roughly 75% of the flowering of winter-sown oilseed rape in the UK occurs over a peak period of about two weeks [32]. However, total flowering duration can extend across five weeks or more and bumble bee colonies may continue to forage on mass-flowering crops throughout their blooming period [46]. Conversely, colonies will vary in the extent to which their development intersects with the blooming period of mass-flowering crops because bumble bee queens emerge from their overwinter sites and initiate colonies at various times in spring [47]. Consequently, colonies of later-emerging queens may develop after the crop's bloom has largely or completely declined and could broadly escape neonicotinoid effects.

Third, our study may underestimate the severity of imidacloprid's effects. For example, we focus primarily on brood production, but there are other demographically important endpoints such as mortality. A diet dosed with imidacloprid at realistically high levels (10 ppb) appears to raise mortality in colonies by increasing the risk that workers become lost whilst foraging and in addition exposed foragers tend to return to the nest with less pollen less often [8]. If these impacts also occur at lower dosages (<10 ppb), which are more typically found in environmental nectar and pollen [21], they could certainly add to the stress on wild bumble bee colonies and diminish their reproductive output. Additionally, while the amount of brood and workers produced in a bumble bee colony can influence the quantity of new queens and males that are produced [30,31], the quality of sexual offspring produced may also be critical for colony fitness. For example, body mass predicts whether a young queen will survive diapause [48] and body size may impact on a male's mating success [49]. Furthermore, wild colonies are likely to be under additional stresses from pathogens [50], parasites [51] and other agrochemicals [8], which could augment the severity of a neonicotinoid's impact and the potential for recovery. Additive [8] and synergistic [52] effects of certain neonicotinoids and other agrochemicals have been reported for bees, but further study into combinatorial effects of neonicotinoids and other potential stressors is necessary. Finally, under laboratory conditions winter honey bees appear to be less sensitive to imidacloprid than summer honey bees [53]. Although winter active bumble bees have been observed at latitudes as far north as southern England [54], unlike winter honey bees they are unlikely to be social foragers because bumble bee colonies typically perish in the autumn before newly mated queens enter hibernation [55]. Therefore, if seasonal differences in sensitivity exist in wild bumble bees, foragers from spring and late summer colonies would have to be compared. Commercially bred bumble bees, which were used in autumn and winter in our current study, are produced throughout the year. As these bees are reared under standardised conditions, it is unlikely that they would show seasonal variation in sensitivity to imidacloprid. However, the effects reported here could be more severe in wild colonies and in future work it would be important to compare the sensitivity of commercially reared and wild bumble bees.

Conclusions

Our study provides further evidence that dietary neonicotinoid pesticides in the environmentally realistic range can have detrimental effects on bumble bee health, specifically by repressing brood production and nutritional intake in queenright colonies. We also show, however, that bumble bees may be somewhat resilient to a pulsed exposure because they exhibit dose-dependent recuperation of brood production when exposure ends. We acknowledge that to interpret the environmental relevance of our findings for wild bumble bee colonies additional studies are necessary. These should seek to establish whether recuperation from pulsed exposure to neonicotinoids occurs during extended exposures and for other demographically important endpoints besides brood production. Finally, the severity of imidacloprid's impact on bumble bees appears to be highly sensitive to its dietary level even within the currently recognized environmentally realistic range [21]. Unfortunately, this range is based on scant published data [56] and more widespread surveys of residues in crops and colonies, such as those recently begun in the USA [25], are therefore urgently required.

Acknowledgments

We thank Dr. H Florance, Dr. CJ Pook and James Smith for their LCMS, SPE and Bayesian modelling expertise, respectively.

Author Contributions

Conceived and designed the experiments: IL JEC. Performed the experiments: IL. Analyzed the data: IL JEC. Contributed reagents/materials/analysis tools: IL JEC. Wrote the paper: IL JEC.

References

1. Potts SG, Biesmeijer JC, Kremen C, Neumann P, Schweiger O, et al. (2010) Global pollinator declines: trends, impacts and drivers. Trends Ecol Evol 25: 345–353.

2. Burkle LA, Marlin JC, Knight TM (2013) Plant-pollinator interactions over 120 Years: Loss of species, co-occurrence and function. Science 339: 1611–1615.

3. Shardlow M (2012) A review of recent research relating to the impact of neonicotinoids on the environment. Buglife website. Available: http://smallbluemarble.org.uk/wp-content/uploads/2012/12/Buglife-A-review-of-recent-research-relating-to-the-impact-of-neonicotinoids-on-the-environment.pdf. Accessed 8 September 2013.

4. Maxim L, van der Sluijs J (2013) Seed-dressing systemic insecticides and honeybees. In: EEA (European Environment Agency), editors. Late lessons from early warnings: science, precaution, innovation. Available: http://www.eea.europa.eu/publications/late-lessons-2. Accessed 8 September 2013.

5. Tomizawa M, Casida JE (2003) Selective toxicity of neonicotinoids attributable to specificity of insect and mammalian nicotinic receptors. Annu Rev Entomol 48: 339-364.

6. Decourtye A, Devillers J (2010) Ecotoxicity of neonicotinoid insecticides to bees. In: Thany SH, editor. Insect nicotinic acetylcholine receptors. New York: Springer. 85-95.

7. Cresswell JE (2011) A meta-analysis of experiments testing the effects of a neonicotinoid insecticide (imidacloprid) on honey bees. Ecotoxicology 20: 149-157.

8. Gill RJ, Ramos-Rodriguez O, Raine NE (2012) Combined pesticide exposure severely affects individual-and colony-level traits in bees. Nature 491: 105-108.

9. Whitehorn PR, O'Connor S, Wackers FL, Goulson D (2012) Neonicotinoid pesticide reduces bumble bee colony growth and queen production. Science 336: 351-352.

10. Henry M, Béguin M, Requier F, Rollin O, Odoux J-F, et al. (2012) A common pesticide decreases foraging success and survival in honey bees. Science 336: 348-350.

11. Cresswell JE, Thompson HM (2012) Comment on "A common pesticide decreases foraging success and survival in honey bees". Science 337: 1453.

12. Henry M, Béguin M, Requier F, Rollin O, Odoux J-F, et al. (2012) Response to comment on "A common pesticide decreases foraging success and survival in honey bees". Science 337: 1453.

13. Williams PH, Osborne JL (2009) Bumblebee vulnerability and conservation world-wide. Apidologie 40: 367-387.

14. Cresswell JE, Desneux N, vanEngelsdorp D (2012) Dietary traces of neonicotinoid pesticides as a cause of population declines in honey bees: an evaluation by Hill's epidemiological criteria. Pest Manag Sci 68: 819-827.

15. USDA (2012) Acreage. United States Department of Agriculture website. Available: http://www.usda.gov/nass/PUBS/TODAYRPT/acrg0612.pdf. Accessed 8 September 2013.

16. Statistics Canada (2012) 2011 Census of Agriculture. Statistics Canada website. Available: http://www.statcan.gc.ca/daily-quotidien/120510/dq120510a-eng.pdf. Accessed 8 September 2013.

17. DEFRA (2012) Agriculture in the United Kingdom 2011. Department for Environment, Food & Rural Affairs website. Available: https://www.gov.uk/government/publications/agriculture-in-the-united-kingdom-2011. Accessed 8 September 2013.

18. Elbert A, Haas M, Springer B, Thielert W, Nauen R (2008) Applied aspects of neonicotinoid uses in crop protection. Pest Manag Sci 64: 1099-1105.

19. FERA (2013) Pesticide usage surveys. The Food and Environment Research Agency. Available: http://pusstats.fera.defra.gov.uk. Accessed 8 September 2013.

20. Rortais A, Arnold G, Halm M-P, Touffet-Briens F (2005) Modes of honeybees exposure to systemic insecticides: estimated amounts of contaminated pollen and nectar consumed by different categories of bees. Apidologie 36: 71-83.

21. EFSA (European Food Safety Authority) (2012) Statement on the findings in recent studies investigating sub-lethal effects of some neonicotinoids in consideration of the uses currently authorised in Europe. EFSA J 10: 2752.

22. Akçakaya HR, Stark JD, Bridges TS (2008) Demographic toxicity: methods in ecological risk assessment. New York: Oxford University Press. 3-19.

23. Tasei J-N, Lerin J, Ripault G (2000) Sub-lethal effects of imidacloprid on bumblebees, Bombus terrestris (Hymenoptera: Apidae), during a laboratory feeding test. Pest Manag Sci 56: 784-788.

24. Moncharmont FXD, Decourtye A, Hennequet-Hantier C, Pons O, Pham-Delègue MH (2003) Statistical analysis of honeybee survival after chronic exposure to insecticides. Environ Toxicol Chem 22: 3088-3094.

25. Rennich K, Pettis J, vanEngelsdorp D, Bozarth R, Eversole H, et al. (2012) 2011-2012 National honey bee pests and diseases survey report. USDA website. Available: http://www.aphis.usda.gov/plant_health/plant_pest_info/honey_bees/downloads/2011_National_Survey_Report.pdf. Accessed 8 September 2013.

26. Mommaerts V, Reynders S, Boulet J, Besard L, Sterk G, et al. (2010) Risk assessment for side-effects of neonicotinoids against bumblebees with and without impairing foraging behavior. Ecotoxicology 19: 207-215.

27. Cresswell JE, Page CJ, Uygun MB, Holmbergh M, Li Y, et al. (2012) Differential sensitivity of honey bees and bumble bees to a dietary insecticide (imidacloprid). Zoology 115: 365-371.

28. Laycock I, Lenthall KM, Barratt AT, Cresswell JE (2012) Effects of imidacloprid, a neonicotinoid pesticide, on reproduction in worker bumble bees (Bombus terrestris). Ecotoxicology 21: 1937-1945.

29. Lopez-Vaamonde C, Koning W, Brown RM, Jordan WC, Bourke AFG (2004) Social parasitism by male-producing reproductive workers in a eusocial insect. Nature 430: 557-560.

30. Owen RE, Rodd FH, Plowright RC (1980) Sex ratios in bumble bee colonies: complications due to orphaning? Behav EcolSociobiol 7: 287-291.

31. Müller CB, Schmid-Hempel P (1992) Correlates of reproductive success among field colonies of Bombus lucorum: the importance of growth and parasites. Ecol Entomol 17: 343-353.

32. Hoyle M, Hayter K, Cresswell JE (2007) Effect of pollinator abundance on self-fertilization and gene flow: application to GM canola. Ecol Appl 17: 2123-2135.

33. Goulson D, Darvill B (2004) Niche overlap and diet breadth in bumblebees; are rare species more specialized in their choice of flowers? Apidologie 35: 55-63.

34. Ramirez-Romero R, Chaufaux J, Pham-Delègue M-H (2005) Effects of Cry1Ab protoxin, deltamethrin and imidacloprid on the foraging activity and the learning performances of the honeybee Apis mellifera, a comparative approach. Apidologie 36: 601-611.

35. He Y, Zhao J, Zheng Y, Desneux N, Wu K (2012) Lethal effect of imidacloprid on the coccinellid predator Serangium japonicum and sublethal effects on predator voracity and on functional response to the whitefly Bemisia tabaci. Ecotoxicology 21: 1291-1300.

36. Nauen R (1995) Behaviour modifying effects of low systemic concentrations of imidacloprid on Myzus persicae with special reference to an antifeeding response. Pestic Sci 44: 145-153.

37. He Y, Zhao J, Wu D, Wyckhuys KAG, Wu K (2011) Sublethal effects of imidacloprid on Bemisia tabaci (Hemiptera: Aleyrodidae) under laboratory conditions. J Econ Entomol 104: 833-838.

38. Azevedo-Pereira HMVS, Lemos MFL, Soares AMVM (2011) Behaviour and growth of Chironomus riparius Meigen (Diptera: Chironomidae) under imidacloprid pulse and constant exposure scenarios. Water Air Soil Pollut 219: 215-224.

39. Lunn DJ, Thomas A, Best N, Spiegelhalter D (2000) WinBUGS - A Bayesian modelling framework: concepts, structure, and extensibility. Stat Comput 10: 325-337.

40. Ihaka R, Gentleman R (1996) R: A language for data analysis and graphics. J Comput Graph Stat 5: 299-314.

41. EFSA (2013) EFSA Guidance Document on the risk assessment of plant protection products on bees (Apis mellifera, Bombus spp. and solitary bees). EFSA J 11: 3295.

42. Suchail S, De Sousa G, Rahmani R, Belzunces LP (2004) In vivo distribution and metabolisation of ^{14}C-imidacloprid in different compartments of Apis mellifera L. Pest Manag Sci 60: 1056-1062.

43. Cresswell JE, Robert F-XL, Florance H, Smirnoff N (2013) Clearance of ingested neonicotinoid pesticide (imidacloprid) in honey bees (Apis mellifera) and bumble bees (Bombus terrestris). Pest Manag Sci DOI: 10.1002/ps.3569.

44. Plowright RC, Jay SC (1968) Caste differentiation in bumblebees (Bombus Latr.: Hym.) I.-The determination of female size. Insectes Soc 15: 171-192.

45. Vogt FD, Heinrich B, Plowright C (1998) Ovary development in bumble bee queens: the influence of abdominal temperature and food availability. Can J Zool 76: 2026-2030.

46. Westphal C, Steffan-Dewenter I, Tscharntke T (2009) Mass flowering oilseed rape improves early colony growth but not sexual reproduction of bumblebees. J Appl Ecol 46: 187-193.

47. Pyke GH, Inouye DW, Thomson JD (2011) Activity and abundance of bumble bees near Crested Butte, Colorado: diel, seasonal, and elevation effects. Ecol Entomol 36: 511-521.

48. Beekman M, Van Stratum P, Lingeman R (1998) Diapause survival and post-diapause performance in bumblebee queens (Bombus terrestris). Entomol Exp Appl 89: 207-214.

49. Amin MR, Bussière LF, Goulson D (2012) Effects of male age and size on mating success in the bumblebee Bombus terrestris. J Insect Behav 25: 362-374.

50. Genersch E, Yue C, Fries I, de Miranda JR (2006) Detection of Deformed wing virus, a honey bee viral pathogen, in bumble bees (Bombus terrestris and Bombus pascuorum) with wing deformities. J Inverteb Pathol 91: 61-63.

51. Brown MJF, Schmid-Hempel R, Schmid-Hempel P (2003) Strong context-dependent virulence in a host-parasite system: reconciling genetic evidence with theory. J Anim Ecol 72: 994-1002.

52. Iwasa T, Motoyama N, Ambrose JT, Roe RM (2004) Mechanism for the differential toxicity of neonicotinoid insecticides in the honey bee, Apis mellifera. Crop Prot 23: 317-378.

53. Decourtye A, Lacassie E, Pham-Delègue M-H (2003) Learning performances of honeybees (Apis mellifera L) are differentially affected by imidacloprid according to the season. Pest Manag Sci 59: 269-278.

54. Stelzer RJ, Chittka L, Carlton M, Ings TC (2010) Winter active bumblebees (Bombus terrestris) achieve high foraging rates in urban Britain. PLoS One 5: e9559.

55. Heinrich B (2004) Bumblebee economics. Cambridge: Harvard University Press. 7-21.

56. Goulson D (2013) An overview of the environmental risks posed by neonicotinoid insecticides. J Appl Ecol DOI: 10.1111/1365-2664.12111.

23

A Multi-Criteria Index for Ecological Evaluation of Tropical Agriculture in Southeastern Mexico

Esperanza Huerta[1]*, Christian Kampichler[2,3], Susana Ochoa-Gaona[4], Ben De Jong[4], Salvador Hernandez-Daumas[1], Violette Geissen[5,6]

1 El Colegio de la Frontera Sur, Unidad Campeche, Dpto. Agroecología, Campeche, México, 2 Universidad Juárez Autónoma de Tabasco, División de Ciencias Biológicas, Villahermosa, Tabasco, México, 3 Sovon Dutch Centre for Field Ornithology, Natuurplaza (Mercator 3), Nijmegen, The Netherlands, 4 El Colegio de la Frontera Sur, Unidad Campeche, Dpto. Sustainability Sciences, Campeche, México, 5 University of Bonn - INRES, Bonn, Germany, 6 Wageningen University and Research Center – Alterra, Wageningen, Gelderland, Netherlands

Abstract

The aim of this study was to generate an easy to use index to evaluate the ecological state of agricultural land from a sustainability perspective. We selected environmental indicators, such as the use of organic soil amendments (green manure) versus chemical fertilizers, plant biodiversity (including crop associations), variables which characterize soil conservation of conventional agricultural systems, pesticide use, method and frequency of tillage. We monitored the ecological state of 52 agricultural plots to test the performance of the index. The variables were hierarchically aggregated with simple mathematical algorithms, if-then rules, and rule-based fuzzy models, yielding the final multi-criteria index with values from 0 (worst) to 1 (best conditions). We validated the model through independent evaluation by experts, and we obtained a linear regression with an $r^2 = 0.61$ ($p = 2.4e{-}06$, $d.f. = 49$) between index output and the experts' evaluation.

Editor: Yong Deng, Southwest University, China

Funding: Financial support was obtained from the Conacyt-SEMARNAT Project "Uso sustentable de los recursos naturales en la frontera sur de México" (Sustainable use of natural resources on the southern border of Mexico) (code SEMARNAT-2002-C01-1109). El Colegio de la Frontera Sur provided infrastructural resources. The funders had no role in study design, data collection and analysis, decision to publish, or preparation of the manuscript.

Competing Interests: The authors have declared that no competing interests exist.

* Email: ehuerta@ecosur.mx

Introduction

In the past 60 years, degradation and deforestation of tropical forests worldwide have occurred much faster and more extensively than in any other period in history [1], [2]. Furthermore, countries like Mexico have been undergoing drastic land use changes. In the tropics of Mexico large parts of its lowland rainforest areas have been converted into pasture and cropland. In the state of Tabasco, for example, only 3.4% of the state is covered with original forest [3], whereas 76.4% of the surface was used for cattle production in 2000 [4] and 15.6% was used for agriculture, principally sugarcane and fruit plantations [3]. The ecological consequences of these land-use changes in Tabasco are well documented. Soil losses in hilly regions are very high up to 200 t ha^{-1} year^{-1} [5]; high pesticide and fertilizer inputs to crops that have replaced forests have caused considerable environmental contamination [6], [7], and soil fertility is decreasing [8], [5], [9].

It is of the utmost importance to identify sustainable land use strategies which are economically attractive for the region's farmers and which may also reconcile the need for food production with that of soil conservation. In order to assist a variety of stakeholders at the local and regional level in making land use decisions, simple evaluation tools are needed. This is even more needed since a high percentage of the population consists of immigrants from other Mexican states, who are unfamiliar with the conditions of the humid tropics and use intensive techniques

for farming the land. This is mostly due to large-scale agricultural development projects, such as "Plan Balancán-Tenosique," named after the two municipalities involved, in which, in the 1970 s, over 1100 km^2 of lowland rain forest was destroyed and converted into crop and pasture land. Ecological values of the 1970 s were very different from those of the present, and government representatives were willing to deforest in order to grant land to farmers [10].

The direct impact of farming is difficult to measure due to methodological difficulties (impossibility of measurement, complexity of the system) or practical reasons (time, costs) [11]. Therefore, the use of indicators appears to be an alternative way of guiding land use decisions [12], [13]. However, the "indicator explosion" [14], that is, the use of an exaggerated number of indicators aimed at assessing environmental impacts of agricultural activities, has been of little use to local decision-makers. Particularly in the tropics, land use decisions are still based on the informal opinion of local experts rather than on implementation of Decision Support Systems for environmentally sound resource management [15]. On the one hand, this is due to farmers' restricted access to modern communications and information technologies; on the other, application of indicators is often beyond the capability of local farmers. For example, the agricultural sustainability index proposed by Nambiar et al. [16], which aims to measure agricultural sustainability as a function of biophysical, chemical, economic, and social indicators, would require considerable training of government stakeholders and farmers in order to be applied, and such training is rarely available.

Thus, any tool which local farmers or regional decision makers may use to support their decision making must be as simple as possible. Agroecosystems (like any other ecosystem) are too complex to be precisely measured and evaluated [17]. We agree with the view of Darnhofer et al. [18] in favour of developing less precise rules of thumb which may be used by farmers as well as to guide local land-use decisions toward a more environmentally friendly system of agriculture.

This index may provide farmers and others involved with a tool to evaluate the sustainability of their management of crop land which is oriented toward diminishing soil damage and conserving soil fertility. The index, exclusively based on terms which describe the environmental conditions of the crop system, is accessible to most farmers, and may be calculated using a simple internet application. In this paper we present a simple, easy-to-use index in order to evaluate the ecological state of farms in south-eastern Mexico. We applied the indicator system to 52 crop production systems in south-eastern Mexico and compared the results of the indicator system with expert opinions.

Human knowledge of how to efficiently and sustainably manage complex systems (including agricultural systems) is incomplete and much of what is thought to be known about this topic is actually incorrect. Yet, decisions must be made by policy makers, agricultural extension agents, and farmers despite uncertainty and knowledge gaps [19]. Therefore, tools to support local decision-makers must be flexible, should not enter into too much detail or precision, and should allow for an adaptive strategy which promotes "learning through management" [19]. Consequently, our rationale for developing an index which aids farmers in making environmentally friendly land-use decisions is based on basic, simplified ecological concepts, i.e. the presence of trees, since trees within an agroecosystem enhance soil microclimate in terms of radiation partitioning (shading), evapotranspiration partitioning, and rain interception/redistribution [20]. These factors all help to retain soil moisture. Branches, bark, roots, and living and dead leaf surfaces provide shelter [21] for soil micro-, meso-, and macro-invertebrates. Tree cover for instance, enhances above- and below-ground diversity, serving to support agricultural sustainability [22].

Materials and Methods

1. Rationale of index composition

We define conventional agriculture as a cropping system, typically promoted by government development programs, that is "capital-intensive, large-scale, highly mechanized agriculture with monocropping and extensive use of synthetic fertilizers, and pesticides" [23], [24]. Furthermore, we acknowledge that farming systems are sustainable only if "they minimize the use of external inputs and maximize the use of internal inputs already present on the farm" [25], [26]. The strategy most frequently linked to sustainability is reduction or elimination of agrochemicals, particularly chemical fertilizers and pesticides [25], [27], [28], [29], [30], [31]. Another key to sustained productivity of agricultural systems is the maintenance of soil functions, such as organic matter and nutrient cycling [32], based on organic inputs [33], above-and below-ground biodiversity [22], and diversifying crop systems with nitrogen-fixing legumes [34]. The principal role of the index we propose is to characterize methods of tillage, external inputs, and crop structure.

2. Primary indicators

We chose 12 field variables as primary indicators related to the above mentioned aspects of ecologically sound agricultural land use based on farmer's practices. These are easy to evaluate in the field and characterize plot structure (primary indicators: tree cover, tree density, tree diversity), crop structure and crop conditions (primary indicators: crop type, crop rotation, crop density, crop colour), tillage (primary indicators: type of tillage, timing of tillage), the use of fertilizers (chemical versus organic) and pesticide application.

2.1 Tree cover. Tree cover is defined as the canopy of trees, measured in the field, and recorded as percentage classes of tree cover in three height classes (trees >15 m, 10–15 m, <10 m). Thus, this variable characterized one aspect of agroecosystem management: the farmers' decision to maintain the canopy of the trees in his or her agroecosystem.

2.2 Tree density. Tree density is defined as the number of trees per area. To measure this, we distinguished three categories of tree density: high density (abundant), medium density, and low density (isolated or no trees). A high number of trees per area guarantee carbon sequestration [35] while a stable microclimate is maintained. This variable, measured in the field, is one the variables that characterize the effect of trees in the agroecosystem.

2.3 Tree diversity. This variable was measured in the field by counting the number of trees species within the plot. In agroecosystems, biodiversity may; (i) contribute to constant biomass production and reduce the risk of crop failure in unpredictable environments, (ii) restore disturbed ecosystem services such as water and nutrient cycling, and (iii) reduce risks of pests and diseases through enhanced biological control or direct pest control [36], [20], [22].

2.4 Crop type. This variable indicates whether the crop is annual, seasonal, or perennial. We obtained this information by observing the type of crop. Annual crops in general have higher environmental impacts, ie: greenhouse emissions, and nutrient leaching, than perennial crops [37].

2.5 Crop rotation. In sustainable farm systems leguminous crops are increasingly used in crop rotations as a source of nutrients, particularly nitrogen for crop growth [38], [39], nitrogen-fixing legumes, contribute to maintaining biodiversity above and in the soil, contribute nitrogen to the soil/plant system, and help avoid the build-up of pest populations [34]. In this study, we asked farmers whether they planted another crop before planting the main crop and whether they practice crop rotation, as crop rotation may assist with weed and pest control [40], [34]. According to Bellon [41], an activity which leads into the maintenance or increase of renewable resources in agroecosystems, is considered as an ecological technology. This variable helped us to characterize the technology used in the agroecosystem.

2.6 Crop density. Crop density is defined as the number of plants (individuals) per area. Three categories were recorded in the field: abundant (high density: 3,000 plants/ha), medium density (1000–1600 plants/ha), and sparse (<1000 plants/ha). This variable also indicated the level of technology applied to the crop, as less intensive techniques typically yield lower densities [42].

2.7 Crop colour. The colour of a crop indicates the nutritional status of the plants; green plants generally have sufficient nutrients, while yellow plants lack nitrogen [43].

2.8 Type of tillage. Type of tillage was categorized into no tillage, manual tillage, and mechanical tillage using machinery, the latter of which generally indicates high disturbance of the soil surface and rapid loss of soil organic carbon and other nutrients

[44]. In the field, we asked the farmers how they prepared the land for crop planting.

2.9 Frequency of tillage. This variable indicates frequency of tillage - every year or every 2 years. With this information, it was possible to estimate the frequency of soil disturbance due to tillage.

2.10 Chemical input. Searching for sustainable productions, it is recommended no or low or use of inorganic fertilizers and pesticides [45], [46]. Long term use of some pesticides, as glifosate can decrease earthworm species number, density and biomass [47], [48]. This information allowed us to estimate the amount of chemical fertilizers and pesticides applied within a given area. These variables (chemical fertilizers and pesticides) are included in the indicator for chemical disturbance.

2.11 Green manure. Is defined as the presence or absence of leguminous crops mixed with the principal crop, generally used to increase total soil nitrogen content. Green manures should always be intercropped, as it has been proven that growing legumes with cereal crops decreases N_2O emissions [49], and therefore is a sustainable, environmentally friendly practice. Examples of green manure use have been observed in traditional Mesoamerican cultures, for example, intercropping beans, as well as other edible plants, within the *milpa* or traditional maize cropping system [50].

3. Index development

Primary indicators were hierarchically aggregated into higher levels, forming intermediate variables, which in turn are structured into a single index that evaluates the ecological condition of a given plot on a scale from 0 (worst) to 1 (best). An index close to zero either would mean that a more environmentally friendly farming techniques need to be implemented or that the plot should be subjected to a fundamental change in land-use, e.g., reforestation, in order to return to a more ecologically sound state. An index close to one, on the other hand, would indicate an ecologically sound land-use. Methods applied in indicator aggregation were (i) simple mathematical operations, ii) sets of if-then rules, and (iii) sets of if -then rules combined with fuzzy logic.

3.1. Aggregation through mathematical operations. Mathematical operations include calculating averages, weighed averages, minimum values, and maximum values. For example, if primary indicator A has the value a and primary indicator B has the value b, then the value x of the intermediate variable X is determined as $x = (a+b)/2$, $x = (w_1*a+w_2*b)$, where $w1$ and $w2$ are weights, or $x = \min(a, b)$ or $\max(a, b)$.

3.2. Aggregation by IF-THEN rules. If primary indicator A can have the discrete values a_1, a_2, and a_3, and primary indicator B can have the discrete values b_1, b_2, and b_3, then the value x of the intermediate variable X is determined by a set of nine (number of levels of A x number of levels of B) rules. For example, IF $A = a_1$ and $B = b_1$ THEN $X = x_1$.

3.3. Aggregation by IF-THEN rules and fuzzy logic. We used small fuzzy rule-based models for aggregation in the case of non-linear interactions among indicators using a continuous numerical scale or an ordinal scale with a large number of possible values. In classic set theory, an object can either be a member (membership = 1) or not (membership = 0) of a given set. The central idea of fuzzy set theory is that an object may have a partial membership of a set, which consequently may possess all possible values between 0 and 1. The closer an element is to 1, the more it belongs to the set; the closer the element is to 0, the less it belongs to the set. To apply the fuzzy set theory, three steps are involved in calculating the model's output. First, for any observed value of the primary indicators, its corresponding membership value in the fuzzy set domain is calculated (*fuzzification*); second,

the memberships of the intermediate variable X are calculated, applying the rules in the fuzzy set theory (*fuzzy inference*); third, the fuzzy results are converted into a discrete numerical output (*defuzzification*; see Wieland 2008 for an introduction to fuzzy models). Fuzzy rule-based models have become popular in ecological modelling [51], [52], and several examples exist of its usefulness in the context of ecosystem evaluation, bioindication, and sustainable management [15], [53], [54]. Here, if both primary indicators A and B can have numerical values from 0 to 1, then the value x of the intermediate variable X is determined by a series of fuzzy set rules representing the linguistic variables "low A", "medium A", "high A", and "low B", "medium B", and "high B", as well as the output "low X", "medium X", and "high X". The value of the intermediate variable X is determined by nine levels of A x the number of levels of B; for example: If A = low and B = low Then X = [low, medium, or high].

To maintain the number of rules as well as their complexity as low as possible, we aggregated only two variables at a time. A simple example shows the reasoning behind this decision; if there are three primary variables, A, B, and C with three categories for each variable, a single rule node requires $3*3*3 = 27$ If-Then rules of the type "If $A = a_1$, a_2, or a_3 and If $B = b_1$, b_2, or b_3 and if $C = c_1$, c_2, or c_3 Then…", whereas if an additional intermediate variable X is introduced to the model, only 18 rules are needed: $3*3 = 9$ rules to aggregate A and B to X (If $A = a_1$, a_2, or a_3 and If $B = b_1$, b_2, or b_3), and $3*3 = 9$ rules to aggregate X and C (If $X = x_1$, x_2, or x_3 and If $C = c_1$, c_2, or c_3).

4. Study area and application of the index

The state of Tabasco in south-eastern Mexico is characterized by a humid tropical climate with a mean annual rainfall between 1200 and 4000 mm and a mean annual temperature of 27°C [55]. Predominant soils are Gleysols and Fluvisols over alluvial sediments in the plains, Vertisols, Cambisols, Luvisols, and Acrisols over Miocene or Oligocene sediments, and Leptosols and Regosols over limestone mountains [9], [56]. We chose the municipalities Balancán and Tenosique in western Tabasco (17°81′50″–18°81′00″ N, 91°80′10″–91°84′60″ W) as a study area (Figure 1), we worked with private, *ejidal* (multipurpose land, where owners can or cannot sell the property according to their legal status in the National Agrarian File), and communal lands (coordinates of each plot are shown in Table 1). The region is mainly a plain towards the North (67% of area has an elevation <20 m. a. s. l.) with hills (29%, 20–200 m. a. s. l.), and mountains (4%, max. 640 m. a. s. l.) in the South, comprising a total area of 5474 km^2. These municipalities have undergone a high degree of land use change over the past 40 years. Until the early 1970 s, this region was still covered by lowland rain forest. The principal form of land use is pastureland, and covers 60% of the land [57]. An additional 30% is cropland, mainly cultivated under small-medium holder systems with seasonal conventional agriculture using high levels of agrochemical inputs (Table 2 & 3). Common crops are maize (*Zea mays*), a variety of hot peppers (*Capsicum* sp), cucumbers (*Cucurbita argyrosperma*), watermelon (*Citrullus lanatus*), perennial fruit crops such as papaya (*Carica papaya*), and biannual crops such as sugar cane (*Saccharum officinarum*) [58].

We chose 52 farms in the study area (Table 1), and selected those farms whose main economic activity (100%) is agriculture and chose one agricultural plot from each farm for evaluation. There were annual and biannual crops with or without trees (Table 2). Average plot size was 32.4±55.1 ha, and the average time that the plot had been used for a given crop system was 2.5±3.0 years.

Table 1. Characterization of plots.

Plot #	Municip	Lat	Long	Altitud (mls)	Plot size (ha)	Type of property	Original-vegetation
1	Bal	2007880	652445	43	18	Ejidal	A
2	Bal	1964368	642505	10	0.5	Private	TRF
3	Bal	1970936	647701	11	160	Private	TRF
4	Bal	1957459	667496	14	2	Ejidal	TRF
5	Bal	1964107	660656	14	7	Private	PT
6	Bal	1960297	658989	22	15	Private	TRF
7	Bal	1976300	676424	40	30	Private	TRF
8	Bal	1990326	665947	40	28	Private	TRF
9	Bal	1964590	667264	11	3	Private	RV
10	Bal	1972260	672382	28	2.5	Ejidal	TRF
11	Bal	1950339	666409	49	40	Private	TRF
12	Bal	1966112	682994	40	80	Private	PT
13	Ten	1931754	704474	65	14	Private	TRF
14	Ten	1912951	690381	130	2	Communal	TRF
15	Ten	1932644	665521	15	3	Ejidal	TRF
16	Ten	1926163	678608	56	68	Private	TRF
17	Ten	1933663	674227	59	35	Private	A
18	Ten	1926570	669486	35	8.5	Ejidal	TRF
19	Ten	1942431	663885	21	1.5	Ejidal	TRF
20	Ten	1942068	671156	31	7.25	Ejidal	TRF
21	Ten	1943167	650736	82	14	Ejidal	A
22	Ten	1923838	669841	118	0. 25	Private	TRF
23	Bal	1973301	695548	105	40	Ejidal	TRF
24	Bal	1982392	701383	74	20	Ejidal	A
25	Bal	1965354	709053	44	11	Ejidal	A
26	Bal	1960496	706464	51	44	Ejidal	TRF
27	Bal	1942484	709731	66	10	Ejidal	A
28	Ten	1944148	673673	16	320	Ejidal	TRF
29	Ten	1930350	659650	29	20	Private	TRF
30	Ten	1924549	656910	41	53	Ejidal	TRF
31	Ten	1926226	652978	65	3	Ejidal	TRF
32	Bal	1955194	688843	45	44	Private	TRF
33	Bal	1960980	679265	49	29	Ejidal	TRF
34	Bal	1942552	696222	48	10	Private	TRF
35	Ten	1938120	686470	49	5	Private	TRF

Table 1. Cont.

Plot #	Municip	Lat	Long	Altitud (mls)	Plot size (ha)	Type of property	Original-vegetation
36	Bal	1968750	695595	60	2	Private	A
37	Bal	1977546	699417	53	200	Ejidal	TRF
38	Bal	1949770	708854	53	23	Ejidal	TRF
39	Bal	1951216	697890	46	2	Ejidal	TRF
40	Ten	1931041	667544	31	17	Private	TRF
41	Ten	1927447	664223	3	3	Ejidal	TRF
42	Ten	1928226	664551	31	3	Ejidal	TRF
43	Ten	1930579	691785	54	15	Communal	TRF
44	Ten	1931219	677895	59	63	Private	TRF
45	Ten	1931525	676500	60	6.75	Ejidal	TRF
46	Ten	1908445	675040	140	40	Ejidal	TRF
47	Ten	1910875	669750	222	45	Ejidal	TRF
48	Ten	1924222	704702	42	30	Ejidal	A
49	Ten	1935076	712227	88	20	Ejidal	TRF
50	Ten	1914876	686285	113	4	Ejidal	TRF
51	Ten	1911365	684878	389	22	Ejidal	TRF
52	Ten	1911503	698606	139	10	Ejidal	TRF

Municip: municipality, Bal: Balancan, Ten: Tenosique. TRF: tropical rain forest, PA: pasture, A: acahual (secondary vegetation). Ejidal: multipurpose land, where owners can or cannot sell the property according to their legal status in the National Agrarian File.

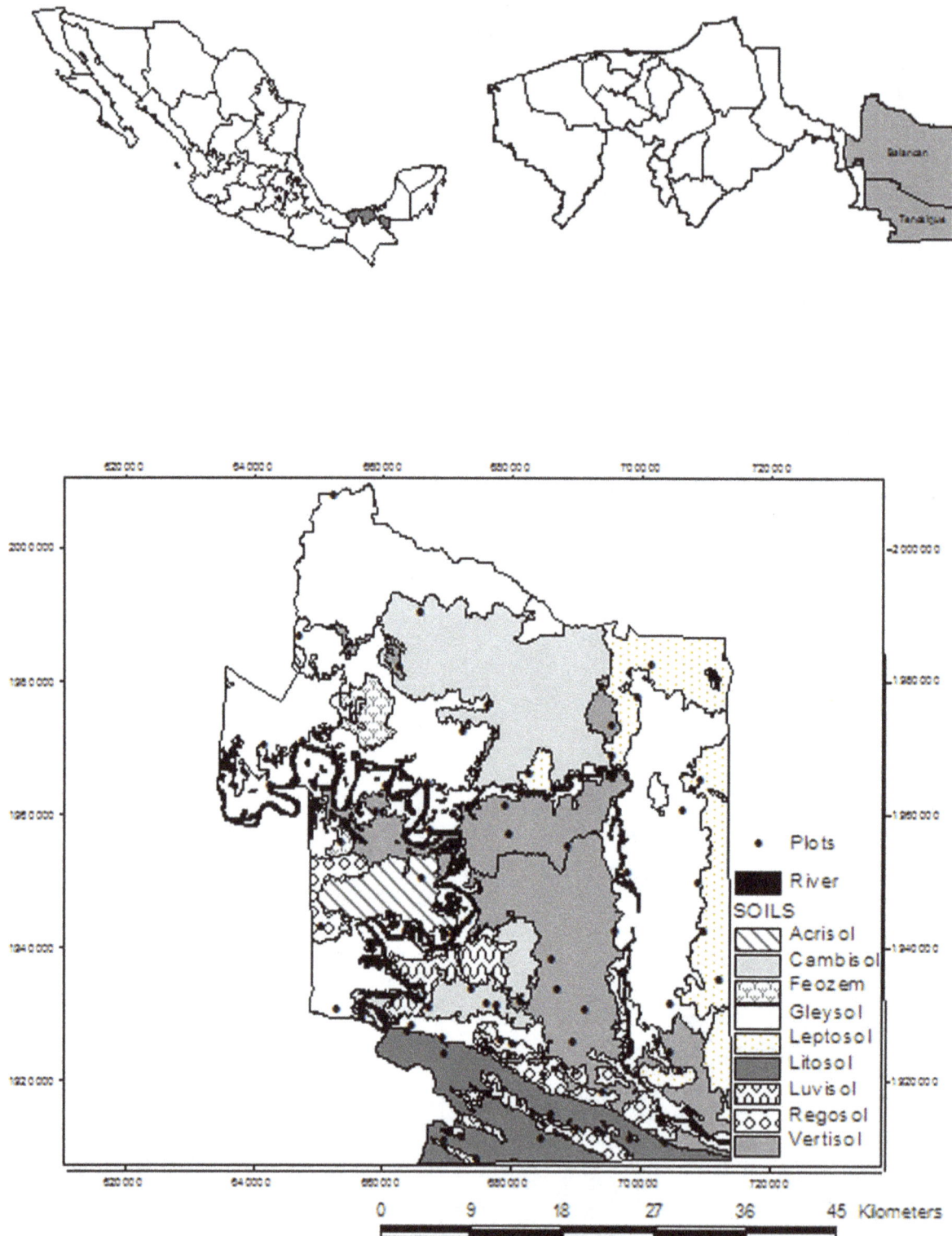

Figure 1. Distribution of 52 evaluated agricultural plots in tropical South-East Mexico.

We questioned farmers as to frequency, amount and type of chemical fertilizers and pesticides applied per area of cropland. After their verbal consent, farmers gave us their complete name and signed next to their name on the record sheet, validating all the information. Due to the fact that we only asked about the use of fertilizers and management of the land, the procedure approval from the Ethics Committee was not required.

Between March and October 2004, for each plot, the values of the primary indicators were determined and the index of ecological condition was calculated. Prior to index calculation, the plots were also evaluated by experts (2 scientists, each with

Table 2. Crop characterization.

Plot	Cycle	Main seasonal or perennial crop	Trees
1	s-s	Zm, Pv, Cs, Cl	Tr, Pa, Cpa
2	s-s	Zm, Pv, Cs, Cl	Sh, Dg
3	w-s	Sv	Sa
4	a-w	Zm, Pv, Cpa	0
5	Y	Zm, Pv, Cl, Cm	0
6	Y	Zm	Cpa, Eg, Tr
7	s-a	Ca	Tr
8	Y	Cp	Cp
9	Y	Zm	0
10	Y	Zm	0
11	w-s	Zm, Pv, Cl	Sm, Cpa
12	Y	Co	Co
13	s-s	Zm	0
14	s-s	Le, Se	0
		Cpa, Zm, Pv	
15	s-s	Cl, Ta, Mp,	Fruit trees
16	Y	Jj	Gs, Hc, Cc, Tr
17	Y	So	0
18	Y	So	0
19	Y	So	0
20	Y	Zm	Mi
21	s-s	Zm	0
22	Y	Csi, Cli, Js	Co, Pa, Bc, Cn, Mi
23	s-s	Zm, Cpa	Cpe
24	s-s	Zm	Cpa
25	s-s	Zm	0
26	s-s	Zm, Cpa	0
27	s-s	Zm, Ca	Sm
28	Y	So	0
29	s-s	Zm, Pv, Cs, Cl	0
30	Y	Zm, Pv, Cs, Cl, Ta	0
31	Y	Zm, Pv, Cs	0
		Cl, Os, Le	
32	s-s	Zm, Pv, Cpa, Cl, Os	Tr
33	Y	Zm, Pv	Sm, Bg
34	s-s	Zm, Pv, Os	Cpa
			Tr
35	Y	Zm, Cl	Tr
36	Y	Zm, Pv, Cpa, Cl, Ca	0
37	s-s	Zm, Pv, Ca	Sm, Pa, Sh
38	Y	Zm, Pv, Cpa, Cl, Ta	Co, Sh
39	Y	Zm, Pv, Cpa, Cl, Ca	
		Cs, Cm	
40	Y	So	Tr, Cpe
			Co, D a
41	Y	So	Tr, Cpe
			Co, D a
42	Y	So	Tr, Cpe
			Co, D a

Table 2. Cont.

Plot	Cycle	Main seasonal or perennial crop	Trees
43	Y	So	Tr, Cpe
			Co, Da
44	Y	So	Tr, Cpe
			Co, Da
45	Y	So	Tr, Cpe
			Co, D a
46	Y	Zm, Pv, Cpa, Me	Sm
47	Y	Zm, Pv, Me, Ib	Bc, Mi, Dg, Cs, Ll
48	Y	Zm, Pv, Cpa, Cl	0
49	Y	Zm, Pv, Ca	Tr, Gs
50	Y	Zm, Pv, Cpa, Cl	Sm, Co, Ma
		Cs, Ib, Me	0
51	Y	Zm, Pv, Cpa, Cs	Fruit trees
52	Y	Zm, Pv, Cpa, Cl, Cs	Co, Sh, Bc
		Mp, Ca, Me	0

Cycle: s–s: spring-summer, w-s: winter-summer, s-a: summer-autumn, Y: all the year. **Main seasonal or perennial crop**: Ca: *Cucurbita argyrosperma* (cushaw pumpkin); Cp: *Carica papaya* (papaya); Cl: *Citrullus lanatus* (watermelon); Cli: *Citrus limon* (lemon); Cm: *Cucumis melo* (muskmelon); Cpa: *Cucurbita pepo* (squash); Cs: *Capsicum* sp. (pepper); Csi: *Citrus sinensis* (orange tree); Ib: *Ipomoea batatas* (sweet potato); Jj: *Jarthropha jurcas* (oil palm); Js: *Jobo spondia* (plum); Le: *Lycopersicon esculentum* (tomato); Me: *Manihot esculenta* (cassava); Mp: *Musa paradisiaca* (banana); Os: *Oryza sativa* (rice); Pv: *Phaseolus vulgaris* (bean); So: *Saccharum officinarum* (sugarcane); Se: *Sechium edule* (pear squash); Sv: *Sorghum vulgare* (milo); Ta: *Triticum aestivum* (wheat); Zm: *Zea mays* (maize). **Trees**: Bc: *Byrsonima crassifolia* (nanche); Cc: *Crescentia cujete* (calabash tree); Co: *Cedrela odorata* (Mexican cedar); Cpa: *Carludovica palmate* (toquilla palm); Cpe: *Ceiba pentandra* (kapok); Cn: *Cocos nucifera* (Coconut); Dg: *Dialium guianense* (wild tamarind); Eg: *Eucalyptus grandis* (Eucalyptus); Gs: *Gliricidia sepium* (Cocoite, gliricidia, cacao de nance); Ll: *Leucaena leucocephala* (white lead tree, jumbay); Ma: *Mammea americana* (mamey apple); Mi: *Mangifera indica* (mango); Pa: *Persea americana* (avocado); Sa: *Sterculia apetala* (camoruco, manduvi tree); Sh: *Swietenia humilis* (small mahogani); Sm: *Spondias mombin* (Yellow plum, bai makok); Tr: *Tabebuia rosea* (savannah oak, maculis).

over 15 years of experience in agroecology) on a scale from 0 (poor condition) to 1 (good condition) in order to test the correlation between the experts' opinions and the index.

Data was normalized for carrying out multiple regression with plot size, altitude of plot site, or previous vegetation cover and the ecological index.

Finally we located the index in this public web address: http://201.116.78.102/~modelo/Index.html, where farmers in the near future can introduce independent data sets and evaluate or monitor their own agro ecosystems.

Results

1. Index structure

Ten nodes were used to aggregate primary indicators to the final index of ecological conditions of agricultural systems (Figure 2). Six of these used simple mathematical operations; two were based on rule sets, and two on fuzzy rule-based models (Table 4).

2. Index application

2.1. General characterization of sampled plots. A total of 67% of the plots were cultivated after cutting of primary lowland forest, 24% after cutting secondary forests of various ages, and 1.5% after cutting riparian vegetation, the rest conversion from pastureland (livestock production) to crop system. On 63% of the farms, maize, beans and pumpkins were cultivated, 13% sugar cane, 12% watermelon, 9% rice and 3% pepper. On 68.5% of the plots, trees were scattered among the crops. Conventional tillage was used on 55.5% of the plots, and pesticides were used on 79.4%, green manure was used on 21% of the plots and chemical

fertilizers on 67%. All farmers understood the meaning of the variables evaluated for their plots (for a complete list of plots, see Table 5).

2.2. Plot evaluation with the index. Ecological condition of the plots ranges from 0.0 to 0.8125 (Figure 3; Table 5). One plot which was intercropped with timber and fruit trees presented the highest value. This site was characterized by an absence of agrochemical use, use of green manure, presence of annual crop rotation, and high tree diversity. We found 18 plots with index values of 0–0.25, 7 plots with values of 0.3–0.5, and 26 plots with values of 0.5–0.7. Thus, the majority of the plots evaluated in this study had an intermediate index value. 50% of the plots with this intermediate index are *ejidal* property, 38% private land, and 12% is under another type of ownership (smallholder or communal). 80% of these intermediate plots were lowland rain forest before being converted into agricultural land, 12% were secondary vegetation, and 8% were pastureland (Table 1). One might believe that a prior forest condition implies a more environmentally friendly agroecosystem that allows for preserving a more diverse system. However, plots with low index values were also lowland rain forest before being turned into agricultural land (Table 5). In this case, the land managers or owners decided to deforest the land to subsequently plant annual crops. 13 plots were larger than 40 ha (Table 1). All of these were previously covered with lowland rain forest; the ecological index ranges from 0.5–0.56 for 46% of these plots, and another 23% have an ecological index of 0.37–0.43. 10 plots fell under the smallest size category (0.25–3 ha) and had an ecological index of 0.5–0.68 (4 plots), 0 to 0.18 (5 plots), and 0.39 (1 plot).

Carrying out multiple regression with normalized data, we did not observe significant correlations between plot size, altitude of

Table 3. Crop land preparation and inputs characterization.

Plot	Tillage	Chemical Fertilization	Pesticides Use	Gm	Oa
1	Ma & Me	NPK 17:17:17 & U	Endosulfan**	0	0
2	Ma	0	Chlorpyrifos*	0	0
3	Me	Pholiar	0	0	c s m
4	Ma	0	0	0	0
5	Ma & Me	NPK 17:17:17	Chlorpyrifos**	Y	Y
6	Ma & Me	NPK 17:17:17	Chlorpyrifos*	0	0
7	Me	0	ID	0	0
8	Me	ID	ID	Y	0
9	Ma & Me	0	Carbofuran*	0	0
10	Me	NPK 18-46-0	ID	0	0
11	Me	NPK 18- 46- 0, U & P	ID	Y	0
12	Me	NPK 18-46-0	ID	0	0
13	Me	NPK 17:17:17 & U	Chlorothalonil** Chlorpyrifos*	0	0
14	Ma	0	Endosulfan*	0	0
15	Me	0	0	Y	0
16	Ma	Urea	0	0	0
17	Me	NPK 19-19-19	ID	0	0
18	Me	NPK 19-19-19	ID	0	0
19	Me	Urea	ID	0	bg
20	Me	0	Methilic*	0	0
21	Me	NPK 17:17:17 & U	Zeta**	0	0
22	Ma	0	0	0	0
23	Me	NPK 17:17:17 & U	Chlorpyrifos**	0	lcf
24	Me	Urea & P	Zeta*	Y	lcf
25	Me	Urea	(2,4-D AMINA)*	Y	0
26	Me	NPK 17:17:17 & U	(Z)-(1R,3R)**	0	0
27	Me	0	Zeta*	N	0
28	Me	NPK 17:17:17	(RS)*	0	0
29	Me	NPK 17:17:17 & U	Chlorpyrifos**	0	0
30	Ma	0	Chlorpyrifos**	Y	0
31	Me	NPK 17:17:17 & U	Chlorpyrifos*	0	0
32	Me	0	0	Y	0
33	Me	0	0	0	0
34	Me	0	0	0	0
35	Me	NPK 17:17:17 & U	0	0	0
36	Me	NPK 17:17:17 & U	Chlorpyrifos*	Y	0
37	Me	0	Urea	0	0
38	Me	NPK 17:17:17 & U	Carbofuran*	Y	Y
39	Me	NPK 17:17:17 & U	Chlorpyrifos**	0	0
40	Me	NPK 17:17:17	(Z)-(1R,3R)*	0	0
41	Me	NPK 17:17:17 & U	(Z)-(1R,3R)*	0	0
42	Me	NPK 17:17:17 & U	(Z)-(1R,3R)*	0	0
43	Me	NPK 17:17:17 & U	(Z)-(1R,3R)*	0	0
44	Me	NPK 17:17:17 & U	(Z)-(1R,3R)*	0	0
45	Me	NPK 17:17:17 & U	(Z)-(1R,3R)*	0	0
46	Ma	0	0	Y	0
47	Me	U	0	0	0
48	Ma &Sb	Urea	Chlorpyrifos**	0	0

Table 3. Cont.

Plot	Tillage	Chemical Fertilization	Pesticides Use	Gm	Oa
49	Me	Urea & P	Zeta*	0	0
50	Ma	Urea	Endosulfan*	0	0
51	Ma	0	ID	0	0
52	Me	NPK 17:17:17 & U	Chlorpyrifos*	0	0

Tillage: Ma: Manual; Me: Mechanized. **Pesticides**: Endosulfan: 6,7,8,9,10,10-Hexachlor- 1,5,5a,6,9,9a-Hexahidro-6,9-metane-2,4,3-benzodioxatiepin-3-oxide; Chlorpyrifos: 0,0-dimetil 0-(3,5,6-trichlore-2-piridinil) fosforotioate (33.8%), Permetrine: 3-fenoxibenzil (1RS)-cis, trans-3-(2,2 diclorovinil)-2,2 dimetil ciclopropane-carboxilate (4.8%); Carbofuran: 2,3 Dihidro-2,2-dimetil-7-benzofuranil metil carbamate; Chlorothalonil: Tetrachloroisoftalonitrile; Zeta: Zeta-cipermetrine a-ciano-3-(fenoxifenil) metil (±) cis-trans; (Z)-(1R,3R): (Z)-(1R,3R)-3-(2-chloro-3,3,3-trifluoroprop-1-enil)-2,2-dimetilcichlopropanecarboxilate de (S)-α-ciano-3-fenoxibencile & (Z)-(1S,3S)-3-(2-chloro-3,3,3-trifluoroprop-1-enil)-2,2-dimetilcichlopropanecarboxilate de (R)-α-ciano-3-fenoxibencile; (RS): (RS)- alfa- ciano-3-fenoxybencil(1RS)-cis-trans-3-(2,2-dichlorvinil)- 1,1-dimetilcichlopropanecasrboxilate. * Once per year, ** 2 per year. **Chemical Fertilization**: U: Urea, P: Phosphorus. **Gm**: Green manure, Y: Yes, 0: null application. **Oa**: Organic amendments: csm: cow, sheep manure, bg: burned grass, lcf: last crop fallow, 0: null application.

plot site, or previous vegetation cover and the ecological index (Kendall's Tau, T = 0.016 p = 0.98, T = 0.11 p = 0.90, T = −0.31 p = 0.75, respectively). Therefore, in this study, it seems that neither size nor location of the plot determines the type of plot management; rather, plot management is likely determined by government development programs and traditional farming techniques.

3. Correlation between index and expert opinion

We obtained a Pearson correlation coefficient of $r^2 = 0.61$ ($p = 2.4e$-06, $d.f. = 49$) between the index and the values determined by independent experts, indicating a satisfactory correspondence (Figure 4). However, the experts systematically awarded higher scores to the plots than did the index. Moreover, they suggested to include additional variables to the index which would yield better information regarding (i) type of organic inputs to the crop system, (ii) types of pest and disease control used, (iii) number of native plant species among the crop, (iv) origin of crop seeds, (v) vegetation surrounding the crop, (vi) presence of vertebrate fauna, and (vii) diversity of soil macroinvertebrates.

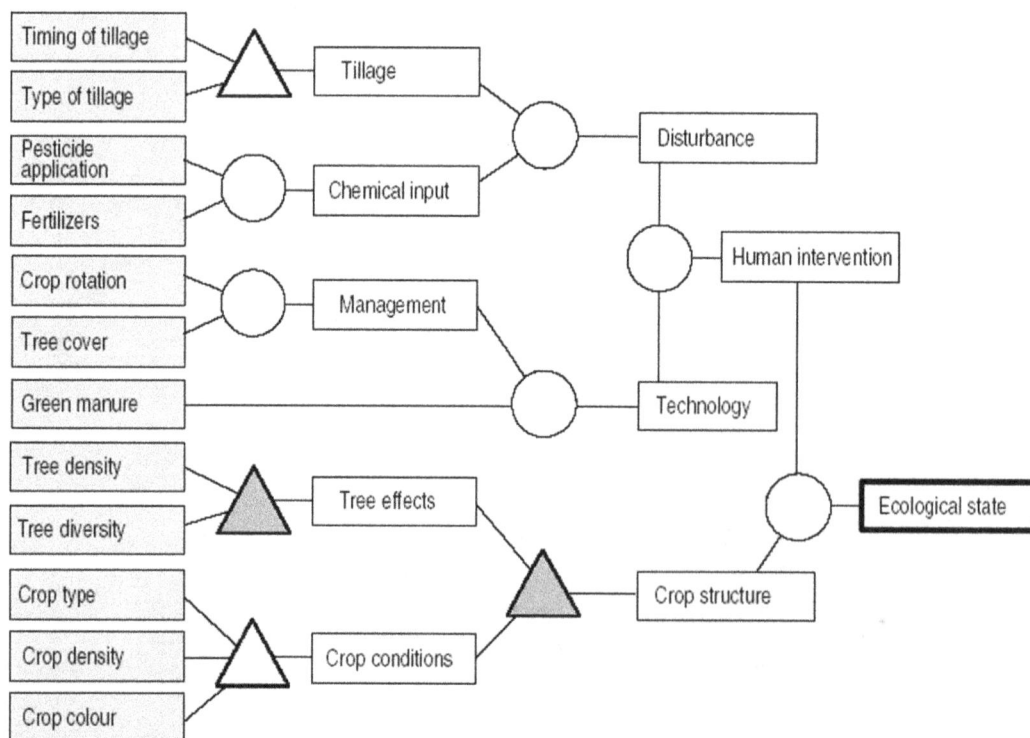

Figure 2. Structure of the Index of ecological condition of tropical agroecosystems. Primary indicators are shaded in grey. Circles represent simple mathematic algorithms, white triangles represent rule sets, and grey triangles represent rule sets, and grey triangles represent rule sets based on fuzzy logic. This index was presented together with other indexes within a frame of Indicators of environmentally sound land use in the humid tropics [15].

Table 4. Description of the measurement levels of the variables.

Groupal variables	Field data variables	Best	Intermediate (5 levels)			Worst
		1	0.75	0.5	0.25	0
Management	Tree cover	all year	presence	present	rare	null
Tree effect	Tree density	high	medium	low	isolated	null
		1	0.66	0.33		0
Crop structure	Tree diversity	>4 species	4 species	2 species		1 species
Crop conditions	Crop type	perennial polyculture	perennial monoculture	biannual		annual
		1	0.33			0
Plough (tillage)	Tillage time	null	frequent			constant
	Tillage form	null	manual			technical
Disturbance						
Chemical inputs	Pesticides	null	frequent			constant
	Fertilizers	null	frequent			constant
Management	Crop rotation	constant	frequent			constant
Technology	Green manure	constant	frequent			null
		1	2 levels			0
Crop conditions	Crop density	abundant				disperse
	Crop coleur	green				yellow

Table 5. Normalized Ecological Index for plots evaluated (52).

Plot	TT	TF	P	F	GM	RC	CA	DA	DivA	CT	CD	Ccol	Ev	Index
1	n	m	c	c	n	n	>1	l	>4	pm	a	g	0.24	0.56
2	n	n	c	n	n	n	>1	i	>4	a	a	g	0.35	0.5
3	c	m	n	c	c	n	>1	i	>4	a	a	g	0.26	0.56
4	n	n	n	n	n	n	n	n	1	a	a	g	0.01	0.37
5	n	n	c	n	c	c	>1	i	>4	a	a	g	0.09	0.62
6	n	m	c	n	c	n	>1	l	>4	a	a	g	0.22	0.5
7	c	t	c	n	n	n	>1	i	>4	a	d	b	0	0.15
8	c	t	c	n	c	c	n	n	1	pp	d	g	0.1	0.12
9	n	t	c	n	c	n	n	n	1	a	a	g	0.09	0.25
10	c	t	c	c	n	n	n	n	1		d	b	0	0
11	c	t	c	c	n	c	>1	i	>4	pm	ID	ID	0.03	0.21
12	c	t	c	c	n	n	>1	h	>4	ba	a	g	0.2	0.5
13	c	t	c	c	n	n	>1	i	>4	a	a	g	0.03	0.5
14	n	n	n	c	n	n	n	n	1	pp	a	g	0.29	0.37
15	n	t	n	c	n	c	>1	h	>4	pp	a	g	0.2	0.81
16	n	n	n	c	n	n	>1	l	>4	ba	a	g	0.25	0.5
17	c	t	n	c	n	n	n	n	1	a	a	g	0	0.25
18	c	t	c	c	n	n	n	n	1	a	d	g	0	0
19	c	t	c	c	n	n	>1	i	1	a	d	g	0	0.06
20	n	t	c	c	n	n	n	i	>4	a	a	g	0.1	0.5
21	c	t	c	c	n	n	n	n	1	a	a	g	0	0.25
22	n	n	n	c	n	n	>1	i	>4	pp	d	g	0.33	0.62
23	n	t	c	c	n	n	>1	i	>4	a	a	g	0.15	0.56
24	n	t	c	n	n	c	>1	l	>4	a	a	g	0.2	0.62
25	n	n	c	c	n	n	n	n	1	a	d	g	0.01	0.06
26	c	m	c	n	n	n	n	n	1	a	a	g	0.25	0.25
27	n	t	c	n	n	n	>1	i	>4	a	a	g	0.01	0.5
28	n	t	c	c	n	n	n	n	1	a	a	g	0	0.25
29	c	t	c	c	n	n	n	n	1	pm	a	g	0.01	0.37
30	c	m	c	n	n	c	n	n	1	pm	a	g	0.1	0.43
31	c	t	c	c	n	n	n	n	1	a	a	g	0.01	0.25
32	c	t	n	c	n	c	n	n	>4	a	d	y	0.09	0.34
33	c	t	n	n	n	n	>1	h	>4	a	a	g	0.15	0.62
34	c	t	n	n	n	n	>1	i	>4	ba	a	g	0.13	0.62
35	c	t	n	c	n	n	>1	l	>4	a	a	g	0.01	0.5

Table 5. Cont.

Plot	TT	TF	P	F	GM	RC	CA	DA	DivA	CT	CD	Ccol	Ev	Index
36	c	t	c	c	n	c	n	n	1	a	a	g	0	0.31
37	c	t	c	n	n	n	>1	i	>4	pm	a	y	0.01	0.15
38	c	t	c	c	n	c	>1	h	>4	pm	a	g	0.2	0.68
39	c	t	c	c	n	n	>1	l	>4	a	a	b	0.06	0.18
40	c	t	c	c	n	n	>1	l	>4	a	a	g	0.19	0.5
41	c	t	c	c	n	n	>1	l	>4	a	a	g	0.19	0.5
42	c	t	c	c	n	n	>1	l	>4	a	a	g	0.17	0.5
43	c	t	c	c	n	n	>1	l	>4	a	a	g	0.2	0.5
44	c	t	c	c	n	n	>1	l	>4	a	a	g	0.2	0.5
45	c	t	c	c	n	n	>1	l	>4	a	a	g	0.2	0.5
46	n	n	n	n	n	c	>1	l	>4	pp	a	b	0.6	0.37
47	n	t	n	n	n	n	>1	l	>4	pp	d	g	0.18	0.5
48	c	n	c	c	n	n	n	n	1	a	a	b	0	0
49	c	t	c	c	n	n	>1	l	>4	a	a	b	0.07	0.18
50	c	n	c	c	n	n	>1	l	>4	pm	a	g	0.2	0.56
51	c	n	c	n	n	n	>1	l	>4	pm	a	g	0.15	0.56
52	c	t	c	c	n	n	>1	l	>4	pm	a	ID	0.24	0.18

TT: Tillage Time: c: >once per year; n: once per year; TF: Tillage Form: n: null, m: manual, t: technical; P. Pesticide use: c: constant, n: null, F: Use of chemical Fertilizers: c: constant, n: null; GM: Green Manure, c: constant, n: null; RC: Crop Rotation: c: constant, n: null; CA: tree cover: >1 year, n: null; DA: tree density: h: high, l: low, i: isolated, n:null; DivA: tree diversity: >4 species, 1 species, CT: crop type: a: annual, ba: biannual, pp: perennial polyculture, pm: perennial monoculture; CD: crop density: a: abundant, d: disperse; Ccol: crop colour: g: green, b: brown, y: yellow; EV: evaluation.

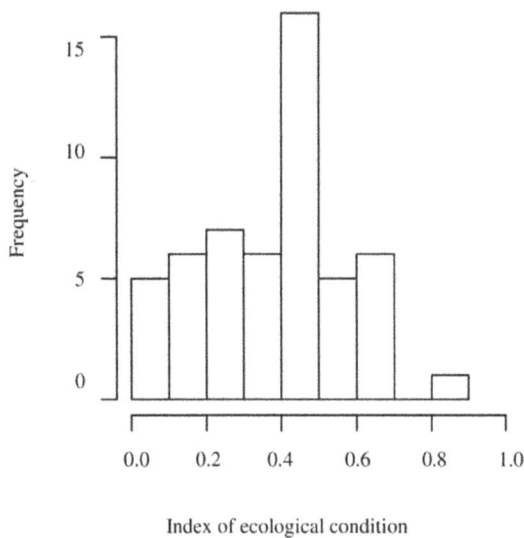

Figure 3. Frequency histogram of values of the index of ecological condition applied to 52 plots in South-East Mexico.

Discussion

Since the Rio Earth Summit, there has been a concerted effort to construct indicators to monitor progress toward sustainable development [59]. Most of these indicators have been developed in Europe (10) and Asia (2) [60]. In Latin America, sustainable

indicators have been developed by Astier et al. [61]; this index mainly focuses on subsistence level agriculture, and evaluates the sustainability of a system.

Our index is geared toward small and medium scale producers who principally grow for market and have a fairly large crop area (32.4 ha on average). According to Bockstaller and Girardin [62], indicators must be elaborated according to a scientific approach, and one of the important steps in this elaboration is validation.

Our index was developed according to the consensus of a group of scientists, with knowledge in agroecology, and was validated independently by 2 scientists, each with over 15 years of experience in agroecology. The evaluation included 3 important steps: design validation, output validation, and end use validation [62].

In previous studies, only seven indicators have been used to evaluate farm systems: crop diversity, crop succession, pesticide use, nitrogen level, phosphorus level, soil organic matter, and irrigation methods [63]. All of these practices depend on the farmer's decision, and to a large extent they impact the environment.

Our index is based on qualitative and quantitative concrete data and includes most of these 7 indicators, except nitrogen and phosphorus, both of which are observed indirectly via plant health, through the crop colour indicator (according to whether plants have a greenish or yellowish colour); organic matter, which is indirectly characterized by the technology applied in the system (green manure indicator, Figure 2); and irrigation, which in this study was not evaluated, given that all plots evaluated were only used for seasonal rainfed agriculture, according to local rainfall patterns.

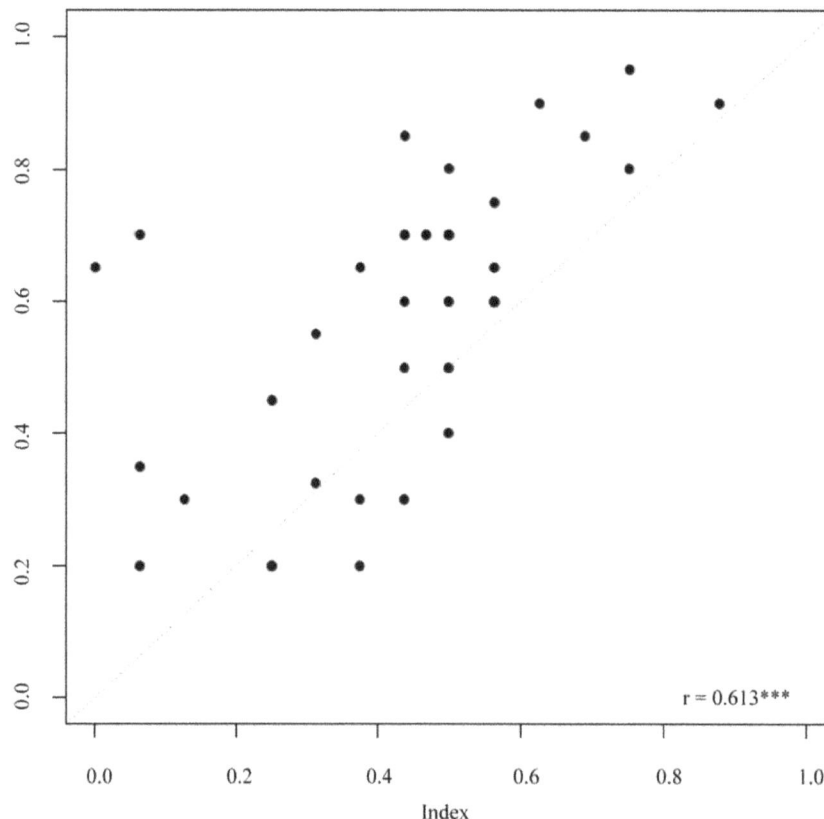

Figure 4. Scatterplot of the values of the index of ecological condition applied to 52 plots in South-East Mexico (x-axis) versus quality values between 0 (worst) and 1 (best) assigned by experts to the same plots (y-axis).

At the farm level there are indicators that evaluate the environmental impact of the agricultural practices and indicators that evaluate the effect of those practices at the local and global level [60].

Our index includes both types of indicators; evaluating agricultural practices: those variables taken in the field: type and frequency of tillage, pesticide application, fertilizer use, crop rotation, tree cover and density, green manure, crop density (see Table 2 & 3). Effect indicators used were disturbance (mainly soil disturbance) measured by tillage (frequency and type) and chemical inputs, technology used, and crop structure (see index, Figure 2). Some existing indexes focus on evaluation or evolution of environmental performance, thus encouraging environmentally sound practices [60], such as crop rotation, organic fertilizers, and no-tillage. Meanwhile, our index identifies the indicator that has the highest environmental impacts in each plot evaluated, with the idea that the farmer could potentially improve these with a given practice, ie. to monitor the soil ecological condition via the use of a soil macroinvertebrates index [64], where the lack of macroinvertebrates informs of a severe negative activity as pollution or conventional tillage.

In developing the variables to be included in our index, we reviewed bibliographic studies and carried out field work obtaining data which we hoped would reflect the negative and environmentally friendly practices commonly used in agro ecosystems of southeastern Mexico. The index can be used by farmers, using the following web address: http://201.116.78.102/~modelo/Index.html.

However, we did not evaluate certain indicators such as water use, and water quality, as did - for example - the index (monitoring tool) of ecological indicators used on a Flemish dairy farm [65]. Nor did we evaluate environmental impacts due to energy consumption [66]. Within a tropical framework, in south-eastern Mexico, the priorities were to identify those practices that were soil perturbing and environment polluting, practices that can be modified by the farmers, by an attitude changing. Nambiar et al. [16], proposed an agricultural sustainability index (ASI) to measure sustainability as a function of biophysical, chemical, economic, and social indicators, our index only measures the ecological state of the agroecosystem, and provides easy to use tools for improving those practices which negatively impact the environment. Van der Werf and Petit [60], state that indicators based on farmer practices cost less in data collection but do not allow for an actual evaluation of environmental impact. In the case of our study, the experts' evaluation correlated satisfactorily with the index, although the index rendered more penalizing scores than did the experts. Some improvements must be made to our index, relating quality and quantity of the applied inputs for instance; the index should specify the kind of manures used and then to evaluate their effect when added to the systems. The consulted experts found important to integrate this information into the index, they also found that the possible relations among different crops and environmental effects of using cow manure, vermicompost, or traditional compost have to be considered. Another variable which the experts suggested should be added to the index is the presence of natural vegetation surrounding the crop. Farmers see advantages of having crops surrounded by secondary forest, diversity in the agricultural area can be increased, ie when different pollinators arrive.

The advantages of using this index is that the common agricultural practices (mechanized land preparation and use of common pesticides ie. Carbofuran, Chlorpyrifos), evaluated in this study as indicators are used throughout the world; over time, through practice, the index may be improved. Farmers and other land owners may realize which of the practices they use are disturbing the environment, due to the fact that these practices generate a value in each of the evaluated indicators. The variables obtained in the field contribute to the information of each of the indicators, and the index is the compendium of all the indicators. Our index doesn't give a sustainability measure, because it does not include socioeconomic indicators of the farms. Further studies are required in order to observe the acceptance of this index by farmers in a regional scale.

Acknowledgments

We thank Lorena Soto and Guillermo Jimenez for participating in the evaluation of the plots. We thank Simon Hernandez de la Cruz, Alejandra Sepulveda Lozada, Lauritania Ibarra Hernandez, and Marcelo Rodriguez Ricardes, who helped to collect data in the field. We are also thankful to Lorena Reyes for bibliographic support. El Colegio de la Frontera Sur provided infrastructural resources. Language revision was done by Ann Greenberg.

Author Contributions

Conceived and designed the experiments: EH CK SOG BDJ SHD VG. Performed the experiments: EH CK. Analyzed the data: EH CK. Contributed reagents/materials/analysis tools: EH CK SOG BDJ SHD VG. Contributed to the writing of the manuscript: EH CK SOG BDJ VG. Figures elaboration: CK SOG.

References

1. Houghton RA (1994) The world wide extent of land-use change. BioScience 44: 305–313.
2. Tilman D, Lehman C (2001) Human-cause environmental change: impacts on plant diversity and evolution. Proceedings of the National Academy of Sciences 98: 5433–5440.
3. INEGI (2008) Cuaderno Estadístico Municipal de Tenosique. Villahermosa, Tabasco: Gobierno del Estado de Tabasco.
4. Grande D, de Leon F, Nahed J, Perez-Gil F (2010) Importance and Function of Scattered Trees in Pastures in the Sierra Region of Tabasco, Mexico. Research Journal of Biological Sciences 5: 75–87.
5. Geissen V, Morales-Guzman G (2006) Fertility of tropical soils under different land use systems-a case study of soils in Tabasco, Mexico. Appl Soil Ecol 841: 1–10.
6. Melgar C, Geissen V, Cram S, Sokolov M, Bastidas P, et al. (2008) Pollutants in drainage channels following long-term application of Mancozeb to banana plantations in southeastern Mexico. Journal of Plant Nutrition and Soil Science 171: 597–604.
7. Aryal DR, Geissen V, Ponce-Mendoza A, Ramos-Reyes RR, Becker M (2012) Water quality under intensive banana production and extensive pastureland in tropical Mexico. Journal of Plant Nutrition and Soil Science 175(4): 553–559.
8. Ortiz SM, Anaya G, Estrada BW (1994) Evaluación, Cartografía y Políticas Preventivas de la Degradación de la Tierra. Chapingo, Mexico.
9. Geissen V, Sanchez-Hernandez R, Kampichler C, Ramos-Reyes R, Sepulveda-Lozada A, et al. (2009) Effects of land-use change on some properties of tropical soils – An example from Southeast Mexico. Geoderma 151: 87–97.
10. Moreno-Unda AA (2011) Environmental effects of the National Tree Clearing Program, Mexico, 1972–1982 Cologne: Cologne University of Applied Sciences 119 p.
11. Bockstaller CGL, Keichinger O, Girardin P, Galan MB, Gaillard G (2009) Comparison of methods to assess the sustainability of agricultural systems. A review. Agron Sustain Dev 29: 223–235.
12. Mitchell GMA, McDonald A (1995) PICABUE: a methodological framework for the development of indicators of sustainable development. Int J Sust Dev World 104–123.

13. Bockstaller C, Guichard L, Makowski D, Aveline A, Girardin P, et al. (2008) Agri-Environmental Indicators to Assess Cropping and Farming Systems: A Review Sustainable Agriculture. In: Lichtfouse E, Navarrete M, Debaeke P, Véronique S, Alberola C, editors. Springer Netherlands. 725–738.

14. Riley J (2001) The indicator explosion: local needs and International challenges. Agr Ecosyst Environ 87: 119–120.

15. Kampichler C, Hernández-Daumás S, Ochoa-Gaona S, Geissen V, Huerta-Lwanga E, et al. (2010) Indicators of environmentally sound land use in the humid tropics: The potential roles of expert opinion, knowledge engineering and knowledge discovery. Ecological Indicators 10: 320–329.

16. Nambiar KK, Gupta AP, Fu Qinglin, Li S (2001) Biophysical, chemical and socio-economic indicators for assessing agricultural sustainability in the Chinese coastal zone Agriculture Ecosystems and Environment 87: 209–214.

17. Cabell JF, Oelofse M (2012) An indicator framework for assessing agroecosystem resilience. Ecology and Society 17: 18 http://dx.doi.org/10.5751/ES-04666-170118.

18. Darnhofer I, Bellon S, Dedieu B, Milestad R (2010) Adaptiveness to enhance the sustainability of farming systems. A review. Agronomy for Sustainable Development 30: 545–555.

19. Allen CR, Fontaine JJ, Pope KL, Garmestani AS (2011) Adaptive management for a turbulent future. Journal of Environmental Management 92: 1339–1345.

20. Malezieux E, Crozat Y, Dupraz C, Laurans M, Makowski D, et al. (2009) Mixing plant species in cropping systems: concepts, tools and models. A review. Agron Sustain Dev 29: 43–62.

21. Jones CG, Lawton J, Shachak M (1997) Positive and negative effects of organisms as physical ecosystem engineers. Ecology and Society 78: 1946–1957.

22. Brussaard L, Caron P, Campbell B, Lipper L, Mainka S, et al. (2010) Reconciling biodiversity conservation and food security: scientific challenges for a new agriculture. Current opinion in Environmental sustainability 2: 34–42.

23. Knorr D, Watkins TR (1984) Alterations in Food Production Knorr DW, Watkins, T.R., editor. New York: Van Nostrand Reinhold.

24. Seufert V, Ramankutty N, Foley JA (2012) Comparing the yields of organic and conventional agriculture. Nature 485: 229–232.

25. Carter H (1989) Agricultural sustainability: an overview and research assessment. Calif Agric 43: 1618–1637.

26. Tellarini V, Caporali F (2000) An input/output methodology to evaluate farms as sustainable agroecosystems: an application of indicators to farms in central Italy. Agriculture, Ecosystems and Environment 77: 111–123.

27. Stinner BR, House GJ (1987) Role of ecology in lower-input, sustainable agriculture: an introduction. Am J Alternative Agric 2: 146–147.

28. Lockeretz W (1988) Open questions in sustainable agriculture. Am J Alternative Agric 3: 174–181.

29. Hauptli H, Katz D, Thomas BR, Goodman RM (1990) Biotechnology and crop breeding for sustainable agriculture. In: Edwards CA, Lal R, Madden P, Miller RH, House G, editors. Sustainable Agricultural Systems Soil and Water Conservation Society: Ankeny, Iowa. 141–156.

30. Madden P (1990) The economics of sustainable low-input farming systems. In: Francis CA, Flora CB, King LD, editors. Sustainable Agriculture in Temperate Zones. New York: John Wiley & Sons. 315–341.

31. Dobbs TL, Becker DL, Taylor DC (1991) Sustainable agriculture policy analyses: South Dakota on-farm case studies. J Farming Systems ResExt 2: 109–124.

32. Blair GJ, Lefroy RD, Lisle L (1995) Soil carbon fractions based on their degree of oxidation, and the development of a carbon management index for agricultural systems. Aust J Soil Res 46: 1459–1466.

33. Ouédraogo E, Mando A, Zombré NP (2001) Use of compost to improve soil properties and crop productivity under low input agricultural system in West Africa. Agriculture, Ecosystems and Environment 84: 259–266.

34. Pretty J, Toulmin C, Williams S (2011) Sustainable intensification in African agriculture. International Journal of Agricultural Sustainability 9: 5–24.

35. Lal R (2004) Soil carbon sequestration impacts on global climate change and food security. Science 304: 1623–1627.

36. Gurr GM, Wratten SD, Luna JM (2003) Multi-function agricultural biodiversity: pest management and other benefits. Basic Appl Ecol 4: 107–116.

37. Börjesson P (1999) Environmental effects of energy crop cultivation in Sweden I: Identification and quantification. Biomass and Bioenergy 16: 137–154.

38. Rommelse R (2001) Economic assessment of biomass transfer and improved fallow trials in western Kenya In: ICRAF Natural Resource Problems PaPP, editor. Natural Resource Problems. Nairobi (Kenya): International Centre for Research in Agroforestry (ICRAF).

39. Nyende P, Delve R (2004) Farmer participatory evaluation of legume cover crop and biomass transfer technologies for soil fertility improvement using farmer criteria, preference ranking and logit regression analysis. Exp Agric 40: 77–88.

40. Koocheki A, Nassiri M, Alimoradi L, Ghorbani R (2009) Effect of cropping systems and crop rotations on weeds. Agronomy for Sustainable Development 29: 401–408.

41. Bellon M (1995) Farmers 'Knowledge and Sustainable Agroecosystem Management: An Operational Definition and an Example from Chiapas, Mexico. Human Organization 54: 263–272.

42. Belalcázar S, Espinosa J (2000) Effect of Plant Density and Nutrient Management on Plantain Yield. Better Crops International 14: 12–15.

43. Shaahan MM, El-Sayed AA, Abou El-Nour EAA (1999) Predicting nitrogen, magnesium and iron nutritional status in some perennial crops using a portable chlorophyll meter. Scientia Horticulturae 82: 339–348.

44. Agbede TM (2008) Nutrient availability and cocoyam yield under different tillage practices. Soil Tillage Research 99: 49–57.

45. Edwards CA (1989) The Importance of Integration in Sustainable Agricultural Systems. Agriculture, Ecosystems and Environment 27: 25–35.

46. Edwards CA, Grove TL, Harwood RR, Pierce Colfer CJ (1993) The role of agroecology and integrated farming systems in agricultural sustainability. Agriculture, Ecosystems and Environment 46: 99–121.

47. García-Pérez JA, Alarcón-Gutiérrez E, Perroni Y, Barois I (2014) Earthworm communities and soil properties in shaded coffee plantations with and without application of glyphosate. Applied Soil Ecology 83: 230–237.

48. Correia FV, Moreira JC (2010) Effects of Glyphosate and 2,4-D on Earthworms (Eisenia foetida) in Laboratory Tests. Bull Environ Contam Toxicol 85: 264–268.

49. Dick J, Kaya B, Soutoura M, Skiba U, Smith R, et al. (2008) The contribution of agricultural practices to nitrous oxide emissions in semi-arid Mali Soil. Use and Management 24: 292–301.

50. Morales H, Perfecto I (2000) Traditional knowledge and pest management in the Guatemalan highlands. Agriculture and Human Values 17: 49–63.

51. Li BL, Rykiel EJ (1996) Introduction. Ecological Modelling 90: 109–110.

52. Salski A (1996) Introduction Fuzzy Logic in Ecological Modelling. Ecological Modelling 85: 1–2.

53. Mendoza GA, Prabhu R (2003) Fuzzy methods for assessing criteria and indicators of sustainable forest management. Ecological Indicators 3: 227–236.

54. Kampichler C, Platen R (2004) Ground beetle occurrence and moor degradation: modelling a bioindication system by automated decision-tree induction and fuzzy logic. Ecological Indicators 4: 99–109.

55. INEGI (2000) Cuaderno Estadístico Municipal de Tenosique. Gobierno del Estado de Tabasco; INEGI, editor. Villahermosa, Tabasco.

56. INEGI (1985) Carta Edafológica, Villahermosa In: E15-8, editor. 1: 250,000. Aguascalientes, México: Instituto Nacional de Estadística, Geografía e Informática.

57. Manjarrez-Muñoz B (2008) Ordenamiento territorial de la ganadería bovina en Balancán y Tenosique, Tabasco. Villahermosa Tabasco: El Colegio de la Frontera Sur. 105 p.

58. Isaac-Márquez R (2008) Análisis del Cambio de Uso y Cobertura del Suelo en los Municipios de Balancán y Tenosique, Tabasco, México. Villahermosa, Tabasco, México: El Colegio de la Frontera Sur.

59. Rigby D, Woodhouse P, Young T, Burton M (2001) Constructing a farm level indicator of sustainable agricultural practice. Ecological Economics 39: 463–478.

60. van der Werf HMG, Petit J (2002) Evaluation of the environmental impact of agriculture at the farm level: a comparison and analysis of 12 indicator-based methods. Agriculture, Ecosystems and Environment 93: 131–145.

61. Astier M, Speelman S, López-Ridaura S, Masera O, Gonzalez-Esquivel CE (2011) Sustainability indicators, alternative strategies and trade-offs in peasant agroecosystems: analysing 15 case studies from Latin America. International Journal of Agricultural Sustainability 9: 409–422.

62. Bockstaller C, Girardin P (2003) How to validate environmental indicators. Agr Syst 76: 639–653.

63. Bockstaller C, Girardin P, Van der Werf HGM (1997) Use of agroecological indicators for the evaluation of farming systems. Eur J Agron 7: 261–270.

64. Huerta E, Kampichler C, Geissen V, Ochoa-Gaona S, de Jong B, et al. (2009) Towards an ecological index for tropical soil quality based on soil macrofauna. Pesq agropec bras 44 (8): 1056–1062.

65. Meul M, Nevens F, Reheul D (2009) Validating sustainability indicators: Focus on ecological aspects of Flemish dairy farms. Ecological Indicators 9: 284–295.

66. Pervanchon F, Bockstaller C, Girardin P (2002) Assessment of energy use in arable farming systems by means of an agro-ecological indicator: the energy indicator. Agricultural Systems 72: 149–172.

PERMISSIONS

LIST OF CONTRIBUTORS

Yufeng Sun and Kongming Wu
State Key Laboratory for Biology of Plant Diseases and Insect Pests, Institute of Plant Protection, Chinese Academy of Agricultural Sciences, Beijing, People's Republic of China

Hao Yu
Department of Entomology, Henan Institute of Science and Technology, Xinxiang, People's Republic of China

Jing-Jiang Zhou and John A. Pickett
Department of Biological Chemistry and Crop Protection, Rothamsted Research, Harpenden, Hertfordshire, United Kingdom

En-Cheng Yang
Department of Entomology, National Taiwan University, Taipei, Taiwan

Graduate Institute of Brain and Mind Sciences, National Taiwan University, Taipei, Taiwan

Hui-Chun Chang and Wen-Yen Wu
Department of Entomology, National Taiwan University, Taipei, Taiwan

Yu-Wen Chen
Department of Animal Science, National Ilan University, Ilan, Taiwan

Rafal Tokarz, Cadhla Firth, Craig Street and W. Ian Lipkin
Center for Infection and Immunity, Mailman School of Public Health, Columbia University, New York, New York, United States of America

Diana L. Cox-Foster
Department of Entomology, The Pennsylvania State University, University Park, Pennsylvania, United States of America Abstract

Benjamin Dainat and Laurent Gauthier
Swiss Bee Research Centre, Agroscope Liebefeld-Posieux Research Station ALP, Bern, Switzerland

Jay D. Evans and Yan Ping Chen
Bee Research Laboratory, United States Department of Agriculture- Agricultural Research Service, Beltsville, Maryland, United States of America

Peter Neumann
Swiss Bee Research Centre, Agroscope Liebefeld-Posieux Research Station ALP, Bern, Switzerland

Department of Zoology and Entomology, Rhodes University, Grahamstown, South Africa

R. Scott Cornman, Jay D. Evans, Yanping Chen, Dawn Lopez and Jeffery S. Pettis
Bee Research Laboratory, Agricultural Research Service, United States Department of Agriculture, Beltsville, Maryland, United States of America

David R. Tarpy and Lacey Jeffreys
Department of Entomology, North Carolina State University, Raleigh, North Carolina, United States of America

Dennis vanEngelsdorp
Department of Entomology, University of Maryland, College Park, Maryland, United States of America

David S. Khoury
School of Mathematics and Statistics, The University of Sydney, Sydney, New South Wales, Australia

Mary R. Myerscough
School of Mathematics and Statistics, The University of Sydney, Sydney, New South Wales, Australia
Centre for Mathematical Biology, The University of Sydney, Sydney, New South Wales, Australia

Andrew B. Barron
Department of Biology, Macquarie University, Sydney, New South Wales, Australia

Fredrik Granberg
Department of Biomedical Sciences and Veterinary Public Health (BVF), Swedish University of Agricultural Sciences (SLU), Uppsala, Sweden
The OIE Collaborating Centre for the Biotechnology-based Diagnosis of Infectious Diseases in Veterinary Medicine, Uppsala, Sweden

Marina Vicente-Rubiano, Deborah Kukielka, Josè Manuel Sánchez-Vizcaíno and Consuelo Rubio-Guerri
Animal Health Department, Faculty of Veterinary, Complutense University of Madrid, Madrid, Spain

Oskar E. Karlsson
Department of Biomedical Sciences and Veterinary Public Health (BVF), Swedish University of Agricultural Sciences (SLU), Uppsala, Sweden
The OIE Collaborating Centre for the Biotechnology-based Diagnosis of Infectious Diseases in Veterinary Medicine, Uppsala, Sweden

SLU Global Bioinformatics Center, Department of Animal Breeding and Genetics (HGEN), SLU, Uppsala, Sweden

Sándor Belák
Department of Biomedical Sciences and Veterinary Public Health (BVF), Swedish University of Agricultural Sciences (SLU), Uppsala, Sweden
The OIE CollaboratingCentre for the Biotechnology-based Diagnosis of Infectious Diseases in Veterinary Medicine, Uppsala, Sweden
Department of Virology, Immunobiology and Parasitology, VIP, National Veterinary Institute (SVA), Uppsala, Sweden

Rodolphe Sabatier and Muriel Tichit
INRA, UMR 1048 SADAPT, Paris, France
AgroParisTech, UMR 1048 SADAPT, Paris, France

Luc Doyen
CNRS, UMR 7204 CERSP, MNHN, Paris, France

Roy M. Francis, Steen L. Nielsen and Per Kryger
Department of Agroecology, Science and Technology, Aarhus University, Slagelse, Denmark

David J. Hawthorne and Galen P. Dively
Department of Entomology, University of Maryland, College Park, Maryland, United States of America

Judy Y. Wu, Carol M. Anelli and Walter S. Sheppard
Department of Entomology, Washington State University, Pullman, Washington, United States of America

David S. Khoury
School of Mathematics and Statistics, The University of Sydney, Sydney, New South Wales, Australia

Andrew B. Barron
Department of Biological Sciences, Macquarie University, Sydney, New South Wales, Australia

Mary R. Myerscough
School of Mathematics and Statistics, The University of Sydney, Sydney, New South Wales, Australia
Centre for Mathematical Biology, The University of Sydney, Sydney, New South Wales, Australia

Jonathan Ivers, Christopher Quock, Travis Siapno, Seraphina DeNault, Christopher D. Smith, John Hafernik and Andrew Core
Department of Biology, San Francisco State University, San Francisco, California, United States of America

Joseph DeRisi and Charles Runckel
Department of Biochemistry and Biophysics, University of California, San Francisco, San Francisco, California, United States of America

Brian Brown
Entomology Section, Natural History Museum of Los Angeles County, Los Angeles, California, United States of America

Ricardo A. Scrosati, Ruth D. Patten and Randolph F. Lauff
Department of Biology, Saint Francis Xavier University, Antigonish, Nova Scotia, Canada

Staša Milojević
School of Library and Information Science, Indiana University, Bloomington, Indiana, United States of America

Gunter J. Sturm, Werner Aberer and Bettina Kranzelbinder
Division of Environmental Dermatology and Venerology, Department of Dermatology, Medical University of Graz, Graz, Austria

Wolfgang Hemmer
Floridsdorf Allergy Center, Vienna, Austria

Chunsheng Jin and Friedrich Altmann
Department of Chemistry, University of Natural Resources and Applied Life Sciences, Vienna, Austria

Eva M. Sturm and Akos Heinemann
Institute of Experimental and Clinical Pharmacology, Medical University of Graz, Graz, Austria

Antonia Griesbacher
Division of Biostatistics, Center for Medical Research, Medical University of Graz, Graz, Austria

Jutta Vollmann and Karl Crailsheim
Institute of Zoology, University of Graz, Graz, Austria

Margarete Focke
Institute of Pathophysiology, Medical University of Vienna, Vienna, Austria

Xinqi Zheng, Tian Xia, Tao Yuan and Yecui Hu
School of Land Science and Technology, China University of Geosciences (Beijing), Beijing, China

Xin Yang
Research Center for Operation and Development of Beijing, Institute of Policy and Management, Chinese Academy of Sciences, Beijing, China

Matt I. Betti and Lindi M. Wahl
Department of Applied Mathematics, Western University, London, Ontario, Canada

Mair Zamir
Department of Applied Mathematics, Western University, London, Ontario, Canada
Department of Medical Biophysics, Western University, London, Ontario, Canada

Rodolphe Sabatier, Kerstin Wiegand and Katrin Meyer
Department of Ecosystem Modelling, Büsgen-Institute, Georg-August-University of Göttingen, Göttingen, Germany

James R. Welch and Carlos E. A. Coimbra Jr
Escola Nacional de Saúde Pública, Fundação Oswaldo Cruz, Rio de Janeiro, Rio de Janeiro, Brazil

Eduardo S. Brondízio
Department of Anthropology, Indiana University, Bloomington, Indiana, United States of America
Anthropological Center for Training and Research on Global Environmental Change, Indiana University, Bloomington, Indiana, United States of America

Scott S. Hetrick
Anthropological Center for Training and Research on Global Environmental Change, Indiana University, Bloomington, Indiana, United States of America

Junyuan Ren, Abigail Cone, Rebecca Willmot and Ian M. Jones
School of Biological Sciences, University of Reading, Reading, United Kingdom

Ian Laycock
College of Life & Environmental Sciences, Biosciences, University of Exeter, Exeter, United Kingdom

James E. Cresswell
College of Life & Environmental Sciences, Biosciences, University of Exeter, Exeter, United Kingdom
Centre for Pollination Studies, University of Calcutta, Kolkata, India

Esperanza Huerta and Salvador Hernandez-Daumas
El Colegio de la Frontera Sur, Unidad Campeche, Dpto. Agroecología, Campeche, México

Christian Kampichler
Universidad Juárez Autónoma de Tabasco, División de Ciencias Biológicas, Villahermosa, Tabasco, México

Sovon Dutch Centre for Field Ornithology, Natuurplaza (Mercator 3), Nijmegen, The Netherlands

Susana Ochoa-Gaona and Ben De Jong
El Colegio de la Frontera Sur, Unidad Campeche, Dpto. Sustainability Sciences, Campeche, México

Violette Geissen
University of Bonn - INRES, Bonn, Germany
Wageningen University and Research Center – Alterra, Wageningen, Gelderland, Netherlands

Index

www.ingramcontent.com/pod-product-compliance
Lightning Source LLC
Chambersburg PA
CBHW080628200326

41458CB00013B/4552